29th International Congress and Exhibition on Condition Monitoring and Diagnostic Engineering Management (COMADEM 2016)

Xi'an, China
20 - 22 August 2016

ISBN: 978-1-5108-6843-4

Printed from e-media with permission by:

Curran Associates, Inc.
57 Morehouse Lane
Red Hook, NY 12571

Some format issues inherent in the e-media version may also appear in this print version.

Printed by Curran Associates, Inc. (2016)

For permission requests, please contact COMADEM International at the address below.

COMADEM International
c/o Prof. Dr. Raj B. K. N. Rao
307 Tiverton Road, Selly Oak
Birmingham B29 6DA. UK

Phone: +44(0) 121 472 2338

rajbknrao@btinternet.com

Additional copies of this publication are available from:

Curran Associates, Inc.
57 Morehouse Lane
Red Hook, NY 12571 USA
Phone: 845-758-0400
Fax: 845-758-2633
Email: curran@proceedings.com
Web: www.proceedings.com

TABLE OF CONTENTS

Multi-mode Tuned Mass Damper Design to Reduce Centrifugal Air Compressor Gearbox and Outlet Pipeline Vibrations

Hang-ling Hu[1], Li-dong He[2*], Jin-ji Gao[3] and Zhen-kun Zhang[4]

[1] Ministry of Education Engineering Research Center of Chemical Safety, Beijing University of Chemical Technology, Beijing 100029, P. R. China, e-mail: huhangling@126.com

[2] Ministry of Education Engineering Research Center of Chemical Safety, Beijing University of Chemical Technology, Beijing 100029, P. R. China, e-mail: 1963he@163.com

[3] Ministry of Education Engineering Research Center of Chemical Safety, Beijing University of Chemical Technology, Beijing 100029, P. R. China, e-mail: gaojinji@263.net

[4] Hebei New Energy Development Co., Ltd., Zhangjiakou 075000, Hebei Province, China, e-mail: niuniu3014@126.com

Abstract. To solve the vibration problems of centrifugal air compressor gearbox and outlet pipeline in a petrochemical plant, the effect of a tuned mass damper (TMD) on the vibration control of a large and complex compressor was investigated. Based on the principle of TMD vibration suppression, the natural frequencies of the compressor and TMD subsystem were calculated using the finite-element analysis software. According to the theory of continuous-structure multi-mode vibration suppression, the natural frequency range of the TMD subsystem should cover the natural frequency range of the main system. Thus, the designs of three different types of TMD were optimized, and their effects on vibration suppression were simulated. Using the correlation between pipeline and gearbox, the TMDs were installed on the pipelines. The vibration of the pipeline and gearbox significantly reduced after the installation, confirming that a TMD can effectively reduce the vibration of a compressor and contribute to safe and stable operation.

Keywords: Centrifugal air compressor, Tuned mass damper, Multi-mode vibration suppression, Finite-element simulation, Harmonic response analysis

1 Introduction

A parallel-axis centrifugal air compressor has strong coupling characteristics. The rotor system of the compressor is a geared rotor system. The vibrations of rotors are no longer independent, and the modes of the rotor system are denser. However, the dynamic design of the rotor system in the compressor industry generally involves the independent design of each rotor. Thus, in practical operations, the compressor often shows abnormal vibrations that can crack the shaft [1–3]. Moreover, the case is coupled with a diffuser, a volute, a chassis, and an outlet pipeline that are connected together, forming an integral system. Although a centrifugal air compressor has a very compact design, it suffers from vibration coupling [4]. Strong coupling characteristics cause complex vibration and difficulty in handling.

The studied compressor severely vibrated. A vibration monitoring and fault diagnosis found that the gearbox and outlet pipeline had overproofed vibrations. The fluid in the compressor channel was chaotic, and the dynamic disturbance was large, particularly at the first and third stages of outlet pipelines. However, at each stage, the rotor operated stably and was dynamically balanced; the sliding bearing oil film was stable. Consequently, rotor dynamic balancing, the traditional method to control vibrations, was not applicable to this case, and a tuned mass damper (TMD) method was investigated for vibration reduction.

TMD is a dynamic vibration absorber and has many advantages such as simple manufacturing, high reliability, and low maintenance costs; it is widely used in the vibration control of architectural structures, bridges and roads, and vehicles [5, 6]. Den Hartog [7] first derived the formula for the designing of an optimal value of single degree of freedom absorber (SDOF-TMD) parameters. Based on this study, several studies [8–11] perfected the design of SDOF-TMD and used in engineering applications. However, although the SDOF-TMD is tuned to a particular structural frequency, the compressor is a continuous large complex system with multiple natural frequencies. The measured vibration contains multiple frequency components that should be suppressed, and thus the SDOF-TMD cannot achieve multi-mode vibration suppression.

To solve the problem, Xu [12] proposed the concept of multiple TMD (MTMD). According to the multiple vibration modes of the primary system, the use of multiple SDOF-TMDs, where the natural frequency of each SDOF-TMD corresponds to a vibration mode, controls multi-mode vibration. Rice [13] investigated two SDOF-TMDs to reduce the vibration of a cantile-

* Corresponding author. Tel.: 010-64443098; fax: 010-64443098; E-mail: 1963he@163.com

ver beam. Nishida [14] built a series of cantilever beam-type TMDs with different frequencies radially assembled in a hub. The resulting MTMD was used to suppress the vibration of the structure. Many studies confirmed that compared to SDOF-TMD, MTMD has a wider frequency band and better vibration suppression efficiency. Because of limited space, however, too many TMDs could not be installed on the compressor. Moreover, the use of too many TMDs with a large mass ratio may change the dynamic characteristics of the compressor.

To solve the problem of vibration suppression of a large number of compressor frequencies, Zhang [15] proposed continuous-structure multi-mode vibration suppression. In this structure, the mass is designed as a continuous structure with multiple natural frequencies, and its modal frequency range covers the frequency range of the main system. At the same time, a large damping material dissipates energy. The application of the proposed structure to a XH 2725 five-axis cooperating machining center well suppressed the spindle vibrations. Based on the theory of continuous-structure multi-mode vibration suppression, the following sections extend the theory to more complex compressor systems to control the vibrations of the gearbox and pipeline.

2 Vibration condition of compressor

2.1 Technical data of compressor

The petrochemical plant has a centrifugal compressor (1000 tons/day capacity) used for compressing various components of air to liquid. The compressor is an integrally geared system produced by Siemens. And its model is STC-GV.

Because of compressor operation, the first and third stage outlet pipelines had severe vibration, particularly that at the third stage. The vibration fractured a fixed bolt on the connecting flange. The reinforcing hoop is a protection for flange. However, it also failed to resolve the fundamental problems of vibration. Long-term vibration also caused damage to other equipment. During the first overhaul downtime, the inspection found that the gear was severely worn, and two tooth surface defects (almost 10 mm large) were discovered. Thus, the replacement of the rotor was unavoidable and produced unexpected costs.

2.2 Analysis of vibration causes

The analysis of vibration detection and fault diagnosis showed that the compressor unit vibration was not caused by the coupling vibration of the gear-rotor system, but closely related to the strong fluid dynamic disturbance within the pipeline. Because of load changes, the centrifugal air compressor should adjust the inlet guide vane frequently. When the compressor flow decreased to a certain value, the inlet positive incidence angle was too large and caused a severe airflow boundary layer separation on the nonworking blade surface. Many air vortexes first emerged in one or several blades and then gradually spread to the adjacent blades. The reverse rotation of the vortexes led to the rotating stall that generated gas pulsation. When pulse airflow met with a varying diameter pipe, elbow, and valve, it struck the pipeline, generating vibrations. Fig. 1 shows the first and third stage outlet pipelines. The two pipelines are shorter and have varying diameters and elbows, thus dramatically changing the air flow velocity and flow direction and causing pipeline vibration [16–18].

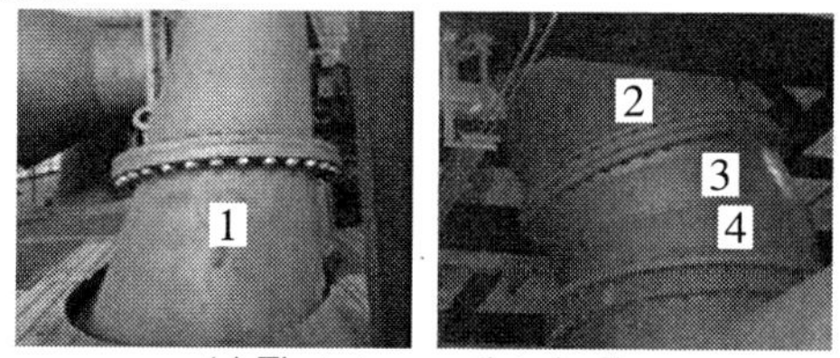

(a) First stage outlet pipeline

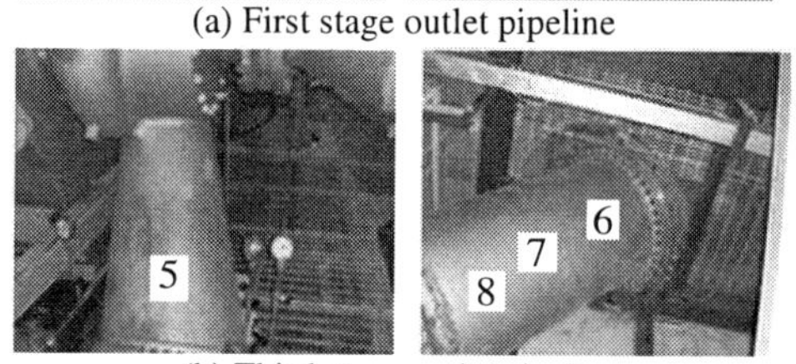

(b) Third stage outlet pipeline

Fig. 1. First and third stage outlet pipelines of compressor

Moreover, because the diffuser, volute, chassis, and outlet pipeline are connected together to become an integral system, the strong piping vibration inevitably caused the compressor unit housing vibration, thus increasing the gearbox vibration. According to the abovementioned analysis and considering the compressor as an integral system with a strong correlation, TMDs were installed on the pipeline. The TMDs efficiently reduced the pipeline vibration. Moreover, the TMDs indirectly reduced the gearbox vibrations.

3 Theory of continuous-structure multi-mode vibration suppression

A TMD consists of a mass, spring, and damper. It (subsystem) is connected to the compressor (the main system) at the location where a significant vibration occurs. By properly selecting the structure parameters, the dynamic characteristics and connection help the subsystem frequency correspond to the main system frequency. A part of the vibration energy is transferred from the main system to the subsystem, and then the vibration of the main structure is reduced. Further, the damper of the subsystem can also absorb some vibration energy [6].

3.1 SDOF-TMD operating principle

A traditional SDOF-TMD is a typical second-order mass-spring-damping system. Fig. 2 shows its operating principle. The equations of motion can be expressed as follows [7]:

$$M\ddot{x}+Kx+c(\dot{x}-\dot{x}_1)+k(x-x_1)=F\sin(\omega t) \quad (1)$$

$$m\ddot{x}_1+c(\dot{x}_1-\dot{x})+k(x_1-x)=0 \quad (2)$$

where M and K are the mass and stiffness of the main system, respectively; m, k, and c are the mass, stiffness, and damping of the TMD subsystem. Generally, the main system damping is much smaller than the TMD subsystem damping; therefore, it can be neglected. x is the main system displacement; x_1 is the TMD subsystem displacement. $F\sin(\omega t)$ is the external harmonic exciting force.

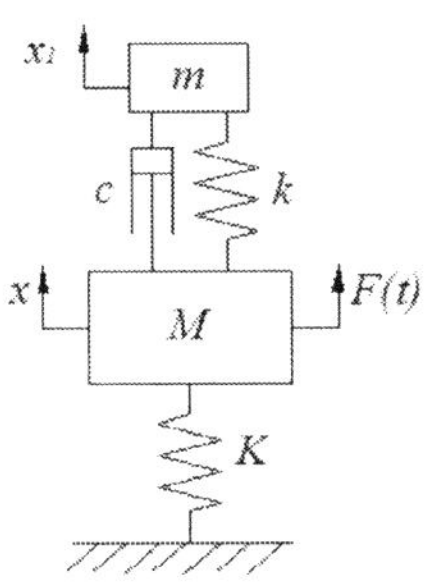

Fig. 2. Operating principle of SDOF-TMD subsystem

3.2 Multi-mode TMD operating principle

By applying the Fourier transform, the SDOF-TMD motion equation becomes:

$$(K-\omega^2M+j\omega C)X(\omega)=F(\omega) \quad (3)$$

or

$$[Z]\{X\}=\{F\} \quad (4)$$

where

$$[H]=[Z]^{-1}=([K]-\omega^2[M]+j\omega[C])^{-1} \quad (5)$$

$[H]$ is the mechanical admittance matrix or transfer function matrix. According to the theory of continuous-structure multi-mode vibration suppression [16], the trend of $[H]$ with the change of k/K can be described as follows: When $k/K<10^{-2}$ (i.e., the TMD subsystem stiffness is much smaller than the main system stiffness) transfer characteristic almost does not change, and it can be assumed that TMD is equal to the surface damping. When $k/K>10^5$ (i.e., the TMD subsystem stiffness is much larger than the main system stiffness) transfer characteristic almost also does not change, it can be assumed that the TMD is completely rigid, i.e., a SDOF-TMD. When $10^{-2}<k/K<10^5$, the stiffness of TMD subsystem and main system is close. The TMD is a continuous structure with multiple modal, and it can perform the multi-mode vibration control of the main system.

4 TMD structure design and simulation

4.1 Compressor modal analysis

To determine that the TMD's frequency band can cover the compressor's natural frequency range, the modal of compressor was analyzed. A solid model of the compressor system was established. Fig. 3 shows the finite-element (FE) model after the meshing. Table 1 lists the natural frequency of the compressor system calculated using the FE software (ANSYS). The compressor system mode shapes at 480 Hz and 550 Hz show that the deformations of the pipeline and gearbox are consistent with the actual vibration.

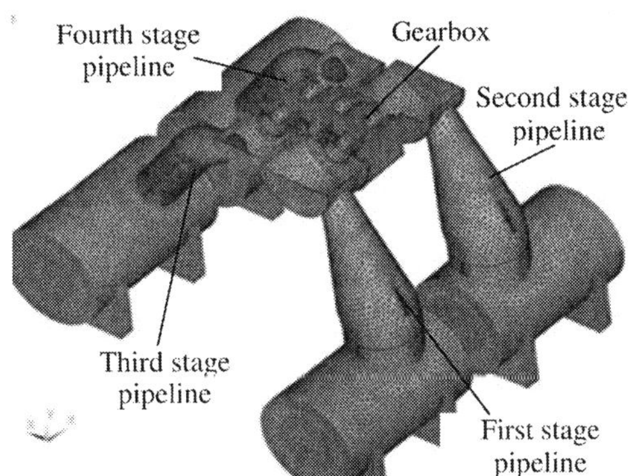

Fig. 3. FE model of compressor

Table 1. Natural frequencies of compressor

Mode number	1	5	10	15	20	25	30
Frequency (Hz)	163	201	289	336	403	426	480
Mode number	40	50	60	70	80	90	100
Frequency (Hz)	550	615	657	703	755	814	870

4.2 TMD structure design

Three types of continuous-structure TMDs were designed. The overall structure of the TMDs was divided into three parts. The front part was the pipe clamp that delivered the vibration energy. The middle part was the elastic connection, and the rear part was a hollow cylindrical mass containing a large damping material. The piston rod was connected to the base, and it was damped by stirring the damping material when the mass vibrated. The compressor vibration was transferred from the clamp and middle connection to the mass and increased the vibration of the mass. Thus, the mass-absorbed part of the compressor vibration energy and damping material dissipated energy, finally decreasing the compressor vibration. The types of TMD's elastic connection were different, and the clamp of the third TMD was reinforced.

Within 0–1500 Hz, the natural frequency range of the three TMDs covered the compressor's natural frequency range. Figs. 4 (a)–(c) show the FE models of the three TMD, and Table 2 lists their natural frequencies.

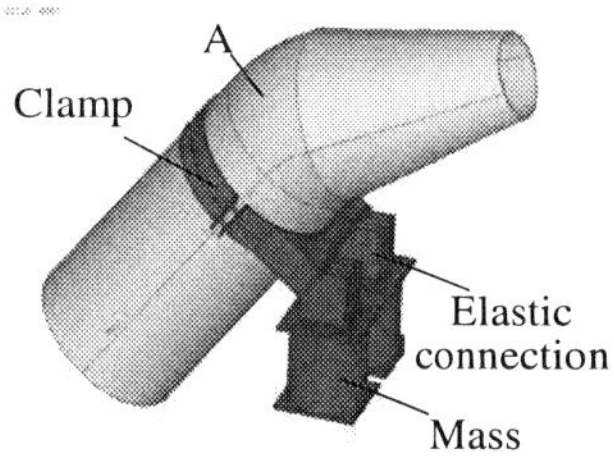

(a) FE model of the first TMD

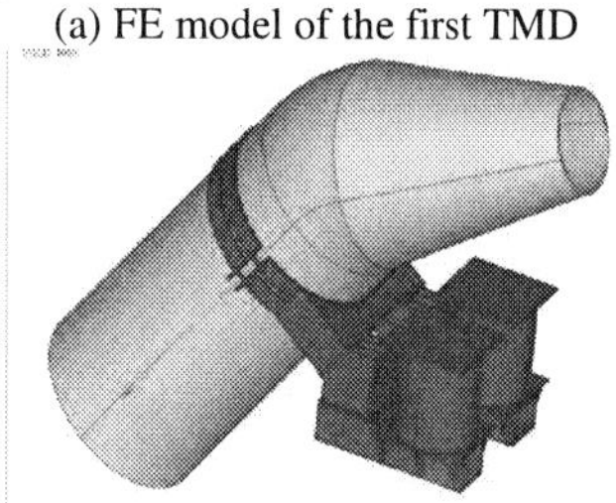
(b) FE model of the second TMD

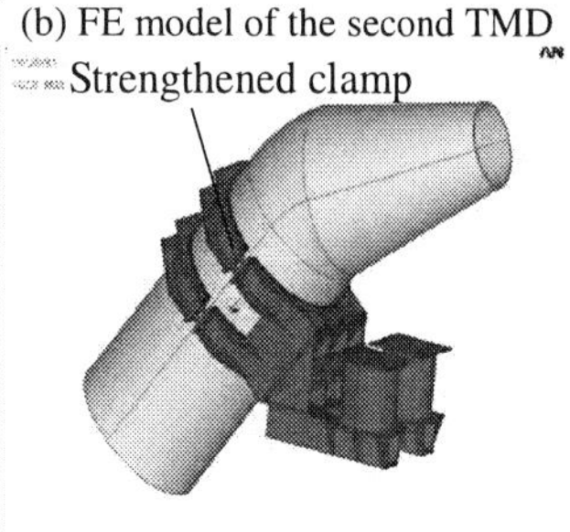

(c) FE model of the third TMD

Fig. 4. FE models of the three types of TMDs

Table 2. Natural frequency of TMDs

Mode number	Frequency (Hz)		
	First TMD	Second TMD	Third TMD
1	117	81	140
2	118	82	143
3	127	87	282
4	130	88	286
5	282	199	383
6	283	200	388
7	496	211	617
8	500	213	641
9	521	353	675
10	523	354	698
11	1111	871	1112
12	1115	876	1130
13	1123	1204	1214
14	1137	1211	1235

The structure vibration reduction effects of the three types of TMDs were simulated by ANSYS harmonic response analysis. At point A (Fig. 4), harmonic forces were applied along the x-, y-, and z-axis. The frequency range of the harmonic forces varied within 0–1500 Hz, and the amplitude was set to 10000 N. Ax, Ay and Az were the displacement responses along the x-, y-, and z-axis. The equation represents the structure displacement response of the pipeline under harmonic forces. Because of the displacement, the frequency domain waves of the pipeline were obtained before and after the installation of the three TMDs as shown in Figs. 5(a)–(c).

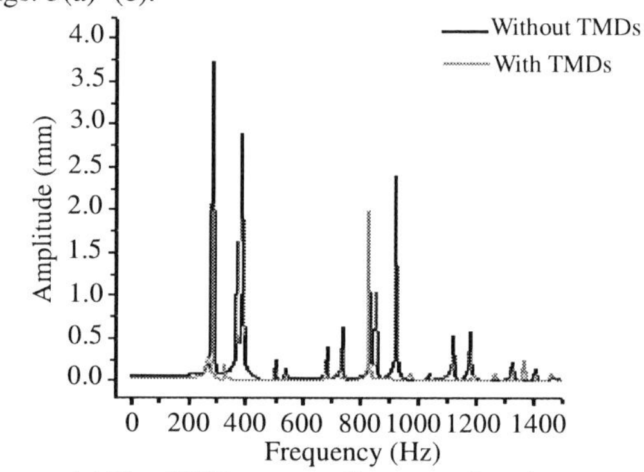

(a) First TMD structure frequency domain wave

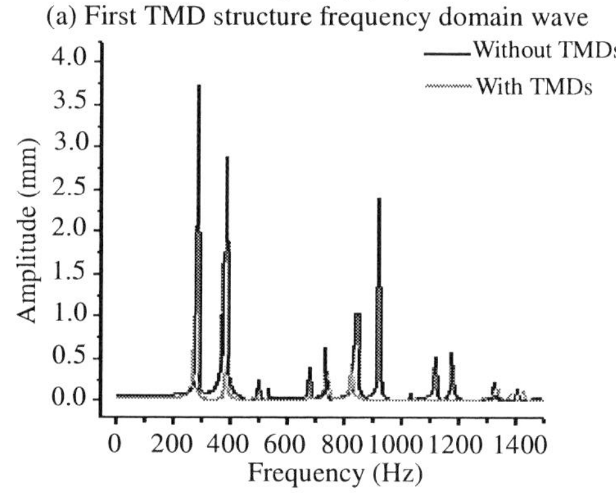

(b) Second TMD structure frequency domain wave

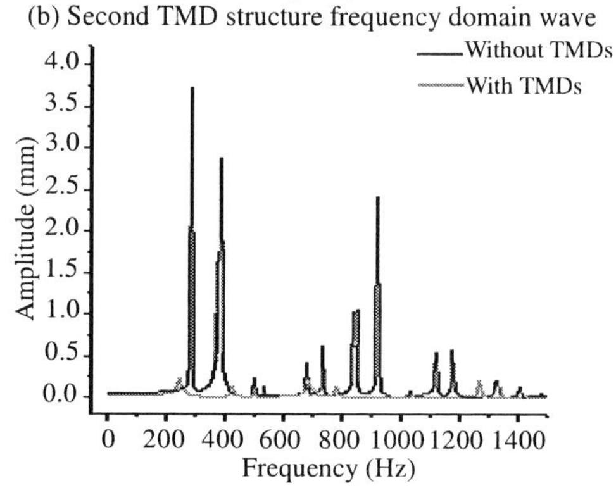

(c) Third TMD structure frequency domain wave

Fig. 5. With and without three types of TMDs frequency domain waves of pipeline

The results show that the third TMD has a better vibration suppression effect reducing amplitude peak of more than 80% on average as the damping material dissipates much more energy. The stiffness ratio of k/K was 9.19, within the range $10^{-2} < k/K < 10^5$, meeting the requirement of a continuous-structure multi-mode vibration suppression. Moreover, the third TMD used a reinforcing clamp, difficult to loose and deform, thus better transferring the vibration. In conclusion, the third TMD structure with a reinforcing clamp is preferred.

4.3 Simulation of vibration suppression

The vibration of the entire compressor system was simulated by harmonic response analysis. The previous analysis conducted on compressor vibrations highlighted that the excitation source is in the pipeline. Harmonic forces were applied at point A (Fig. 4) along the x-, y-, and z-axis to simulate pipeline vibration. The frequency range of the harmonic forces was 0–1500 Hz, and the amplitude was set to 10000 N. The result shows that the pipeline vibration coupled with the compressor case caused the gearbox vibration. It also confirmed the analysis of vibration causes.

According to the analysis of vibration causes, four groups of TMDs were installed on the outlet pipelines, two groups at the first stage and two groups at the third stage. After installing TMDs on the solid model, the same harmonic forces applied in the same positions excited the pipeline, and the vibration of the entire compressor was studied. The result shows that the displacement response of the compressor gearbox equipped with TMDs was reduced than the situation without TMDs. Fig. 6 shows the frequency domain curves of the gearbox before and after installing TMDs and shows the vibration suppression effect on the gearbox vibration.

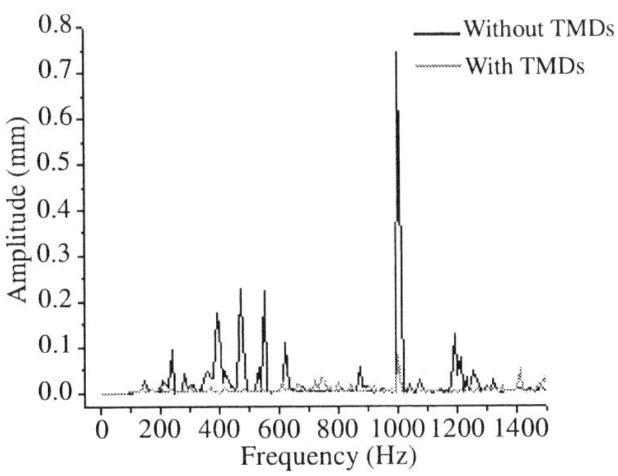

Fig. 6. Frequency domain waves of the compressor gearbox with and without TMDs

The amplitude peak decreased from 0.8 mm to 0.1 mm, and the average vibration decreased by about 80%. This result indicates that TMD not only reduced the pipeline vibration, but also suppressed the vibration of the gearbox. Therefore, the use of TMDs is feasible and effective to ensure the safe operation of a compressor.

5 Practical applications of TMDs and their effects on vibration suppression

According to the vibration conditions, available actual space, and implementation feasibility, the installation positions of TMDs were determined. To avoid installing TMDs in a cone position, two groups of TMDs were placed in the inclined pipe section of the third stage outlet pipeline, and two groups of TMDs were placed in the straight section and tilt flange of the first stage outlet pipeline. Thus, four groups of TMDs were installed in the entire compressor system.

After the installation, the pipeline vibration was measured, and the results show that the effect of damping was remarkable. The measurement points are shown in Fig. 1. Among them, measurement points 1–4 were on the third stage outlet pipeline; measurement points 5–8 were on the first stage outlet pipeline. Table 3 compares the vibration amplitudes of each measurement point with and without TMDs.

Table 3. Vibration amplitudes of the compressor pipeline with and without TMDs

Measurement point	Without TMDs (mm/s)	With TMDs (mm/s)	Decrement (%)
1	154.3	94.3	39.9
2	180.9	108.0	40.1
3	205.8	138.4	32.9
4	192.4	89.2	53.7
5	150.8	58.2	61.5
6	207.4	89.8	57.3
7	259.1	86.3	66.7
8	185.6	92.5	50.2

The comparison highlights that the vibration amplitude of the first stage outlet pipeline decreased by on average 41.7%, whereas that of the third stage outlet pipeline decreased by on average 58.9%. Table 4 compares the vibration amplitudes of the compressor gearbox with and without TMDs.

Table 4. Vibration amplitudes of the compressor gearbox with and without TMDs

Measurement point	Without TMDs (mm/s)	With TMDs (mm/s)	Decrement (%)
Horizon	8.48	2.83	66.7
Vertical	6.01	3.34	44.5
Axial	9.75	3.50	64.2

The comparison shows that the vibration amplitude of the compressor gearbox decreased by on average 58.5%. According to the ISO 10816-3 vibration standards, the vibration grade changed from grade D to grade B.

6 Conclusions

(1) A parallel-axis centrifugal air compressor has strong coupling characteristics, making it difficult to solve the vibration. According to the vibration detection and fault diagnosis, the gas pulsation caused by the rotating stall is the main source of vibration. By considering the compressor as an integral system with a strong correlation, TMDs were installed on the pipelines to reduce the pipeline vibration. The TMDs reduced the gearbox vibration as an indirect effect.

(2) Because a compressor has many variable frequencies of vibrations to be suppressed, a continuous-structure TMD subsystem was designed. The main system's multifrequency vibration was controlled using the TMDs' multi-modes. Thus, the continuous-structure multi-mode vibration suppression method is feasible. Compared to other methods such as SDOF-TMD or MTMD, it is more suitable for actual engineering applications.

(3) The TMDs were optimized using the FE software ANSYS. The modes of the entire compressor system and TMD subsystem were calculated, and the TMD subsystem's natural frequency range covered the natural frequency range of the compressor system. By harmonic response analysis, the effect of vibration reduction was simulated, and the optimal structure was investigated from the perspective of strength. The vibration condition was improved after the practical assembling.

(4) A practical application indicates that the multi-mode TMDs vibration suppression method can reduce the vibration of pipelines and gearbox and help compressor to operate safely.

References

1. Xia B Q, Yu L, Xie Y B. Dynamics analysis of geared rotor-bearing system of DH type turbine Compressor [J]. Journal of Vibration Engineering, 2003, 16(2): 251-255.
2. Xie H J. Research on rotor system dynamic properties of DH centrifugal compressor [D]. Beijing: China University of Petroleum, 2009.
3. Li M, Sun T, Hu H Y. Review on dynamics of geared rotor-bearing systems [J]. Journal of Vibration Engineering, 2002, 15(3): 249-256.
4. Mu S. Finite element analysis of dynamic characteristics of large welding casing [D]. Xi'an : Xi'an University of Technology, 2008.
5. Chen S R, Cai C S. Coupled vibration control with tuned mass damper for long-span bridges [J]. Journal of Sound and Vibration, 2004, 278: 449-459.
6. Liu Y Z, Yu D L, Zhao H G, et al. Review of passive dynamic vibration absorbers [J]. Chinese Journal of Mechanical Engineering, 2007, 43(3): 14-21.
7. Den H. Mechanical vibration. 4th edition [M]. New York: McGraw-Hill, 1956.
8. Nishihara O, Asami T. Closed-form solutions to the exact optimizations of dynamic vibration absorbers (minimizations of the maximum amplitude magnification factors) [J]. ASME Journal of Vibration and Acoustics, 2002, 124(4): 576-582.
9. Asami T, Nishihara O. Analytical and experimental evaluation of an air damped dynamic vibration absorber: design optimizations of the three-element type model [J]. ASME Journal of Vibration and Acoustics, 1999, 121(3): 334-342.
10. Ren M Z. A variant design of the dynamic vibration absorber [J]. Journal of Sound and Vibration, 2001, 245(4), 762-770.
11. Pennestri E. An application of Chebyshev's Min-Max criterion to the optimal design of a damped dynamic vibration absorber [J]. Journal of Sound and Vibration, 1998, 217(4): 757-765.
12. Xu K, Igusa T. Dynamic characteristics of multiple substructures with closely spaced frequencies [J]. Earthquake Engineering and Structural Dynamics, 1992, 21: 1059-1070.
13. Rice H J. Design of multiple vibration absorbers systems using modal data [J]. Journal of Sound and Vibration, 1993, 160: 378-385.
14. Nishida E, Koopmann G H. A method for designing and fabricating broadband vibration absorbers for structural noise control [J]. ASME Journal of Vibration and Acoustics, 2007, 129(4): 397-405.
15. Zhang C L. The mechanism study of broadband frequency tuned mass damper and its applications [D]. Nanjing: Southeast University, 2007.
16. Li Y, Jiang P Z. Process fluid machinery. second edition[M]. Beijing: Chemical Industry Press, 2008.
17. Zhang X H, Xu Z C, Duan L G, et al. Analysis and precaution for surge of centrifugal air compressor [J]. Hebei Metallurgy, 2013, 1: 49-51.
18. Song X H, Fan D H, Qiao J. Piping vibration analysis and measures of vibration damping of centrifugal compressor [J]. Science Technology and Engineering, 2009, 9(17): 4930-4933.

WEAK FAULT ENHANCEMENT METHOD FOR BLADE CRACK BY USING STOCHASTIC RESONANCE

Changbo He[1] and Hongkun Li[2]

1 School of Mechanical Engineering, Dalian University of Technology, Dalian 116024
1019894427@qq.com
2 School of Mechanical Engineering, Dalian University of Technology, Dalian 116024
lihk@dlut.edu.cn

Abstract. Centrifugal compressor is a piece of key equipment for petrochemical factories. For centrifugal compressor, blades as the most important part suffer periodic vibration and flow induced excitation mechanism, which are prone to failure. As the blade failure has a serious impact on the operation of a compressor, which may lead to catastrophes, the method for the incipient weak fault determination is urgently needed. Although FFT and envelope analysis have been widely used for rotating equipment, they are not very helpful for the determination of incipient crack of a blade as the signal modulation and noise interference. In this research, stochastic resonance is used for the pressure pulsation signal which is acquired by pressure sensors to extract the characteristic frequency of crack. Comparing the result with FFT, it can be concluded that crack characteristics can be determined by stochastic resonance effectively although the information is weak.

Keywords: Centrifugal compressor, blade crack, feature extraction, stochastic resonance

1 Introduction

During the operation of a centrifugal compressor, the blades suffer the combined effects of centrifugal force, unsteady flow, vibration, etc [1]. Therefore, as the most sensitive part in the compressor, blades exposed to periodic vibration and fluid excitation for a long time [2]. At the same time, with the development of turbomachinery, the working environment for the impeller is more harsh, the result is that the blades are more susceptible to cracks and other faults. How to detect crack fault of blades dynamically with a relatively simple but accurate method is one of the difficult problems in fault diagnosis of engineering and academic circle. Usually, displacement sensors are used to detect the shaft vibration [3-4], but using shaft vibration signal to identify the crack is always very difficult. Compared to the shaft crack, the identification for blade crack is more difficult. Many scholars studied blades with different ways. Rao and others proposed a method which is extracting the characteristic information from the vibration signal of gas turbine to recognize the blade status [5]. WITEK studied the develop process of the blade crack using vibration signal obtained from the laboratory [2]. EGUSQUIZA and others studied the blade failure mechanism of turbine pump, which provides the theoretical support for blade feature extraction [6]. In addition, acoustic emission signal is also used to study the classification of turbine blades [7]. Support vector machine has very good performance on blade crack classification and prediction, so it is used to detect the blade damage of helicopter [8]. At the same time, König made a series of research on centrifugal compressor impeller failure mechanism through pressure pulsation signal, which has good results [9]. Pressure pulsation signal is used for the extraction of the crack of a blade in this paper.

2 Stochastic resonance system

Stochastic resonance model usually consist of three basic elements, which are weak input signal, noise and nonlinear system used for signal processing separately. Figure 1 shows the structure of stochastic resonance system. The weak input signal can be periodic signal, non-periodic signal, deterministic signal or a random signal, and other types of signal. The noise, which is actually the random signal satisfying a certain requirement of statistical characteristic, can be inherent noise of the system or external noise, such as white noise, gauss noise or non gauss noise. The mixed signal of useful signal and noise is selected to be the input of the system, then the processed signal can be obtained through the system.

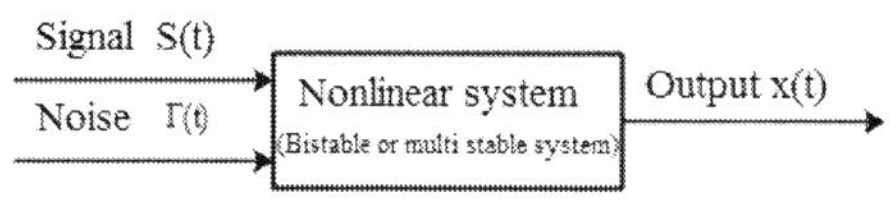

Fig.1 The general structure block diagram of stochastic resonance

2.1 Bistable stochastic resonance model

Bistable system is studied by many scholars in the study of the effect of noise on nonlinear systems, as its typical effect. The bistable system subjected to random noise and external force can be described by Langevin equation(LE), which is shown as follows

$$\begin{cases} \dot{x} = -V'(x) + s(t) + \eta(t) \\ \langle \eta(t) \rangle = 0, \quad \langle \eta(t), \eta(0) \rangle = 2D\delta(t) \end{cases} \tag{1}$$

Wherein, $V(x)$ is the nonlinear bistable potential function, and

$$V(x) = -\frac{a}{2}x^2 + \frac{b}{4}x^4 \qquad a > 0, b > 0 \tag{2}$$

$s(t)$ represents the external force, $\eta(t)$ is the Gaussian white noise with zero mean value and D intensity.

Take formula (2) into formula (1), then formula (3) can be obtained, which is shown as follows.

$$\dot{x} = ax - bx^3 + s(t) + \eta(t) \tag{3}$$

x is the output signal of Stochastic resonance system. a, b represent the structure parameters of this nonlinear bistable system, both of them are real number greater than zero. The differential equation above describes the over damped motion of unit particle subjected to both external force and noise in double-well potential. The height of potential barrier is $\Delta V = \frac{a^2}{4b}$, when there is no the effect of external force and noise. The potential function takes the minimum value at $x_m = \pm\sqrt{a/b}$, and takes the maximum value at $x_b = 0$.When there is only external periodic excitation signal, the bistable potential well changes periodically driven by this excitation signal. However, the transition switch of particles will occur according to Kramers transition rate when there only exists noise. The periodic signal will introduce periodic variation to the switch of potential well of the stochastic resonance system, which can effectively synchronize with the transition caused by the noise, when the system is affected by both weak periodic signal and noise. The output signal of the system will synchronize with weak periodic signal when the signal, noise and the system's nonlinearity reach a certain matching state. As a result, the weak signal can be enhanced as the action of bistable stochastic resonance system.

3 Application for the weak fault signal

3.1 The experiment condition

The bistable stochastic resonance system is used to extract weak characteristic frequency of a centrifugal compressor. The impeller is a semi-open one which is used for test with 13blades and 800 mm diameter. The rotating speed of the impeller is 5000r/min in this experiment, so the corresponding shaft frequency and blade passing frequency are 83.3 Hz and 1083.3 Hz respectively. A notch is machined on a single blade of the normal semi-open impeller to simulate the crack. Collecting the pressure pulsation signal which can reflect the crack information during the operation of the compressor. As the thickness of the blade is very large, the information of the blade crack is relatively weak which means the useful frequency is difficult to be identified.

Fig. 1. The impeller with crack

3.2 Pressure pulsation signal analysis

Pressure pulsation is generated due to the interaction between the rotor and the stator inside the compressor, so the characteristic frequency is mainly blade passing frequency and its harmonic frequency. When there exits a crack in the blade, the characteristic frequency of crack will be modulated to blade passing frequency as it belongs to low frequency componet. Analyzing the sampled data by FFT, only the twice frequency of shaft frequency can be clearly found seen as Fig 2. However, there is no information about crack frequency as the strong interference of noise, which indicates FFT is not enough for the detection of weak fault signal. So, envelope analysis and bistable stochastic resonance system are used to process the acquired signal respectively, the result are shown as Fig 3 and Fig 4. It can be seen from these two figures that the characteristic frequency 53 Hz of crack is through stochastic resonance more clear than the result of envelope analysis. Comparing the results of FFT, envelope and bistable stochastic resonance, it can be found stochastic resonance method is very helpful for the frequency determination of weak fault.

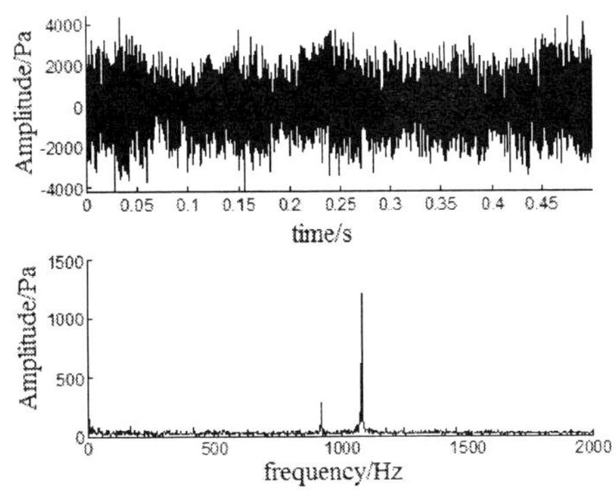

Fig. 2. Time domain and frequency domain of pressure signal

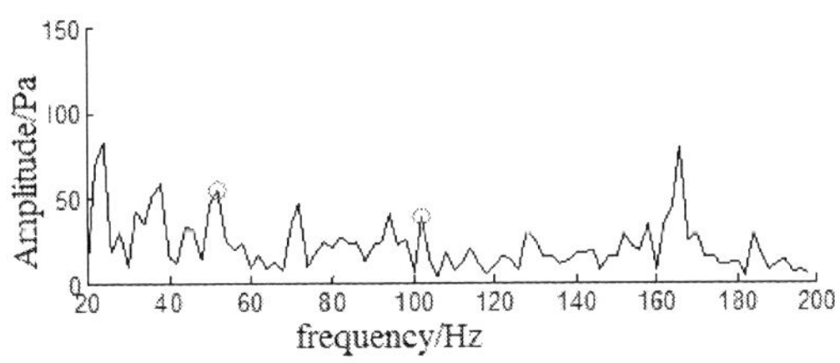

Fig. 3. Envelope analysis for pressure signal

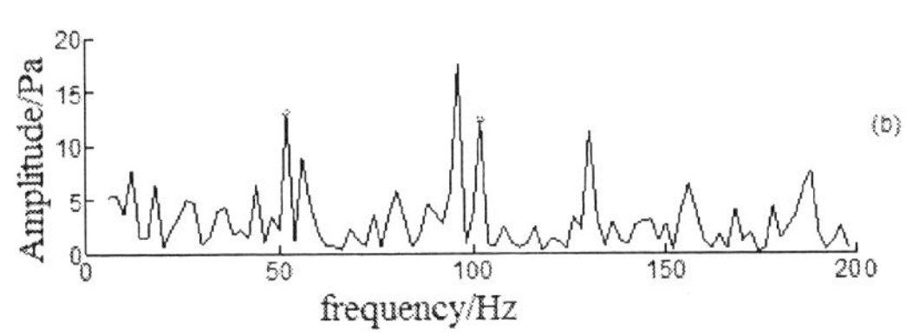

Fig. 4. The result of stochastic resonance

4 Conclusion

1. The vibration of blades will be affected when there exists crack, and the blade's vibration will affect pressure pulsation of airflow further. Therefore, the pressure pulsation signal contains crack information when the blade is faulted. But as the crack information belongs to early weak fault, it is difficult to be identified by traditional methods.
2. The analysis result indicates that bistable stochastic resonance system is helpful for the determination of crack frequency comparing with FFT and envelope analysis.

References

1. EISINGER F L. Acoustic fatigue of impellers of rotating machinery[J]. Journal of Pressure Vessel Technology, 2002, 124(2): 54-160.
2. WITEK L. Experimental crack propagation and failure analysis of the first stage compressor blade subjected to vibration [J]. Engineering Failure Analysis, 2009, 16(7): 2163-2170.
3. ELBHBAH K, SINHA J K. Vibration-based condition monitoring of rotating machines using a machine composite spectrum [J]. Journal of Sound and Vibration, 2013, 332(11): 2831-2845.
4. SARAVANAN K, SEKHAR A. Crack detection in a rotor by operational deflection shape and kurtosis using laser vibrometer measurements [J]. Journal of Vibration and Control, 2012, 0(0): 1-13.
5. RAO A R, DUTTA B. Vibration analysis for detecting failure of compressor blade [J]. Engineering Failure Analysis, 2012, 25: 211-218.
6. EGUSQUIZA E, VALERO C, HUANG X, et al. Failure investigation of a large pump-turbine runner [J]. Engineering Failure Analysis, 2012, 23: 27-34.
7. JOOSSE P, BLANCH M, DUTTON A, et al. Acoustic emission monitoring of small wind turbine blades [J]. Journal of Solar Energy Engineering, 2002, 124(4): 446-454.
8. PAWAR P M, JUNG S N. Support vector machine based online composite helicopter rotor blade damage detection system [J]. Journal of Intelligent Material Systems and Structures, 2008, 19(10): 1217-1228.
9. KÖNIG S, PETRY N, WAGNER N G. Aeroacoustic phenomena in high-pressure centrifugal compressors—a possible root cause for impeller failures [C]. Proceedings of the 38th Turbomachinery Symposium, 2009.

Research on Performance Degradation Assessment Technology for Ball Screw

Xiaochen Zhang[1,*], Dongxiang Jiang[1], Hongli Gao[2]

[1] State Key Lab of Power Systems, Department of Thermal Engineering, Tsinghua University, Beijing 100084, China
zhangxch2008@hotmail.com, jiangdx@mail.tsinghua.edu.cn
[2] School of Mechanical Engineering, Southwest Jiaotong University, Chengdu 610031, China
hongli_gao@home.swjtu.edu.cn

Abstract. To evaluate the performance of ball screw, performance degradation assessment technology based on dynamic fuzzy neural network (DFNN) is proposed. A kind of screw test platform designed for accelerated test is introduced. A three-way accelerometer and two one-way accelerometers are installed to monitor the real-time acceleration signals of screw. With the method of principal component analysis (PCA), high dimension features are mapped to low dimensional space and stored into knowledge base together with screw expected residual lifetime. Meanwhile, screw performance degradation model which is mainly made up of two DFNNs receives real-time feature vector. By means of empirical mode decomposition (EMD), screw performance degradation tendency is extracted from the output of the online DFNN and treated as screw residual lifetime. When the output of the online DFNN appears a large deviation compared with the expected residual lifetime, the standby DFNN will copy network parameters of the old online DFNN and be the new online DFNN. The old online DFNN would be optimized by artificial fish swarm algorithm (AFSA) and retrained to get a good prediction result. The experimental results show that the proposed performance degradation model could meet the need of screw performance assessment.

Keywords: degradation assessment, ball screws, fuzzy neural networks, real time systems, artificial fish swarm algorithm

1 Introduction

The essence of ball screw assessment technology is predicting the screw performance degradation trend and calculating the screw residual life[1-2]. A screw condition monitoring system in accordance with manufacturing requirements will contribute to improving the product quality and reducing the manufacturing cost. In recent years, scholars and enterprises have made great efforts on the research of screw, screw condition monitoring technologies based on multi-sensor fusion and artificial intelligence have been widely used and achieved some progress[3-4].

At present, acceleration sensors are adopted to monitor screw vibration condition. Artificial intelligence modeling methods, such as artificial neural network and grey system, are used to establish screw performance degradation model, while model parameters are determined by training samples. Only when enough training samples are obtained and used to train the degradation model offline, the prediction accuracy of model can be guaranteed. In addition, as the key component of NC machine tool, screw motion system is a kind of typical nonlinear system. Variations of NC machine tool structure parameters have great impacts on the output of the acceleration sensors, so the degradation models are not easy to apply on different types of NC machine tools[5-6].

This paper presents a kind of screw performance degradation model based on dynamic fuzzy neural network (DFNN), which contains two parallel life calculating models to realize model online training, predicting and parameters updating. DFNN can accomplish the structure identification and the parameter updating at the same time, which is adapted to system online modeling. The experiment results show that the proposed screw performance degradation model can accurately predict the residual lifetime of screw, which has a good industrial prospect.

2 Screw Condition Monitoring System

2.1 Screw Accelerated Test

Fig. 1. Screw accelerated test platform

The screw accelerated test platform is shown in Figure 1. A Beckhoff servo motor is used to drive a ball screw, meanwhile the screw nut is connected with the back surface of a sliding platform. A gear rack mechanism is installed on the upper surface of the sliding platform. The gear and a magnetic powder brake are connected by a torque sensor. With the help of the gear rack mechanism, the magnetic powder brake can impose axial force on ball screw. By controlling the output torque of the magnetic powder brake, the axial force of ball screw can be adjusted. One three-way accelerometer is installed on screw nut while two one-way accelerometers are installed on two bearing chocks. Three Beckhoff EL3632 modules connected with the Beckhoff industrial PC are responsible for accelerometer signal acquisition. Industrial PC stores the vibration data, while the sampling frequency is 5kHz. Ball screw works in horizontal installation state. The installation method of the ball screw is one end fixed while the other end floated. Ball screw reaches 40mm in axis diameter and 5mm in lead.

2.2 Mathematical Morphology Filter

Mathematical morphology filter is an important nonlinear filter developed in recent years, which can eliminate some interferential components of the signal without corrupting the signal shape[7-9]. Considering characteristics of the rotating machinery vibration signal, which not only has random noise but also disturbed by impulse signal, mathematical morphology filter is used to reduce noise. The transformation form of the mathematical morphology filter mainly consists of corrosion , expansion, open operation and closed operation. The one-dimensional discrete signal can be expressed as $f(n)$, while $F=(0,1,...,N-1)$ is the domain of $f(n)$. $g(m)$ is the structure element with a domain $G=(0,1,...,M-1)$. On the premise of $N>>M$, corrosion and expansion can be defined as follows:

$$(f\Theta g)(n) = \min[f(n+m) - g(m)] \qquad m \in G \tag{1}$$

$$(f \oplus g)(n) = \max[f(n-m) + g(m)] \qquad m \in G \tag{2}$$

Open operation and closed operation can be expressed as follows:

$$(f \circ g)(n) = (f\Theta g \oplus g)(n) \tag{3}$$

$$(f \bullet g)(n) = (f \oplus g\Theta g)(n) \tag{4}$$

Open-closed transformation form and closed-open transformation form can be built as follows:

$$F_{oc} = (f \circ g \bullet g)(n) \tag{5}$$

$$F_{co} = (f \bullet g \circ g)(n) \tag{6}$$

In order to achieve better filtering effect, combination transformation form which is used in this paper can be formulated as:

$$y(n) = \frac{1}{2}(F_{oc} + F_{co}) \tag{7}$$

3 Screw Performance Degradation Assessment

3.1 Dynamic Fuzzy Neural Network

As is shown in Figure 2, DFNN is a kind of expansion radial basis function neural network which is consist of five layers[10]. In Figure 2, the first layer is the input layer where the input variables $x_1, x_2, \ldots, x_r$ are input to the nodes; the second layer is the membership function layer where MF_{ij} is the membership function; the third layer is the t-norm layer and the fourth layer is the normalized layer. System output y and weight w_j constitute the fifth layer. u is the number of the rules.

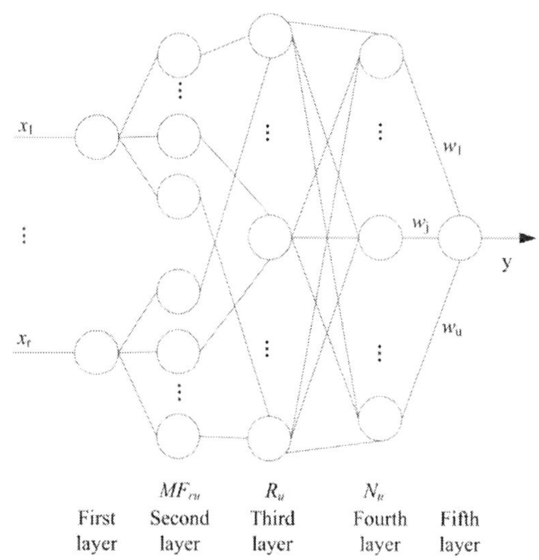

Fig. 2. The structure diagram of DFNN

Each layer can be expressed as follows:

(1) The first layer: each node is responsible for a variable.

(2) The second layer: the jth membership function u_{ij} of the input variable x_i can be defined as:

$$\mu_{ij}(x_i) = \exp\left[-\frac{(x_i - c_{ij})^2}{\sigma_j^2}\right] \tag{8}$$

Where $i=1,2,\cdots,r$; $j=1,2,\cdots,u$. c_{ij} is the center while σ_j is the width of u_{ij}.

(3) The third layer: nodes are used to represent fuzzy rules. The output of the *j*th node R_j can be expressed as:

$$\phi_j = \exp\left[-\frac{\sum_{i=1}^{r}(x_i - c_{ij})^2}{\sigma_j^2}\right] \tag{9}$$

Where $j=1,2,\cdots,u$.

(4) The fourth layer: the output of *j*th node N_j is defined as:

$$\psi_j = \frac{\phi_j}{\sum_{k=1}^{u}\phi_k} \tag{10}$$

Where $j=1,2,\cdots,u$.

(5) The fifth layer: the system output can be calculated as:

$$y(\mathbf{X}) = \sum_{k=1}^{u} w_k \cdot \psi_k \tag{11}$$

Where w_k is the weight of *k*th fuzzy rule.

For DFNN, to get the compact structure, rule activeness detection is very important. Inactive rules should be rejected on time. Though DFNN can keep a relative compact structure through prune technology, it is still easy to fall into local optimum since network parameters are initialized randomly. Artificial fish swarm algorithm (AFSA) is applied to search the optimal initialization network parameters in this paper. Figure 3 shows the algorithm flowchart of DFNN and ASFA. DFNN initialization network parameters including minimum debugging standard ε_{min}, maximum debugging standard ε_{max}, minimum output error e_{min}, maximum output error e_{max}, overlap factor of the radial based function k, width renewal factor k_w, rule threshold k_{err}, decay constant γ, convergence constant β, function width of the first rule σ_0.

The AFSA fitness function can be expressed as:

$$f = \frac{1}{\sum_{p=1}^{s}\left|y'_p - t_p\right|} \tag{12}$$

Where t_p is the true value and y'_p is the predictive value. The number of imitating prediction samples is s.

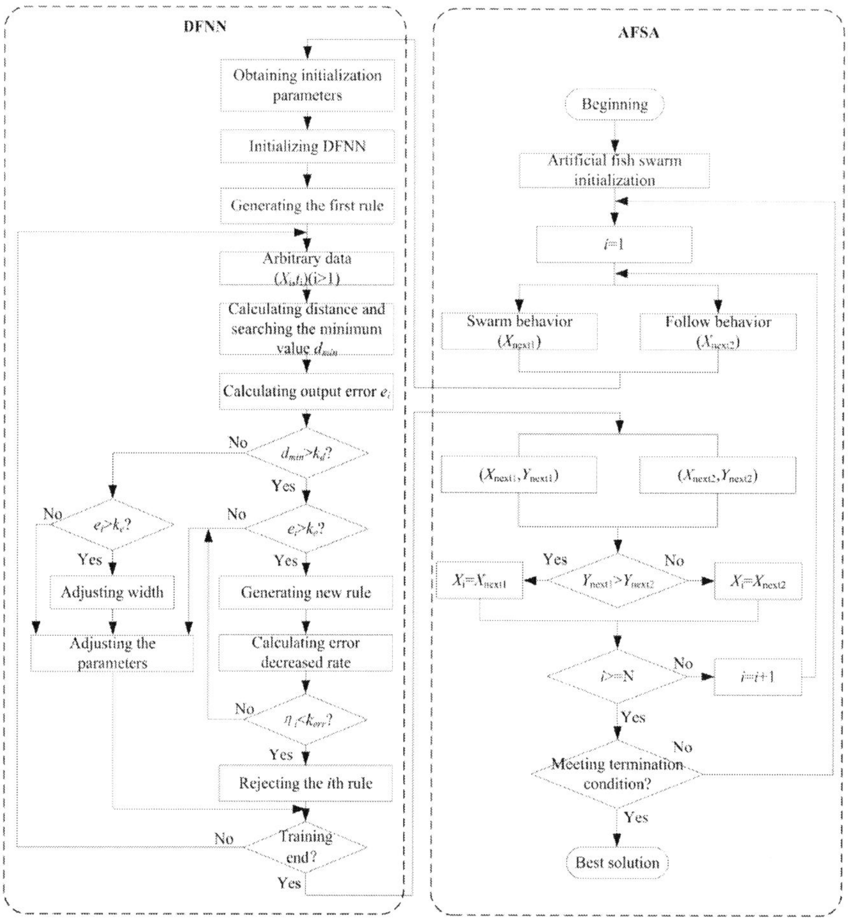

Fig. 3. Algorithm flowchart of DFNN and AFSA

3.2 Screw Performance Evaluation Index

As an important performance evaluation index, residual lifetime is adopted to express screw performance degradation. Screw lifetime is defined as follows:

$$L = \left(\frac{C_a}{F_m} \right)^3 \times 10^6 \tag{13}$$

Where L is the rated revolution of ball screw, C_a is the rated dynamic load and F_m is the axial load. Then the ball screw rated lifetime L_h can be expressed as:

$$L_h = \frac{L}{60 \times n_m} \tag{14}$$

Where n_m is the rotational speed of ball screw. Thus

$$L_e = L_h - L_w \tag{15}$$

Where L_e is the expected residual lifetime of ball screw and L_w is the servicing lifetime.

The expected residual lifetime of ball screw presents a kind of slow descending tendency, which means that the desired output of DFNN is a kind of trend curve. Therefore, this paper adopts empirical mode decomposition (EMD) to search the slow tendency from DFNN output.

3.3 Screw Performance Degradation Model

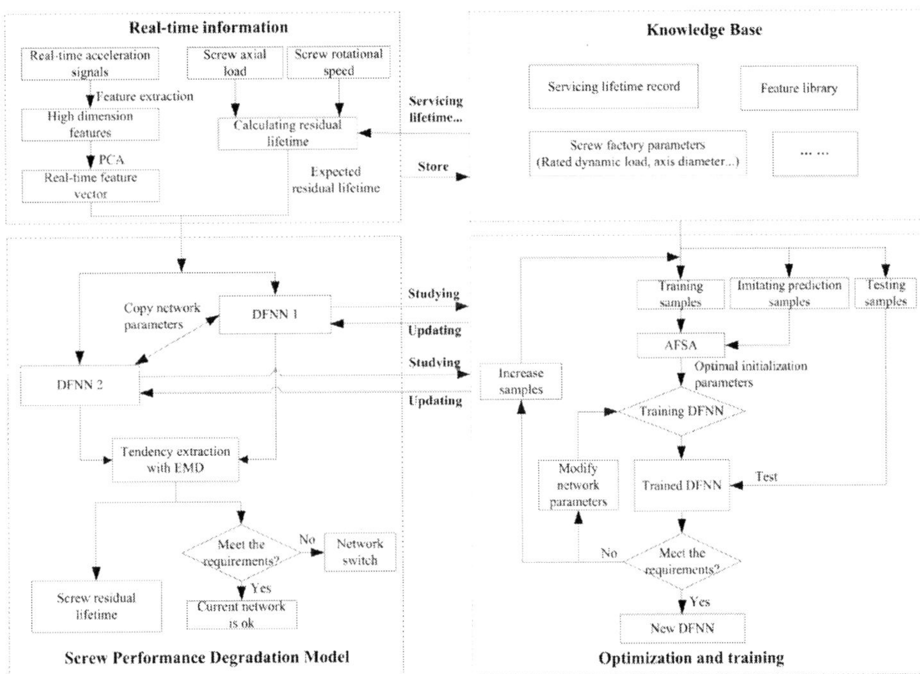

Fig. 4. Flowchart of building screw performance degradation model

Figure 4 shows the flowchart of building screw performance degradation model. From Figure 4, we can know that screw performance degradation model is made up of two DFNN. One DFNN is responsible for predicting screw residual lifetime online, while the other one is on standby status. This kind of structure ensures that one DFNN can be real-time working at any time.

When the output of the online DFNN appears a large deviation compared with the expected residual lifetime, the standby DFNN will copy network parameters of the old online DFNN and be the new online DFNN. The old online DFNN would be optimized by AFSA and retrained to get a good prediction result. Then the new retrained DFNN will be the online working network again while the new network parameters would be assigned to the other DFNN.

4 The Experiment Results

In this paper, high dimension features are made up of following parts. The acceleration signal is decomposed into 5 levels selecting wavelet 'db1', while wavelet packet energies are calculated; Time domain or frequency domain features. Considering the certain correlation among the high dimensional features, principal component analysis (PCA) is applied to eliminate redundant information and get the feature vector.

In the process of optimization, 100 samples are treated as imitating prediction samples to avoid over optimized, while 100 training samples and 100 testing samples are used to train and test the screw performance degradation model. Parameters of the AFSA are selected as follows: population scale is 20; the number of the iterations is 50, the initialization parameters interval of DFNN is [0, 3], visual distance is 1, while the congestion factor is 0.618.

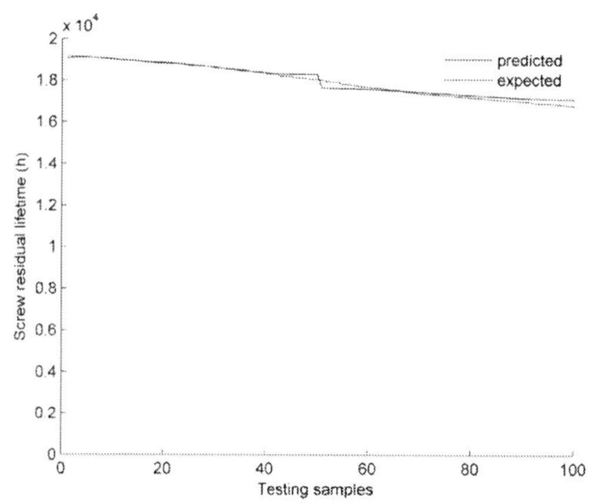

Fig. 5. Predicted screw residual lifetime

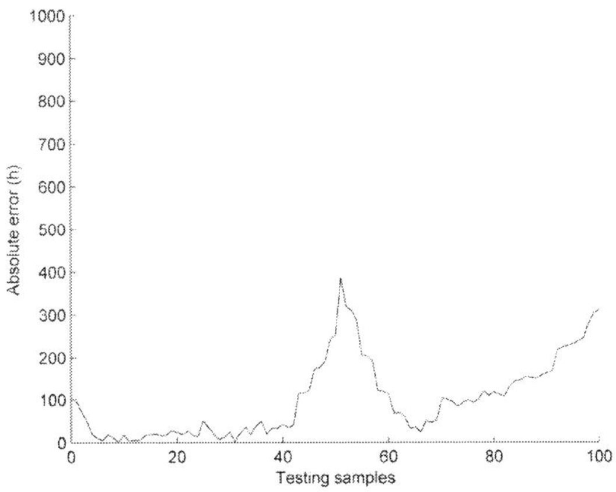

Fig. 6. Absolute error of the predicted results

Figure 5 shows the predicted screw residual lifetime, while Figure 6 shows the absolute error of the predicted results. It is clear that the mean error of the predicted results is 101 hours and the maximum error is 387 hours. Compared with the ball screw rated lifetime, the predicted error of screw performance degradation model is less than 3%, which means that the model accuracy is acceptable. The results indicate that screw performance degradation model could assess screw performance effectively.

5 Conclusion

Screw performance degradation assessment technology for ball screw is studied in this paper. Some conclusions could be obtained as follows:

(1) Acceleration signals can be used to monitor screw performance;

(2) As an important performance evaluation index, residual lifetime can effectively express screw performance degradation degree;

(3) Proposed screw performance degradation model presents an acceptable accuracy and is suitable for screw performance degradation assessment.

References

1. Z. Z. Xu, X. J. Liu, H. K. Kim, et al. Thermal error forecast and performance evaluation for an air-cooling ball screw system[J]. International Journal of Machine Tools & Manufacture, vol. 51, pp. 605-611, April 2011.
2. X. C. Zhang, H. L. Gao, H. F. Huang. Screw performance degradation assessment based on quantum genetic algorithm and dynamic fuzzy neural network[J]. Shock and Vibration, vol. 2015, Article ID 150797, 11 pages, doi:10.1155/2015/150797, 2015.

3. Y. C. Huang, J. L. Chang. Prognostic diagnosis of ball screw preload loss for machine tool through the hilbert-huang transform and multiscale entropy measure[J]. Journal of the Chinese Society of Mechanical Engineers, vol. 32, pp. 349-358, August 2011.
4. M. C. Chang, J. L. Liou, C. C. Wei, et al. Fractal analysis for vibrational signals created in a ball-screw machine operating in short- and long-range tribological tests[J]. Journal of Tribology, vol. 135, pp. 031101:1-11, July 2013.
5. G. Q. Fu, J. Z. Fu, Y. T. Xu, et al. Accuracy enhancement of five-axis machine tool based on differential motion matrix: geometric error modeling, identification and compensation[J]. International Journal of Machine Tools and Manufacture, vol. 89, pp. 170-181, February 2015.
6. A. W. Khan, W. Y. Chen. A methodology for systematic geometric error compensation in five-axis machine tools[J]. International Journal of Advanced Manufacturing Technology, vol. 53, pp. 615-628, March 2011.
7. D. H. Xiao, T. He, Q. Pan, et al. Impulsive noise cancellation of acoustic emission signal based on iterative mathematical morphology filter[J]. Journal of Vibroengineering, vol. 15, pp. 1752-1764, December 2013.
8. A. J. Hu, G. J. Tang, L. S. An. De-noising technique for vibration signals of rotating machinery based on mathematical morphology filter[J]. Chinese Journal of Mechanical Engineering, vol. 42, pp. 127-130, April 2006.
9. X. C. Zhang, H. L. Gao, H. F. Huang, et al. Optimization design of mathematical morphology filter based on quantum genetic algorithm[J]. Journal of Southwest Jiaotong University, vol. 49, pp. 462-469, June 2014.
10. M. J. Er, T. P Tan, S. Y. Loh. Control of a mobile robot using generalized dynamic fuzzy neural networks[J]. Microprocessors and Microsystems, vol. 28, pp. 491-498, May 2004.

Multiwavelet denoising with the data-driven block threshold for pipeline defect detections

Hailiang Sun, Jingzhi Cui, Fei Ma, Qilin He, Haoyang Zhou, Yihua Mo

Beijing Institute of Astronautical Systems Engineering, Beijing, 100076, China

Tel: +86-188-1099-9730, Email: hailiang41@live.cn

Abstract. The ultrasonic guided wave (UGW) has several advantages of less attenuation and longer distance on the propagation path than conventional non-destructive testing methods. It can monitor the pipelines in a wide range and full length. It is significant to extract incipient fault features as early as possible. However, the weak features in pipelines are always immersed in noises of the environment and the equipment. In addition, the dispersion of UGW make it more difficult to detect incipient defects in pipelines. Wavelet denoising is a useful tool for incipient defect detection and its effect mainly depends on the feature separation and the noise elimination. Multiwavelets have two or more multiscaling functions and multiwavelet functions. They possess the properties of orthogonality, symmetry, compact support and high vanishing moments simultaneously. The deficiencies of conventional block threshold denoising are the invariant block size and the global threshold; hence it cannot accurately represent local concentrated energy of the signals in engineering applications. The data-driven block threshold selected the optimal block length and threshold at different decomposition levels by using the minimum Stein's unbiased risk estimate. A multiwavelet denoising technique with the data-driven block threshold was proposed in this paper. The detection of the pipeline defect indicated that the proposed method successfully detected the weak features of incipient faults.

Keywords: ultrasonic guided wave; multiwavelet denoising; data-driven block threshold; pipeline; defect detection

Introduction

Fault detection plays an important role to reduce unscheduled machine downtime and avoid catastrophic accidents [1]. It is significant to extract incipient fault features as early as possible, which is a complex and challenging task that requests advanced analytical methods. The ultrasonic guided wave (UGW) has been proven very efficient for inspecting plate- and pipe-like components because of their capability of propagating along the plane of plate- and pipe-like structures with little propagation loss and their sensitivity to both surface and subsurface features [2]. However, the measured UGW signals are usually complex and non-stationary due to variable operational conditions and defects. What is more, weak features of incipient faults are always immersed in noises generated by the environment and the equipment. Consequently, the main challenge for fault detection is how to accurately extract weak features of incipient faults.

Up to now, researchers have extensively studied advanced signal processing methods for fault feature detections in pipelines. Alleyne et al. [3] presented an experimental time-reversal procedure which was able to regenerate a simple waveform at the receiver position. Prada and Fink [4] and Ing and Fink [5] adopted the time-reversal mirror (TRP) to compensate for the dispersion of Lamb waves. Gangadharan et al. [6] investigated the feasibility of time reversal for resolution improvement under broadband signal excitation. In summary, the researches on pipeline defect detection only focused on feature detections and wave dispersion compensation. Studies on weak feature detections of incipient faults are rare. Actually, UGW signals in pipelines are non-Gaussian and nonstationary, and the defect features in the signals are usually weak.

Because of the multi-resolution analysis of wavelet transform, wavelet denoising technique has obtained a good result in signal denoising and been widely used in engineering applications [7]. For the feature detection, multiwavelets possess more than two different multiscaling functions and multiwavelet functions, which can possess such properties as compact support, symmetry, orthogonality and high vanishing moment simultaneously. One of the earliest and most popularly used multiwavelets with multiplicity 2 is GHM multiwavelet, which are effective for feature detections [8,9]. For the noise elimination, the conventional term-to-term denoising techniques neglect the correlation between wavelet coefficients of fault features, which cannot accurately extract the weak fault features. Based on the correlation between neighboring wavelet coefficients, Cai [10] discussed the James-Stein rule for the block length selection. The length $\log n$ agreed with the compromise between global and local adaptation.

The result of multiwavelet denoising with the block threshold mainly depends on the block length and the threshold. However, the conventional block threshold denoising techniques cannot effectively solve the problems: (1) The threshold is selected according to the prior experience, which generates a large deviation in signal denoising. (2) The universal threshold is adopted at all decomposition levels, which is not the optimal one for each level. To improve conventional wavelet denoising techniques, Cai [11] estimated the data-driven block length and threshold by using the minimum SURE. Further studies are still necessary to verify the optimal selection of block lengths and thresholds.

This paper proposed a multiwavelet denoising technique with the data-driven block threshold for the pipeline defect detection. The structure of this paper is organized as follows. In Section 2, the theory of the block threshold and the data-driven block threshold is given. In Section 3, the proposed method is applied to the pipeline defect detections. Conclusions are given in Section 4.

2 Multiwavelet denoising with the data-driven block threshold

2.1 Multiwavelet denoising with block threshold

Suppose the measured signal $s(n)$ is

$$s(n) = f[n] + \sigma r[n] \tag{1}$$

where $r(n)$ is independently distributed as $N(0,1)$, σ is the variance of the distribution, $f[n]$ is the unknown features to recover. The purpose of signal denoising is to eliminate noise and recover the unknown features $f[n]$ from the noisy signal $s(n)$.

For multiwavelet denoising techniques, the preprocessing is necessary to transform the one-stream signals into multiple-stream ones. Eq. (1) can be expressed as $\boldsymbol{S} = \boldsymbol{F} + \boldsymbol{E}$, where $\boldsymbol{S} = \left[s^{(1)}, s^{(2)}\right]^{\mathrm{T}}$ is the sampled vector signal after preprocessing, $\boldsymbol{F}$ is the unknown vector features corresponding to $f[n]$, $\boldsymbol{E}$ has a multivariate normal distribution $N(0,\boldsymbol{V}_j)$. $\boldsymbol{V}_j$ is the covariance matrix for high-frequency multiwavelet coefficients at the resolution scale j. By using the standard transform $\theta_{j,k} = \boldsymbol{D}_{j,k}^{\mathrm{T}} \boldsymbol{V}_j^{-1} \boldsymbol{D}_{j,k}$, we obtain a positive scalar value that in the absence of any signal component has a χ_M^2 distribution, where M is the number of elements in $\boldsymbol{S}$.

A robust covariance estimation method is used to estimate $\boldsymbol{V}_j$ directly from the observed coefficients [12]. The pseudo-code for the robust covariance estimation is listed [13]:

$$\mathrm{mad}(y)=1.4826 * \mathrm{median}(|y - \mathrm{median}(y)|) \tag{2}$$

where the function $\mathrm{median}(\cdot)$ is used to find out the median value in a given sequence. Float a_1, a_2, b_1, b_2, $\boldsymbol{V}_j$ are obtained by Eq. (3)

$$\begin{aligned}
&a_1 = 1/mad(row_1) \\
&a_2 = 1/mad(row_2) \\
&b_1 = mad(a_1 \cdot row_1 + a_2 \cdot row_2) \\
&b_2 = mad(a_1 \cdot row_1 - a_2 \cdot row_2) \\
&\boldsymbol{V}_j = \begin{bmatrix} 1/a_1 \cdot a_1 & b_1 - b_2/[(b_1 + b_2) \cdot a_1 \cdot a_2] \\ b_1 - b_2/[(b_1 + b_2) \cdot a_1 \cdot a_2] & 1/a_2 \cdot a_2 \end{bmatrix}
\end{aligned} \tag{3}$$

In the first row y is a row vector, and the definition of $\mathrm{mad}(y)$ is given. This definition is applied to calculate a_1, a_2, b_1, b_2 by replacing y with row_1 and row_2. Here row_1 and row_2 are the two rows of multiwavelet coefficients at the resolution scale j.

For multiwavelet block threshold denoising, the incorporation of the multiwavelet coefficients of the block k at the resolution scale j is represented as below

$$Sum_{j,k}^2 = \sum_{n=0}^{L} \theta_{j,k+n}^2, \tag{4}$$

where L is the block length.

The shrinkage rule for multiwavelet denoising with block threshold is expressed as:

$$\hat{\boldsymbol{D}}_{j,k} = \begin{cases} \boldsymbol{D}_{j,k}(1 - \mu^2 / Sum_{j,k}^2), & Sum_{j,k}^2 \geq \mu^2 \\ 0, & Sum_{j,k}^2 < \mu^2 \end{cases} \tag{5}$$

where $\mu = 2\log n_j$, n_j is the length of the multiwavelet coefficients at the scale j.

2.2 Multiwavelet denoising with the data-driven block threshold

The data-driven block threshold chooses the optimal block length and threshold by minimizing SURE [11,14]. Suppose the multiwavelet coefficients at the $m-th$ sub-band are $w_m = \{s_{i,j,k},\ i,j,k \in \text{the } m \text{ subband}\}$, the unknown feature without noise is $f_m = \{f_{i,j,k}\}$, where i denotes the $i-th$ branch of multiwavelet decomposition, j denotes the decomposition level and k denotes the translation of the multiwavelet coefficients at the level j. It was proved that the expected loss or risk of the estimator $\hat{f}_j$ for the true feature f_j could be unbiased estimated based on multiwavelet coefficients w_m. Generally, suppose the variance is $\sigma = 1$, then

$$E\{\|\hat{f}_j - f_j\|_2^2\} = N_m + E\{\|g(f_j)\|_2^2 + 2\nabla g(f_j)\} \tag{6}$$

where $g(f_j)=\{g_n\}_{n=1}^{N_m}=\hat{f}_j-f_j$, $\nabla g=\sum_n \frac{\partial g_n}{\partial f_j}$, it is obtained

$$E\|g_n(f_n)\|_2^2=\left\|\hat{f}_j-f_j\right\|_2^2=\begin{cases} f_n^2 \cdot \mu^4 \Big/ (Sum_{j,k}^2)^2, & Sum_{j,k}^2>\mu^2 \\ f_n^2 & , Sum_{j,k}^2\le\mu^2 \end{cases} \tag{7}$$

$$\nabla\frac{\partial g_n}{\partial f_j}=\begin{cases} -\mu^2\left(Sum^2{}_{j,k}-2f^2{}_n\right) \Big/ (Sum_{j,k}^2)^2, & Sum^2{}_{j,k}>\mu^2 \\ -1 & , Sum^2{}_{j,k}\le\mu^2 \end{cases} \tag{8}$$

$$\text{SURE}(w_m,\mu,L)\text{=}N_m+\sum_n\|g_n(f_n)\|_2^2+2\sum_n\frac{\partial g_n}{f_n} \tag{9}$$

Then $E\{\left\|\hat{f}_j-f_j\right\|_2^2\}=E\{\text{SURE}(w_m,\mu,L)\}$ is the unbiased estimator of the expected loss. The minimum $\text{SURE}(w_m,\mu,L)$ can be obtained by selecting the optimal block length L^m and threshold μ^m, that is

$$(\mu^m,L^m)=\arg\min_{\mu,\,L}\text{SURE}(\text{w}_m,\mu,L) \tag{10}$$

As mentioned above, the block length and threshold are calculated by assuming $\sigma=1$. Therefore, for the collected signals with non unit variance, the standard deviation $\hat{\sigma}$ of the signal should be estimated and the variance of the wavelet coefficients should be normalized. A good estimator for the variance is obtained by the median value of the absolute values of the diagonal sub-band coefficients at the finest decomposition level.

$$\hat{\sigma}=\frac{median(|f_j|)}{0.6745} \tag{11}$$

The estimator is constructed through a hybrid method. Set $T_d = n_j \sum (x_i^2 - 1)$, $\gamma_d = n_j^{-1/2} \log_2^{3/2} n_j$ and $\mu^F = 2L\log n_j$. (μ^*, L^*) denotes the optimal block length and threshold which have the minimum SURE, under the additional constraint condition in the searching region.

$$(\mu^*, L^*) = \underset{\max\{L-2,0\} \le \mu \le \mu^F,\ 1 \le L \le n_j^{1/2}}{\arg\min} SURE(x, \mu, L) \tag{12}$$

The estimator $\hat{f}^*(x)$ for f is

$$\begin{cases} \underline{\hat{f}}^*_b = \underline{\hat{f}}_b(\mu^*, L^*) & T_d > \gamma_d \\ \hat{f}_i^* = (1 - {2\log nn}/{x_i^2})_+ x_i & T_d \le \gamma_d \end{cases} \tag{13}$$

The estimator is called SureBlock estimator. When $T_d \le \gamma_d$, it degenerates to James-Stein estimator whose block length is 1. The hybrid scheme is used to guard against the situations of extreme sparsity of a mean vector.

As shown in Fig. 1, multiwavelet denoising with the data-driven block threshold can be generalized as below.

1. Preprocess the measured signal and perform l-level decomposition.
2. Select the optimal block length and threshold by minimizing SURE at each decomposition level.
3. Threshold the detail coefficients at each decomposition level according to Eq. (13) using the optimal block length and threshold, the purified signal $\hat{d}_{j,k}$ is obtained.
4. Reconstruct the purified signal $\hat{d}_{j,k}$ at each decomposition level and the approximation signal to obtain the multi-stream denoising result $\tilde{s}_{0,n}$.

5. Postprocess the multi-stream denoising result $\tilde{s}_{0,n}$ to obtain the one-stream result x.

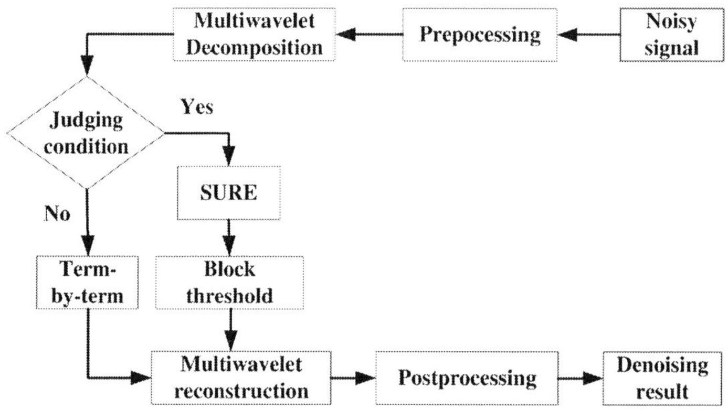

Fig. 1. Multiwavelet denoising with data-driven block threshold.

3 Pipeline defect detections

In this experiment, the pipeline with artificial defect is shown in Fig. 2, the length of the pipe is 1200mm, the diameter is Ø300mm and the wall thickness is 1mm. The excitation PZT sensors and the receiving sensors are placed on the left end of the pipe. The distance between the artificial crack and receiving sensors is 250mm. The depth of the crack is only 0.3mm.

The excitation signals are windowed tonebursts with a precise center frequency and a limited bandwidth, which helps to prevent the excitation of undesired modes at other frequencies and reduce the effect of dispersion on the propagation of the desired mode. In this experiment, a Hanning-windowed toneburst centered at 80 kHz is used as the original signal. The UGW signals pass through the monitoring area and are captured by the PZT sensors. They are acquired at a sampling rate of 2.5 MHz.

The receiving UGW signals are shown in Fig. 3, the incident wave packet and boundary-reflected wave packet are clear. The defect-scattered wave packet is interfered by the noise. In the Hilbert transform of the receiving signals, the envelope of the damage-scattered wave packet is also immersed in the noise.

Fig. 4 shows the result of the proposed denoising method. The incident wave packet and boundary-reflected wave packet are more distinct than the original signals. In addition, the defect-scattered wave packet is also distinct. In the Hilbert transform of the denoising result, the distance between the envelope of incidence and that of defect is equal to the distance between the crack and receiving sensors.

Fig. 5 shows the result of multiwavelet denoising using neighbouring coefficients. The incident wave packet and boundary-reflected wave packet are distinct. However, the defect-scattered wave packet is not as distinct as the result in Fig. 4. Moreover, the Hilbert envelope of the defect-scattered wave packet is also immersed in the noise. In addition, there was a small additional impulse as denoted by the ellipse in Fig. 5, which interfered the feature detection of the pipe defect.

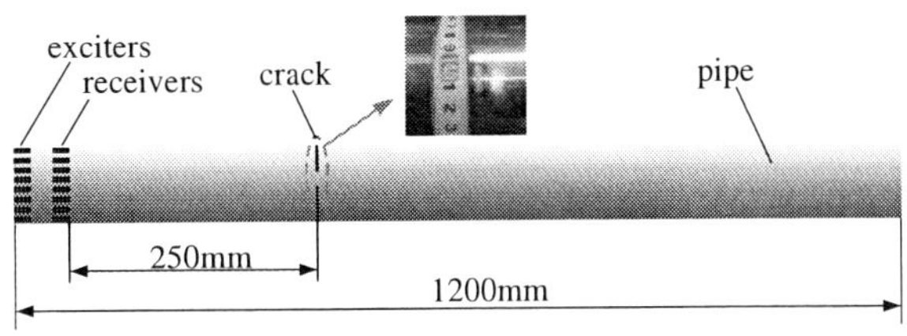

Fig. 2. The experiment pipeline with artificial damage.

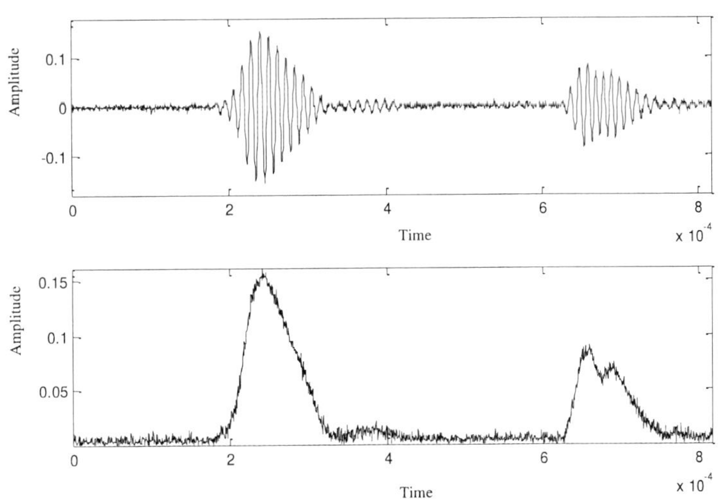

Fig. 3. The original UGW signals and its Hilbert Transform.

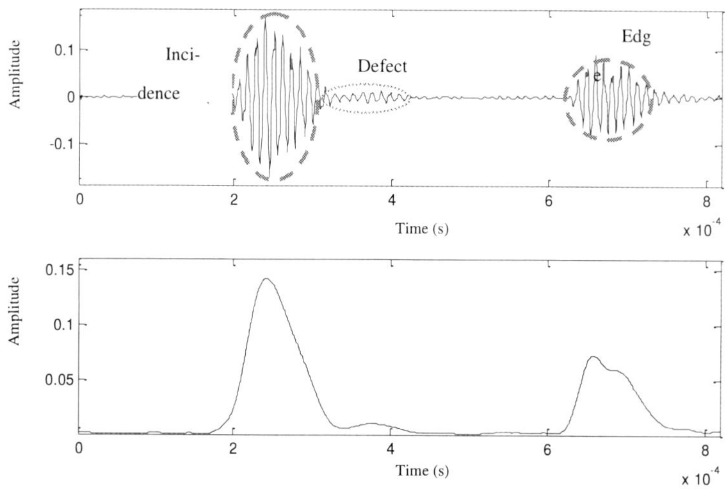

Fig. 4. The denoising result of the proposed method.

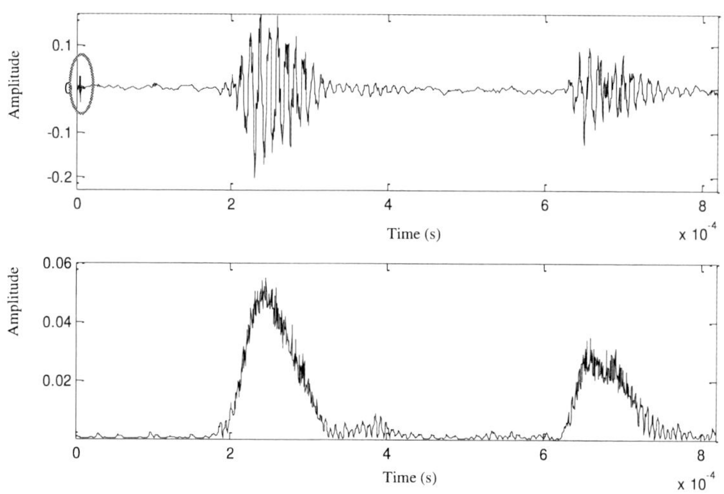

Fig. 5. The result of multiwavelet denoising using neighboring coefficients.

4 Conclusion

The ultrasonic guided wave is efficient for inspecting defect in pipelines. The data-driven block threshold technique selects the optimal block length and threshold by using the minimum Stein's unbiased risk estimate when evaluating the unknown features. Multiwavelets have more than two multiscaling functions and multiwavelet functions, which can effectively match variant fault features caused by the transmission path or the equipment resonance. They possess excellent properties of orthogonality, compact support, symmetry and high vanishing moment simultaneously, which are important for accurate incipient fault extractions. The proposed technique combines the advantages of GHM multiwavelets and the data-driven block threshold. The defect detection in pipelines indicates that the proposed method can effectively detect the transient features of incipient faults in pipelines.

Acknowledgments

This research was financially supported by the project of National Natural Science Foundation of China (No. 51505036). Special thanks are expressed to Dr. Yong Chang for helping with describing and analysing the experimental data.

References

1. R.Q. Yan, R.X. Gao, Rotary machine health diagnosis based on Empirical Mode Decompositon, ASME Transactions on Journal of Vibration and Acoustics 130 (2) (2008) 021007.
2. X. Wang, P. W. Tse, and A. Dordjevich, "Evaluation of pipeline defect's characteristic axial length via model-based parameter estimation in ultrasonic guided wave-based inspection," Meas. Sci. Technol., vol. 22, no. 2, art. no. 025701, 2011.
3. D. N. Alleyne, T. P. Pialucha, and P. Cawley, "A signal regeneration technique for long-range propagation of dispersive Lamb waves," Ultrasonics, vol. 31, no. 3, pp. 201-204, 1993.
4. C. Prada and M. Fink, "Separation of interfering acoustic scattered signals using the invariants of the time-reversal operator. Application to Lamb waves characterization," J. Acoust. Soc. Am., vol. 104, no. 2, pp. 801-807, 1998.
5. R. K. Ing and M. Fink, "Self-focusing and time recompression of Lamb waves using a time reversal mirror," Ultrasonics, vol. 36, no. 1–5, pp. 179-186, 1998.
6. R. Gangadharan, C. R. L. Murthy, S. Gopalakrishnan, and M. R. Bhat, "Time reversal technique for health monitoring of metallic structure using Lamb waves," Ultrasonics, vol. 49, no. 8, pp. 696-705, 2009.

7. D.L. Donoho, De-nosing by soft-thresholding, IEEE T. Inform. Theory. 41(3) (1995) 613-627.
8. J. Yuan, Z.J. He, Y.Y. Zi, and et al., Adaptive multiwavelets via two-scale similarity transforms for rotating machinery fault diagnosis, Mech. Syst. Signal. Pr. 23(5) (2009) 1490-1508.
9. H.L. Sun, Y.Y. Zi, Z.J. He, and et al. Multiwavelettransformanditsapplicationsinmechanical fault diagnosis – A review, Mech. Syst. Signal. Pr. 43 (2014) 1-24.
10. T. Cai, Adaptive wavelet estimation: A block thresholding and oracle inequality approach, Ann. Stat. 27(1999) 898-924.
11. T. Cai, H. Zhou, A data-driven block thresholding approach to wavelet estimation, Ann. Stat. 37(2009) 569-595.
12. J. Yuan, Z.J. He, Y.Y. Zi, Gear fault detection using customized multiwavelet lifting schemes, Mech. Syst. Signal Pr. 24(5) (2010) 1509-1528.
13. X.D. Wang, Y.Y. Zi, Z.J. He, Multiwavelet denoising with improved neighboring coefficients for application on rolling bearing fault diagnosis, Mech. Syst. Signal. Pr. 25(2011) 285-304.
14. H.L. Sun, Y.Y. Zi, Z.J. He, Wind turbine fault detection using multiwavelet denoising with the data-driven block threshold. Applied Acoustics, 77 (2014) 122-129.

A Hybrid Decision Making Method for Underwater Vehicle Anti-sink Recovery Maneuver

Deyu He[1,2], Niaoqing Hu[1,2], Lei Hu[1,2], Yiping Guo[3], Lun Zhang[1,2]

1. Laboratory of Science and Technology on Integrated Logistics Support, National University of Defense Technology, Changsha 410073, China; 2. College of Mechatronics Engineering and Automation, National University of Defense Technology, Changsha 410073, China; 3. The Jiujiang Branch of 707 Research Institution, China Shipbuilding Industry Corporation, Jiujiang 332007, China

hedeyu@nudt.edu.cn

Abstract. Once stuck rudder fault happens to the underwater vehicle, it's very likely to lead to accident of sinking to seabed. Thus it is significant to take appropriate anti-sink recovery maneuvers to recover the underwater vehicle. A hybrid decision making method based on fuzzy safety assessment and event tree is proposed in this paper. The key parameters involved in the safety station of the anti-sink progress were extracted. The current safety level of underwater vehicle was assessed by the evaluated parameters based on fuzzy safety assessment theory. Event tree of stuck rudder fault accident was illustrated based on the anti-sink recovery maneuvers. Reverse deduction was made according to the safety level and the anti-sink maneuvers were thus generated. Dynamic simulation model of anti-sink maneuvers of underwater vehicle was established. Stuck rudder faults were injected in several typical cases of different running scenes. Anti-sink maneuvers were generated according to the proposed method. Simulation results showed that the underwater vehicle could all be successfully recovered. The proposed method may provide an available choice for other decision situation.

Keywords: Fuzzy safety assessment, event tree, underwater vehicle, recovery maneuver, decision making

1 Introduction

Sailing safety of underwater vehicle (UV) is the basis of all its regular tasks, including training and cruising. Besides fire, collision and ocean condition, faults of itself is a significant factor that causes accident. Steering system is a pivotal subsystem which controls the moving path and gesture of UV. Once it burns into fault, rudders are likely to be out of control, which is called "stuck rudder fault". When this situation happens, collision to the seabed might be finally caused if improper decision is made, which will lead to huge loss of stuff and economy. Therefore, it's of great significance to perform accurate safety assessment and proper anti-sink recovery maneuvers decision making. However, traditional decision making methods for anti-sink recovery maneuvers are usually based on experience, which is too subjective to be trusted.

As for anti-sink recovery maneuvers, WANG etc. analyzed the effect of several recovery maneuvers after the stern rudder burned into stuck fault in 2009. Simulation calculation was performed and effective recovery maneuvers under different running scenes were accordingly proposed [1]. In 2010, DING etc. applied fuzzy assessment to the analysis of risk level when stuck rudder fault occurred [2].The above works involves litter about the systemic decision making method for anti-sink recovery maneuvers.

To solve the above problem, a hybrid decision making method for UV anti-sink recovery maneuver based on fuzzy safety assessment (FSA) and event tree analysis (ETA) was proposed in this paper. In section 2, theory foundation of the proposed methodology will be introduced. In section 3, the performance process of the proposed method will be illustrated. In section 4, five case studies will be presented to verify the proposed method. Conclusion will be drawn in section 5.

2 Theory Foundation

In the process of system safety assessment, there are many fuzzy concepts and variables that need to be investigated. Using fuzzy mathematics, these fuzzy and qualitative indicators could be expressed by quantitative numbers. As for fuzzy safety assessment, S.M. Miri Lavasani etc. applied fuzzy risk assessment to the safety assessment of oil and gas offshore wells in 2011 [3]. In 2012, G.S. Beriha etc. applied fuzzy logic approach to the prediction of different types of accidents (fatal to minor) of Indian industries in an uncertain environment [4]. Golam Kabir etc. applied fuzzy Bayesian belief network to the safety assessment of oil and gas pipelines in 2015 [5].

Event tree analysis (ETA) is to analyze the results of event series leaded by initiative event. ETA is commonly used in the accident and reliability analysis of system safety. Every step (node) of any event from initiative event to result could come into two branches: successful (normal) or unsuccessful (fault). As for ETA, Refaul Ferdous etc. utilized fuzzy set and evidence theory to handle the data uncertainties of ETA in 2009. They applied the proposed method to the risk assessment of LPG release [6]. In 2013, Tony Rosqvist etc. applied ET to the flood protection in Finland [7]. Nahid Ramzali used fuzzy ETA to the barrier analysis of offshore drilling system [8].

3 Hybrid Decision Making Method

The basic principle of the proposed hybrid decision making method is illustrated as Fig.1. After stuck fault happened to the stern rudder of UV, the threat/safety level could be determined by safety mark which was obtained from fuzzy safety assessment. In the event tree of stuck rudder accident, slowing down, bow rudder and direction rudder are usually used to perform dynamic anti-sink recovery [9]. Once the above maneuvers are all unsuccessful, disastrous accident will be resulted in. Different anti-sink maneuvers are supposed to correspond to different threat level. Therefore, reverse deduction could be made according to the threat level and the anti-sink maneuvers were thus generated as in Table 1.

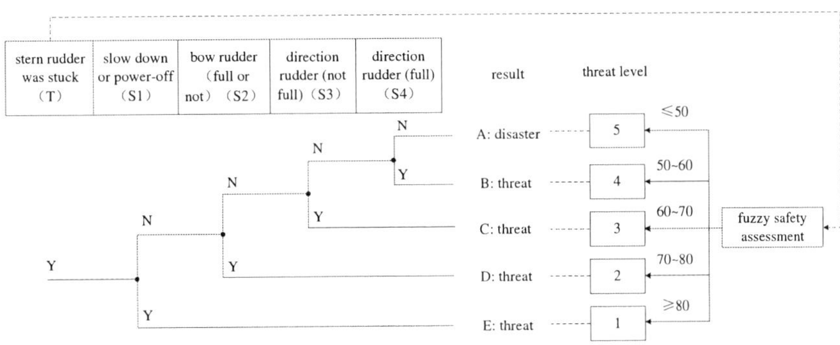

Fig. 1. Basic principle of hybrid decision making method

Table 1. The generated anti-sink maneuvers according to safety level

Threat level	anti-sink maneuvers
1	slow down;
2	slow down; positive bow rudder;
3	slow down; full positive bow rudder (25°); rotating direction rudder
4	slow down; full positive bow rudder (25°); full rotating direction rudder (35°)

One-level fuzzy safety assessment was adopted and the detailed procedures are stated as following.

1. The establishment of indicator set

The possibility of colliding to the seabed because of stuck rudder fault is mainly determined by four parameters: depth to the seabed Δh (m), trim δ (°), changing rate of depth α_h (m/s) and the changing rate of trim α_δ (°/s). These four parameters were extracted to establish the indicator set as illustrated in Fig.2.

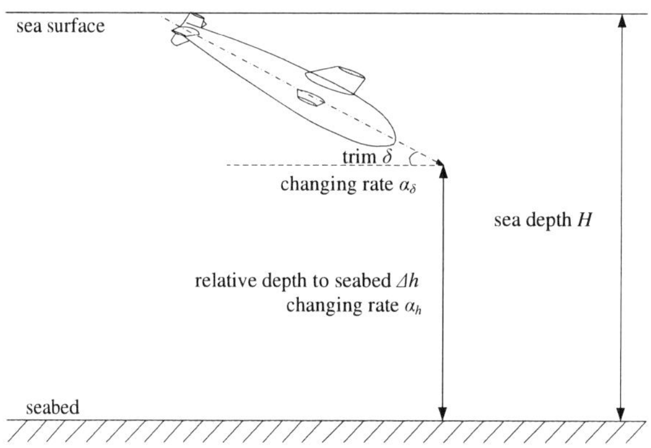

Fig. 2. Illustration of the indication set

2. The determination of weight set.

The weight set of the above indicator set was straightly given as [0.6 0.2 0.2 0.1] according to the experts' experiences.

3. The Establishment of the assessment matrix
4. Calculation of integrated assessment matrix
5. Getting the overall marks of system safety

6. Synthetic assessment of the system safety level

4 Case Study

In order to verify the proposed hybrid decision making method, dynamic simulation model of anti-sink maneuvers of UV was established. Five typical simulation cases were studied. Imagine that the initial speeds and depths of the five cases are all 12 knots and 30m respectively. UV dived at different sea depths with different stern rudder angles. Then stuck rudder faults happened to stern rudders and decision making was performed according to the proposed method. The simulation results are listed in Table 3. All the UVs in the five cases could be successfully recovered in the end. It could be inferred from the results that the proposed hybrid decision making method is valid under different sea depth and stuck rudder angle condition.

Table 2. Details and results of the five cases study

case	sea depth	stuck rudder angle	threat level	anti-sink recovery maneuvers	succeed or not
1	100	18	4	slow down; full positive bow rudder; full rotating direction rudder	yes
2	150	24	4	slow down; full positive bow rudder; full rotating direction rudder	yes
3	200	30	4	slow down; full positive bow rudder; full rotating direction rudder	yes
4	250	30	3	slow down; full positive bow rudder; rotating direction rudder (30°)	yes
5	300	30	3	slow down; full positive bow rudder; rotating direction rudder (20°)	yes

5 Conclusion

This paper proposed a hybrid decision making method based on fuzzy safety assessment and event tree. The threat/safety level of UV was assessed based on fuzzy safety assessment theory. Event tree of stuck rudder fault accident was illustrated based on the basic anti-sink recovery maneuvers. Reverse deduction was made according to the threat level and the anti-sink maneuvers were thus generated. Dynamic simulation model of anti-sink maneuvers of UV was established. Anti-sink maneuvers were generated according to the proposed method. Simulation results showed that the UV could all be successfully recovered. The proposed method may provide an available choice for other decision situation.

Acknowledgement. This work was supported by the National Natural Science Foundation of China [grant number 51475463].

References

1. WANG Jing-qi, SHI Sheng-da, and ZHANG Wei-kang, Safe Maneuvering Technology for Ensuring Submarine's Survivability. Journal of Naval University of Engineering, 2009. 21(1): 63-67. (in Chinese)
2. DING Lei, LIN Li, GUO Yi-ping, et al, Safe Maneuvering The Research and Analysis of Emergency Control on Jam of Submarine Based on the Fuzzy Assessment Theory. Ship Science and Technology, 2010. 32(1): 67-70. (in Chinese)
3. S.M. Miri Lavasani, Z. Yang, J. Finlay, J. Wang. Fuzzy risk assessment of oil and gas offshore wells, Process Safety and Environmental Protection, 89 (2011)277–294
4. G.S. Beriha, B. Patnaik, S.S. Mahapatra, S. Padhee. Assessment of safety performance in Indian industries using fuzzy approach, Expert Systems with Applications, 39 (2012) 3311–3323
5. Golam Kabir, Rehan Sadiq and Solomon Tesfamariam. A fuzzy Bayesian belief network for safety assessment of oil and gas pipelines, Structure and Infrastructure Engineering, 2015(1-16)
6. Refaul Ferdous, Faisal Khan, Rehan Sadiq, Paul Amyotte, Brian Veitch. Handling data uncertainties in event tree analysis, Process Safety and Environmental Protection, 87(2009):283–292
7. Tony Rosqvist, Riitta Molarius, Hanna Virta, Adriaan Perrels. Event tree analysis for flood protection- An exploratory study in Finland, Reliability Engineering and System Safety, 112(2013): 1–7
8. Nahid Ramzali, Mohammad Reza Miri Lavasani, Jamal Ghodousi. Safety barriers analysis of offshore drilling system by employing Fuzzy Event Tree Analysis, Safety Science, 78 (2015): 49–59
9. ZHAO Tingdi. Safety Design, Analysis and Verification. Beijing: National Defense Industry Press; 2011. (in Chinese)

Wear Testing Methods and Online Monitoring Study of Wet Clutch Friction plate

Wang Liyong[1,2], Li Hao[1], Li Le[1],Chen Xi[1]
(1Beijing Information Science and Technology University, The ministry of education key laboratory of modern measurement and control technology, 2Collaborative Innovation Center of Electric Vehicles in Beijing, Beijing100192,China)

Abstract: For the friction plate wear testing of wet clutch and the on-line monitoring about the size and number of wear particle, the test system and method of the wet clutch friction plate wear amount was designed, on-line monitoring of friction plate temperature and oil wear particles were achieved. The a hydraulic flexible connection system is used as the dynamic input, on-line oil monitoring sensor was located on the pressure oil and lubricating oil circulation circuits to test the size and quantity of the wear particles in oil. The clutch and hydraulic components were protected from particles damage by online monitoring. The particles were generated because of excessive wear of friction. To protect the frictional parts of clutch against multiple engagement-induced abnormal damage, the temperature sensors were used to test the temperature in the clutch steel; and monitor the operating condition of friction plate to assures the clutch reliability. By the experimental testing, wear online testing system can be effectively used for friction plate wear amount test.

Key words: wet clutch; flexible connection; wear; real-time monitoring

1 Introduction

The wet Clutch is one of the core components of the automatic transmission system and hydraulic mechanical transmission. It is necessary to study friction wear test methods of the clutch plates and wear particles on-line oil monitoring, in order to meet the vehicle transmission demand for high-speed, heavy, high-power-density, high-reliability of the transmission system.[1]

The friction pair of wet shifting clutch is composed of the internal toothed friction plate fabricated from copper-based powder and the external toothed dual dish-shaped steel plate. The wear degree of friction plate has immediate effect on the reliability and performance of integrated transmission [1~2]. Clutch wear is a process of continuous damage to surface material of clutch friction pair. Conventionally, the wear loss of clutch friction plate is determined through direct weighing or thickness measurement after disassembly. Despite high accuracy, this method may frequently impair the original match. Some scholars launched studies on the computing method of clutch wear loss, which are of extremely high theoretical guiding significance and practical value for overcoming the clutch failure resulting from excessive wear of friction plate [3]. Most scholars' studies on wear loss focus on theoretical calculation, and few scholars pay attention to the measurement of friction plate wear loss based on actual simulation of vehicle motion state. By means of direct drive on clutch input shaft, the test rig provided with no transitional flexible protection may frequently cause damage to equipment. Existing test rigs are equipped with no on-line detection and sensing system, and most of them employ off-line analysis method for wear and oil test, making it impossible to monitor in real time if friction plate is in normal working condition during clutch multiple engagement test [4]. Hence, the study on wear measurement design method for wet clutch friction plate is of great importance for clutch reliability test.

2 Operating Principle of Wet Clutch

Clutch transfers power through the friction between friction plate and dual steel plate, and the clutch friction pair is in a liquid &

---

[1] Address correspondence to Wang Liyong, The ministry of education key laboratory of modern measurement and control technology, Beijing Information Science and Technology University, Beijing 100192, China. E-mail: wangliyong@bistu.edu.cn

semi-liquid friction state at frictional contact surface during the process of friction. During the process of friction, the friction surface is isolated by an oil film, which assures stable friction coefficient and moderate wear loss of friction pair under extremely high pressure. Fig. 1 shows the fundamental structure of wet multi-plate friction clutch.

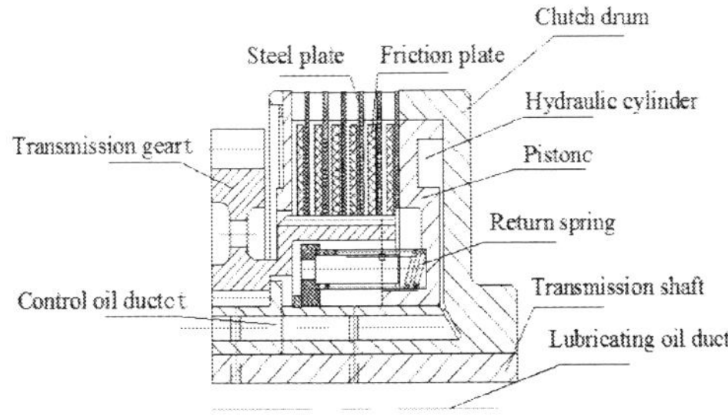

Fig. 1 Structure diagram of wet multi-plate friction clutch.

The clutch consists of a few dual steel plates circularly arranged in an alternating manner and friction plates. Its engagement is hydraulically controlled: When clutch engagement is needed to transfer power, the hydraulic pump outputs high pressure oil that pushes the piston to the left for the mutual compression between dual steel plate and friction plate; due to the speed difference there-between, the sliding friction under pressing force generates a lot of friction heat until there is no relative sliding between dual steel plate and friction plate; then, both rotate at the same speed for torque transmission. For the disengagement of clutch, the control oil duct drains oil, when the piston moves to the right under the action of reset spring; the disappearance of axial pressure on friction pair leads to the instant reduction in frictional force; as a result, the clutch is disengaged, and the transmission is interrupted. Lubricating oil passes through lubricating oil duct and flows across the friction pair contact surface to bring about lubricating and cooling effect, which assures the high-speed smooth operation of clutch [5].

3 Wear Test Rig Structure Design

The test rig is composed of power mechanism, control mechanism, test object & data acquisition mechanism, and brake mechanism..

Power mechanism includes oil tank, motor, hydraulic pump, hydraulic motor, and adjustable inertia group 1. . The motor output side connects with hydraulic pump, hydraulic motor and adjustable inertia group 1 in turn; the control mechanism regulates the motor speed and controls the operation of hydraulic pump; the input side of hydraulic pump is connected to oil tank. The output side of adjustable inertia group 1 is connected with the test object & data acquisition mechanism..

The test object & data acquisition mechanism comprises two speed & torque sensors, a clutch test kit, an on-line oil monitoring sensor, a radiator, a temperature sensor, a pressure sensor, a collector ring, and a data acquisition card; the clutch test kit is provided with a cross-compound arrangement, where a pair of reduction gears are used for transmission connection between input shaft and output shaft, of which both are hollow. The output side of adjustable inertia group 1 is coaxially connected to the input shaft of clutch test kit, and the output shaft of clutch test kit is connected with brake mechanism; a speed & torque sensor is connected respectively to the input shaft and output shaft of clutch test kit; both speed & torque sensors transmit the speed & torque signal collected to the control mechanism. Temperature sensor is arranged on the friction plate of clutch test kit, while the pressure sensor is arranged on the pressure disc of clutch test kit. The output sides of both sensors are connected to the collector ring mounted at the end of input shaft of clutch test kit via the data cable in hollow shaft of clutch test kit; the data collected is transferred to data acquisition card through the collector ring. Hydraulic auxiliary system is designed to feed pressure oil and cooling oil to clutch test kit; the oil duct between hydraulic auxiliary system and clutch test kit is furnished with on-line oil monitoring sensor that transmits the oil wear particle number signal collected to data acquisition card, which transfers the received data to control mechanism.

The brake mechanism is designed for braking purpose of clutch test kit output shaft, and comprises adjustable inertia group 2 and brake. One side of adjustable inertia 2 is connected with the clutch test kit output shaft, while the other end to the brake, which is connected to and driven by the hydraulic auxiliary system. The control mechanism comprises frequency converter, host PC, manual operation control cabinet, and displacement controller. The data acquisition card transfers the received data to the host PC, and the two speed & torque sensors transmit the collected speed & torque signal to host PC and manual operation control cabinet, respectively. The host PC controls the operation of frequency converter, displacement controller and hydraulic auxiliary system based on the received data; it regulates motor speed through frequency converter, and hydraulic pump through displacement controller. The manual operation control cabinet is connected to hydraulic auxiliary system, so the hydraulic auxiliary system could be controlled with manual operation control cabinet. Fig. 2 shows the diagram of test rig, and Fig 3 represents the structure diagram of hydraulic auxiliary system.

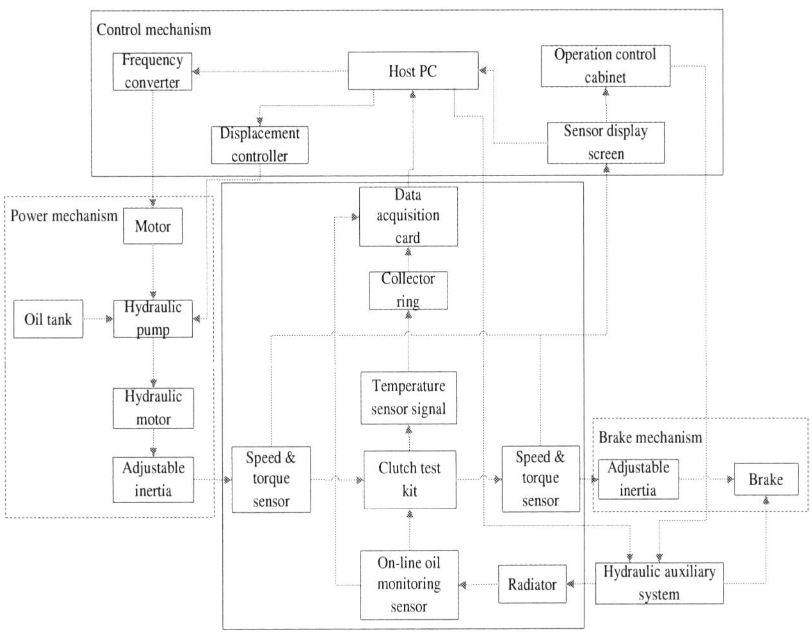

Fig. 2 Test rig diagram

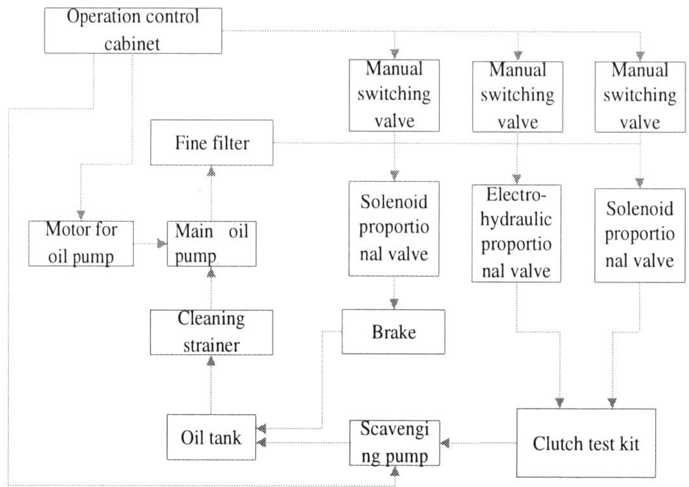

Fig.3 Structure diagram of hydraulic auxiliary system

4 Temperature sensor arrangement scheme

The temperature signal at the contact surface of the friction plates and steel was tested. There is oil at contact surface and it is unsuitable for non-contact measurement, temperature sensor use the high-precision, high sensitivity of K-type thermocouple. If the sensor is arranged on the friction plate, the mechanical structure and the material properties of friction sheet would be influenced. Taking into account the actual vehicle in less wear and tear steel, and the main cause of failure is the thermal deformation, so the sensor is arranged in the even of steel. Figure 4 show that sensors were arranged in depth and location along radial direction of steel friction.A1 ~ A5 radially arranged at different depths.A3, B1, C1, D1 are arranged along the circumferential direction at the same depth. the through-holes between A2 and A3, A3 and A4 to provide the conditions arranged the sensor in the latter part. Figure 5 is the physical map after steel perforation. In order to achieve the fastest heat exchange between sensor and steel, the diameter holes embedded width of the steel as large as sensors, which are 1mm.

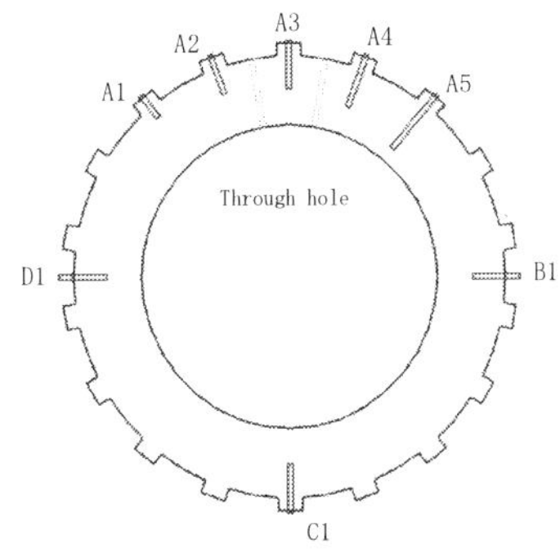

Fig 4Thermocouple sensor arrangement schematic

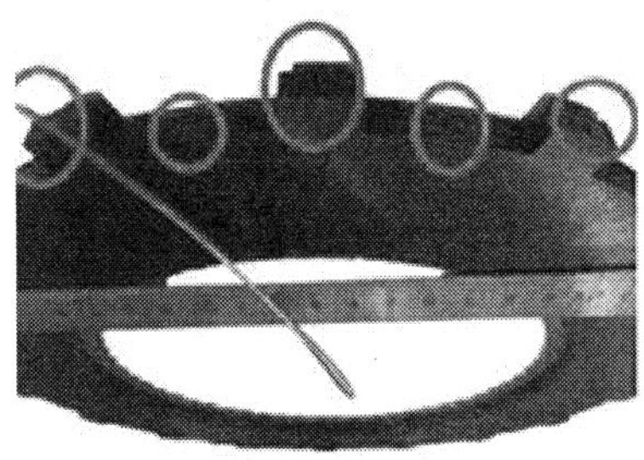

Fig 5Dual steel machining holes schematic

Using the temperature measurement program, the temperature of A1 ~A5 position in wet clutch dual steel were measured. It was shown in Fig 6.

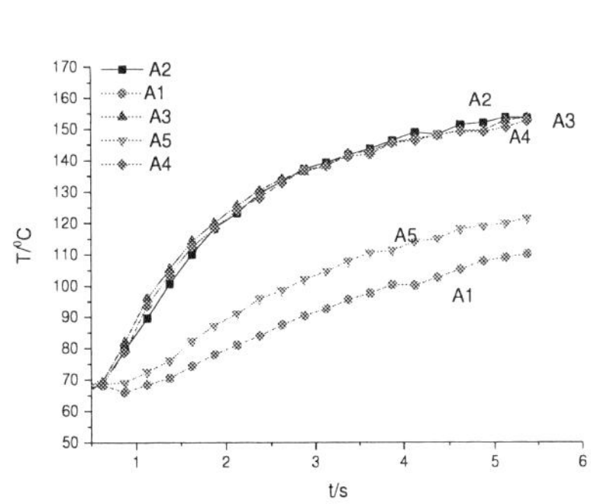

Fig 6temperature curve of A1 ~A5

5 Wear Loss Testing Method

The testing methods for wear loss of wet clutch friction plate include the test for determining the dynamic and static friction torques of friction plate, the dynamic and static friction coefficients, and the test on wear loss of clutch friction plate.

5.1 Test on Static Friction Torque and Static Friction Coefficient.

The host PC controls frequency converter to keep the motor operating at rated speed, and regulates the hydraulic pump displacement through displacement controller so that it operates at low speed, in which case the high motor speed and the low speed and large torque of hydraulic motor are realized for the measurement of wet clutch's static friction torque and static friction coefficient. Wear-in the wet clutch so that the contact area of friction plate reaches 80%; have the wet clutch engaged, brake the output side of wet clutch, and slowly regulate the output speed of hydraulic motor; when the output speed of hydraulic motor is so regulated that the wet clutch input shaft slips, the torque and pressure are the static friction torque and positive pressure to be determined, respectively.

5.2 Determination of Dynamic Friction Torque and Dynamic Friction Coefficient.

Keep the motor running at rated speed, increase the hydraulic motor displacement, and drive adjustable inertia group 1 to rotate with high speed, in which case the torque is low, and the rotation speed is high - it's suitable for determining the dynamic friction torque and dynamic friction coefficient of wet clutch. Regulate the hydraulic motor displacement to make the rotation speed of wet clutch input shaft reach 3,000 r/min; engage and keep the wet clutch running at high speed; turn off the power source of motor, and apply the brake to wet clutch output shaft; under the action of adjustable inertia group 1, the input shaft keep on rotating to cause sliding friction of friction plate; the torque measured at this point is the dynamic friction torque determined with the wet clutch output side fixed when the input side is subjected to forced sliding friction with different rotation speeds.

5.3Test on Wear Loss of Clutch Friction Plate.

(1) Start the motor, and control the hydraulic motor through host PC so that the wet clutch could be subjected to multi-engagement test (30s/cycle on an average) at different rotation speeds; make adjustment based on the pre-determined engagement pressure with time taken into consideration. The wear loss may vary depending on engagement pressure and sliding friction time: The engagement pressure is directly proportional to sliding friction time..

(2) Check if the engagement pressure is the pre-determined pressure: If yes, regulation the speed; if not, regulate the speed after adjusting the engagement pressure to the set value. In this embodiment, 800 cycles of engagement are performed for each measurement of wear loss on wet clutch friction plate. To better simulate the operating state of clutch, the 800 cycles of engagement are divided into four groups (200 cycles/group) for tests at different engagement pressures and rotation speeds..

The engagement pressures set are: 0.3Mpa for Group 1; 0.6Mpa for Group 2; 1.2Mpa for Group 3; and 1.4Mpa for Group 4. The rotation speeds under fixed engagement pressure are: 800r/min for Group 1; 1,200r/min for Group 2; 1,600r/min for Group 3; and 2,000r/min for Group 4.

(3) The host PC controls the displacement of hydraulic motor, adjusts the output speed to 800r/min, and controls the engagement of wet clutch. After 200 cycles of engagement, if the online oil monitoring sensor gives an alarm, it' s necessary to turn off the motor and check the wet clutch kit; if no alarm is given, it' s necessary to take measurement of oil sample and adjust the output speed of wet clutch kit to 1,200r/min for the test on Group 2. Perform tests on Groups 3 and 4 in a similar way.

(4) Upon the completion of 800 test cycles, it' s recommended to disassemble the wet clutch kit, change the friction plate, adjust the

pressure (to a value that is different from the pressure value set in previous step), and perform the test. Such a single-variable test on wet clutch friction plate could better describe the reliability and safety of friction plate and wet clutch. The operating procedure is as shown in Fig. 7.

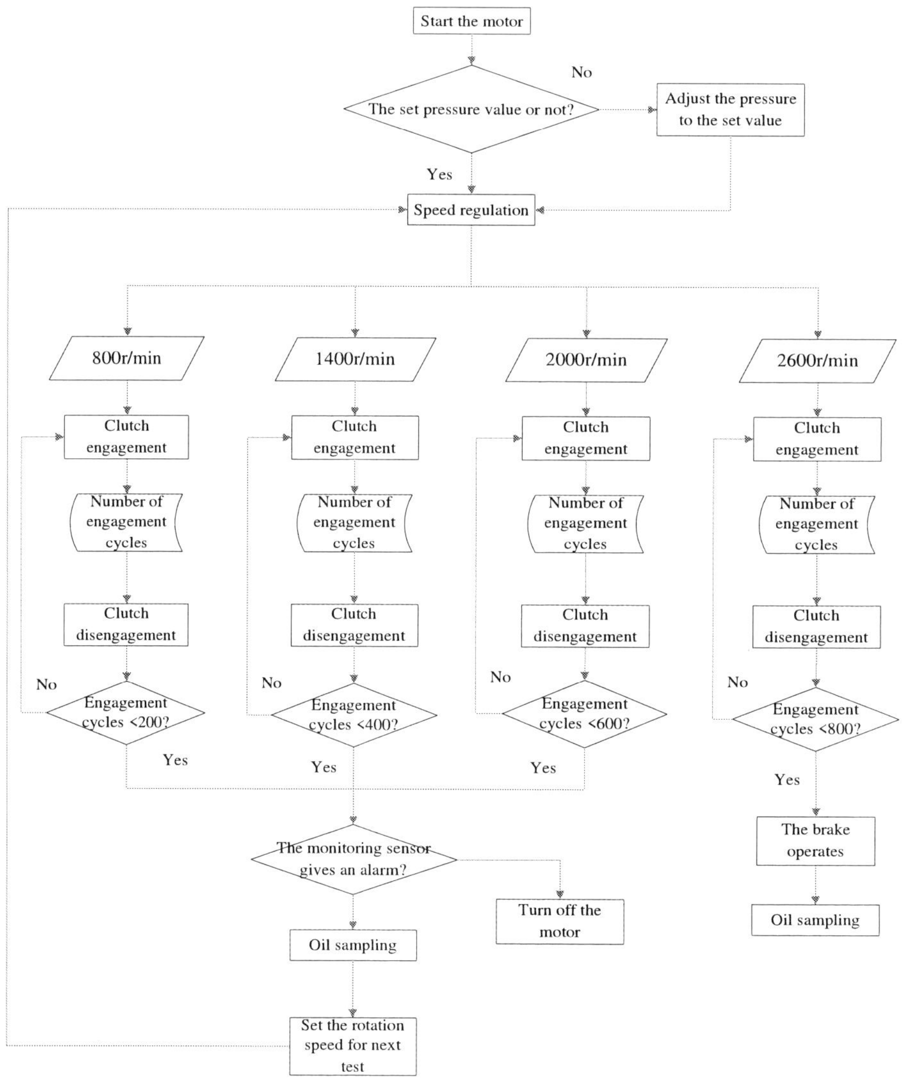

Fig. 7Test method & flow diagram

6Conclusion

(1) Provide requisite conditions for the tests on various properties of wet clutch by using motor as the source of power, taking hydraulic motor as the vehicle engine, and using the inertia and brake to simulate the operating status of vehicle at start, climbing, and braking.

(2) The connection between motor and hydraulic pump and between hydraulic pump and hydraulic motor assures the operation safety of motor with the clutch is braked. Such a hydraulic connection and high contributes to excellent engagement flexibility; in case of static friction, the output side is in a braking state, while the rotation speed at input side would drop to zero rapidly. The hydraulic motor could protect the motor against the braking-induced damage.

(3)Multiple locations insteel disc of the measured wet clutchequipped with thermocouple sensors which measured the temperature, andreal-time monitoring the work status of friction plate.The test temperature result was referenced, which provided an experimental basis for researching steel disc surface temperature field.

(4)The clutch tested and the circulating oil duct of pump station are furnished with online metal particles sensors for real-time monitoring of metal particle content in hydraulic oil during the multiple engagements of clutch so as to prevent repeated engagement-induced frictional damage to frictional parts of clutch, and to assure the reliability of clutch test.

Acknowledgements:

This work was completed with the support of Beijing finance found of science and technology planning project.(KZ201611232032), （KM201611232004）, and young prominent talentprojects. (No.2014000026833ZK25）(No. 2014000020124G096)（CIT&TCD201404112）

References

[1] MA Biao. A Study on Hydraulic Buffering Valve Dynamic PerformanceSimulation of Vehicular Power Shift Transmission.Beijing:BEIJING INSTITUTE OF TECHNOLOGY,1999;

[2] Li Guoqiang.Study on torque characteristics of mul-trplate wet clutches during engagement.Beijing:BEIJING INSTITUTE OF TECHNOLOGY,2014;

[3]Wang Liyong.Research on Wear Calculation Method for Friction Disc of Wet Shifting ClutchBeijing:BEIJING INSTITUTE OF TECHNOLOGY 2008

[4] Lu Guodong. Thermal Analysis of Friction Pair of Wet Multi-disc Clutch . ChongQing: CHONGQING UNIVERSITY 2011;

[5]Zhang Xueyong, A Study on the Wear-rules of the Friction Disk[J] . 2002(5) : 44-45.

[6]Li Jun. Theoretical Calculation of Wear Valueof Clatch Friction Disc for Truck SXZ1032[J] . 1999, 12(4) : 48-51.

CASE-BASED REASONING (CBR) IN CONDITION MONITORING & DIAGNOSTIC ENGINEERING MANAGEMENT (COMADEM): A LITERATURE SURVEY

B.K.N. Rao
(COMADEM International, UK)

Abstract

It is the universal truth that experience is the best teacher. All civilizations have evolved by learning from past experiences of many previous generations from different cultural backgrounds. The primary sources of these past experiences are to be found in many historical, mythological, literary, cultural and spiritual documents/case studies. Past experiences still remain the primary source of reference from which the new knowledge/wisdom is generated, discovered, disseminated & from which civilized societies are constantly evolving. It is a basic human survival instinct to learn from the past to resolve day-to-day socio-economic problems and to continuously improve the quality of life. It is embedded in our genes. As a scientific quest, Case-Based Reasoning (CBR) was introduced as a powerful diagnostic and prognostic pro-active management strategy in its own right. Its philosophy is that all reasoning is based on past cases personally experienced. As a scientific endeavour CBR is based on the hypothesis that similar problems have similar solutions & new problems can be solved by revisiting or adapting solutions of past problems. If used judiciously, it can be a powerful method for computer reasoning and for better decision making. There are several potential advantages of fully exploring & exploiting CBR in the proactive and holistic discipline of Condition Monitoring and Diagnostic Engineering Management (COMADEM). This paper reviews this aspect through a brief literature survey.

Keywords: case-based reasoning, artificial intelligence, condition monitoring, diagnostic engineering, proactive management

1. Introduction

The CBR approach is an important area of Artificial Intelligence (AI) to solve problems from the previously accumulated knowledge of known scenarios that can provide a solution to new problems arise. It was Albert Einstein who said that "The only source of knowledge is experience". Knowledge is at the heart of knowledge management. Knowledge is associated with three components. These are (a) Logical Reasoning, (b) Intuition and (c) Wisdom. CBR is closely associated with knowledge management. Multiple thinking strategies are needed for high-quality decision-making while dealing with case-based reasoning. There are different types of reasoning. These are a) Deductive reasoning, b) Inductive reasoning, c) Abductive reasoning, d) Analogical reasoning, e) Fallacious reasoning and f) Automated reasoning. Intuition is central to discussions about the nature of scientific and philosophical reasoning and what it means to be human. Intuition, a phenomenon of the mind, describes the ability to acquire knowledge without inference or the use of reason. It is worth remembering great spiritual scientist/philosopher Sri Madhvacharya's statement that "knowledge without wisdom is dangerous and irresponsible". Wisdom in interpreted as knowledge gained by having many experiences in life. It is the natural ability to understand things that most other people cannot understand.

2. What is Case Based Reasoning (CBR)?

A short definition of case-based reasoning is that it is a methodology for solving problems by utilizing previous experiences. It involves retaining a memory of previous problems and their solutions and, by referencing these, solve new problems. Generally, a case-based reasoner will be presented with a problem. It may be presented by either a user or another program or system. The case-based reasoner then searches its memory of past cases (the case base) and attempts to find a case that has the same problem specification as the current case. If the reasoner cannot find an identical case in its case base, it will attempt to find the case or cases in the case base that most closely match the current query case. Case Based Reasoning is also called Case-Based Prediction (CBP). CBR was founded on the belief that human memory is memory of autobiographical events (related to times, places, associated emotions, and other contextual who, what, when, where and why) that can be explicitly stated. This episodic memory comprises human knowledge accumulated from past experiences and stored in deep memory of CNS. When faced with a new problem, a human often intuitively relates the problem to one or more memory episodes and arrives at a solution from these episodes. CBR is a computational intelligence tool whose knowledge source is composed of a memory of important past cases. In the real world it is observed from the analysis about tendencies that similar problems usually have similar solutions. Besides, it is very possible that past situations happen again in a similar way. It is reasonable to provide new solutions by recovering and adapting past similar cases. CBR has been applied to a variety of complex engineering and manufacturing systems.

Case-based reasoning has been formalized as a six-step process [Thomas Roth–Berghofer and Ioannis Iglezaki]. These are as follows:

Step 1: Retrieve: Given a target problem, retrieve from memory cases relevant to solving it. A case consists of a problem, its solution, and, typically, annotations about how the solution was derived.

Step 2: Reuse: Map the solution from the previous case to the target problem. This may involve adapting the solution as needed to fit the new situation.

Step 3: Revise: Having mapped the previous solution to the target situation, test the new solution in the real world (or a simulation) and, if necessary, revise.

Step 4: Retain: After the solution has been successfully adapted to the target problem, store the resulting experience as a new case in memory.

Step 5: Review: This step consists of two tasks: measure and monitor

Step 6: Restore: This step is described by the tasks select and modify.

CBR systems store knowledge in four different compartments: 1) Vocabulary, 2) Case Base, 3) Similarity Aeessment, 4) Solution Adaptation. There are two kinds of CBR systems; (1): problem-solving systems and interpretive systems. A problem-solving system focuses on the construction of solutions suited to the new case by modifying previous case solutions. (2) An interpretive system evaluates and justifies new cases based on the similarities or differences with the previous cases.

3. CBR Architecture, Frameworks & Tools

CREEK is an architecture for knowledge-intensive CBR, which has been developed at the Norwegian University of Science and Technology and focuses on open and weak-theory domains. jCOLIBRI is an object-oriented framework for building CBR systems that has been developed by the Department of Software Engineering and Artificial Intelligence at University of Madrid. myCBR is an open source CBR tool developed at the DFKI, which focuses on domain and similarity modelling for case retrieval. IUCBRF is also an open source framework for CBR system development, which is implemented in Java and developed at the Indiana University. On the commercial side the probably most widely used tool in Europe is the Information Access Suite (named e:IAS) from empolis. Another use of the CBR methodology is its inclusion in multi-agent-systems. Today the internet is not only used for information gathering, but also for social networking and sharing personal views, experiences and media. This development is usually subsumed by the term Web 2.0. Both these new fields, the Semantic Web as well as the Web 2.0, benefit from the application of case-based reasoning (http://www.wi2.uni-trier.de/publications/2009_KI_CBR.pdf). The most used CBR tools: CBR Shell, FreeCBR, jCOLIBRI, myCBR and eXiTCBR.

4. Why use Case Based Reasoning?

When used in the appropriate situations, case-based reasoning offers many advantages. The following is a brief summary of appropriate situations reported: (a) Reduction of the knowledge acquisition task, (b) Avoid repeating mistakes made in the past, (c) Graceful degradation of performance, (d) Able to reason in domains that have not been fully understood, defined or modelled, (e) May be able to make predictions as to the probable success of a proffered solution, (f) Learn over time. (g) Reason in a domain with a small body of knowledge, (h) Reason with incomplete or imprecise data and concepts, (h) Avoid repeating all the steps that need to be taken to arrive at a solution, (i) Provide a means of explanation, (j) Can be used in different ways, (k) Can be applied to a broad range of domains, (l) Reflects human reasoning.

5. CBR as a Diagnostic & Prognostic Strategy

In general, the constitutive elements of CBR are summarized as follows: A case library which contains features describing the problem; plus outcomes and solutions, methods used and assessments of the efficacy of those methods employed. A coding mechanism used to code information contained in cases. An adaptation scheme which is used to infer the "best" solution to a new problem from the stored cases. A method to organize the cases in the case library in meaningful and cohesive clusters. A Dynamic CBR paradigm contains time stamped cases and relates events, outputs, etc. along with the temporal domain. Dynamic Case Based Reasoning (DCBR) is a powerful technique for Fault Diagnosis and Prognosis. DCBR allows the accumulation of experience from inclusion of the new cases and therefore accommodates learning. Model Based Reasoning (MBR) techniques for fault diagnosis of dynamic systems consisting of interconnected components/subsystems have been based primarily on qualitative reasoning methods. Matlab system has been developed for multi-agent fault diagnosis for predictive health monitoring.

6. Case Based Reasoning in COMADEM – A Brief Literature Survey

The holistic philosophy of Condition Monitoring and Diagnostic Engineering Management (COMADEM) is to continuously improve and enhance the quality, reliability, safety, availability, maintainability and performance of all assets (both physical and human) for as long as possible and to derive maximum benefits with minimum risk. This means continuous knowledge discovery, generation and dissemination and proactively and holistically managing all assets. Its aim is to bring industries and academia to work in harmony with a view to shape our future and create value for the benefit of all mankind.

Plenty of published information is available on the application of CBR in COMADEM covering theory and practice in condition-based intelligent machine fault diagnosis and failure prognosis including state-of-the-art algorithms, methodologies, and contributions from leading experts, including cost -benefit analysis tools and performance assessment techniques - Covers theory and practice in a way that is rooted in industry research and experience. In the following, CBR has been successfully exploited in predictive maintenance, e-manufacturing environment, failure diagnosis and prognosis of engineering systems such as chiller system, product development, integrated structural health monitoring, drilling engineering, automotive systems, robotics, aircraft maintenance, Aircraft Landing Gear Extension and Retraction Control System, rotating machinery, fleet vehicles, bearings, Air Traffic Control systems, Transportation infrastructure, especially highway bridges, business processes.

The CBR is principally applied in the area of predictive maintenance that assist the user in the diagnosis of failures and support in the decision-making equipment maintenance in industrial environments [Kim et al. (2009), Yee et al. (2009), Na et al.(2008)]. Edinburgh Research & Innovation based in UK has developed a Case-Based Reasoning (CBR) Shell. It is an effective software tool for guiding diagnosis and fault finding in present instances by identifying patterns and knowledge implicit in historic information databases.[http://www.researchinnovation.ed.ac.uk/Opportunities/Case-Based-Reasoning#page=overview]. Mobile Intelligence Corporation, an advanced research and development firm specializing in applying artificial intelligence to solve complex problems. Their expertise is focused in artificial intelligence, autonomous robotics, command/control of groups of autonomous robots, stereo-vision, case-based reasoning (CBR), and software agent architectures. Luis Fernando Alape Realpe, Alberto Jose Álvares, Edgar Jhonny Amaya Simeon, Iván Patricio Moreno Marcos (2012) proposes the application of the approach of Case-Based Reasoning (CBR) to build a knowledge-base of an intelligent system for maintenance predictive of hydropower equipment. Hamid Berenji, Yan Wang and Abhi Saxena have applied Dynamic Case-based Reasoning in the fault diagnosis and prognosis. Klinger, Stottler and LeClair (1992) have applied CBR in manufacturing applications. Kai Cheng and Bateman (2008) have explored the potential benefits of CBR in e-Manufacturing environment. George Vachtsevanos et al (2006)). Olsson & Funk (2012) used Dynamic Case-based Reasoning (DCBR) as a powerful technique for fault diagnosis and prognosis. The authors have developed DCBR as a dynamic model of a chiller system. Belecheanu, Pawar, Barson, Bredehorst, & Weber, (2003) have applied case based reasoning to decision support in new product development. Olsson, Funk et al (2015) have developed CBR based fault diagnosis via fusion of information from a case stream.; Ann Pattersson-Hine et al (2005) have published a review on Diagnostic Techniques for Integrated Structural Health Monitoring (ISHM) Applications. Samad Valipour Shokouhi, Pål Skalle & Agnar Aamod have overviewed applications of case-based reasoning in drilling engineering. Funk (2015) and Begum Shahina etal (2011) have applied hybrid case-based reasoning & case-based reasoning strategies to the future of health science & health care systems. In their paper, Balakrishnan & Vasant Honavar (1996) examines and compares several different approaches to the design of intelligent systems for diagnosis applications. These include expert systems (or knowledge based systems), truth (or reason) maintenance systems, case=based reasoning systems, and inductive approaches like decision trees, ANNs (or artificial connectionist systems), and statistical pattern connection systems. Each of these approaches is demonstrated through the design of a system for a simple automobile fault diagnosis task. Pous, Colomer and Melendez (2004) extended a Fault Dictionary towards a Case-Based Reasoning System for Linear Electronic Analogue Circuits Diagnosis. Olsson, Funk & Bengtsson (2004) applied Case-Based Diagnosis to Industrial Robots based on acoustic signals. Erik Olsson, Peter Funk & Ning Xiong of Malardalen University, Sweden used Case-based reasoning to diagnose faults in industry using sensor readings. Fault diagnosis of industrial equipment becomes increasingly important for improving the quality of manufacturing and reducing the cost for product testing. Developing a fast and reliable diagnosis system presents a challenge issue in many complex industrial scenarios. The major difficulties therein arise from contaminated sensor readings caused by heavy background noise as well as the unavailability of experienced technicians for support. A novel method for diagnosis of faults by means of case-based reasoning and signal processing has been proposed. The received sensor signals are processed by wavelet analysis to filter out noise and at the same time to extract a group of related features that constitutes a reduced representation of the original signal. The derived feature vector is then forwarded to a classification component that uses case-based reasoning to recommend a fault class for the probe case. This recommendation is based on previously classified cases in a case library. Case-based diagnosis has attractive properties in that it enables reuse of past experiences whereas imposes no demand on the size of the case base. The proposed approach has been applied to fault diagnosis of industrial robots at ABB Robotics and the results of experiments are very promising. To diagnose failures in aircraft much knowledge is required. This knowledge is distributed over various knowledge sources. Pascal Reuss, Alexander Hundt, Klaus-Dieter Althoff, Wolfram Henkel, and Matthias Pfeiffer of the German Research Centre for Artificial Intelligence Kaiserslautern, Germany, present an approach for a decision support system for diagnosis and maintenance within the OMAHA research project. The authors compare their approach to other diagnostic approaches for technical diagnosis and describe the case-based agents within the decision support system in more detail. Jackson, Austin, Fletcher, & Jessop (2003) of the University of York have produced a technical report on Delivering a grid enabled distributed aircraft maintenance environment (dame). Yang Yang (2011) of Cranfield University has published an M.Sc Thesis on Aircraft Landing Gear Extension and Retraction Control System Diagnostics Prognostics and Health Management. The turbomachinery systems on oil platforms are extremely sensitive, as their failure can cause total shutdown of the platform's production activity. Scheduled stoppages for preventive maintenance are effective, since qualified personnel and replacement parts are available and ready. On the other hand, shutting down a turbo machine ceases petroleum production. Not rarely, maintenance people conclude the machine could have worked much longer. Monitoring the machines' behaviours to predict the best time to stop has been an industry trend. Failure prognosis has been performed using statistical methods and Artificial Intelligence techniques, such as Neural Networks, Fuzzy Systems and Expert Systems. Inhaúma Neves Ferraz, Ana Cristina Bicharra Garcia presents the results obtained by developing an expert system based on case-based reasoning for turbomachinery failure prognostics. Feret & Glasgow (1997) combined case-based and model-based reasoning for the diagnosis of complex devices. Saxena. Wu & Vachtsevanos (2005) integrated diagnosis and prognosis architecture for fleet vehicles using dynamic case-based reasoning. Dnyanesh Rajpathak, Rahul Chougule & Pulak Bandyopadhyay (2010) developed an application association mining and CBR Integrated framework for effective service and repair in the automotive Domain. In this paper, the authors propose a novel integrated framework combining association rule mining, case-based-reasoning and text mining that can be used to continuously improve service and repair in an automotive domain. The developed framework enables identification of anomalies in the field that cause customer dissatisfaction and performs root cause investigation of the anomalies. It also facilitates identification of the best practices in the field and learning from these best practices to achieve lean and effective service. Association rule mining is used for the anomaly detection and the root cause investigation, while case-based-reasoning in conjunction with text mining is used to learn from the best practices. The integrated system is implemented in a web based distributed architecture and has been tested on real life data. Developing Knowledge-Based Systems: Car Failure Detection using Expert System. Adsavakulchai (2014) of the School of Engineering, University of the Thai Chamber of Commerce, Thailand developed an expert system describing the knowledge-base of the car failure detection. 19 rule-based of car fault diagnosis expert system using the Visual Basic and Microsoft Access as tools for helping inexperienced mechanic was employed as the decision support system. During the test phase the system behaved well according to the rules used. It was concluded that car failure detection expert system was helpful although it might not give a complete guidance and help as a human expert do. Xiaoli Qin (1999) developed a CBR design system to assist human designers in solving bearing design problems in the mechanical engineering domain. It retrieves previously designed cases from a case-base, using certain adaptation techniques to repair them and their solutions to satisfy the current problem requirements. His approach combines parametric adaptations and constraint satisfaction adaptations. The technique of parametric adaptation considers not only parameter substitution, but also the interrelationships between the problem definition and its solution. The technique of constraint satisfaction adaptation provides a method to globally check the design requirements to assess case adaptability. Currently, the system can be used and tested in the Rolling Bearing design domain. Previous attempts to use Artificial Intelligence in Air Traffic Control (ATC) have never attained the level of confidence necessary for controllers to effectively use it in the real world. This lack of success is due in large measure to knowledge engineering difficulties in modelling ATC decision making. In this thesis Andrea Bonzono (1998) describe the successful application of CBR to this problem. He describe what was required to make CBR work and asses the knowledge engineering impact of CBR. The novelty of the approach presented in this thesis is in the manner that artificial intelligence is used as an intelligent assistant rather than an expert system, and in the technique used, which is CBR instead of the standard rule-based systems (RBS). The integration of rule-based and case-based reasoning methods has shown a substantial improvement with regards to performance over the individual methods. Verdande Technology As wants to try integrating the rule based reasoning method with an existing case based system. In this research project Azeb Bekele Eshete (2009) focuses on designing, implementing and testing of a demo system that demonstrates the development of a rule based reasoning component and integrating it with the existing case based system of Verdande Technology As. Transportation infrastructure, especially highway bridges, is characterized by their growing deterioration rate. Increased traffic loads, severity of environmental conditions, and deferred maintenance decisions are the main reasons for this rapid deterioration. Transportation agencies have developed Bridge Management Systems (BMS's) to select optimal maintenance actions and to predict future funding needs. BMS's require bridge deterioration models to predict the future condition of bridges and to estimate their remaining service life. Most of the state-of-the-art BMS's employ Markov Decision Process (MDP) in modelling bridge deterioration. The main shortcomings of Markovian models are that they assume state independence, neglect improvement actions taken in the past, and overlook the interaction of bridge components. Moreover, these models are difficult to update when new data are obtained or a different rating system is used. In this research, a Case-Based Reasoning (CBR) approach was introduced by Morcous, George (2000) to provide BMS's with a realistic, accurate, and generic deterioration model that eliminates the previous shortcomings. This approach predicts the future condition of a query bridge case by retrieving bridge cases stored in the BMS database that are similar to the query case in their physical features, environmental and operating conditions, and previous inspection and maintenance records. To implement this approach, a CBR shell called CBRMID was developed to be used in modelling the deterioration of any infrastructure facility. This shell was designed to meet a set of requirements, such as hierarchical decomposition of cases, representation of time-dependent data, data accumulation, case adaptation, data derivation, system modularity, versatility, and extensibility. CBRMID consists of three main modules: (i) Case Based Reasoner, for case retrieval and adaptation; (ii) System Maintainer, for case and knowledge accumulation and modification; and (iii) Database, for case and knowledge representation. CBRMID was implemented using the object-oriented programming language C++, the object-oriented database management system ObjectStore 6.0, and the expert system programming language CLIPS 6.10. Cases obtained from the Ministry of Transportation of Quebec (MTQ) database were employed in developing a "proof of concept" CBR application for modelling the deterioration of concrete bridge decks. Comparison between the solutions provided by the developed application and the actual solutions of real-world cases indicated that the CBR approach has a great potential in modelling bridge deterioration and, consequently, may have a serious impact on our transportation system. In his thesis Kapetanakis Stylianos (2012) presents an approach towards the effective monitoring of business processes using Case-Based Reasoning (CBR). The rationale behind this research was that business processes constitute a fundamental concept of the modern world and there is a constantly emerging need for their efficient control. They can be efficiently represented but not necessarily monitored and diagnosed effectively via an appropriate platform. Motivated by the above observation this research pursued to which extent there can be efficient monitoring, diagnosis and explanation of the workflows. Workflows and their effective representation in terms of CBR were investigated as well as how similarity measures among them could be established appropriately. The monitoring results and their following explanation to users were questioned as well as which should be an appropriate software architecture to allow monitoring of workflow executions. Time series analysis covers many different task types including classification, indexing, clustering, prediction etc., which deals with the data represented as a sequence of data points, measured at successive time periods. For a long time, time series analysis was based solely on statistical methods. Recently, some of the research was done in the area of applying different Machine Learning and Data Mining techniques in time series analysis. However, there are only a few papers on applying the CBR in time series analysis. In his thesis Vladimir Kurbalija (2009) describes one promising approach where a problem of time series analysis and prediction was solved by using Case Based Reasoning (CBR) technology. Case-Based Reasoning has become successful technique for knowledge-based systems in different domains. This promising technique is based on the use of previous experience in a form of cases so as to understand better and solve new problems in a particular domain. The main assumption in CBR is that, for many particular domains, similar problems usually have similar solutions. Déchamp, Dutech, Montroig, Qian, Racoceanu, Rasovska, Brézillon, Charpillet, Jaffray, Moine, Morello, Müller, Nguengang, Palluat, & Pelissier used CBR as a prognosis and diagnosis tool in the PROTEUS E-maintenance platform. The PROTEUS project has identified several maintenance tasks and functionalities where Artificial Intelligence tools or methods could bring many benefits. The authors focused on the diagnosis and prognosis of faulty states in equipment. They showed how the generic "PROTEUS AI Web Service" can be instantiated with different AI tools to do diagnosis or prognosis. PROTEUS architecture can work with various data and knowledge sources (like FMECA, SCADA, CMMS, ERP) enabling the use of both maintenance and production data. The characteristics of these "Web Services" were then analyzed so as to specify the features of a "Meta-Tool" that, for a given diagnosis or prognosis task, would help decide which tools are best suited. This Meta-Tool itself could be built as an instance of the generic "PROTEUS AI Web Service". Vahid Julali & David Leake (2015) have published a CBR based case study of large-scale adaptation rule generation by using big data. Additional references are also included in the list of references.

7. Discussion & Conclusion

It is important to study the past to define the future. CBR is based on the past case histories of human experiences gathered from various disciplines. CBR is based on a memory-centered cognitive model. This raises a number of following questions. a) While Retrieval based purely on similarity remains the mostly used technique, important limitations of similarity-based retrieval have come to light with the emergence of new and more demanding applications of CBR. [http://www.iiia.csic.es/~mantaras/Retrieval_Overview.pdf]. (b) How accurate, credible and trustworthy are the past case history? c) How to trust the past case history? d) How to extract the 'truth' from the case history? e) What role human centred reliability play in CBR? f) How relevant it is to adapt past solutions to new (and challenging) problems to modern high speed complex human-machine systems which are designed to operate in unfamiliar/harsh environments? g) Is CBR culture-based? h) How to secure the information against the increasing threat of cyberattacks? i) Human experiences are subjective and are prone to contamination & misinterpretations. .History of past experiences can be unreliable because of errors in comprehension, recall, evaluation, expression and personal beliefs, culture and environment. CBR's reliability is highly dependent upon the trust-worthiness, reliability and the state of the human mind. No one fully knows the psychology and ontology of human mind. These are some of the least understood problems in CBR.

Also, many organizations do not have cybersecurity plans to combat increasingly dangerous cyberattacks. The Internet of Things, cloud computing and big data are prone to the cyberattacks. Decisions based on Truth, Wisdom, Honesty, Trust, Openness & Welfare of the Society will, no doubt, enhances the "Quality of Life". Knowledge without wisdom is dangerous and irresponsible. CBR based on holistic knowledge, rich real-life experiences and collective wisdom from different sections of society over a lengthy period will result in a trust - worthy and reliable diagnostic and prognostic tool which will benefit mankind.

References

1. Aamodt, A., Plaza, E.: Case-Based Reasoning: Foundational Issues, Methodological Variations, and System Approaches. AI Communications, 7(1), 39-59 (1994)
2..Aamodt,A.: Explanation-Driven Case-Based Reasoning. LNCS, pp.274-274(1994)
3..Julie Main, Tharam Dillon & Simon Shiu. A tutorial on Case-based reasoning
4..Kolodner, J. Case-based reasoning. Morgan Kaufman Publications. (1993).
5..Althoff, K.D., Hanft, A., Schaaf, M.: Case factory - maintaining experience to learn. Advances in Case-Based Reasoning Lecture Notes in Computer Science 4106/2006, 429–442 (2006)
6..Riesbeck, C, Schank, R.: Inside Case-Based Reasoning. Lawrence Erlbaum Associates, Mahwah (1989)
7..Reuss, P.: Concept and implementation of a Knowledge Line - retrieval strategies for modularized, homogeneous topic agents within a multi-agent-system (in German). Master's thesis, University of Hildesheim (2012)
8..Reuss, P., Althoff, K.D.: Explanation-aware maintenance of distributed case-based reasoning systems. In: LWA 2013. Learning, Knowledge, Adaptation. Workshop Proceedings. pp. 231–325 (2013)
9..Althoff, K.D., Reichle, M., Bach, K., Hanft, A., Newo, R.: Agent based maintenance for modularised case bases in collaborative mulit-expert systems. In: Proceedings of the AI2007, 12th UK Workshop on Case-Based Reasoning (2007).
10..Bach, K.: Knowledge Acquisition for Case-Based Reasoning Systems. Ph.D. thesis, University of Hildesheim, dr. Hut Verlag Mnchen. (2013).
11. Amjad Abou Assali, Dominique Lenne,and Bruno Debray Case Retrieval in Ontology-Based CBR Systems; University of Technology of Compiègne.
12..Kim, H.J., Bae, C.H., Kim, S.H., Lee, H.Y., Park, K.J., Suh, M.W. "Development of a Knowledge-Based Hybrid Failure Diagnosis System for Urban Transit", International Journal of Automotive Technology, Vol. 10, No. 1, pp. 123–129. (2009).
13..Yee, P.S., Kiong, L.S., Song L.W. "Adaptive Case-based Reasoning for Fault Diagnosis", International Conference of Soft Computing and Pattern Recognition. (2009).

14..Luis Fernando Alape Realpe, Alberto Jose Álvares, Edgar Jhonny Amaya Simeon, Iván Patricio Moreno Marcos, ABCM Symposium Series in Mechatronics - Vol. 5. A Methodology Based In Case-Based Reasoning To Build A Knowledge-Base Applied To Failure Diagnosis System Of Hidrogenerators Machinery. (2012).
15..Hamid Berenji, Yan Wang, Abhi Saxena http://icsl.gatech.edu/aa/images/2/28/FUZZIEEE05_DCBR.pdf
16..George Vachtsevanos, Frank L. Lewis, Michael Roemer, Andrew Hess, Biqing Wu
Intelligent Fault Diagnosis and Prognosis for Engineering Systems, ISBN: 978-0-471-72999-0, 456 pages, October 2006, John Wiley
17. T Olsson and P Funk. Case-based reasoning combined with statistics for diagnostics and prognosis
25th International Congress on Condition Monitoring and Diagnostic Engineering Management. (2012).
18..R. Belecheanu, K.S. Pawar, R.J. Barson, B. Bredehorst, F. Weber, "The application of case based reasoning to decision support in new product development", Integrated Manufacturing Systems, Vol. 14 Iss: 1, pp.36 – 45. (2003).
19..Fault diagnosis via fusion of information from a case stream
Tomas Olsson, P.Funk, et al. 23rd International Conference on CBR 2015 held in Frankfurt am Main, Germany, Sept. 28-30. (2015).
20..Ann Patterson-Hine, Gordon Aaseng, Gautam Biswas and Sriram Narasimhan, "A Review of Diagnostic Techniques for ISHM Applications", NASA, USA. (2005).
21..Samad Valipour Shokouhi, Pål Skalle & Agnar Aamod. An overview of case-based reasoning applications in drilling engineering
http://www.idi.ntnu.no/~agnar/publications/air-12.pdf
22..P. Funk. School of Innovation, Design & Engineering, Malardalen University, Sweden. Why hybrid case-based reasoning will change the future of health science and healthcare. (2015).
23..Begum Shahina et al. Case-Based reasoning systems in the health sciences: a survey of recent trends and developments. Systems, Man and Cybernetics, Part C, Applications and Reviews, IEEE Transactions on 41, 4 (2011), 421-434.
24..Karthik Balakrishnan, & V. Vasant Honavar. Iowa University. Intelligent Diagnosis Systems, (1996).
25..Pascal Reuss, Alexander Hundt, Klaus-Dieter Althoff, Wolfram Henkel, and Matthias Pfeiffer. Case-based agents within OMAHA Project. German Research Center for Artificial Intelligence Kaiserslautern, Germany http://www.dfki.de 2 Institute of Computer Science, Intelligent Information Systems Lab University of Hildesheim, Hildesheim, Germany http://www.uni-hildesheim.de 3 Airbus Kreetslag 10 21129 Hamburg, Germany.
26..Jackson, T., Austin, J., Fletcher, M., Jessop, M.: Delivering a grid enabled distributed aircraft maintenance environment (dame). Tech. rep., University of York (2003).
27..Yang Yang. Aircraft Landing Gear Extension And Retraction Control System Diagnostics Prognostics And Health Management. MSc Thesis. Cranfield University. (2011).
28..Inhaúma Neves Ferraz, Ana Cristina Bicharra Garcia
Turbo Machinery Failure Prognostics, Chapter in Modern Advances in Applied Intelligence, Volume 8481 of the series Lecture Notes in Computer Science pp 349-358 http://link.springer.com/chapter/10.1007%2F978-3-319-07455-9_37
29..Feret, M., Glasgow, J.: Combining case-based and model-based reasoning for the diagnosis of complex devices. Applied Intelligence 7, 57–78 (1997).
30. Saxena, A., Wu, B., Vachtsevanos, G.: Integrated diagnosis and prognosis architecture for fleet vehicles using dynamic case-based reasoning. In: Autotestcon (2005).
31..Dnyanesh Rajpathak, Rahul Chougule, Pulak Bandyopadhyay An Integrated Framework for Effective Service and Repair in the Automotive Domain: An Application of Association Mining and Case Based Reasoning, International Journal of Computers in Industry, (2010)
32..Adsavakulchai, S. Developing Knowledge-Based Systems: Car Failure Detection using Expert System, School of Engineering, University of the Thai Chamber of Commerce, Thailand www.iiste.org ISSN 2224-5758 (Paper) ISSN 2224-896X (Online) Vol.4, No.12, (2014)
33..Xiaoli Qin, A case-based reasoning system for bearing design. MSc thesis, Drexel University. June (1999).
34..Andrea Bonzano, ISAC: a Case-Based Reasoning System for Aircraft Conflict Resolution. PhD Thesis. University of Dublin, April (1998).
35..Azeb Bekele Eshete, Integrated Case Based and Rule Based Reasoning for Decision Support. Master in Information Systems Thesis, Norwegian University of Science and Technology, July (2009).
36.Morcous, George, Case-based reasoning for modeling bridge deterioration, PhD thesis, Concordia University, (2000).
37..Kapetanakis, Stylianos, Intelligent monitoring of business processes using case-based reasoning. PhD thesis, University of Greenwich. (2012).
38..Vladimir Kurbalija, Time series analysis and prediction using Case Based Reasoning Technology. PhD Thesis, UNIVERZITET U NOVOM SAD, (2009)

39..L. Déchamp, A. Dutech, T. Montroig, X. Qian, D. Racoceanu, I. Rasovska, B. Brézillon, F.Charpillet, J-Y. Jaffray, N. Moine, B. Morello, S. Müller, G. Nguengang, N. Palluat, L.Pelissier. On the Use of Artificial Intelligence for Prognosis and Diagnosis in the PROTEUS E-maintenance platform. PROTEUS ITEA European Project URL : http://www.proteus-iteaproject.com
40..Vahid Julali & David Leake. CBR meets Big Data: A case study of large-scale adaptation rule generation. Proceedings of the 23rd International Conference on CBR held in Frankfurt am Main, Germany during 28-30th Sept. (2015).
41..Thomas Roth–Berghofer and Ioannis Iglezakis. Six Steps in Case–Based Reasoning: Towards a maintenance methodology for case–based reasoning systems. http://www.iglezakis.net/downloads/papers/gwcbr2001-2.pdf
42..The following two web sites provide many useful information and links to other CBR resources: • Ian Watson at the University of Salford maintains a site with URL http://www.ai-cbr.org • Ralph Bergmann, Ivo Vollrath and Sascha Schmitt at the University of Kaiserslautern maintain a site with URL http://www.cbr-web.org
43..http://www4.comp.polyu.edu.hk/~csckshiu/pdf/shiu01scbrb2.pdf
44..http://citeseerx.ist.psu.edu/viewdoc/download?doi=10.1.1.126.1155&rep=rep1&type=pdf
45..http://citeseerx.ist.psu.edu/viewdoc/download?doi=10.1.1.454.8651&rep=rep1&type=pdf
46..https://home.cc.gatech.edu/dil/uploads/Design,%20Innovation%20and%20Case-Based%20Reasoning.pdf
47..http://www.es.mdh.se/pdf_publications/1075.pdf
48..http://www.es.mdh.se/pdf_publications/811.pdf
49..https://www.aaai.org/Papers/Symposia/Spring/1999/SS-99-04/SS99-04-011.pdf
50..https://www.researchgate.net/publication/200443873_Assumption-Based_Network_Reasoning_and_Fault_Diagnosis
51..http://people.unipmn.it/mino/pubs/ECCBR06.pdf
52..http://www.emeraldinsight.com/doi/abs/10.1108/09576060310453335
53..https://macsphere.mcmaster.ca/bitstream/11375/8491/1/fulltext.pdf

Research on Equipment Fault Prediction Method Based on Full Vector Spectrum

Lei Chen[1,2], Jie Han[1], Wenping Lei[1], Lingjun Li[1] and Liya Wang[1]

[1] Institute of Vibration Engineering, Zhengzhou University, 450001 Zhengzhou, PRC,
[2]School of Chemical Engineering and Energy, Zhengzhou University,
450001 Zhengzhou, PRC, chenlei@zzu.edu.cn

Abstract. Fault prediction is one of the key factors of the predictive maintenance. Researches on mechanical fault prediction, currently, are mainly focused on the vibration trend prediction and the equipment residual life evaluation. Since not enough researches have been done on the prediction of the spectrum structure of the dynamic signals, there is still lack of effective methods on the prediction of the potential fault types of the equipment. Therefore, a time series equipment fault prediction model was proposed to solve the above problem. The proposed model was built on the basis of the adaptive prediction and full vector spectrum. A fault prediction system based on time series analysis was established in this study. The proposed fault prediction model included three levels of content. Firstly, a vibration strength prediction model was constructed based on vibration feature of amplitude domain. The model was used to predict the vibration intensity changes and determine the deterioration trend of equipment. Secondly, a fault spectrum structure prediction model was built. The inputs of the model were full vector spectrums of time series. The model was used to predict the possible fault properties. Thirdly, a multi-dimensional full vector prediction model was established. The inputs of the model were full vector spectrums of the multiple measurement positions of the machine. Several related spectrum structures were calculated and predicted. By analysis of the predicted spectrum structures, the machine running state and possible fault can be determined. The data fusion method used in the prediction models is on the basis of the methodology of full vector spectrum. The uncertainty of the spectrum structure can be eliminated by the designed data acquisition and fusion method. The reliability of the diagnosis on fault character was improved. The proposed prediction model provides a feasible method for the prediction of the spectrum structure and the potential equipment fault. It has vital significance to the reasonable arrangement of equipment maintenance.

Keywords: Fault prediction; Full vector spectrum; Data acquisition; Information fusion; AR model.

1 Introduction

Mechanical equipment maintenance is generally divided into three categories: breakdown maintenance, preventive maintenance and predictive maintenance [1]. Fault prediction is to predict the running state and forcast the potential failure of the machine in the future period of time according to the machine's running conditions[2].

Researches and applications of fault prediction provides effective means for the machine maintenance. By these means, one may now determine a potential fault such as cracked rotor, bad bearing, seal instability, etc. and make a reasonable arrangement of equipment maintenance. The research and application can bring considerable economic and social benefits for the enterprise.

Currently, most equipment failure prediction research is based on a single vibration parameter. Researches focused on the aspects of equipment health, fault severity, vibration amplitude, and so on [3]. By this means, the tendency of the numerical changes of the parameters are figured out and the residual lifes of the machines are assessed. But there are lack of methods on the prediction of fault locations, fault characters and evaluation of the whole machine. This is due the fact that the traditional spectrum analysis is based on a single transducer, although the test system using one or more transducers. At the same interface of rotating machine, different vibration spectrum will be obtained because of the different installation direction of the transducers. It is obvious that the information collected by a single transducer is not comprehensive. To predict the spectrum with a single transducer cannot satisfy the requirement of reliability. In response to these circumstances, information fusion was made based on the theory of full vector spectrum [4, 5]. A data-driven fault prediction system was established. The possible change of the vibration amplitude, spectrum structure and the overall running state of the machine can be predicted.

2 Full Vector Spectrum Based Prediction Method

2.1 Principle of Full Vector Spectrum

Usually, the vibration of the rotating machine is measured by non-contact eddy current transducers. The transducers are installed with the X-Y layout or V installation [2]. The rotor vibration collected by the two transducers of x and y directions are a combination of motion orbits for some harmonic waves w_i (i = 1, 2, 3...). Movement equation of x, y direction can be expressed as

$$\begin{cases} x = \sum_{i=1}^{\infty} X_i \cos(w_i t + \emptyset_{xi}) = Re(\sum_{i=1}^{\infty} \bar{X}_i e^{jw_i t}) = Re[\sum_{i=1}^{\infty}(x_{ci} + jx_{si})e^{jw_i t}] \\ y = \sum_{i=1}^{\infty} Y_i \cos(w_i t + \emptyset_{yi}) = Re(\sum_{i=1}^{\infty} \bar{Y}_i e^{jw_i t}) = Re[\sum_{i=1}^{\infty}(y_{ci} + jy_{si})e^{jw_i t}] \end{cases} \quad (1)$$

In the above formula, $\bar{X}_i$, $\bar{Y}_i$ represent the complex amplitudes of the vibration signals in the direction x, y of the harmonic i (i= 1, 2, 3...); the trajectories of each harmonic are an ellipse, that is

$$(Y_{ci}^2 + Y_{si}^2)x_i^2 + (X_{ci}^2 + X_{si}^2)y_i^2 - 2(X_{ci}Y_{si} + X_{si}Y_{ci})x_i y_i = (X_{si}Y_{si} - X_{ci}Y_{ci})^2 \quad (2)$$

When measuring the same signal with transducer in different directions, different frequency spectrums will be obtained. That is to say, whether the transducer mounts in x direction or in y direction, the acquired signal in one transducer is one-sided. In order to obtain the reliable fault prediction, it is required to overcome the one-sidedness of the vibration obtained by a single transducer. In this study, full vector signal is fused and the full vector spectrum will be extracted for the purpose of fault predication.

2.2 Structure of Prediction System

The structure of prediction method is shown in figure 1, which is mainly composed of three layers: data layer, prediction layer and evaluation layer.

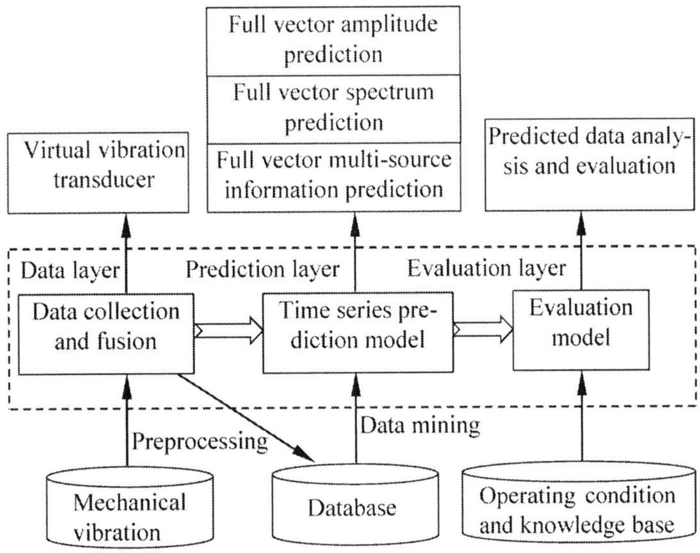

Fig 1. Diagram of Full Vector Fault Prediction

In the data layer, machine fault diagnoses with only one transducer's signal may produce misjudgment. So information fusion is needed. The full vector spectrum is the synthesis of two homologous transducers' information. After completing the data acquisition at the same interface with x-direction and y-direction transducers synchronously, the information fusion was carried out based on the theory of full vector spectrum. The fused data's spectral structure can reflect the operation state of the machine comprehensively.

In the prediction layer, vibration data was classified and mined in accordance with the machine operating conditions. The prediction model was build up based on the theory of auto regressive (AR). The amplitude was predicted according the data of full vector vibration. The major axis of the vibration ellipse, called main vector, reflects the most significant vibration. The spectrum structure prediction was based on main vector. Amplitudes at different frequencies have different change tendencies, so the prediction models at each frequency point is different. The parameters of AR models are a coefficient matrix. As time goes on, the parameters of AR model need to be revised to adapt to the changes of vibration. So an iterative algorithm for AR models was constructed to correct model coefficient in real time.

In the evaluation layer, the machine running condition was predicted according to the set of the alarm value and risk value. In the domain of predicted results, the fault character and position were judged out according to the spectral structure changes. The overall running state was evaluated by the predicted results of the multi-source vibration.

3. Prediction methods

3.1 Prediction method and amplitude prediction

Equipment fault prediction methods on the whole can be divided into two categories: qualitative and quantitative methods [6-8]. In this paper, the fault prediction model was constructed based on auto regressive (AR) model. AR(n) model was commonly used in time series analysis.

$$x(t) = \emptyset_1 x_{t-1} + \emptyset_2 x_{t-2} + \cdots + \emptyset_n x_{t-n} + \varepsilon_t \quad (3)$$

Equation (3) is an n-order AR model used to describe the vibration amplitude sequence, denoted as AR(n). Where n stands for the order of the model; $\emptyset_i$ stands for the parameters of the model; ε_t stands for the random noise; x(t) stands for the vibration amplitude of day t, x(t-n) stands for the amplitude of day t-n. Figure 2 shows that the predicted values were basically in agreement with the actual ones.

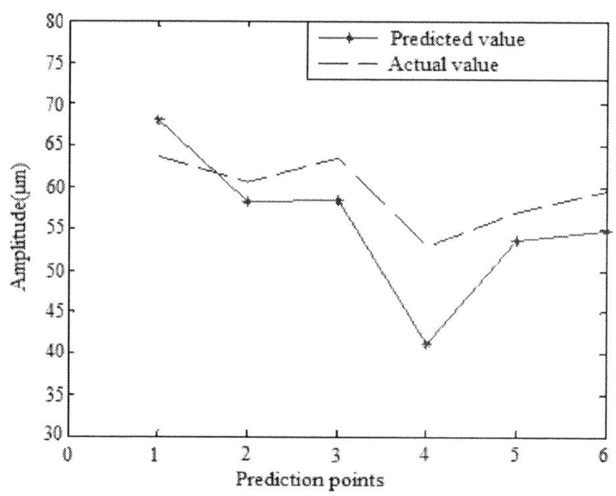

Fig 2. Contrast of actual value and predicted

3.2 Full vector spectrum prediction

Spectrum structure prediction is a feasible method for fault property prediction. Signal obtained with only one single transducer is one-sided for fault prediction. Date fusion must be carried out to enhance the reliability of spectrum. The data fusion is based on the theory of full vector spectrum. The main vector was used for the prediction of spectrum structure. Comparing with the spectrum in one direction only, the main vector spectrum structure can reflect the operation state of the machine comprehensively.

When the spectrum prediction was carried on, vibration amplitude of each frequency point of the signal series was used to construct a AR model. As shown in equation 4.

$$x_1(t) = \emptyset_{11} x_{1(t-1)} + \emptyset_{12} x_{1(t-2)} + \cdots + \emptyset_{1n} x_{1(t-n)} + \varepsilon_{1t}$$
$$x_2(t) = \emptyset_{21} x_{2(t-1)} + \emptyset_{22} x_{2(t-2)} + \cdots + \emptyset_{2n} x_{2(t-n)} + \varepsilon_{2t}$$
$$\cdots$$
$$x_m(t) = \emptyset_{m1} x_{m(t-1)} + \emptyset_{m2} x_{m(t-2)} + \cdots + \emptyset_{mn} x_{m(t-n)} + \varepsilon_{mt} \quad (4)$$

The prediction model is a two-dimensional matrix, and the coefficient matrix of the AR model can be written as

$$A = \emptyset_{ij} = \begin{bmatrix} \emptyset_{11} & \emptyset_{12} & \cdots & \emptyset_{1n} \\ \emptyset_{21} & \emptyset_{22} & \cdots & \emptyset_{2n} \\ \vdots & \vdots & \ddots & \vdots \\ \emptyset_{m1} & \emptyset_{m2} & \cdots & \emptyset_{mn} \end{bmatrix} \tag{5}$$

Where i = 1, 2,..., m, stands for the number of frequency point, j = 1,2,...,n stands for the order of AR model. The result of spectrum structure prediction with AR(13) is shown in figure 3. The potential fault property can be determined by analyzing the spectrum structure in the prediction domain.

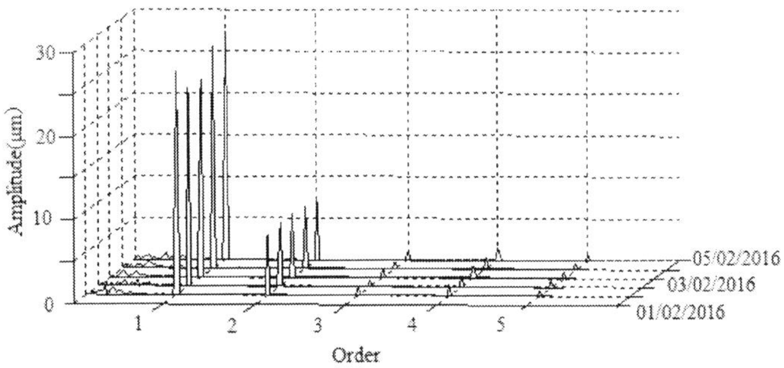

Fig 3. Full vector spectrum prediction with AR(13).

3.3 Multi-dimensional prediction and overall evaluation

Evaluation of vibration status of a machine often requires measurement results of multiple measuring sections. That is to say, it need to predict not only the spectral structure of one measurement section of the machine, but also need to predict spectra structures of multiple sections. The overall evaluation can be made in accordance with the prediction results.

$$Y = AX + \varepsilon \tag{6}$$

Where, $A = \emptyset_{ikj}$ =

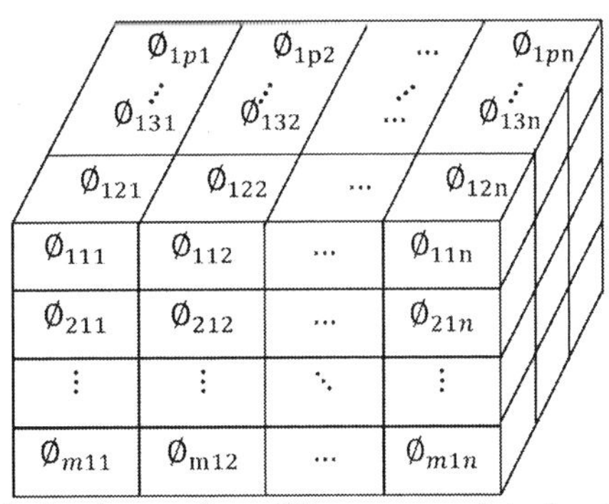

Where i = 1,2,...,m, stands for the number of frequency points, k = 1,2,...,p stands for the number of sections; j=1,2,...n stands for the order of AR model.

Comprehensive prediction model of multiple measurement cross sections can be set up based on equation (6). The potential machine fault position and property can be determined by the prediction model.

4. Conclusions

Feature extraction and fault prediction are the key technologies of predictive maintenance. Due to the shortcomings of one-sidedness of single transducer measurement result of rotating machinery, homologous information fusion was carried out based on the theory of full vector spectrum. The AR(n) prediction model was set up based on the fusion vibration signal series; The constructed vibration amplitude prediction model can be used for determining the deterioration trend of equipment. The built spectrum structure prediction model based on the theory of full vector spectrum can be used for the prediction of the possible fault properties. The established multi-dimensional full vector prediction model based on the signal of multiple measurement positions can be used for the prediction of machine running state and possible fault position. The data fusion method used in prediction models are all on the basis of the methodology of full vector spectrum. The uncertainty of the spectrum structure can be eliminated by the designed data acquisition and fusion method. The proposed prediction model provides a feasible method for the prediction of the spectrum structure and the potential equipment fault. It provides a feasible means for the reasonable arrangement of equipment maintenance.

5. Acknowledgements

This project is supported by the Guidance Program of the Key Projects of Scientific and Technical Research of Department of Education of Henan Province (13B603970.0) and the Opening Project of Key Laboratory of Precision Manufacturing Technology and Engineering, Henan Polytechnic University (PMTE201302A).

References

1. J.Ma, X.L.Xu, Z.D.Hua, Survey of Fault Predication Methods for Rotating Machinery, Process Automation Instrumentation, 32,1-3 (2011)
2. L.Chen, J.Han, W.P.Lei, et al. Full Vector Signal Acquisition and Information Fusion for the Fault Predication, International Journal of Rotating Machinery, 2016, 1-7(2016)
3. Y.Y.Li, J.Han, L.Chen, et al. Research on Condition Trend Prediction Method of Mechanical Equipment Based on AR-Kalman, Manufacturing Automation, 12, 95-96, 120(2014)
4. Donald E. Bently, Charles T. Hatch, Bob Grissom, Fundamentals of Rotating Machinery Diagnostics. Bently Pressurized Bearing Press, Minden Nevada (2002)
5. J.Han, L.D.Shi. Full Vector Spectrum Technology and Its Engineering Application, China machine press, Beijing (2008)
6. D.H. Zhou, Y.Y.Hu, Fault Diagnosis Techniques for Dynamic System, Acta Automatica Sinica, 35, 748-755,(2009)
7. X.L. Xu, T. Wang, The Research and Realization of the Fault Reasoning Intelligent Expert System in the Rotor of Rotating Machine, 31, 12-13, 2004
8. Z.Y. Song, G.Y. Zhang, Y.Q. Su, One Fusion Algorithm of Equipment Fault Prediction Based on Rough Set Theory and Grey Model, Applied Mechanics and Materials, vols. 687-691, 1377—1379(2014)

EEMD-Based Adaptive Multi-Scale Noise-Controlled Stochastic Resonance for Mechanical Fault Diagnosis

Jimeng Li[1*], Jinfeng Zhang[2], Yungang Zhang[1] and Ping Xie[1]

[1] College of Electrical Engineering, Yanshan University, Qinhuagndao 066004, PR China
[2] College of Liren, Yanshan University, Qinhuagndao 066004, PR China
[*]Corresponding author: Telephone: +86 0335-8057041; E-mail: xjtuljm@163.com;
Address: No. 438, West Part of Hebei Avenue, Qinhuagndao 066004, PR China

Abstract: Effective extraction of weak features submerged in strong noise that are indicative of structural defects has remained a major challenge. Therefore, an adaptive multi-scale noise-controlled stochastic resonance (SR) method for mechanical fault diagnosis is proposed by using the adaptive decomposition ability of ensemble empirical mode decomposition (EEMD) and the unique advantage of SR using noise to enhance weak signal feature. According to the oscillation characteristics of signal itself, EEMD can adaptively decompose the fault signals into different frequency bands and optimize the data analysis flow. Therefore, through EEMD, the noise intensity of different scales can be adjusted to realize the SR detection of weak signal. The validity and practicability of the proposed method is verified by engineering application.

Keywords: Stochastic resonance; Multi-scale; EEMD; Adaptive; Fault diagnosis

1 Introduction

In view of the inherent link between machine operation and vibration, vibration signals collected from mechanical equipment carry rich information on machine health conditions. Therefore, vibration-based condition monitoring and fault diagnosis methods [1-3] have been extensively investigated. Unlike traditional denoising techniques, stochastic resonance (SR) [4-5] is a kind of typical noise-assisted data processing method, which realizes the detection of weak signal by utilizing noise instead of eliminating noise, and the weak signal features are not weakened or destroyed but enhanced. Consequently, by virtue of the unique advantage of using noise to enhance weak signals, SR has been widely studied in machinery fault diagnosis field [6-9].

Although SR is an effective method to extract weak information features from the strong background noise, the detection effect of SR for the complicated and nonstationary signal is not satisfactory. Therefore, a new ensemble empirical mode decomposition (EEMD)-based adaptive multi-scale noise-controlled SR method for mechanical fault diagnosis is studied. As a noise-assisted adaptive decomposition method, EEMD can eliminate the mode mixing phenomenon in empirical mode decomposition

(EMD) through adding white noise to signal [10], and realizes the adaptive multi-scale decomposition of complex signal according to the oscillation characteristics of signal itself. And then, select the effective intrinsic mode functions (IMFs) to reconstruct the input signal which is used as the input of SR system. Finally, the noise control parameter of reconstructed signal at different scales is optimized and selected by using the genetic algorithm. Engineering application demonstrates that the proposed method is validated to be effective in detecting the weak fault features in mechanical fault diagnosis. The remaining parts of this paper are arranged as follows. Section 2 is mainly dedicated to provide the theoretical background of SR and EEMD. The method of EEMD-based adaptive multi-scale noise-controlled SR method is described in detail in Section 3. Section 4 performs engineering application to verify the effectiveness of the proposed method. Finally, conclusions are drawn in section 5.

2 Theoretical Foundation

2.1 SR Theory

The overdamped SR equation with a bistable model can be expressed as follows:

$$x(t)' = ax(t) - bx(t)^3 + s(t) + n(t) \tag{1}$$

where $x(t)$ denotes the resonance system output, parameters a and b are positive real numbers, $s(t)$ is a input signal, and $n(t)$ is a Gaussian white noise with zero mean and D variance. The potential function is $U(x,t) = -\frac{a}{2}x(t)^2 + \frac{b}{4}x(t)^4$ with two stable points $x = \pm\sqrt{a/b}$ and one critical stable point $x_0=0$. The height of potential barrier is $\Delta U=a^2/b$. From Eq. (1), the system output $x(t)$ is actually the Brownian particle trajectory in the potential function $U(x)$ under the combined action of input signal and noise. And, the crucial process of using SR to detect weak signal is to adjust the noise intensity or system parameters to make the transition rate of the particle caused by noise match the oscillation period of the input signal, thus amplifying the particle movements in single potential well to the transition motions between double potential wells to realize the feature enhancement of input signal.

2.2 EEMD Algorithm

EEMD, as an improve method of EMD, is presented to alleviate the mode mixing problem in EMD. First a white noise is added to a signal to make the signal continuous in different scales, and then the noise-added signal is decomposed by EMD, finally through the ensemble mean of enough trials, the added white noise can be decreased or even completely canceled out and the ensemble mean is treated as the true answer. The EEMD algorithm can be given as follows [10]:

1. Initialize the number of ensemble N, the standard deviation of added white noise σ;
2. Add a Gaussian white noise with the given standard deviation σ to the investigated signal $x(t)$, and decompose the noise-added signal $x_i(t)=x(t)+n_i(t)$ with EMD into m

IMFs $c_{ij}(t)$ and a residue $r_i(t)$, where $n_i(t)$ denotes the ith added white noise, and $x_i(t)$ indicates the noise-added signal of the ith trial, $c_{ij}(t)$ represents the jth IMF of the ith trial, and m is the number of IMFs.

3. Repeat steps (2) with different white noise each time, and calculate the ensemble mean $\bar{c}_j$ of the N trials for each IMF, then conduct the mean $\bar{c}_j$ of each of the m IMFs as the final IMFs.

3 EEMD-Based Adaptive Multi-Scale Noise-Controlled SR Method

According to the oscillation characteristics of signal itself, EEMD algorithm can realize the adaptive multi-scale decomposition for input signal, which provides a new control strategy for realizing SR by adjusting the multi-scale noise. Therefore, a new EEMD-based adaptive multi-scale noise-controlled SR method is proposed to realize the resonance detection of weak signal. The important steps of the proposed method are described in detail below.

3.1 Selection of effective IMFs

EEMD can achieve the adaptive segmentation of frequency band, and the decomposition level/scale n and the bandwidth of each IMF are not set by manual. Therefore, how to select the appropriate IMF components from the n IMFs obtained by EEMD, namely that determines the reconstruction scale L to reconstruct the input signal is the key of realizing the multi-scale noise-controlled SR method.

Firstly, use the statistic characteristics of IMFs in frequency domain that the frequency spectrum distribution of IMFs can be well fitted by Gaussian distribution to calculate the center frequency f_i of each IMF component. The frequency spectrum of each IMF component is sorted by its amplitude values in descending order, and select the top 5 frequency values to calculate an average f_{av}, which is used as the center frequency of the ith IMF component, namely $f_i = f_{av}$.

Then, calculate the difference values between the target signal frequency f_0 and the center frequency f_i of each IMF component, and the decomposition level i of the IMF component corresponding to the minimum difference value is selected and assigned to the reconstruction scale L, thus the first L IMFs are served as the reconstruction components of multi-scale noise control algorithm.

3.2 Construction of band-pass filters

The decomposition process of EEMD indicates that the extraction of IMF is similar to the frequency band partition of dyadic wavelet transform. However, because of the influence of noise, the obtained IMFs are not strictly orthogonal each other. Thus, the statistic characteristics of IMFs in frequency domain are further applied to construct the corresponding filters to filter the L IMFs, so as to improve the orthogonality of adjacent IMFs. First, based on the frequency spectrum distribution of the first IMF component, the parameters of high-pass filter are designed as follows:

$$f_{1s}=(f_1+f_2)/2,\quad f_{1p}=f_{1s}+10 \tag{2}$$

where f_{1s} is the cutoff frequency of passband, f_{1p} is the cutoff frequency of stopband. f_1 and f_2 are the center frequencies of the first and second IMF component respectively.

Similarly, according to the frequency spectrum distribution of each IMF$_i$ (i=2,3,...,L), the parameters of band-pass filter corresponding to each IMF component are constructed as follows:

$$\begin{cases} f_{is_1}=(f_i+f_{i+1})/2, & f_{ip_1}=f_{is_1}+10 \\ f_{is_2}=(f_i+f_{i-1})/2, & f_{ip_2}=f_{is_2}-10 \end{cases} \quad i=2,3,\cdots,L \tag{3}$$

where f_{is1} and f_{is2} are the stopband lower limit frequency and upper limit frequency respectively, f_{ip1} and f_{ip2} are the passband lower limit frequency and upper limit frequency respectively. f_i is the center frequency of the ith IMF component.

3.3 Multi-scale reconstruction with noise control of different scale

The L IMFs are filtered by using the constructed filters, and their corresponding filtered results LB$_i$ are adjusted by introducing a controllable parameter α to realize the noise control of different scale, finally the adjusted results LB$_i$ are employed to reconstruct the input signal. The reconstruction formula is as follows:

$$Rex=\sum_{i=1}^{L}(2^{\alpha+i}\bullet \mathrm{LB}_i)\quad i=1,2,\cdots,L \tag{4}$$

where Rex is the reconstructed signal, and α is a controllable parameter, which denotes the noise intensity of different scale, and LB$_i$ is the filtered ith IMF component.

3.4 Algorithm flow of EEMD-based adaptive noise-controlled SR method

Through above analysis, the procedures of the proposed method are as follows: (1) Vibration signal pre-processing. Vibration signals measured on bearings are demodulated by the Hilbert envelop analysis, and the obtained envelope signals are used as the input of EEMD algorithm. (2) Parameter initialization. Set the SR system parameters a=1 and b=1, the searching range of noise intensity, the initial parameters of genetic algorithm, and the standard deviation of added white noise and the ensemble number of EEMD algorithm. (3) Multi-scale decomposition and filter processing. Using EEMD algorithm to decompose the envelop signal obtained by pre-processing into different frequency components IMFs, and then, the top L IMFs are filtered with the constructed filters. (4) Multi-scale noise control SR. Eq. (4) is adopted to reconstruct the input signal by using the filtered L IMFs, and then the reconstructed signal Rex is entered into SR system. The genetic algorithm is employed to adaptively select and optimize noise intensity α, whose objective function is established by using SNR of resonance output, and thus the best resonance result of input signal is obtained. (6) Fault identification based on SR detection result.

4 Engineering Application

Through the regular detection for wind turbines in a wind farm, it was found that the vibration response of the front bearing of generator of a wind turbine increased. To find out the reasons, the vibration signal needs to be further analyzed. The signals were sampled at 12.8 kHz, and the length of data is 16384. The average rotational speed of the generator is 1406 r/min, and the bearing type of generator is 6324.

Fig. 1 (a) shows the original vibration signal of the front bearing of the generator. The inconspicuous impulse components can be found in the time waveform. In Fig. 1 (b), apart from the rotational frequency of 23.44 Hz, there is not useful feature information relevant to bearing fault in the spectrum.

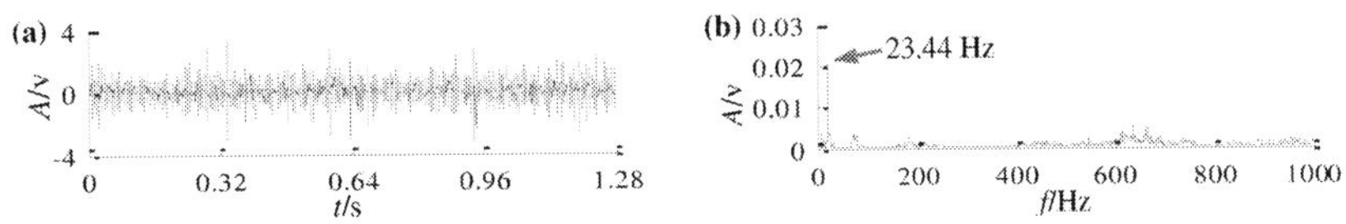

Fig. 1. Bearing signal: (a) the time waveform; (b) the spectrum

For the purpose of detecting the performance of the generator, the proposed method is applied to process the vibration signal. The searching range of noise intensity α is [0.1, 10], the standard deviation of added white noise is 0.08, and the ensemble number is 50. The genetic algorithms parameters settings: the number of initial population is 50, the maximum number of generations is 25 and the precision of the variables is 1e-8, etc. And the optimal noise intensity obtained by genetic algorithm is α=6.7, and the corresponding detection results are displayed in Fig. 2. Obviously, lots of background noise is eliminated, and the high-frequency signal energy is concentrated to target signal through SR system, some useless interferences are greatly decreased, thus making the spectrum peak 113.3 Hz more outstanding in the spectrum as shown in Fig. 2 (b). The decibel value corresponding to the fault characteristic frequency of bearing inner ring calculated by using Shock Pulse Method [11] is 21.64 dB, which indicates there is an early damage on the bearing inner ring. And the bearing of generator should be paid more attention and detected regularly.

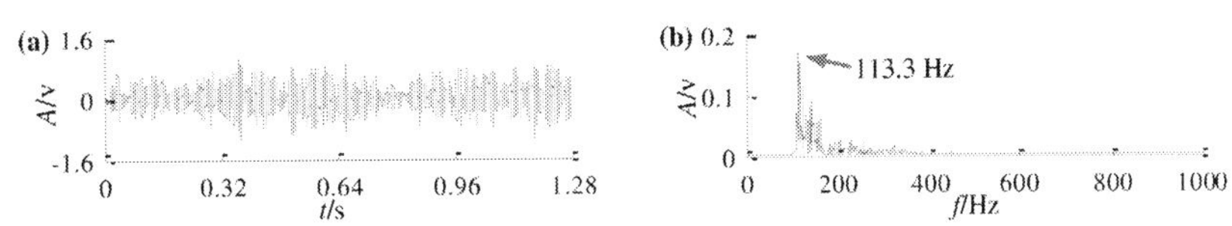

Fig. 2. Detection results: (a) the output waveform of SR; (b) the output spectrum of SR

5 Conclusion

Aiming at the complexity and diversity of fault signals collected from mechanical equipment, a new EEMD-based adaptive multi-scale noise-controlled SR method for weak feature extraction is proposed in the paper. According to the oscillation characteristics of signal itself, fault signals can be adaptively decomposed into different frequency bands with EEMD. And then, by adjusting the noise intensity of different scales with the help of the genetic algorithm, the weak features submerged in strong noise can be enhanced and extracted with SR method. The engineering case indicates that the proposed method is able to extract the fault characteristic information and identify the faults effectively.

Acknowledgements
This work was supported by the National Natural Science Foundation of China (Grant No. 51505415), China Postdoctoral Science Foundation Funded Project (Grant No. 2015M571279) and Science and Technology Support Project of Qinhuangdao City (Grant No 201502A008).

References

1. Yunusa-Kaltungo, A., Sinha, J.K., Elbhbah, K.: An Improved Data Fusion Technique for Faults Diagnosis in Rotating Machines. Measurement 58, 27--32 (2015)
2. Cong, F.Y., Zhong, W., Tong, S.G., Tang, N., Chen, J.: Research of Singular Value Decomposition Based on Slip Matrix for Rolling Bearing Fault Diagnosis. J. Sound Vib. 344, 447--463 (2015)
3. Wang, Y.X., Markert, R., Xiang, J.W., Zheng, W.G.: Research on Variational Mode Decomposition and Its Application in Detecting Rub-Impact Fault of the Rotor System. Mech. Syst. Signal Pr. 60-61, 243--251 (2014)
4. Benzi, R., Sutera, A., Vulpiani, A.: The Mechanism of Stochastic Resonance. J. Phys. A-Math. Gen. 14, 453--457 (1981)
5. Nicolis, C., Nicolis, G.: Stochastic Aspects of Climatic Transitions–Additive Fluctuations. Tellus 33, 225--234 (1981)
6. Lu, S.L., He, Q.B., Kong, F.R.: Stochastic Resonance With Woods–Saxon Potential for Rolling Element Bearing Fault Diagnosis. Mech. Syst. Signal Pr. 45, 488--503 (2014)
7. Zhao, R., Yan, R.Q., Robert, G.: Dual-Scale Cascaded Adaptive Stochastic Resonance for Rotary Machine Health Monitoring. J. Manuf. Syst. 32, 529--535 (2013)
8. Lai, Z.H, Leng, Y.G.: Generalized Parameter-Adjusted Stochastic Resonance of Duffing Oscillator and Its Application to Weak-Signal Detection. Sensors 15, 21327--21349 (2015)
9. Hu, B.B., Li, B.: A New Multiscale Noise Tuning Stochastic Resonance for Enhanced Fault Diagnosis in Wind Turbine Drivetrains. Meas. Sci. Technol. 27, 025017_1--025017_14 (2016)
10. Lei, Y.G., He, Z.J., Zi, Y.Y.: EEMD Method and WNN for Fault Diagnosis of Locomotive Roller Bearings. Expert Syst. Appl. 38, 7334--7341 (2011)
11. Li, Z., He, Z.J., Zi, Y.Y., Chen, X.F.: Bearing Condition Monitoring Based on Shock Pulse Method and Improved Redundant Lifting Scheme. Math. Comput. Simulat. 79, 318--338 (2008)

The Fault Diagnosis of Rolling Bearing Based on A Revised Hilbert–Huang Transform

Yongjian Ji[1], Xibin Wang[2], Zhibing Liu[2,*], Hongjun Wang[3]

[1]School of Mechanical Engineering, Beijing Institute of Technology,
[2]Key Laboratory of Fundamental Science for Advanced Machining,
Beijing Institute of Technology,
No.5 South Zhongguancun Street, Haidian District, Beijing 100081, P.R.China,
[3]Key Laboratory of Modern Measurement & Control Technology,
Beijing Information Science and Technology University,
No.12 East Qinghexiaoying Road, Beijing 100192, P R China

Abstract. Due to the non-linear and high randomness, it is difficult to diagnose the fault type of rolling bearing in linear methods. In the present paper, to diagnose the bearing fault accurately and effectively, a revised Hilbert–Huang transform (HHT) was proposed. The local linear extrapolation method was introduced to suppress end effects and the combination of auxiliary signal and decorrelation operator was introduced to eliminate mode mixing. To verify the effectiveness of the revised HHT, this new HHT method has applied to analyze the signal of inner race fault. The results indicate that the revised method is more reliable than the traditional in the aspect of rolling bearing fault diagnosis.

Keywords: Rolling bearing, fault diagnosis, Hilbert–Huang transform

1 Introduction

Rolling bearings are widely used in mechanical drive systems in manufacturing. The failure of a rolling bearing may lead to the failure of the mechanical system, so find out the causes of fault and eliminate the failure is necessary in industrial production. Due to the high stochastic and nonlinear characteristics, it is difficult to diagnose the fault type of bearings in traditional linear methods. Hilbert-Huang transform (HHT) has big advantages over the traditional linear method in analyzing non-linear and non-stationary signals. Although HHT is a promising approach in the aspect of fault diagnosis, it still has some drawbacks such as end effects and mode mixing. In order to overcome the above drawbacks, a revised HHT method is proposed here.

2 Hilbert-Huang transform (HHT)

HHT is proposed by Huang et al. [1] and it consists of two parts: the empirical mode decomposition (EMD) and the Hilbert spectral analysis (HSA). Firstly, a data is decomposed into a series of intrinsic mode functions (IMF) by EMD method. Secondly,

*Corresponding author.
E-mail address: liuzhibing@bit.edu.cn

with the Hilbert transform, the intrinsic mode functions (IMF) yield instantaneous frequencies as functions of time that give sharp identifications of imbedded structures. Finally, an energy-frequency-time distribution can be obtained. In addition, we can also get the marginal spectrum. The main drawbacks of HHT are end effects and mode mixing, which is harmful to the analysis result.

3 A revised HHT method

3.1 The suppression of end effects by local linear extrapolation

During the process of EMD, the upper and lower envelopes are obtained by cubic spline interpolation. Since the local maximum and the local minimum cannot be determined at the start point and endpoint of the data, serious problems of the spline fitting will occur near the ends, where the cubic spline fitting can have large swings, which called end effects [1]. A new method which is called local linear extrapolation is proposed to eliminate end effects. Taking the determination of right extreme points as an example, this method can be expressed as follows:

1) The determination of right maximum point:

(1) The local maximum value points A and B are determined; (2) A straight line AB is determined by point A and B. Point C is the intersection point of line AB and the time axis which correspond to the right endpoint, as shown in Fig. 1 (a); (3) If the value of point C is less than the value of right endpoint D, the right endpoint D is identified as the maximum value; If the value of point C is greater than the value of point D, as shown in Fig. 1 (b), then the maximum value point is identified in the following situations: If the value of point C is greater than twice of the value of point B, the maximum value point is identified as half of the value of adding C to D, otherwise, the maximum value point is identified as intersection point C.

2) The determination of right minimum point:

(1) The local minimum value points A\` and B\` are determined; (2) A straight line A\`B\` is determined by point A\` and B\`. Point C\` is the intersection point of line A\`B\` and the time axis which correspond to the right endpoint, as shown in Fig. 1 (c); (3) If the value of point C\` is greater than the value of right endpoint D\`, the right endpoint D\` is identified as the minimum value; If the value of point C\` is less than the value of right endpoint D\`, as shown in Fig. 1 (d), then the minimum value point is identified in the following situations: If the value of point C\` is less than twice of the value of point B\`, the minimum value point is identified as half of the value of adding C\` to D\`, otherwise, the minimum value point is identified as intersection point C\`. The method of determining left extreme points is similar to the above.

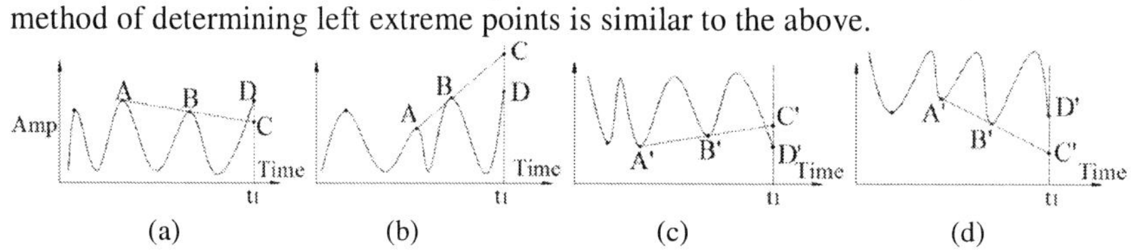

Fig. 1. Local linear extrapolation method legend

In order to verify the effectiveness of local linear extrapolation method in the aspect of restraining end effects, this method is applied to analyzing the simulation signal $y_1(t) = 5\sin(2\pi 15t) + 6\sin(2\pi 80t)$. The IMFs are shown in Fig. 2. The IMFs in Fig. 2 (a) are obtained by the traditional EMD method and the IMFs in Fig. 2 (b) are obtained by the revised EMD. It is obviously that end effects exist in both ends of IMF2 in Fig. 2 (a), but each IMF in Fig. 2 (b) is normal.

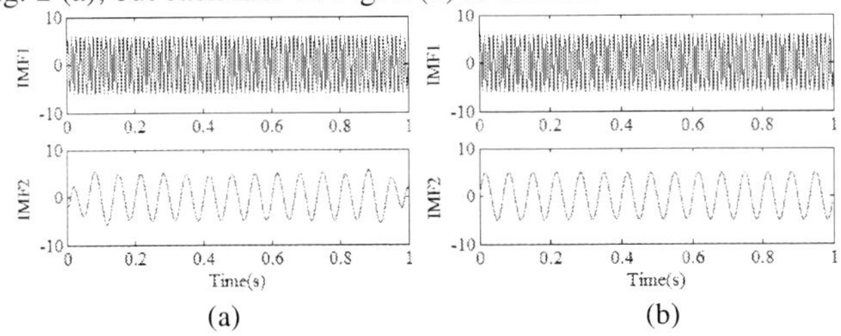

Fig. 2. The IMFs, (a) The first two IMFs obtained by the traditional EMD method, (b) The first two IMFs obtained by the revised EMD method

3.2 The suppression of mode mixing by adding sinusoidal signal and imbedding decorrelation operator

Mode mixing, which is defined as either a single IMF consists of signals of widely disparate scales or a signal of a similar scale residing in different IMF components, it could make the individual IMF lose its physical meaning [2]. The envelope will be changed if a signal mixed with discontinuous signal. It is easy to cause overshoots and undershoots since the envelopes are mutational at the junction of two signals, such as Fig. 3 (a) shows, so mode mixing is inevitable. Extreme points of the signal will be redistributed and the 'mutational event' which caused by singular signal will be unconspicuous if we add high frequency signal to the original signal. Although the envelope will be changed, such as Fig. 3 (b) shows, its trend becomes more in line with the original signal. So we can obtain a better decomposition effects.

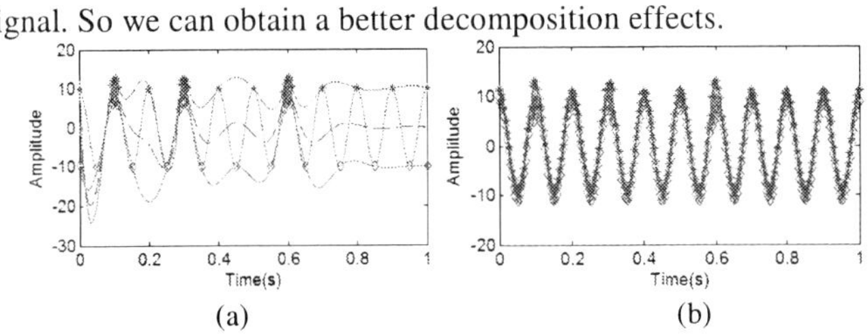

Fig. 3. The envelopes of each signals. (a) The envelopes of containing discontinuous signals, (b) The envelopes after adding high frequency sinusoidal signal

As discussed by Peng et al. [3], the first IMF always contains a wide range of frequencies. Since the arrangement of IMFs is according to the frequency from high to low, the frequency of the added sinusoidal signal should larger than the highest frequency of original signal so that we can ensure the sinusoidal signal is obtained as the first IMF. The added signal and discontinuous signal are mixed together in the first IMF, so we can obtain the discontinuous signal by minus the added signal from IMF1. In order to optimize the distribution of each extreme point, the amplitude of the added high frequency sinusoidal signal usually be chosen as 0.15~0.25 times of the original signal's peak.

Though adding high frequency sinusoidal signal to original signal can eliminate mode mixing, it is still useless to the signal which mixed with small frequency ratio signals. In order to eliminate mode mixing, we should make sure that the IMFs are orthogonal to each other. So the decorrelation operator [4] is introduced during the process of EMD. Correlation coefficient r is defined as Eq. (1)

$$r = \frac{y^T(n) \cdot x(n)}{y^T(n) \cdot y(n)} \tag{1}$$

Obviously, $r \cdot y(n)$ represents the cross-correlation part of $x(n)$ and $y(n)$. The data $v(n)$ which is not correlated to $y(n)$ can be obtained by Eq. (2).

$$v(n) = x(n) - r \cdot y(n) \tag{2}$$

According to the above mentioned, a new method is proposed to eliminate mode mixing, which is expressed as the following:

(1) Add high frequency sinusoidal signal $\hat{\hat{x}}(t)$ to original signal X(t), we can obtain a mixed-signal X'(t), then the new signal X'(t) is processed by EMD, finally a series of IMFs $c_i(t)$ can be obtained.

(2) The first IMF $\overline{c}_1(t)$ which without the added high frequency sinusoidal signal $\hat{\hat{x}}(t)$ can be obtained by Eq. (3)

$$\overline{c}_1(t) = c_1(t) - \hat{\hat{x}}(t) \tag{3}$$

(3) Calculate the correlation coefficient r_1 between $\overline{c}_1(t)$ and $c_2(t)$, then we can get the IMF $\hat{c}_1(t)$ by Eq. (4)

$$\hat{c}_1(t) = \overline{c}_1(t) - r_1 \cdot c_2(t) \tag{4}$$

(4) Reserve $\hat{c}_1(t)$, then we can obtain $\hat{X}(t)$ by the equation $X(t) - \hat{c}_1(t) - \hat{\hat{x}}(t)$, decomposing $\hat{X}(t)$ by EMD, we get a series of IMFs $\tilde{c}_i(t)$. Calculate the correlation coefficient $\overline{r}_1$ between $\hat{c}_1(t)$ and $\tilde{c}_1(t)$, if $\overline{r}_1$ is less than threshold δ, $\hat{c}_1(t)$ is defined as the first IMF, that is IMF1. Otherwise, repeat the decorrelation operator between $\hat{c}_1(t)$ and $\tilde{c}_i(t)$.

(5) Repeat the operation of decorrelation until all of the correlation coefficients is less than the threshold δ, finally we can get a series mutually orthogonal IMFs.

In order to validate the effectiveness of this method, a simulated signal which contains a variety of frequency components and discontinuous signal is taken as an example to be decomposed by the above method. The simulated signal $y_2(t)$ is shown as Eq. (5)

$$y_2(t) = 10\cos(2\pi 10t) + 6\sin(2\pi 50t) + 5\sin(2\pi 80t) + y_3(t) \tag{5}$$

$$y_3(t) = 3\sin(2\pi 200t) \tag{6}$$

Where signal $y_3(t)$ is a discontinuous signal. The added signal is $y_4(t)=4\sin(2\pi 232t)$, the results of EMD are shown in Fig. 4. Fig. 4 (a) shows the first four IMFs which are obtained by the revised EMD, and it is obviously that the four kinds of frequency components are decomposed without mode mixing. However, the IMFs in Fig. 4 (b), which are obtained by the traditional EMD, serious mode mixing exist in each IMF.

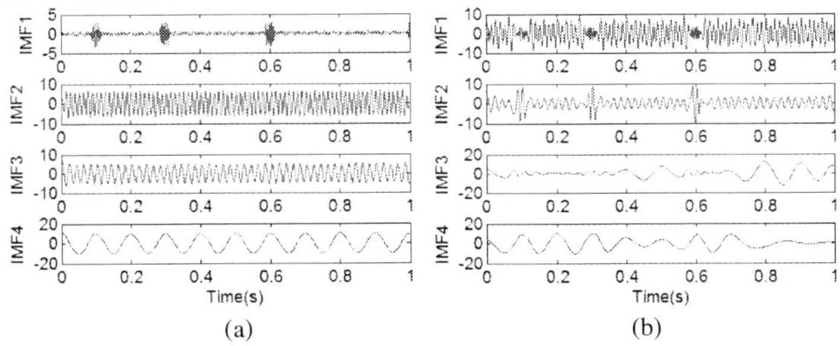

Fig. 4. The IMFs, (a) The first four IMFs obtained by the revised EMD, (b) The first four IMFs obtained by the traditional EMD

4 Fault diagnosis by the revised HHT method

In order to verify the effectiveness of the revised HHT method in the aspect of rolling bearing fault diagnosis, the rolling bearing fault data which obtained from the Case Western Reserve University bearing data center [5] is analyzed by the revised HHT method. The vibration signals of bearing were collected under the condition of inner race fault. The 6205-2RS JEM SKF deep groove ball bearing is utilized in this experiment and its basic parameters are shown in table.1.

Table 1. Bearing size (mm)

Inside diameter	Outside diameter	Thickness	Ball diameter	Pitch diameter	Ball number
25	52	15	7.94004	39.0398	9

The inner race fault frequency f_i can be calculated according to Eq. (6).

$$f_i=\frac{r}{60}\cdot\frac{1}{2}\cdot Z\cdot\left(1+\frac{d}{D}\cdot\cos\alpha\right) \tag{6}$$

Where r=rotate speed (rpm), d=ball diameter, D=pitch diameter, Z=number of balls, α =contact angle, it can be determined as zero when the bearings is running. We can obtain the inner race fault frequency $f_i=162.186Hz$ according to Eq. (6). The length of the selected data is 4096. The process of diagnosis is shown in Fig. 5. The added high frequency sinusoidal signal is determined as $y_5(t)=0.01\sin(2\pi 1222t)$. The marginal spectrum of the fault signal is shown in Fig. 5. Fig. 5 (b) illustrates that the failure frequency of inner race fault that obtained by the revised HHT method is 162.6 Hz, which is closer to the actual failure frequency than the frequency 156.7 Hz that obtained by the traditional HHT method. So it indicates that the revised HHT method is more effective than the traditional.

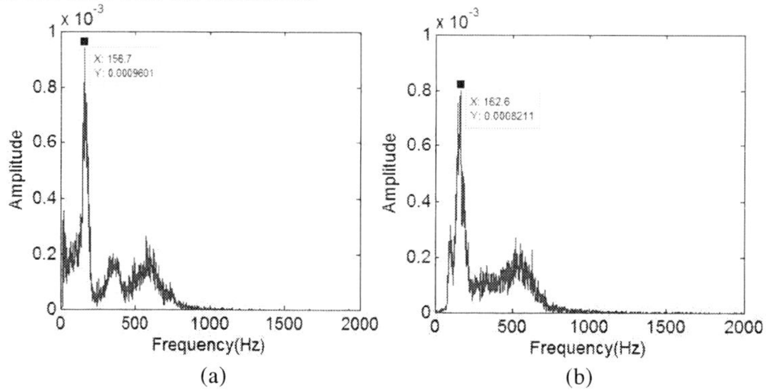

Fig. 5. The marginal spectrum, (a)The marginal spectrum obtained by traditional HHT method, (b) The marginal spectrum obtained by the revised HHT method

5 Conclusions

It is difficult to analyze the fault types of rolling bearing in traditional linear methods. HHT method is effective in the aspect of processing non-stationary and non-linear signals. In order to solve the problems of end effects and mode mixing, a revised HHT method was proposed here. In the proposed method, the local linear extrapolation and high frequency sinusoidal signal, decorrelation operator were introduced to suppress end effects and mode mixing respectively. To test the effectiveness of the proposed method, the acceleration signal of rolling bearing inner race fault which obtained from the Case Western Reserve University bearing data center is disposed by traditional HHT method and the revised method respectively. The results illustrate that the revised HHT method is more accurate than the traditional in the aspect of bearing failure diagnosis. So the revised HHT method is effective in the aspect of fault diagnosis for rolling bearing.

References

1. Huang, N.E., Shen, Z., Long, S,R.: The empirical mode decomposition and the Hilbert spectrum for nonlinear non-stationary time series analysis. Proc. R. Soc. 454, 903-995 (1998)
2. Huang, N.E., Wu, Z.H.: A review on Hilbert-Huang transform: method and its applications to geophysical studies. Rev. Geophys. 21-23 (2008)
3. Peng, Z.K., Tse, P.W., Chu, F.L.: A comparison study of improved Hilbert-Huang transform and wavelet transform: application to fault diagnosis for rolling bearing. Mech. Syst. Signal Process. 19, 974-988 (2005)
4. Xiao, Y., Yin, F.L.: Decorrelation EMD: a new method of eliminating mode mixing. Journal of Vibration and Shock. 2015,04:25-29.(in Chinese)
5. Case Western Reserve University. Bearing data. http: //csegroups. case. edu/ bearing data center/ pages/ download-data-file.

Milling Chatter of Thin-walled Workpieces and its Effect on Surface Topography

Dongqian Wang1, Zhibing Liu2,*, Xibin Wang2, Zhenghu Yan1 and Junqing Wang1

1School of Mechanical Engineering, Beijing Institute of Technology, No.5 South Zhongguancun Street, Haidian District, Beijing 100081, P.R.China,
2Key Laboratory of Fundamental Science for Advanced Machining, Beijing Institute of Technology, No.5 South Zhongguancun Street, Haidian District, Beijing 100081, P.R.China,

Abstract. For the process of thin-walled workpieces milling chatter phenomenon and its effects on the surface topography, the dynamic model of process system is established. Simulation analysis method is adopted to get the cutting stability lobes diagram of dynamic model. The cutting force signal is analyzed in the frequency range, and the chatter frequency is identified. By collecting tool vibration displacement signal of flank milling, the 3D surface morphology prediction model is set up and the MATLAB simulation is applied. The experimental results show that the stability lobes diagram and the surface topography prediction model have a good consistency with the actual machining results. By analysis of the chatter influence on surface morphology, changing the cutting parameters can obtain better surface topography.

Keywords: Chatter, simulation, stability lobes diagram, surface topography

1 Introduction

With the rapid development of aviation, thin-walled workpieces are widely used and the surface topography requirements are quite strict. Regenerative chatter and multi-order modal shapes of the thin-walled workpieces will affect the final processing results. These properties lead to the finished surface leave chatter marks, and even cause the workpieces scrapped[3]. Budak and Altintas [1]built the chatter prediction model to obtain the cutting stability lobe diagram(SLD) that helps technologists choose appropriate parameters to avoid chatter. However the SLD changes constantly in the machining, which makes two dimensional SLD difficult to predict and determine the machining parameters in advance[6]. Zhang Xiaojian and Xiong Caihua [2]considered multi-order modal shapes into the dynamical equation by calculating the maximum transfer function values and got a conservative two-dimensional SLD. However, the efficiency of machine tool's cutting is suppressed.

The mutual interference between tool and workpieces is the main cause of cutting surface topography [5]and introducing the vibration signal to geometric model is a widely used method. K.F.Ehmannh and M.S.Hong [4] assumed that the vibration

E-mail: liuzhibing@bit.edu.cn

signal is random and added it to the contour prediction model. The machining parameters of these models did not distinguish stable region and unstable region. This led us to introduce the position and modal shapes into the dynamic equations, and analyze the surface topography in the stable and unstable regions respectively.

This paper attempts to introduce machining position fitting function and modal shape fitting functions of thin-walled workpieces to the dynamic equation at the same time, and the three-dimensional SLD which varies with different modes and different positions is obtained. Then the three-dimensional SLD of different modal shapes are enveloped to obtain a comprehensive three-dimensional SLD, which was verified by the milling experiment.

2 Modeling of milling dynamics

The process of thin-walled workpieces and tool interaction is a dynamic cutting, and dynamic model can be simplified as a perpendicular two degree of freedom system as shown in figure 1.

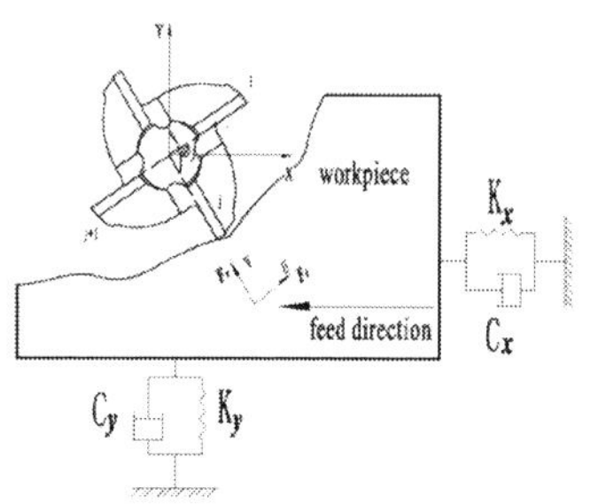

Fig. 1. Dynamic model of end milling

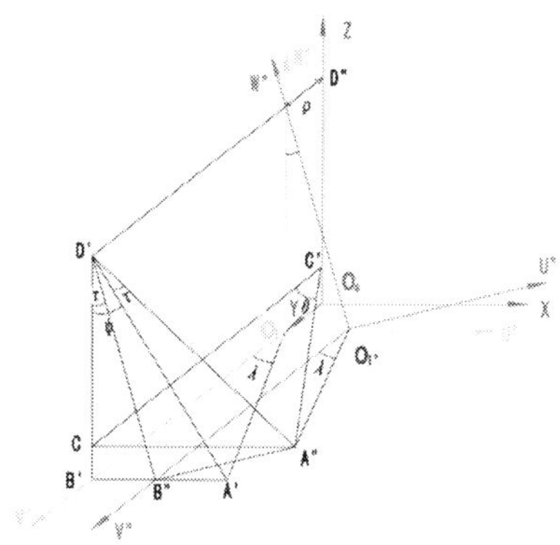

Fig. 2. Actual installation model of tool

Compared with the Y direction, X direction has a greater stiffness, so the system is further simplified as a single degree of freedom (Y direction) milling model.

$$F_y = \frac{1}{2} b K_t a_{yy} \Delta y \quad \text{with } \Delta x \approx 0 \ \Delta y = y(t) - y(t-T) \tag{1}$$

During milling the thin-walled workpieces, the tooth passing frequency, the spindle rotation frequency and the harmonic frequency of thin-walled workpieces will excite the first few modal shapes of thin-walled workpieces, which affects the transfer function analysis of the process system. The modal shapes of the thin-walled workpieces are different with the machining position, and the flexible characters of the thin-walled workpieces can also change the modal parameters along the machining position. Because of the small damping, the amount of change can be ignored. Therefore, the milling dynamic equation in x direction becomes:

$$m(x)\ddot{y} + c\dot{y} + k(x)y = f_y(x) \tag{2}$$

The matrix equation (2) is coupled of multi modal shapes of the thin-walled workpieces, and the modal coordinate transformation is carried out. Using the transformation $y=\Phi q$, equation (2) is written as

$$M(x)\ddot{q}+C\dot{q}+K(x)q=\Phi^T f_y(x) \tag{3}$$

where $M(x)=\Phi^T m(x)\Phi$, $C=\Phi^T c\Phi$, $K(x)=\Phi^T k(x)\Phi$, $\Phi=[\varphi_1(x),\varphi_2(x),...\varphi_n(x)]$

Also y is the function of x and time t, which can be written as $y(x,t)=\sum_{j=1}^{n}\varphi_j(x)q_j(t)$

where $\varphi_j(x)$ is the jth -order modal shape function which is normalized. Due to the different normalized modal shapes is orthogonal, equation (3) can be expressed as

$$M_j(x)\ddot{q}_j+C_j\dot{q}_j+K_j(x)q_j=\frac{1}{2}\varphi_j^T(x)\varphi_j(x)bK_t a_{yy}[q_j(t-T)-q_j(t)] \quad (j=1,2,...,n) \tag{4}$$

Based on the work by Altinats and Budak[1], Define

$$\Lambda=\Lambda_{\mathrm{Re}}+i\Lambda_{\mathrm{Im}}=\frac{1}{\sum_{j=1}^{n}[\varphi_j^2(x)G_j(iw)]} \tag{5}$$

Then the expression of the limit depth is

$$b_{\lim}=\frac{2\pi}{N_t K_t(\Lambda_{\mathrm{Re}}^2+\Lambda_{\mathrm{Im}}^2)}\Lambda_{\mathrm{Re}}(1+\frac{1}{\kappa^2}) \tag{7}$$

with $\omega_c T=-2\varepsilon+2k\pi \quad k=0,1,2,3...$

where $\kappa=\frac{\Lambda_{\mathrm{Im}}}{\Lambda_{\mathrm{Re}}}=\frac{-\sin\omega_c T}{1+\cos\omega_c T}$, $\varepsilon=-2\arctan\kappa$ is the phase shift between the inner and outer modulations, and T is the tooth passing period. Then the spindle rotating speed can be obtained as $\Omega=\frac{60}{N_t T}$ (8)

3 Surface topography modeling of flank milling in stable region

As shown in Fig. 2, the tool is assumed to exist eccentric displacement ρ and tilt angle τ . From the geometric point of view, the interference between cutting edge and the thin-walled workpieces is removed, and the surface topography is composed of the residual height in the cutting area. The flank milling area of thin-walled workpieces is divided into grids with the formation of $m\times n$ (represent X direction and Z direction respectively), and the height of Y direction (h_Y) corresponding to each matrix point is calculated. The micro roughness parameters $R_a=\frac{f_t^2}{32(R+f_t N_f/\pi)}$ [7] of trajectory is introduced into the surface topography model, and the sine wave signal as forced vibration is also added to the surface topography model. Then the expression of Y direction surface topography model is

$$P(x) = H_Y * Ra + A_1 \sin(\omega_1 x + \varphi_1) \tag{9}$$

where $\omega_1 = 2\pi f_1$, f_1 is the spindle rotation frequency, A_1 is the amplitude, φ_1 is the initial phase.

The cutting depth b of cutter is discretized into N_z elements whose thickness is $\Delta z = \frac{b}{N_z}$, and the jth cutting edges of the ith cutting element is calculated in the following form

$$L(i) = L - (i-1)\Delta z, \quad \lambda(j) = \lambda - 2\pi(j-1)/N_f, \quad \varphi(i,j) = arctan[Rsin\lambda(j)/L(i)] \tag{10}$$

$$\theta(i,j) = arctan\{[\sqrt{(Rsin\lambda(j))^2 + L(i)^2} sin(\varphi(i,j)+\tau)]/[\rho + Rcos\lambda(j)]\} \tag{11}$$

where λ is the initial position angle of the cutting edge, L is the tool length, φ is the cutting edge deviation angle, θ is the position angle of Spindle coordinate system, R is the radius of cutter and R_c is the actual rotation radius of cutting edge in A point. Substitute equation (10)- (11) into equation (9) and introduce the offset Y_0 which is between spindle coordinate system and workpieces coordinate system to obtain

$$H_Y(i,j) = R_c(i,j)cos\theta(i,j) + Y_0 \tag{12}$$

Then the equation (12) can be expressed as $P(i,j,x) = H_Y(i,j) * Ra + A_1 \sin(w_1 x + \varphi_1)$ (13)

4 Simulation and experimental results discussion

Along the X direction of the machining, the hammer method was used to fit out the modal frequency function and modal stiffness function. In order to obtain the multi-order modal shapes of the thin-walled workpieces, the modal shapes analysis was carried out by using Finite Element Method (FEM). Using the MATLAB to simulate the calculation process, the results are shown in Fig.3.

Selecting the point F(11000rmp,3mm), at the spindle speed of 7000rmp two-dimensional SLD was got from the three-dimensional SLD. It could be seen that the machining process experienced stable-unstable- stable. In Fig.4, the chatter-stability-chatter phenomenon is observed from the experimental results. The time-domain cutting force signal was transformed into frequency domain by Fast Fourier Transform (FFT) .The frequency distribution was observed from the frequency spectrum, except spindle frequency, cutter tooth passing frequency and its harmonic frequency, and the chatter frequency was also found. In the process of machining, an obvious unstable-stable- unstable phenomenon(C-D-E) was observed, which was in good agreement with the second-order modal shape SLD[2].In the region D, due to the second-order modal shape was close to zero, the first-order modal shape played the main role, and region D was stable in the first-order modal shape. When the tool left middle position, second-order modal shape played a dominant role again, which led machining to the unstable region E. The above results proved the accuracy of three-dimensional SLD. According to the above method, other points on the graph were verified, which was in good agreement with the three-dimensional SLD.

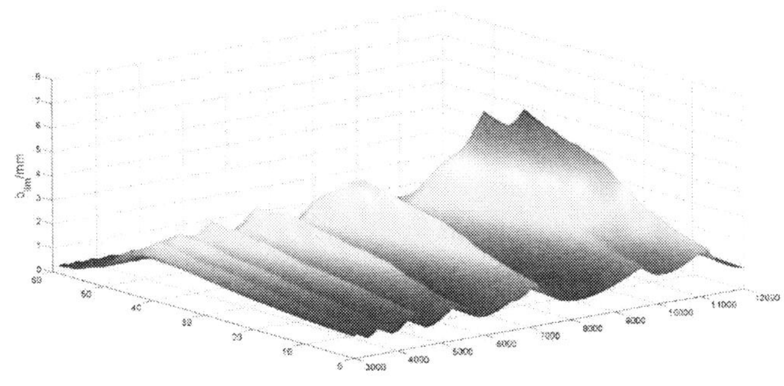

Fig. 3. The comprehensive three-dimensional SLD

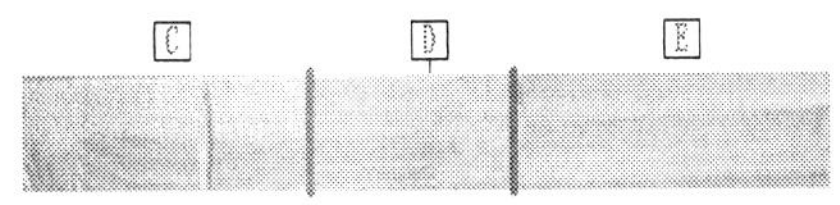

Fig. 4. The machined surface

Selecting point A (7000rmp,2mm) which is in stable region from the three-dimensional SLD, the experiment was carried out. The profile peak height and width of the surface topography prediction model was consistent with the actual machining results. Along the machining direction the surface topography prediction model was intercepted at a certain point, and the roughness value was predicted by the two-dimensional figure. As shown in Fig. 5 and Fig. 6, the measured value of roughness was $Ra = 1.612\mu m$, and the predictive value of roughness was $Ra = 1.676\mu m$ .The error between the predictive value and the measured value is 4% .

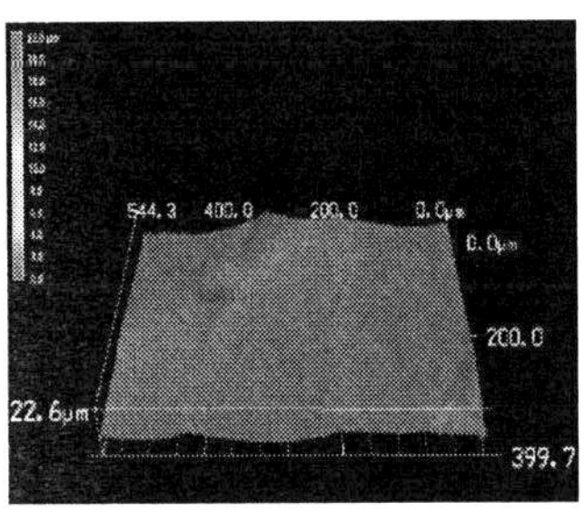

Fig.5. The measured surface topography

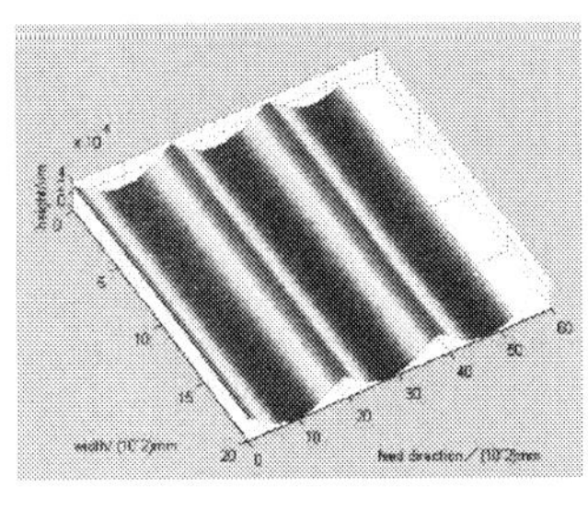

Fig. 6. The predictive surface topography

Selecting B (7000rmp,5mm) from the three-dimensional SLD, the experiment was carried out. Experiments showed that the machined surface topography of thin-wall workpieces is irregular and the prediction model is invalid.

5 Conclusions

Along the machining direction, the fitted modal shape functions and the modal parameters of thin-walled workpieces are introduced to the dynamic equation. The results show that there is a good consistency between the three dimensional SLD and the machining results, which proves the accuracy of the three dimensional SLD. In order to ensure the stability of the machining process, the three dimensional SLD of the different modal shapes are enveloped to obtain the comprehensive three-dimensional SLD which can be more reliable to guide the selection of machining parameters.

The profile peak height and width of the surface topography prediction model in the stable region is consistent with the actual machining result, which verify the correctness of the prediction model. In the unstable region, the surface topography prediction model of the thin-walled workpieces is in poor agreement with the prediction model. This shows that the influence of the vibration on the surface topography is greater than that of the forced vibration, and the main way to obtain a better surface topography is suppressing the vibration.

6 References

1. Altintas Y, Budak E. Analytical prediction of stability lobes in milling. CIRP Ann Manuf Technol, 1995, 44(1): 357-362.
2. Zhang Xiao-jian, Xiong Cai-hua, DING Ye, et al. Stability analysis in milling of thin-walled workpieces with emphasis on the structural effect, Proceedings of the Institution of Mechanical Engineers, Part B: Journal of Engineering Manufacture, 2010, 224(4): 589-608.
3. Schmitz Tony L, Scott Smith K. Machining Dynamics Frequency Response to Improved Productivity. New York: Springer Science + Business Media, LLC, 2009.
4. K. F. Ehmann, M. S. Hong, A generalized model of the surface generation process in metal cutting. Annals of CIRP, 2009, 43(1): 483-486.
5. Hao Jiang, Xinhua Long, Guang Meng. Study of the correlation between surface generation and cutting vibrations in peripheral milling. Journal of materials processing technology. 2008, 208(1-3): 229-238.
6. Aijun Tang, Zhanqiang Liu: Three-dimensional stability lobe and maximum material removal rate in end milling of thin-walled plate. The International Journal of Advanced Manufacturing Technology. 2009, 43(3): 33-39.
7. M. Alauddin, M.A Baradie.Computer-aided analysis of a surface-roughness model for end milling. Journal of Matweials Processing Technology,2005,55(2):123-127.

Simulation Analysis Based on Explicit Dynamics for Dual Rotor Intermediate Bearing

Wei Xiang, Huaqing Wang*, Jingsheng Jiang, Yanliang Ke

College of Mechanical and Electrical Engineering, Beijing University of Chemical Technology, No.15 in East Road of North Third Ring, Chaoyang District, Beijing, 100029, China
Huaqing wang is the corresponding author.
E-mail: hqwang@mail.buct.edu.cn.

Abstract: In order to study the dynamic characteristics of the defect intermediate bearing under complex conditions, in view of the intermediate bearing working characteristics that the low-pressure rotor shaft neck is supported by the outer ring bearing and the high-pressure rotor shaft neck is supported by the inner ring bearing, the dynamic model of the cylindrical roller bearing is established by using ANSYS/LS-DYNA, and dynamic simulation is carried out. In order to avoid signals attenuation through the outer ring raceway. The lower surface center point and the right center point of the intermediate bearing outer ring surface are regarded as the object of study. Signal features of the outer ring fault intermediate bearing are characterized by extracting the acceleration curve of the points. Simulation results show that fault characteristic frequency of intermediate bearing can be found by acceleration responses with spectrum analysis.

Keywords: Intermediate bearing; explicit dynamics; simulation; spectrum analysis

1 Introduction

Intermediate bearing is the key component supporting the high and low pressure rotor of double rotor engine and the inner and outer circles are often used in the same direction and reverse rotation [1]. The intermediate bearing is prone to wear and fatigue spalling because of poor working conditions, large dynamic load, high temperature, and high speed. Once the failure will lead to a sharp increase in engine vibration, and it may cause an accident that the axle shaft broken. Therefore, it is very important to study the dynamic modeling and vibration response of the intermediate bearing, in order to ensure the safety and reliability of the engine [2-3].

In recent years, with the development of computer technology, finite element software simulating the force situation of the intermediate bearing has become the mainstream [4-7]. The finite element method is adapted to dynamic simulation and structure analysis of the defect intermediate bearing in theory and studied the vibration characteristics of defect intermediate bearing. And the dynamic simulation method can be used to prove the correctness of the calculation result, which can effectively save the experimental time, reduce the experimental cost. It has important significance to carry out the research on the on-line monitoring and fault diagnosis of the intermediate bearing.

In this paper, the simulation analysis of the fault intermediate bearing is carried out by using ANSYS/LS-DYNA and based on the explicit algorithm. The model of the intermediate bearing is established by selecting the material model which is in accordance with the actual conditions, the reasonable contact and constraint conditions. The intermediate bearing outer raceway node stress level with the outer ring fault distance variation rule is analyzed and the characteristic frequency of the outer ring fault is extracted.

2 Basic theory

2.1 Basic algorithm of explicit dynamic [8]

In the dynamic finite element analysis, the solution equations of the system are:

$$\boldsymbol{M}\ddot{a}(t) + \boldsymbol{C}\dot{a}(t) + \mathrm{K}a(t) = \mathbf{Q}(t) \tag{1}$$

In the expression, $\ddot{\boldsymbol{a}}$ is the nodal acceleration vector of the system; $\dot{\boldsymbol{a}}$ is the nodal vector of the system; $\boldsymbol{M}$ is mass matrix; $\boldsymbol{C}$ is damp matrix; $\boldsymbol{K}$ is stiffness matrix and $\boldsymbol{Q}$ is the nodal load vector of the system. In LS-DYNA, the central difference scheme of the direct integration method is used to integrate the motion equations and in the central difference method, the acceleration and the velocity can be expressed as:

$$\ddot{a}(t) = \frac{1}{\Delta t^2}(a_{t-\Delta t} - 2a_t + a_{t+\Delta t}) \tag{2}$$

$$\dot{\boldsymbol{a}}(t) = \frac{1}{2\Delta t}(-a_{t-\Delta t} + a_{t+\Delta t}) \tag{3}$$

The recursive formula of each discrete time point solution can be obtained by putting type (2)、(3) into the equation (1):

$$\left(\frac{1}{\Delta t^2}\boldsymbol{M} + \frac{1}{2\Delta t}\boldsymbol{C}\right)a_{t+\Delta t} = \boldsymbol{Q} - \left(\boldsymbol{K} - \frac{1}{\Delta t^2}\boldsymbol{M}\right)a_t - \left(\frac{1}{\Delta t^2}\boldsymbol{M} - \frac{1}{2\Delta t}\boldsymbol{C}\right) \tag{4}$$

The each discrete time point displacement can be solved by using type (4) after giving calculation method and initial conditions.

2.2 Fault characteristic frequency of intermediate bearing

During the rotation of the rolling body, each passing through the defect will produce an impact, and with the different location of the defect, the frequency of the impact is also different. Therefore, the fault type of the intermediate bearing can be determined by extracting the characteristic frequency. In the case of the same direction rotation of the inner and outer rings of intermediate bearing, the out ring fault characteristic frequency calculation formula of each element of the medium bearing are shown as follows:

$$f_e = \frac{z}{2}(1 - \frac{d}{D}\cos\alpha) \times \frac{|n_i - n_e|}{60} \tag{5}$$

In the formula: f_e is the outer fault characteristic frequency; Z is the number of rolling elements; d is ball diameter (mm) and D is the pitch diameter (mm); α is the connect angle; n_i is outer race (rpm) and n_e is inner race (rpm).

3 Modeling

3.1 Element selection and material property setting

Finite elements SOLID164 and SHELL163 are selected and meshed by using ANSYS/DYNA. The model with an arc shape defect of 0.4mm×0.2mm on the outer ring roller has 15297 nodes and 11160 elements, as shown in Figure 1(b), and A points and B points of the bearing outer ring is selected as the observation point, as shown in Figure 1(a). It is assumed that the material of the inner, outer rings and rolling elements are GCr15, the cage is cold-rolled steel. Density of GCr15 is $7.83\times10^{-9} t/mm^3$, elastic modulus is $2.06\times10^5 MPa$, Poisson's ratio is 0.3; Density of cold-rolled steel is $7.83\times10^{-9} t/mm^3$, elastic modulus is $0.93\times10^5 MPa$, Poisson's ratio is 0.346.

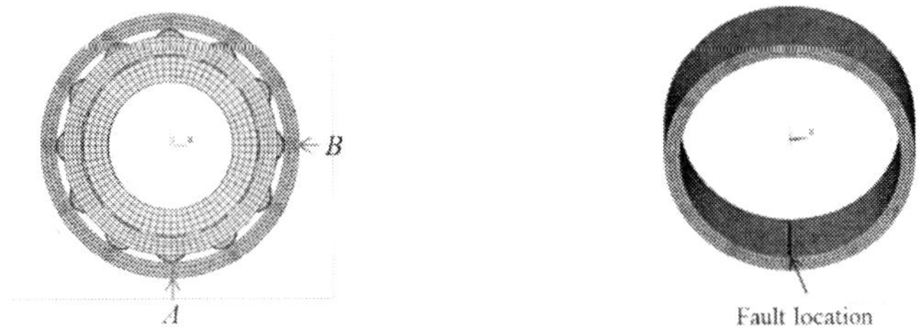

(a) Location of observation points (b) outer ring of intermediate bearing

Fig. 1. Finite element model of the intermediate bearing

3.2 Boundary conditions and loads setting

The inner surface of the inner rings and the outer rings are rigid material with constraints of Z translational degrees of freedom and X, Y rotational degrees of freedom. The speed load is applied to the inner surface of the inner ring and the inner surface of the outer ring and the radial load is applied on the lower half surface of inner ring in form of pressure.

4 Analysis and results

Termination time is determined to be 1s. Time interval is 1000. The inner ring speed is 300rpm and the outer ring speed is 1500rpm. According to the speed and size of the simulation conditions, the theoretical value of the outer ring is 98.46Hz. Binary output file is ANSYS and LS-DYNA. Dynamic characteristics of the intermediate bearing are obtained based on explicit dynamics. The normalized time domain and spectrum diagram of A point with X direction are shown in Fig. 2. Fig. 2(a) is the acceleration time domain diagram of A point with the X direction. Fig. 2(b) and Fig. 2(c) are respectively the spectrum diagrams of the normal and the outer ring fault intermediate bearing of A point with X direction.

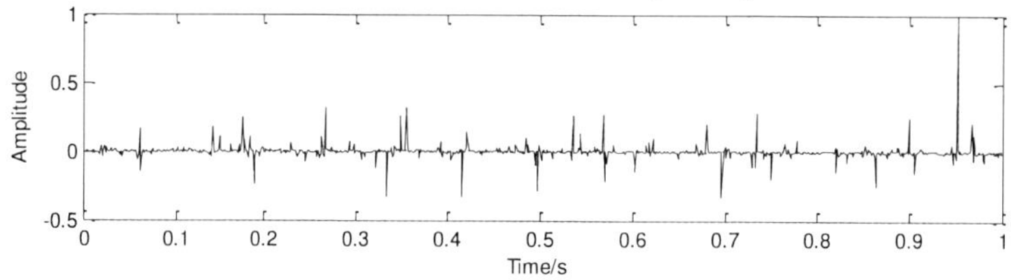

(a) Time domain diagram of acceleration

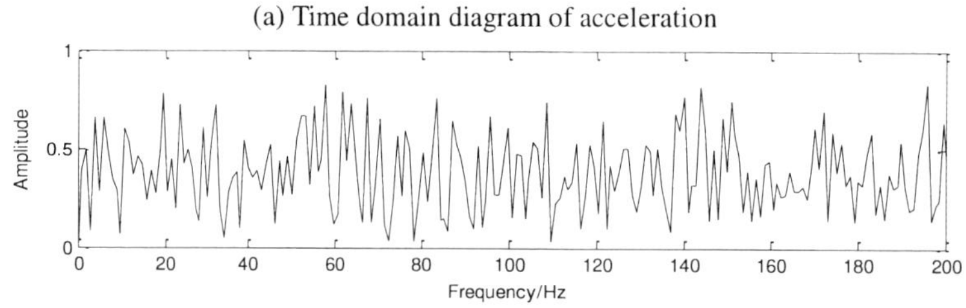

(b) Spectrum diagram of the normal intermediate bearing

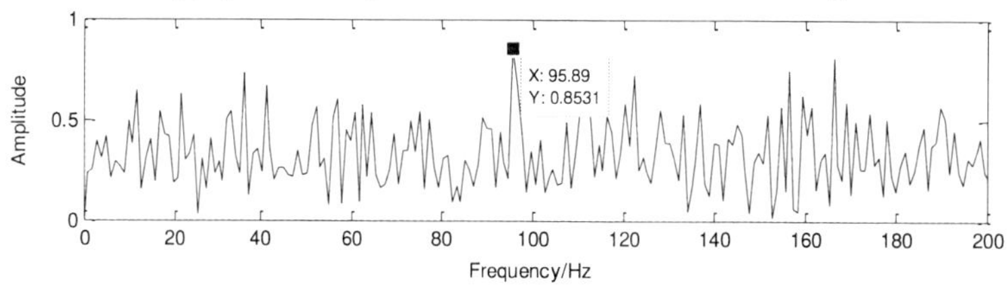

(c) Spectrum diagram of the outer ring fault intermediate bearing

Fig. 2. Time and spectrum diagrams of A point with X direction

Compared Fig. 2(b) and Fig. 2(c), we can see that it is easy to determine the outer ring fault characteristic frequency through the spectrum diagram of the outer ring fault intermediate bearing. However, the characteristic frequency of the outer ring fault can not be seen through the spectrum diagram of the normal intermediate bearing. And Fig. 2(c) shows that the fault feature frequency value obtained through the simulation result is very close to the theoretical value, which can be recognized as the outer ring fault.

Fig. 3(a) is the acceleration time domain diagrams of B point with the Y direction. The spectrum diagrams of the normal and the fault intermediate bearing of B point with Y direction are respectively shown in Fig. 3(b) and Fig. 3(c). The diagrams of Fig.3 are all normalized.

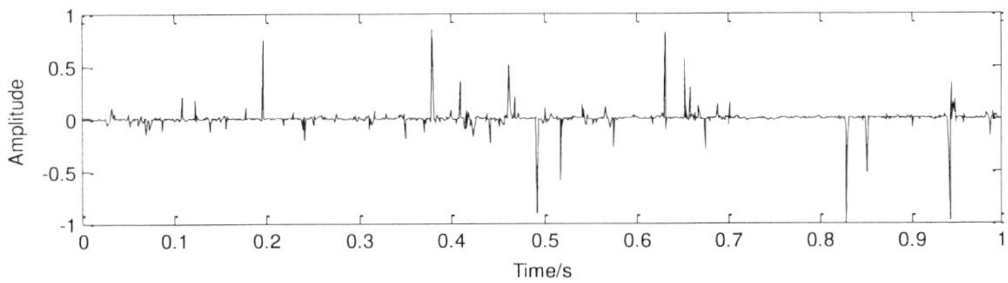

(a) Time domain diagram of acceleration

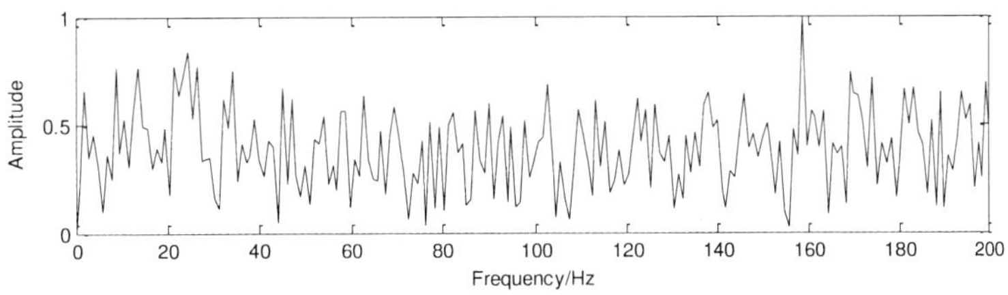

(b) Spectrum diagram of the normal intermediate bearing

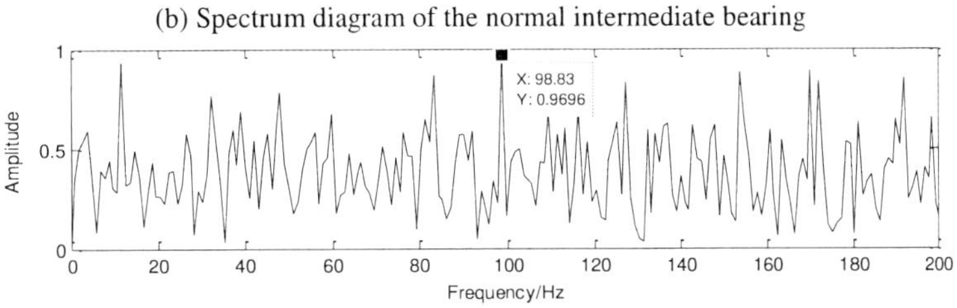

(c) Spectrum diagram of the outer ring fault intermediate bearing

Fig. 3. Time and spectrum diagrams of B point with Y direction.

It can be seen that the amplitude fluctuation of the spectrum curve is small in the vicinity of the outer ring fault characteristic frequency value. On the contrary, we can determine the out ring fault characteristic frequency that shown in Fig. 3(c). Therefore, with the same situation of the fault characteristic frequency diagram of A point, the fault characteristic frequency of B point can obviously reflect the fault characteristic frequency of the outer ring fault intermediate bearing. And compared Fig. 2(c) and Fig. 3(c), we can see that the outer ring fault characteristic frequency with Y direction is closer to the theoretical value than X direction. Therefore, in the fault signal analysis of the intermediate bearing, the acceleration signal in Y direction should be analyzed preferentially.

5 Conclusions

A finite element model of outer ring fault intermediate which considers the speed of inner and outer rings, connect effects and constrains under friction is established successfully by ANSYS/LS-DYNA. And the simulation method that through analyzing its dynamic characteristics with ANSYS/LS-DYNA is feasible. The results show that the outer ring fault characteristic frequency can be easy obtained through doing FFT transform for acceleration simulation signals of intermediate bearing and a feasible idea for further study of the intermediate bearing fault transfer characteristics is provided.

Acknowledgements

This project is supported by the National Natural Science Foundation of China (51375037)

References

1. H. Xuan, G. H. Luo, D. P. Gao. Performance analysis of aero-engine mediation bearing. Journal of Aerospace Power, 2007, 22 (3): 439-443.
2. K. J. Xu, F. H. Dong, H. Q. Qin. Dynamic modeling and vibration response analysis of the inner ring bearing with defect. Journal of Naval Aeronautical Engineering Institute, 2015, 30(1): 36-42.
3. M. F. Liao, Z. G. Ma, Y. Q. Liu. Fault characteristics and diagnosis method of intershaft bearing in aero-engine. Journal of Aerospace Power, 2013, (12):2752-2758.
4. X. Hu, G. H. Chen. An intelligent fault identification method of rolling bearings based on LSSVM optimized by improved PSO. Mechanical Systems and Signal Processing, 2013, 35 (1/2): 167-175.
5. W. S. Su, F. T. Wang, Z. Hu, et al. Rolling element bearings faults diagnosis based on optimal morlet wavelet filter and autocorrelation enhancement. Mechanical Systems and Signal Processing, 2010, 24 (5): 1458-1472.
6. B. Muruganatham, M. A. Sanjith, B. K. Kumar, et al. Roller element bearing fault diagnosis using singular spectrum analysis. Mechanical Systems and Signal Processing, 2013, 35(1/2): 150-166.
7. T. Slack, F. Sadeghi. Explicit finite element modeling of subsurface initiated spalling in rolling contacts. Tribology International, 2010, 43 (9): 1693-1702.
8. X. Wei. H. Q. Wang, T. Gang, H. F Yuan. Simulation analysis of an engine rotor mediation bearing based on explicit dynamics. Journal of Beijing University of Chemical Technology, 2016, 43(3): 91-96.

Flexible Simulator for the Vibration Analysis of Rolling Element Bearings

J. Kansanaho[1], K. Saarinen[2] and T. Kärkkäinen[3]

1 University of Jyväskylä, Department of Mathematical Information Technology, PO Box 35, FI-40014 University of Jyväskylä, Finland; jarno.m.kansanaho@jyu.fi; phone +358400248111

[2] ABB AB, Corporate Research, Forskargränd 8, SE-721 78, Västerås, Sweden; kari.saarinen@se.abb.com

[3] University of Jyväskylä, Department of Mathematical Information Technology, PO Box 35, FI-40014 University of Jyväskylä, Finland; tommi.karkkainen@jyu.fi

Abstract. The detection of incipient faults as early as possible has great economic value in the monitoring of the rolling element bearings in industrial applications. In the early stage, a local fault in the bearing element produces a series of weak impacts at a rate dependent on the bearing geometry. These impacts in turn excite a special type of vibration that is, in principle, possible to detect using various condition monitoring methods.

However, the analysis of vibration measurements taken from a real industrial environment can be challenging because measurements are noisy and periodic phenomena from other external, known and unknown sources may overlap the known behavior of interest, for example, in the spectral representation. For instance, electromagnetic forces in slowly rotating machines driven by frequency converters, may generate low-frequency pulses with low-frequency carriers, which are difficult to filter out and may create false alarms when traditional envelope analysis is used for fault detection. In particular, when the goal is early fault detection, we need to identify various vibration sources and understand their effect on the analysis to detect weak pulses from faulty bearings and at the same time avoid false alarms.

In this paper, we present a flexible simulator for advancing early bearing fault detection. In the simulator, the vibration signals are generated by parametric models of the impulse responses for different vibrating components, with adjustable noise and jitter effects included. The possibility to adjust models and parameters during the simulation allows a more realistic exploration of changes, for example, in different operating conditions. This includes the generation of non-stationary components into the vibration signals, which are not suitable *per se* for frequency domain methods. It is possible to load measured vibration signals to verify the simulation model. The simulator is being developed with LabVIEW™ and MATLAB software products.

The main purpose of the versatile simulation environment is to enable the rigorous testing of existing and to-be-developed analysis tools consisting of digital filters, frequency domain methods (FFT, HFRT), time-frequency methods (STFT, WT) etc. In turn, improved knowledge on the behavior of analysis methods and approaches can be reflected back to the simulation when searching the limits of the early fault detection methods.

Keywords: Rolling element bearings, Early fault detection, Vibration measurements, Spectral analysis, Simulation model

1 Introduction

Rolling element bearings (REBs) are widely used in various machines, motors, wheels, cooling fans etc., to support rotating shafts. Failures of bearings are the most common reason for machine breakdowns. The most common cause of failure is fatigue in REBs [1]. In terms of condition monitoring and system maintenance, research on REB vibrations is essential. An ideal method is a condition-based maintenance that is able to detect incipient faults so early that it is possible to complete maintenance actions before the bearing fails. Compared to the planned maintenance actions, unplanned stoppages are very expensive.

Generally, REB defects are classified into distributed and localized effects. The distributed effects are surface roughness, waviness, misaligned races, and off-size rolling elements. The localized defects include cracks, pits, and spalls caused by fatigue on the rolling surfaces [2]. The resonances of the bearing parts and housing structure emerge when a rolling element strikes a localized defect; an exponentially decaying ringing is generated [2] [3]. These vibrations in the 2 – 20 kHz range, are commonly measured by acceleration sensors. Ultrasound and acoustic emission measurements cover considerably higher frequencies.

The main motivation for this work is that the weak bearing fault signatures experience interference from noise from different external sources and internal mechanical components, which makes bearing fault detection from real measurements a very challenging task. In this paper, we present a vibration signal simulation approach to improve bearing fault detection analysis.

In the simulator, the vibration signals are generated by parametric models of the impulse responses for different vibrating components, with adjustable noise and jitter effects included. The possibility to adjust models and parameters during the simulation allows a more realistic exploration of changes, for example, in different operating conditions. This includes the generation of non-stationary components into the vibration signals, which are not suitable per se for frequency domain methods. It is possible to load measured vibration signals to verify the simulation model. The simulator is being developed with LabVIEW™ and MATLAB software products.

The nature of the vibrations generated by the REBs, a parametric model of the impulse response and the information about the disturbances are explained in the second chapter (theoretical background). The flexible simulator and its demonstration in two different scenarios of the vibration signal simulation and analysis are introduced in the third chapter (simulator). Conclusions and ideas for future development are summarized in the fourth chapter.

2 Theoretical background

Vibrations generated by REBs are the vibration of the raceway ring, the vibration of the cage, rolling element passage vibration, vibration due to waviness, vibration due to flaw, vibration due to contamination, and seal vibration [4]. Other than flaw vibrations, the presented vibrations of the REBs can be considered normal. Bearing fault signals have a deterministic part and a stochastic part. The stochastic part is quasi-cyclostationary and very often small because of the low-pass filtering effect of random jitter, combined with the band-pass filtering effect of resonances that are excited [5]. In some cases, the REB signals are modeled as cyclostationary [6].

The local fault in the bearing element produces a series of weak impacts at a rate dependent on the bearing geometry. These impulses occur when the defective part hits other elements of the bearing. If the shaft is rotating at a constant speed, these impacts will occur periodically with a certain frequency, which can be calculated based on the location of the fault, the bearing geometry, and the shaft speed. These frequencies are known as the bearing fault frequencies and can be determined for a fault on: the outer race (Ball Pass Frequency Outer, BPFO), the inner race (Ball Pass Frequency Inner, BPFI), the ball or roller, (Ball Spin Frequency, BSF), and the gage (Fundamental Train Frequency, FTF) [6]. A ratio of the length to width of the localized defects has a remarkable effect on the generated pulse waveform [7].

The dynamic behavior and features of the rolling element bearings are studied using different dynamic simulation models. Jantunen et al. [8] classified dynamic simulation models into the following categories: bearing contact, clearances, EHL contact, distributed defect, and localised defect. Further, the dynamic simulation of the localised defects is divided into geometrical defect, force defect, and defect function.

In the simulations, we need a model for the impulses generated by the defects that is simple to parametrize and at the same time realistic enough. Randall et al. [1] provided a useful analytical equation for a single degree of freedom impulse response:

$$r(t) = \mathrm{A}e^{\left(\frac{-t}{\tau}\right)}\sin(\omega_r t), \quad (1)$$

where ω_r is the frequency of the excited resonance and τ is the damping time constant(s).

Liang and Faghidi evaluated the signal-to-interference ratio when the fault-bearing signal contains multiple interferences. They present a vibration signal model [9] of the form:

$$r(t) = \mathrm{x(t)} + v(t) = Ae^{-\beta t}\cos(\omega_r t + \varphi) + \textstyle\sum_{k=1}^{K} L_k \cos(\omega_{lk} t), \quad (2)$$

where A is the amplitude of the fault signal, β is the structural damping characteristic, L_k represents the frequency of the kth interference component, L_k denotes the amplitude of the kth interference component, x(t) is the vibration signal containing fault generated impulse, and v(t) contains multiple vibration interferences. Instead of the constant amplitude in equations 1 and 2, we have added the possibility to modulate the amplitude based on rotation speed:

$$A(t) = \mathrm{A}_0 + A_1 \sin(\omega_{rotation} t). \quad (3)$$

where $\omega_{rotation}$ is the rotation frequency. The modulation term is added into bearing model equation (eq.2):

$$r(t) = x(t) + v(t) = A(t)e^{-\beta t}\cos(\omega_r t + \varphi) + \sum_{k=1}^{K} L_k \cos(\omega_{lk} t) \tag{4}$$

The occurrences of the impacts in faulty bearing never reproduce exactly at the same position due the slip of the rolling elements [5]. This random variation is modeled with jitter. Jitter, a measure of uncertainty, is the deviation from the true periodicity of a presumed periodic signal. Jitter can be deterministic, random, or both. Randal et al. presented a model for jitter [6] where the random variable ΔT_i is defined as the difference:

$$\Delta T_i = T_{i+1} - T_i, \tag{5}$$

where T_i is the ith time of occurrence.

The possibility to adjust models and parameters during the simulation allows a more realistic exploration of changes, for example, in different operating conditions. This includes the generation of non-stationary components into the vibration signals, which are not suitable per se for frequency domain methods. Certain components in the rotary machinery, such as rotary blades, frequency converter etc., create vibrations that disturb the analysis. The disturbances can be modeled with a sinusoidal pulse, a parametric pulse model, or a wavelet in the simulator.

3 Simulator

Understanding the nature of the disturbances in complicated vibration signal analysis scenarios will lead to accurate conclusions when selecting and developing vibration signal analysis methods. The flexible simulator offers an interactive tool for the vibration signal simulation and analysis for hard-to-analyze scenarios. The main parts of the simulator are the pulse generator, vibration signal generator and, vibration signal analysis module. The simulation and analysis process is described in Figure 1. The pulse generator produces a (shock) pulse based on the parametric model of the impulse response. The model parameters are adjustable. In addition, the white noise can be added to the pulse signal. A real measured pulse can be loaded and plotted into the same window with the simulated signal. This makes it possible to adjust the simulated pulse to match with the true pulse. The vibration signal generator produces the simulated vibration signal. The simulated shock pulses are added to the base signal based on the adjustable frequency and the first pulse start position. The length of the base signal is given and white noise is added to it. There is often a slight misalignment on the shaft and the rolling element bearings. To simulate this, a sinusoidal low-frequency component is added to the base signal. Signal pre-processing is done with digital filters such as linear continuous-time filter (elliptic), finite impulse response filter (FIR), and wavelet filter. Standard statistical measures, including the root mean square (RMS), peak-to-peak value, crest factor, and kurtosis, are calculated in the analysis module. Mainly frequency domain methods (FFT, HFRT) and time-frequency methods (STFT, WT) are used in the analysis of the vibrations. In addition, with the simulator, it is easy to compare algorithms developed earlier [10] with the newer development (Early SPD) [11]. Improved understanding on the behaviour of the analysis methods and approaches can be reflected back to the simulation, for example, when searching the limitations of the early fault detection methods.

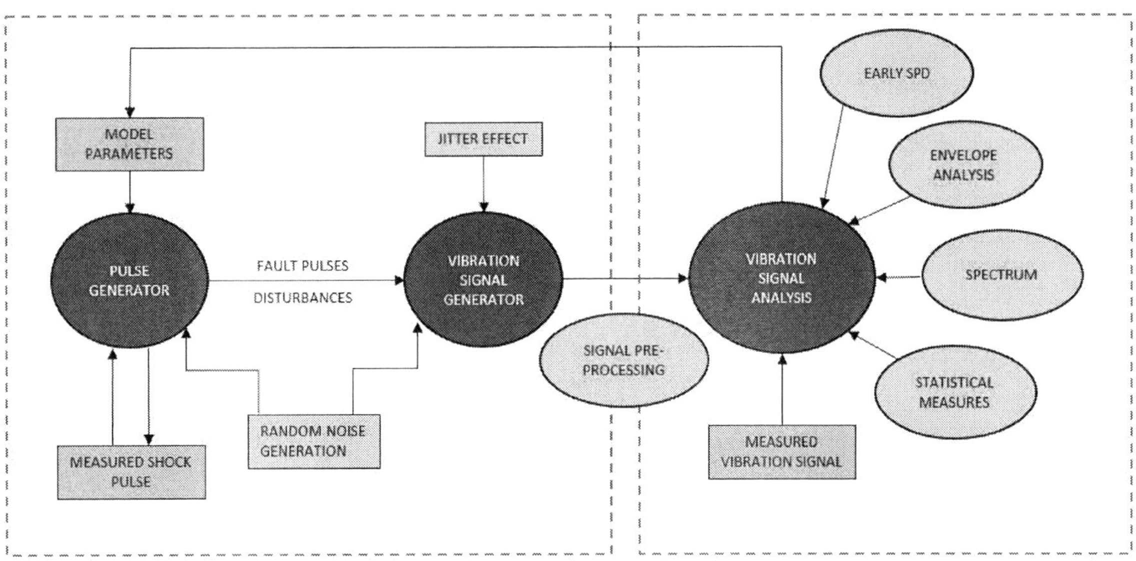

Fig. 1. Process flow of the flexible simulator

4 Results

The goal of the demonstration of the simulator is to show how the vibrations generated by machine components other than rolling element bearings may create false alarms on defects. Two case studies are introduced. The first simulated signal contains bearing fault pulses, sinusoidal disturbance, and noise. The fault pulse is modeled using equation 2 with the resonant frequency of 2500 Hz. The pulse repetition frequency of the fault, i.e., bearing fault frequency is 10.2 Hz and frequency of sinusoidal disturbance is 7.0 Hz. Figure 2 shows the simulated signal (case 1). In the second case (Figure 3), we model disturbances generated by blade and electric motors using pulses with resonant frequencies of 171 Hz (blade) and 4500 Hz (electric motor). The pulse repetition frequencies of the blade and motor disturbance are 5.0 Hz and 12.0 Hz, respectively. There are no fault pulses in the second case; the rolling element bearing is healthy. The simulated signal is shown in Figure 3.

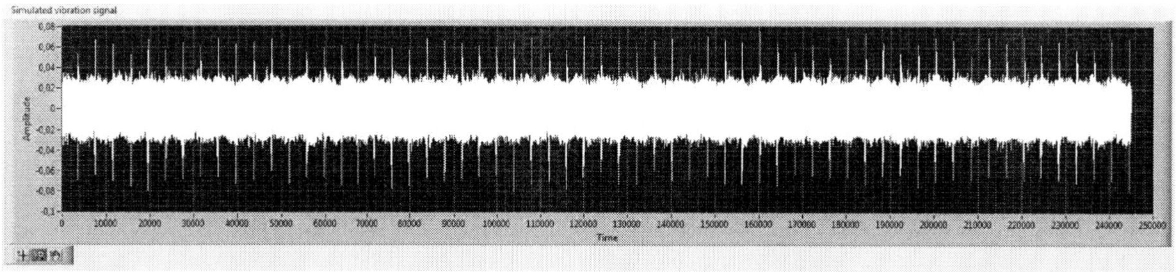

Fig. 2. Simulated vibration signal of a faulty bearing

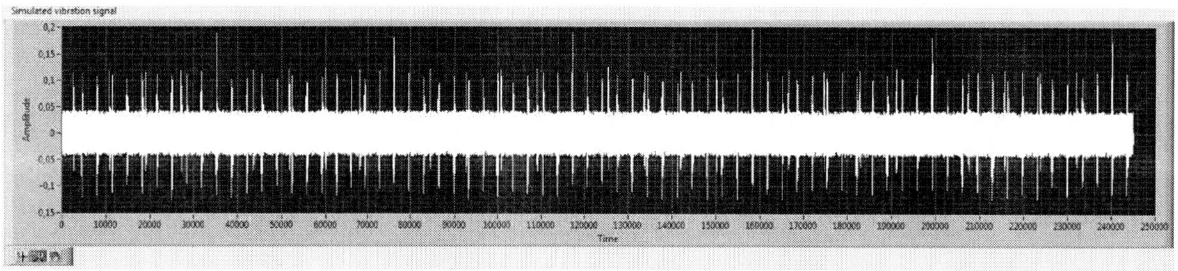

Fig. 3. Simulated vibration signal of healthy bearing

Frequency domain methods, spectrum and envelope spectrum are used in the vibration signal analysis. Like in common envelope analysis, we use the high-pass filter (1500 Hz cut-off-frequency) before calculating the envelope signal and envelope spectrum. Figure 4 incorporates spectra of the first case. The sinusoidal disturbance frequency (7 Hz) and resonant area exited by fault pulses (2500 Hz) appear in the left hand side of Figure 4. The envelope spectrum reveals clearly the fault frequencies of the simulated vibration signal of a faulty bearing. The frequency axis of the envelope spectrum is normalized according to the rotation speed.

Figure 5 incorporates the spectra of the second case. The resonant frequencies of the disturbing components are seen in the spectrum. In addition, there are some leakages to lower frequencies. Harmonics of blade- and motor-generated frequencies appear in the spectrum of the enveloped signal (right hand side). Despite the usage of the high-pass filter (1500 Hz), attenuation on the amplitudes of the blade frequency and its harmonics is small in the envelope spectrum. The second harmonic of the blade frequency (10 RPM order) is very close to the bearing fault frequency (10.2 RPM order) that may lead to a false alarm.

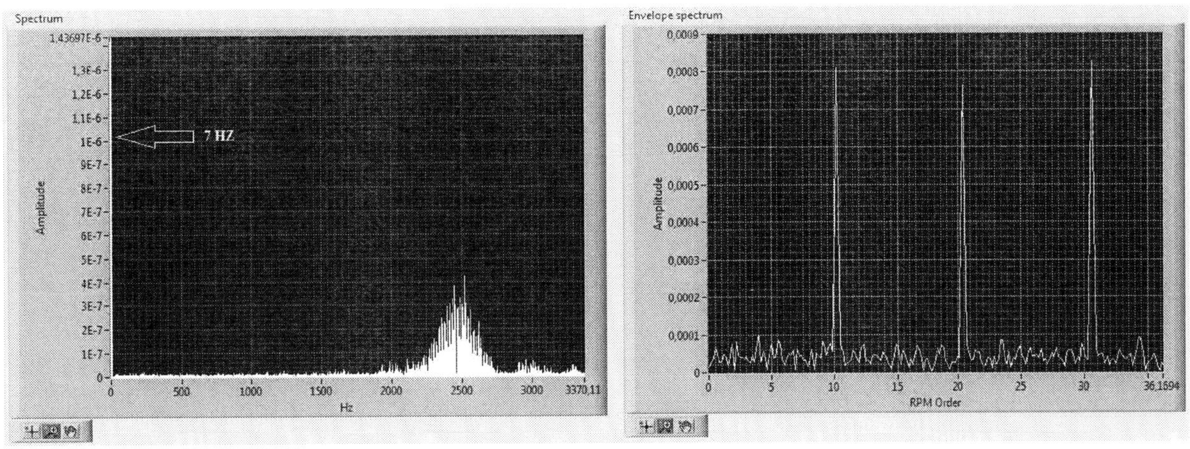

Fig. 4. Case 1: signal spectrum (left) and envelope spectrum (right)

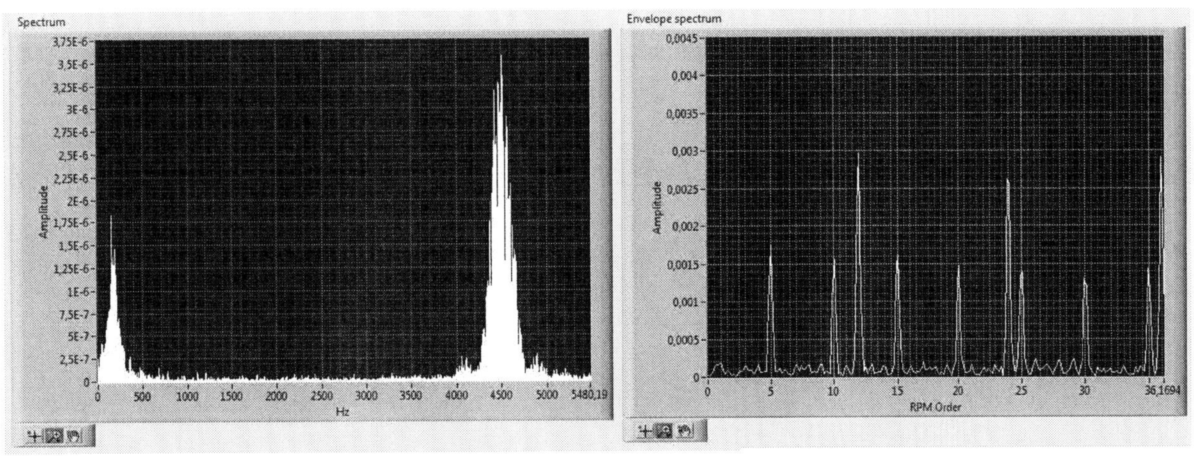

Fig. 5. Case 2: signal spectrum (left) and envelope spectrum (right)

5 Discussion and conclusions

The developed flexible simulator offers a new approach to study vibration signals that are complicated to analyze. The basic idea was to simulate bearing fault pulses and disturbances as they appear in the measured signals. In a perfect simulation scenario, the analysis results should be the same for the simulated and measured signals. The simulator advances the analysis, especially in the cases when there are disturbing non-stationary components in the vibration signals that create false alarms. The power of the simulator was demonstrated in the case study. The results of the case study show how the vibrations of the other known machine components disturb the analysis of the vibrations produced by REB faults and the weakness of the used high-pass filter on removing pulsating low-frequency disturbances from the envelope spectrum. Further development of the simulator will concentrate on the modeling of disturbances in REB study and the development of an advanced wavelet analysis module.

References

1. N. Sawalhi and R.B. Randall, Vibration response of spalled rolling element bearings: Observations, simulations and signal processing techniques to track the spall size, Mechanical Systems and Signal Processing 25 (2011) 846–870

2. N. Tandon and A. Choudhury, An analytical model for the predition of the vibration response of rolling element bearings due to a localized defect, Journal of Sound and Vibration 205(3) (1997) 275–292

3. Z. Kiral and H. Karagulle, Simulation and analysis of vibration signals generated by rolling element bearing with defects, Tribology International 36 (2003) 667–678

4. T. Momono and B. Noda, Sound and Vibration in Rolling Bearings, Motion & Control No. 6 (1999)

5. R. B. Randall, J. Antoni, and S. Chobsaard, The relationship between spectral correlation and envelope analysis in the diagnostics of bearing faults and other cyclostationary machine signals, Mechanical Systems and Signal Processing (2001) 15(5), 945–962

6. R. B. Randall and Jerome Antoni, Rolling element bearing diagnostics — A tutorial, Mechanical Systems and Signal Processing 25 (2011) 485–520

7. J. Liu, Y. Shao, and T. C. Lim, Vibration analysis of ball bearings with a localized defect applying piecewise response function, Mechanism and Machine Theory 56 (2012) 156–169

8. I. El-Thalji and E. Jantunen, A summary of fault modelling and predictive health monitoring of rolling element bearings, Mechanical Systems and Signal Processing 60-61 (2015) 252–272

9. M. Liang and H. Faghidi, Intelligent bearing fault detection by enhanced energy operator, Expert Systems with Applications 41 (2014) 7223–7234

10. J. Toukonen, M. Orkisz, M. Wnek, K. Saarinen, and Z. Korendo, ARMADA – Advanced Rotating Machines Diagnostics Analysis tool for added service productivity. Proceedings of COMADEM2001 conference, (2001) 717–724

11. K. Saarinen and J. Nowak, A method for detection and automatic identification of damage to rolling bearings. EP 2053375

Fault Feature Extraction based on SVD Envelope Demodulation for Wind Turbine Gearbox

Tao Chen [1,2] and Xiao-li Xu [1,2]

[1] Key Laboratory of Modern Measurement and Control Technology, Ministry of Education, Beijing Information Science and Technology University, xuxiaoli@bistu.edu.cn

[2] Measurement and Control of Mechanical and Electrical System, Beijing Key Laboratory, Beijing Information Science and Technology University, chentao@bistu.edu.cn

Abstract: Wind turbines, running in poor conditions and under external dynamic bearing, are prone to faults or failures. In order to guarantee a reliable operation of wind turbines, it is important to conduct fault diagnosis research. Focusing on the vibration signal of wind turbines gearbox with faults containing periodic impact composition and presenting modulation phenomenon, the fault feature is extracted from the signal by demodulation analysis. First, the vibration signal is decomposed into different scales by SVD method, and then envelope demodulation analysis is conducted to get fault features of the entropy. Fault feature extraction of industrial wind turbines shows that this method can effectively suppress noise and extract reliable fault sensitive features, and also provide scientific basis for fault diagnosis.

Keywords: SVD, Envelope demodulation, Wind turbines, Gear box, Fault feature extraction

1 Introduction

Wind turbines are used to generate electricity from the kinetic power of wind in wind farms both onshore and offshore, which are efficient sources of cheap energy. There are two main kinds of wind generators, one with a vertical axis, and the other with a horizontal axis which has been dominant in the current market. As wind turbines run in poor conditions and under external dynamic bearing, they are prone to faults or failures. In order to guarantee a reliable operation of wind turbines, it is important to conduct fault diagnosis research [1,2].

Horizontal-axis wind turbines (HAWT) have a main rotor shaft and an electrical generator at the top of a tower pointed at the direction of the wind, and components of HAWT are shown in Fig1. Most have a gearbox, which turns the slow rotation of the blades into a quicker rotation that is more suitable to drive an electrical generator. Gear box is an important assembly in the turbines and about 40% faults of the wind turbines mainly include gearbox faults and bearing faults [3]. Focusing on the vibration signal of wind turbines gearbox with faults containing periodic impact composition and presenting modulation phenomenon, the fault feature is extracted from the signal by demodulation analysis.[4]

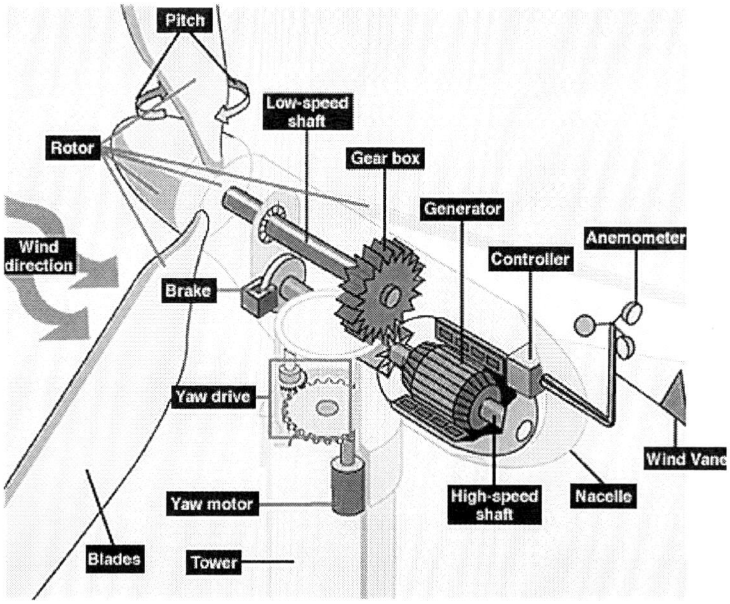

Fig 1 Components of HAWT

2 SVD Envelope Demodulation Method

2.1 Principle of SVD

When a failure happens on the gearbox of a wind turbine, the fault vibration signal is weak with modulation in wide frequency range. Since SVD algorithm has a good stability and invariance, the decomposition of Singular Value can reflect the intrinsic attributes of the fault data monitored. Thus it can be applied to demodulate the vibration signal so as to extract spectrum of fault sensitive characteristics from the decomposition band.

SVD has evident advantages. Firstly, SVD has a singular value stability; the tiny disturbance of matrix singular value before and after the change is not greater than the perturbation matrix's 2-norm. Secondly, SVD has s singular values scale invariance; normalized processing eliminating amplitude cannot essentially change the singular value. Thirdly, SVD has singular values rotation invariance; singular value feature vector not only has the orthogonal transformation, rotation, displacement, mirrored invariance of algebra and geometry, but also possess a good stability and the noise resistance. Therefore, SVD is widely used in pattern recognition and image analysis to get a unique and stable characteristics described and to reduce the dimension of feature space or to improve the ability of resistance to interference and noise [5,6].

In the linear algebra, SVD is an eigen decomposition of a $m \times n$ (usually $m > n$), matrix H via an extension of polar decomposition. SVD can be computed via the following formula:

$$H = U \Sigma V^* \tag{1}$$

where U is an $m \times m$ real or complex unitary matrix, Σ is a $m \times n$ rectangular diagonal matrix with non-negative real numbers on the diagonal, and V is an $n \times n$ real or complex unitary matrix. The diagonal entries σ_i of Σ are the singular values of H. The columns of U and the columns of V are called the left-singular vectors and right-singular vectors of H respectively. The left-singular vectors of H are a set of orthonormal eigenvectors of HH^*. The right-singular vectors of H are a set of orthonormal eigenvectors of H^*H.

$$\Sigma = diag(\sigma_1, \sigma_2, \cdots, \sigma_r) \tag{2}$$

Where $\sigma_1 \geq \sigma_2 \geq \ldots \geq \sigma_r$, $\sigma_i (i = 1, 2, \cdots, r)$ is a nonzero singular value. The first few values are bigger, which contain the most information matrix H. By SVD the matrix H is decomposed into r bands and the eigenvector corresponding to the maximum eigenvalue is the direction of the maximum variance distribution.

As matrix H presents the recording signal composed of noise and the useful signals, the singular values of H can be partitioned into two groups as

$$\Sigma = \begin{bmatrix} \Sigma_s & 0 \\ 0 & \Sigma_n \end{bmatrix} \tag{3}$$

Where Σ_s contains r singular values associated with r source signals and Σ_n contains $m - r$ singular values associated with the noise. Corresponding to this, the matrix U can be partitioned as

$$U = \begin{bmatrix} U_s & U_n \end{bmatrix} \tag{4}$$

Where U_s contains the first r columns of U associated with the signal part and U_n contains the last $m - r$ columns of U associated with the noise part. Consequently, the span of the columns of U_s is called the signal subspace (of dimension r) and the span of the columns of U_n is called the noise subspace (of dimension $m - r$).

Since the signal and noise subspaces are orthogonal to each other, the solution to the signal separation can be presented as an orthonormal projection of the data onto the signal subspace, where by noise contributions are eliminated. Formally, this can be written as

$$S = U^T H = \begin{bmatrix} U_s^T H \\ U_n^T H \end{bmatrix} = \begin{bmatrix} H_s \\ H_n \end{bmatrix} \tag{5}$$

Where H_s is an $(r \times m)$ matrix containing estimates of the r source signals. Fault feature extraction based on SVD is done on the source signals with a better noise reduction effect. And SVD has noticeable advantages of calculation suppleness with good physical meaning.

2.2 Procedure of Fault Feature Extraction based on SVD Envelope Demodulation

When conducting fault diagnosis on wind turbines, fault feature extraction starts from an initial set of monitored data and builds derived features intended to be informative and non-redundant. Fault feature extraction based on SVD is related to dimensionality reduction by decomposing the data in to into different scales[7].

According to principle of SVD, the calculation procedure of fault feature extraction based on SVD is listed below:

(1) Gather monitored data column and change the data column into matrix H_{m*n};

(2) Calculate eigenvalues and eigenvectors of matrix H according to formula (1);

(3) Reorganize the eigenvalues in descending order and adjust the eigenvectors corresponding to the eigenvalues;

(4) Calculate the cumulative contribution rate of the eigenvalues;

(5) Get fault features of the entropy of the main eigenvalue dimension.

To conduct a singular value decomposition of the monitored signal of wind turbines, the signal is firstly transformed into a matrix, generally the signal in Hankel matrix showing the best demodulation decomposition effect. The Hankel matrix of discrete digital signal $x = (x(1), x(2), \cdots, x(N))$ is structured as below:

$$H = \begin{bmatrix} x(1) & x(2) & \cdots & x(n) \\ x(2) & x(3) & \cdots & x(n+1) \\ \vdots & \vdots & \vdots & \vdots \\ x(N-n+1) & x(N-n+2) & \cdots & x(N) \end{bmatrix} \tag{6}$$

Where $1 < n < N$, $m = N - n + 1$ and $H \in R^{m \times n}$ .The Hankel matrix H is called a reconstruction attractor matrix. The complete separation of the signals Hankel matrix requires the maximum number of rows and columns, that is to say, $n = N/2$ and $m = N/2 + 1$.

The fault features of the entropy of the main eigenvalue dimension is calculated via the following formula

$$\Delta E_i = -\left(\frac{\sigma_i}{\sum_j^l \sigma_j}\right) \log\left(\frac{\sigma_i}{\sum_j^l \sigma_j}\right) \tag{7}$$

Like other types of information entropy, the SVD entropy has a similar significance on describing the information quantity of the signal, i.e. the lager the value of E_i, the more complex the signal is in composition.

3 Application of Fault Feature Extraction based on SVD Envelope Demodulation

Selecting wind turbine gearbox as research object, vibration signals are inquired from the sensors installed on the turbine. According to the algorithm principle above, the procedure of SVD feature extraction is written using MATLAB. The signals data is transformed into a Hankel matrix H .Fig2 is the time domain plot of wind turbine gear box fault. Table 1 shows the Eigne Contribution by SVD. The procedure is to extract the decomposed component whose cumulative variance contribution rate is more than 90%. The SVD method can effectively suppress noise.

Fig3 shows the extracted entropy feature analysis of the turbine fault signal and Fig4 shows the extracted entropy feature analysis of the turbine normal signal.

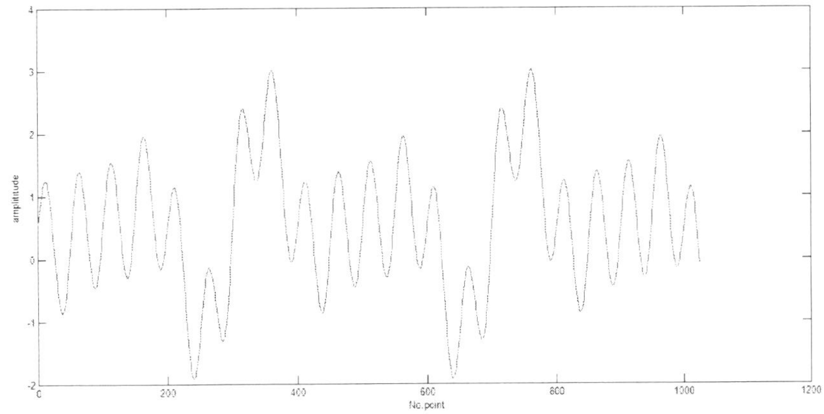

Fig2 Time Domain Plot of Fault

Table 1 Eigne Contribution by SVD

Order	Contribution rate	Cumulative contribution rate
1	16.64%	16.64%
2	16.63%	33.27%
3	15.89%	49.16%
4	15.43%	64.59%
5	8.86%	73.45%
6	8.61%	82.06%
7	7.17%	89.23%
8	7.01%	96.24%
9	3.75%	99%

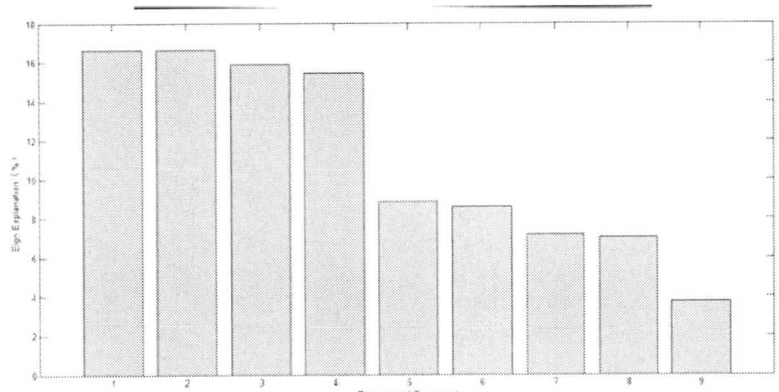

Fig.3 SVD Entropy Analysis of Fault

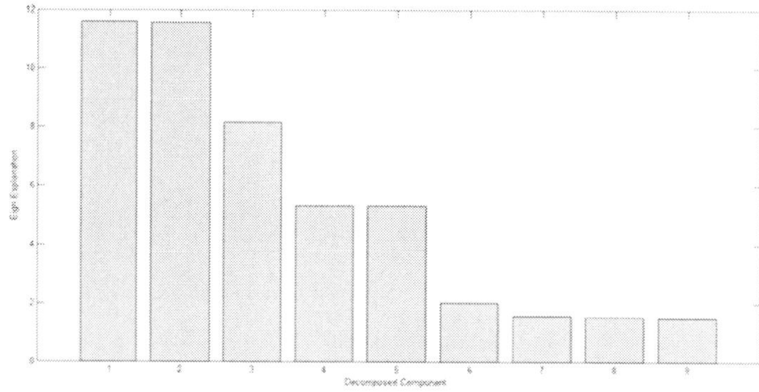

Fig4 SVD Entropy Analysis of Normal

From Fig3 and Fig4, the entropy feature of fault and normal can be seen clearly that the entropy values of fault are obviously larger than those of the normal. Furtherly the decomposed component can be analyzed furtherly to determine the type of the fault. That is to say, SVD can extract reliable fault sensitive features, and also they can provide scientific basis for the further fault pattern recognition and further fault trend prediction.

4 Conclusions

SVD means to extract features from the mass data with a better noise reduction effect. The application of feature extraction shows that SVD has a good performance in feature extraction application for wind turbines. It is feasible to apply this method to the study of similar projects concerning fault diagnosis.

Acknowledgement

This work was supported by the Natural Science Foundation of China（51105041，51275052）. A sincere gratitude is also extended to the anonymous reviewers for their constructive suggestions.

References

[1] M.R. Wilkinson, F. Spianto, M. Knowles, etal. Towards the zero maintenance wind turbine [J]. The 41st International Universities Power Engineering Conference, 2006, 1:74-78.

[2] P. J. Tavner, J. Xiang, F. Spinato. Reliability of Turbines[J]. Wind Energy, 2007,10: 1-18.

[3] R. Johan, B.L. Margareta. Survey of failures in wind power systems with focus on Swedish wind power plants during 1997-2005[J],IEEE Transactions on Energy Conversion, 2007,22(1):167-173.

[4] Zhipeng Feng,Ming Liang,Fulei Chu. Recent advances in time–frequency analysis methods for machinery fault diagnosis: A review with application examples[J]. Mechanical Systems and Signal Processing. 2013 (1)

[5] Reza Golafshann, Kenan Yuce Sanliturk. SVD and Hankel matrix based de-noising approach for ballbearing fault detection and its assessment using artificial faults[J]. Mechanical Systems and Signal Processing,2016(70-71): 36-50

[6] Hamid Hassanpour, Amin Zehtabian, S.J. Sadati.Time domain signal enhancement based on an optimized singular vector denoising algorithm[J]. Digital Signal Processing,2012 (22): 786-794

[7] MengZhi hui,WangChang.Fault Diagnosis for Rolling Bearings Based on SVD De-noising and spectral Kurtosis. Bearing2013,(10):52-55

Improved Ensemble Empirical Mode Decomposition for Compound Fault Diagnosis of Roller Bearings

Gang Tang[1], Ganggang Luo[1], Weihua Zhang[2], Caijin Yang[2], Huaqing Wang[1]

[1] College of Mechanical and Electrical Engineering, Beijing University of Chemical Technology, Beijing 100029, China
[2] State key Laboratory of Traction Power, Southwest Jiaotong University, Chendu 610031, China
E-mail: tanggang@mail.buct.edu.cn

Abstract: Compound fault signal is very common in the running status monitoring of rotating machinery. It usually contains multiple characteristic signals and strong random noise, and hence, it is difficult to extract fault features using conventional methods. Empirical mode decomposition (EEMD) is developed and used to decompose mixed signals and extract fault features. In order to reduce the operating time and improve the running efficiency for compound fault diagnosis, the present paper proposes an improved method based on EEMD with independent component analysis. A vibration signal is first decomposed into intrinsic mode functions by EEMD to obtain multichannel signals. Then, the cross correlation coefficient is calculated and a threshold value is set as stopping criteria for the process of decomposition. Finally, the compound faults are separated effectively with independent component analysis. Experiments validate the effectiveness of the proposed method in compound fault separation, and a comparison experiment shows that the proposed method has higher effectiveness and practicability than the common ensemble empirical mode decomposition method.

Keywords: roller bearing; fault diagnosis; ensemble empirical mode decomposition; independent component analysis

1. Introduction

As one of the most widely mechanical parts components in rotating machines, roller bearing is also one of the most vulnerable components of these machines [1]. Therefore, how to collect enough and useful information of a roller bearing has become more and more important.

Vibration signals are usually used to detect roller bearing faults, which can provide useful information about different fault features. However, in practice, most vibration signals are often compounded by different fault features. In addition, these compound fault features are often buried by noise in the frequency spectrum, which will affect the fault identification through traditional envelope spectrum analysis. In recent years, blind source separation (BSS) proposed by Herault gives a new insight to signal processing for compound fault diagnosis [2]. Independent component analysis (ICA) is one of the main approaches to realizing BSS [3]. By operating in statistics, demixing matrix can be estimated by ICA to separate a set of independent sources from the original mixed signal. Nevertheless, one of the most important precondition of ICA is that the number of sensors must be equal to or greater than the number of original source signals, which is very difficult to be satisfied because sensors are often positioned in a particular number and locations on a roller bearing [4].

To solve this problem, many researchers have employed decomposition strategy to demix signals. Empirical mode decomposition (EMD) [5] is one of the most popular methods. As a typical non-stationary signals, the vibration signal of a roller bearing can be separated to a set of intrinsic mode functions (IMFs) by EMD. And the number of IMFs is considerably more than the number of original source signals [6]. However, the major drawback of EMD is mode mixing [7], which causes some inveracious IMFs. Aiming at solving this problem, ensemble EMD (EEMD) was proposed by Wu and Huang [8] by adding white noise to the original signal. For compound fault diagnosis, the method combing EEMD and ICA has been proved to be effective, in which EEMD is used to separate the signal and ICA is used to extract the fault features [9]. However, it provides a huge running time increasing over the sampling points, which hinders its applications to high-frequency rotating machines.

In order to reduce the running time, in the present paper, an improved method based on EEMD and the cross correlation coefficient is proposed, in order to demix the original signal and filtrate effective components. The rest of this paper is organized as follows. The improved ensemble empirical mode decomposition is described in Section 2. The experimental results are provided in Section 3. Finally, a conclusion is given in Section 4.

2. Improved Ensemble Empirical Mode Decomposition

In order to eliminate the mode mixing, EEMD was proposed by adding finite white noise to the original signal. For compound fault diagnosis, EEMD is usually used to separate the original signal and ICA is used to extract the fault features. However, there are some drawbacks in this method.

a. EEMD provides a huge running time increasing over the sampling points
b. IMFs separated by EEMD don't have to be completely used as the input matrix of ICA. So, many IMFs are redundant.
c. As the input matrix of ICA, the IMFs need to meet the correlation requirements with the original signal. And the screening process has not been found in EEMD.

To overcome the problems, an improved EEMD based on cross correlation coefficient is proposed in this paper. By calculating the cross correlation coefficient between each IMF and the original signal, and setting a threshold to stop the iterations, the proposed method can reduce the running time and filtrate effective components.

The steps of improved EEMD algorithm follows:

STEP 1. Obtain an ensemble signal by adding a set of Gaussian white noise signal $\omega(t)$ to the original signal $x(t)$:

$$X(t) = x(t) + \omega(t)\,. \tag{1}$$

STEP 2. Decompose $X(t)$ with EMD to get IMFs:

$$X(t) = \sum_{j=1}^{n} c_j + r_n\,. \tag{2}$$

STEP 3. Calculate the cross correlation coefficient and set a threshold.

$$\rho_{XY} = \frac{cov(X,x)}{\sqrt{D(X)}\sqrt{D(x)}} \tag{3}$$

STEP 4. Add different Gaussian white noise $\omega_i(t)$, and repeat steps above.

$$X_i(t) = \sum_{j=1}^{n} c_{ij} + r_{in}\,. \tag{4}$$

STEP 5 Eliminate the effect of reference structure at time domain based on the principle of the zero mean in Gaussian white noise spectrum. The IMF $c_n(t)$ can be expressed as:

$$c_n(t) = \frac{1}{N}\sum_{i=1}^{N} c_{i,n}(t)\,. \tag{5}$$

STEP 6 Original signal $x(t)$ can be decomposed as

$$x(t) = \sum_{n=1}^{m} c_n(t) + r_m(t)\,. \tag{6}$$

To improved EEMD method, the cross correlation coefficient is used to identify the correlation of components and filtrate the valid IMFs. And a given threshold is used to stop iteration and output result. Finally, input the IMFs to ICA and the fault features can be extracted from envelope spectrum. The process details of this method for an improved system are shown in Figure 1.

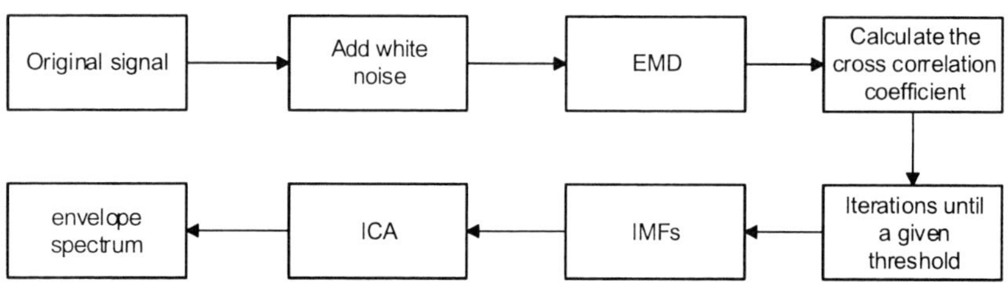

Figure 1. The process details of proposed method

3. Experimental Results

To validate the efficiency of the proposed method in practical applications, a vibration signal is gathered from acceleration sensors of a rotary machinery The main acceleration sensors are mounted on the top and side of the bearing house to measure the vibration signal of the 2 V channel. The compound fault of the bearing outer race and rollers is considered as the research object. For the fault diagnosis test, an outer-race flaw (0.5 mm (width) × 0.15 mm (depth)) and a roller flaw (0.5 mm (width) × 0.15 mm (depth)) are artificially created using a wire-cutting machine.

To fully analyse the signal features and acquire more comprehensive information on the fault diagnosis, a large sampling frequency (100 KHz) is chosen. A vibration signal is measured and a section samples with 5 s is selected for the experiment. The signal at time domain and frequency domain is shown in Figure.2.

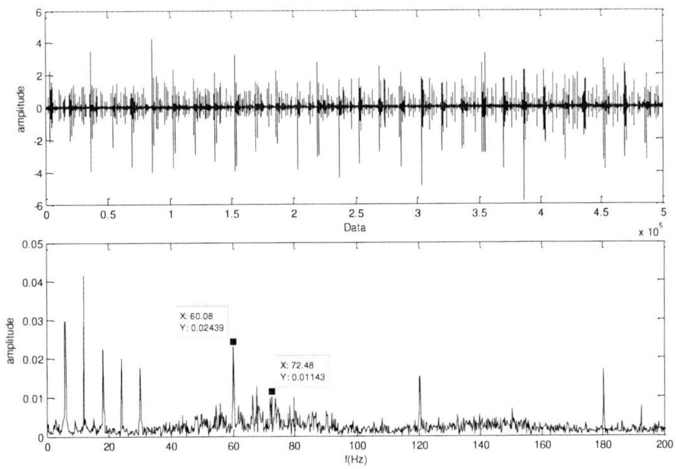

Figure 2. The vibration waveform and its spectrum

Then, the threshold is set at 0.2 and the decomposition signals can be obtained with the improved EEMD method. The decomposition results are shown in Figure 3.

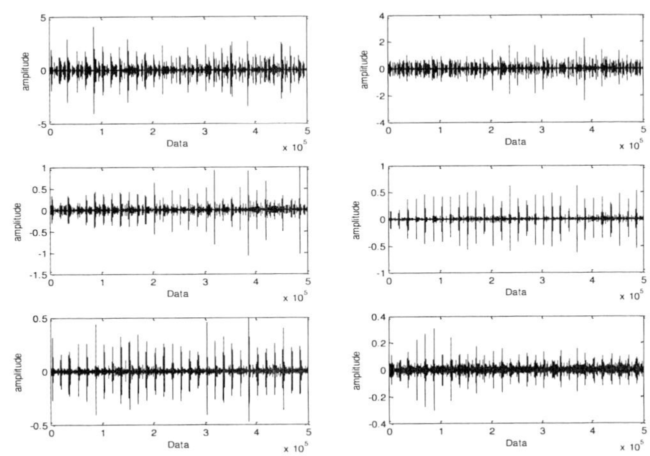

Figure 3. The decomposition results

Finally, the envelope of IMFs is used as the input matrix for ICA to separate and identify the compound faults. The separation results are shown in Figure 4.

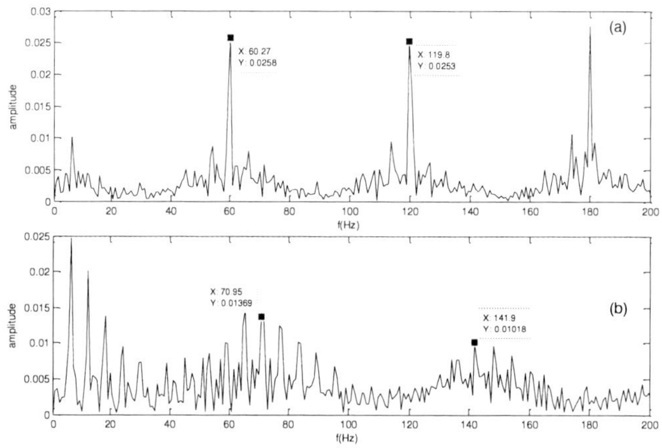

Figure 4. Spectrum of separated signals obtained by proposed method

From Figure 5, we can see that outer-race defect and roller defect can be separated and identified obviously. And this improved EEMD method shows its high efficiency in running time. The comparison with traditional EEMD method in running time is shown in Figure 5.

As mentioned above, the improved EEMD method can be used to extract compound fault features of roller bearings.

The study result shows that the proposed method has higher adaptability and practicability in separating signals.

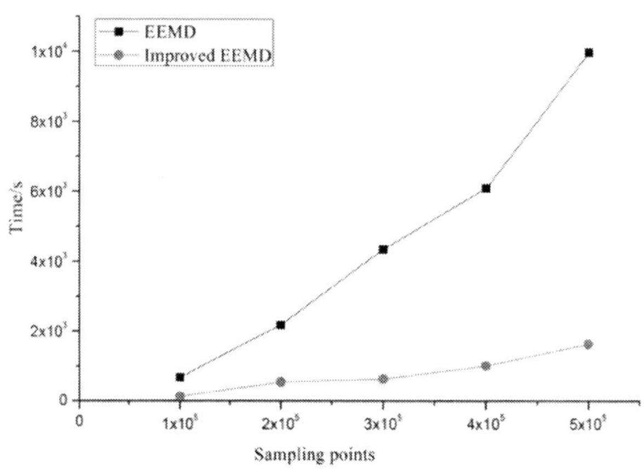

Figure 5. The comparison in running time

4. Conclusion

To solve the problems of big running time in compound fault diagnosis, a new improved EEMD method is proposed in this paper based on cross correlation coefficient. An experiment shows that the compound fault features can be accurately identified in a short amount of time. However, the proposed method is sensitive to noise, which is a drawback currently. Future work will focus on further optimization of the proposed method.

Acknowledgments. This work is partially supported by the National Natural Science Foundation of China (Grant no.51405012, 51375037), the Open Fund of State Key Laboratory, Southwest Jiaotong University (TPL1603) and PetroChina Innovation Foundation (2014D-5006-0104).

References

1. Harris, T.: Rolling Bearing Analysis. John Wiley and Sons, Inc. (United States). 1013 (1991)
2. Zarzoso, V., Nandi, A.K.: Blind source separation. Springer, United States (1999)
3. Hyvärinen, A., Karhunen, J., Oja, E.: Independent Component Analysis. John Wiley & Sons, United States (2004)
4. Shao, R., Hu, W., Li, J.: Multi-fault feature extraction and diagnosis of gear transmission system using time-frequency analysis and wavelet threshold de-noising based on EMD. Shock and Vibration. 20(4), 763-780 (2003)
5. Rilling, G., Flandrin, P., Goncalves, P.: On empirical mode decomposition and its algorithms. IEEE-EURASIP Workshop Nonlinear Signal Image Process. 3, 8–11 (2003)
6. Yang, Q., An, D.: EMD and wavelet transform based fault diagnosis for wind turbine gear box. Advances in Mechanical Engineering. 5, 212836 (2013)
7. Ricci, R., Pennacchi, P., Lombardi, M., Mirabile, C.: Failure diagnostics of a spiral bevel gearbox using EMD and HHT. Proceedings of the ISMA2010 Including USD, Leuven, Belgium 20–2 (2010)
8. Wu, Z., Huang, N.E.: Ensemble empirical mode decomposition: a noise-assisted data analysis method. Advances in adaptive data analysis. 1(01), 1–41 (2009)
9. Wang, H., Li, R., Tang, G., Yuan, H., Zhao, Q., Cao, X.: A compound fault diagnosis for rolling bearings method based on blind source separation and ensemble empirical mode decomposition. PloS one. 9, 1–13 (2014)

De-nosing Method of Rolling Bearing Based on Packing Numbers-Manifold Learning

Jingsheng Jiang, Huaqing Wang*, Hongfang Yuan, Wei Xiang and Yanliang Ke

School of Mechanical & Electrical Engineering, Beijing University of Chemical Technology, Chaoyang District, Beijing, 100029, China
Huaqing Wang is the corresponding author.
E-mail: *hqwang@mail.buct.edu.cn

Abstract. Aiming at the problem that the actual engineering vibration signals are influenced by complex working conditions and strong noise, a de-nosing method based on packing numbers-manifold learning is proposed. Firstly, the original signal containing strong noise is decomposed into a series of intrinsic mode functions (IMFs) by complementary ensemble empirical mode decomposition (CEEMD) method. Secondly, local tangent space alignment (LTSA) method is employed to project IMFs set into low dimensional matrix. To determine reduction dimension, packing numbers is introduced. At last, low dimensional components are inverse projected to the original high dimensional matrix by principal manifold reconstruction. After eliminating the noise, the reconstructed IMFs matrix is formed into a new signal. Thus envelope demodulation is carried on the new signal for fault diagnosis. Experiments show that the proposed method has better de-noising effect than wavelet transform and is more suitable for early fault diagnosis of rolling bearings.

Keywords: packing numbers, LTSA, principal manifold reconstruction, fault diagnosis

1 Introduction

Rolling bearings play an important role in rotating equipment, and the mechanical failure of bearings may result in financial loss, and even death. Therefore, condition monitoring and fault diagnosis of rolling bearings is critically important [1]. The actual engineering vibration signals are interfered by strong noise with strong nonlinear characteristic, so traditional linear filter cannot produce desirable de-noising effect. For instant, wavelet transform is limited by parameters choose [2]. When a rolling bearing fault is weak at an early stage or at a low shaft speed, weak fault features are often embedded in background noise. So it is no easy task to extract the representative features from the original signal [3]. Recently chaotic time series analysis is widely used in engineering field. Phase space reconstruction, as the foundation of chaotic time series analysis, is proposed by Takens [4]. To determine embedding dimension and time delay in phase space reconstruction, many methods have been put forward. But huge computation and complex algorithm are the limitations of these methods.

Furthermore, the results yielded by different methods are widely divergent [5]. To solve the above problems, the paper adapts IMFs set instead of high dimensional phase space. Considering that IMFs set may be distorted in the process of dimension reduction when reduction dimension are chosen at random, a new de-noising method based on packing numbers and local tangent space alignment reconstruction is introduced. Figure 1 shows the process of fault diagnosis. Firstly, the original signal is decomposed into IMFs by CEEMD and the intrinsic dimension of IMFs set is estimated by packing numbers. Secondly, IMFs set is projected to intrinsic dimensional matrix by LTSA. And then a new IMFs set is reconstructed by principal manifold reconstruction. Thus a new signal is obtained summing over these reconstructed IMFs. Finally, envelope demodulation is carried on the new signal for fault diagnosis. Experiments prove that the method has better de-noising effect and is well proper to early fault diagnosis of rolling bearing.

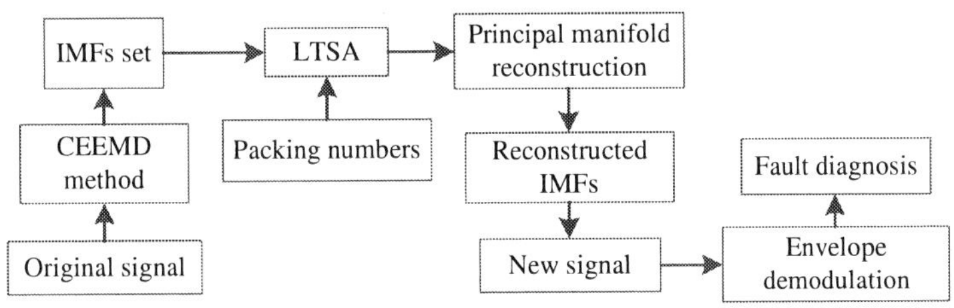

Fig. 1. Process of fault diagnosis

2 Intrinsic Dimension Estimation and Packing Numbers

Informally, the intrinsic dimension of a random vector X is defined as the minimum number of independent free variables needed to model the data set. With correlation dimension and topological dimension defined, packing numbers, one of intrinsic dimension estimation methods is proposed. Early correlation dimension method is usually used to estimate intrinsic dimension of a data set [6]. But when it obeys non-uniform distribution, the intrinsic dimension D_{corr} will be severely underestimated. To overcome the problem, capacity dimension is adopted instead. Compared with correlation dimension method, it does not rely on probability distribution of manifold. But it is noted that capacity dimension method isn't widely used because of computational complexity [7-8]. In this case, packing numbers method is introduced to redefine capacity dimension, and the definition is as follows.

$$D_{cap} = -\lim_{r \to 0} \frac{\log M(r)}{\log r} \quad (1)$$

For a finite sample, the zero limit cannot be achieved. So we need to redefine the capacity dimension in a scale-dependent manner.

The scale-dependent capacity dimension of a finite set $\mathbf{S}_n = \{x_1, \cdots, x_n\}$ is

$$\hat{D}_{cap}(r_1,r_2)=-\frac{\log M(r_2)-\log M(r_1)}{\log r_2-\log r_1} \tag{2}$$

Finding r-packing numbers $M(r)$ for a data set $\mathbf{S}_n=\{x_1,\cdots,x_n\}$ is equal to finding the cardinality of a maximum independent vertex set $\mathbf{MI}(\mathbf{G}_r)$ and graph $\mathbf{G}_r(\mathbf{V},\mathbf{E})$ with vertex set $\mathbf{V}=\mathbf{S}_n$ and edge set $\mathbf{E}=\left\{(x_i,x_j)\middle|d(x_i,x_j)<r\right\}$. Given a data set $\mathbf{S}_n$, we start with an empty centers $\mathbf{C}$. First add x_1 to $\mathbf{C}$ and in an iteration we add to $\mathbf{C}$ data points that at a distant of more than r from all data points in $\mathbf{C}$, till every point in $\mathbf{S}_n$ has been visited. The estimate $\hat{M}(r)$ is the cardinality of $\mathbf{C}$. The bias of $\hat{M}(r)$ is independent of r but the variance of $\hat{M}(r)$ can affect the estimation accuracy, see [6]. To eliminate the ill effects, we arrange all data points at random ten times and take the average of calculated packing numbers as the estimate $\hat{M}(r)$. Similar with r-covering number, given a specific r, r-packing numbers $M(r)$ is in proportion to $r^{-D_{pack}}$. So we established cylindrical coordinates system with $\lg(r)$ as x-axis, $\lg\hat{M}(r)$ as y-axis. Calculate $\hat{M}(r)$ of different distances and plot these points in cylindrical coordinates system. Thus the estimated intrinsic dimension $\hat{D}_{pack}$ is the negative slope of line segment.

3 LTSA De-noising Algorithm

Assume a time series $\mathbf{S}=[s_1,s_2,\cdots,s_n]$, and a set of IMFs are obtained by CEEMD method [9]. Then IMFs are reconstructed into a high dimensional matrix $\mathbf{P}$.

$$\mathbf{P}=[IMF_1\ IMF_2\ \cdots\ IMF_D] \tag{3}$$

The low dimensional coordinate system of the high dimensional embedded space $\mathbf{P}$ is determined by LTSA. Thus the low dimensional global principal manifold with row vector orthogonal is acquired [10-11].

$$\mathbf{T}=[\tau_2,\tau_3,\cdots,\tau_{d+1}]\in R^{d\times n} \tag{4}$$

For a phase point IMF_j, $\mathbf{IMF}_j^k$ is set as neighbor data set. First choose the number k of nearest neighbors and the dimension d of principal manifold to construct adjacency graph $\mathbf{G}$; Compute the local tangent coordinates $\mathbf{\Theta}_i$ and the right singular feature vector $\mathbf{Q}_i$ corresponding to the largest d singular value of centered matrix $\mathbf{P}_j - \bar{\mathbf{P}}_j^k 1_k^T$. Where $\bar{P}_j^k=\frac{1}{k}\sum_{j=1}^{k}P_j$ and the local tangent coordinates $\mathbf{\Theta}_i=\mathbf{Q}_i^T(P_j-\bar{P}_j^k 1_k^T)$. The reconstruction formula is defined as follows [12].

$$y_i = f(\tau_i) = \bar{P}_i + Q_i L_i^{-1}(\tau_i - \bar{\tau}_i) \tag{5}$$

To return to one dimensional signal, it can be recorded as

$$x'(t) = \sum y_i \tag{6}$$

4 Analysis and Discussion

Weak fault features are often embedded in background noise at an early stage. To simulate early fault of rolling bearings, an inner-race fault bearing and an outer-race fault bearing are respectively taken and the bearings' defects were artificially made with the use of a wire-cutting machine. The defects are all 0.2 mm × 0.05mm (width × depth). In the experiment, the spindle speed was set to 600 rpm and the sample frequency was set to 10000 Hz. The inner ring fault feature frequency is 70.91 Hz and the outer ring fault feature frequency is 49.09 Hz by calculating. First the inner ring fault vibration signal is processed, and the time-domain waveform is shown in Fig.2.

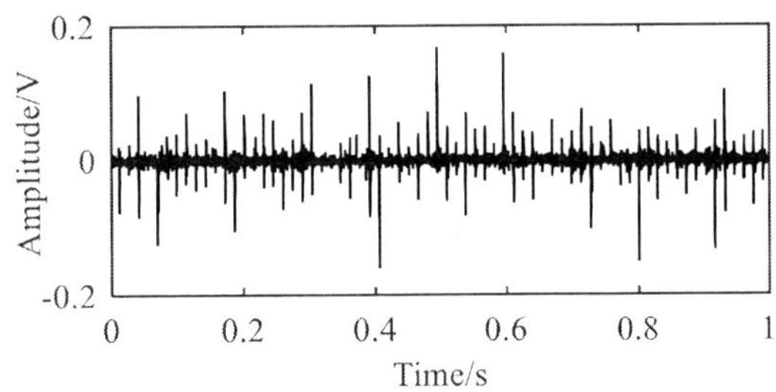

Fig. 2. Inner ring fault vibration signal

The signal is decomposed into a series of IMFs by CEEMD, and IMFs are constructed to high dimension matrix **P**. Then use packing numbers to estimate the intrinsic dimension of **P**. From the result of linear fitting in Fig.3, we can solve estimated dimension $\hat{D}_{pack} = 1$.

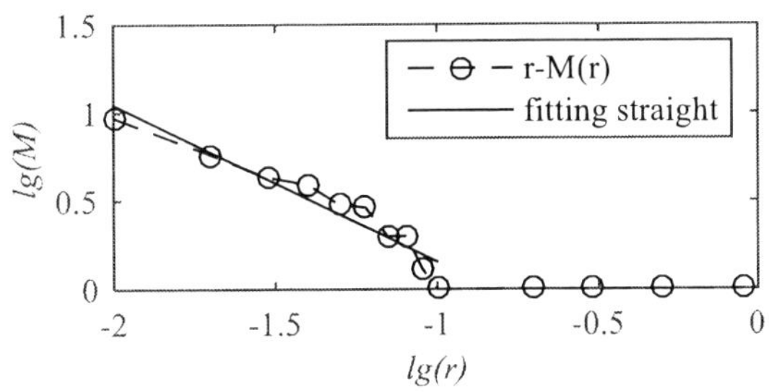

Fig. 3. Linear fitting result between *lg(r)* and *lg(M(r))*

According to LTSA de-noising algorithm, set target dimension to 1, and the new signal is formed whose time-domain waveform is shown in Fig.4. The amplitude of signal has reduced after noise reduction.

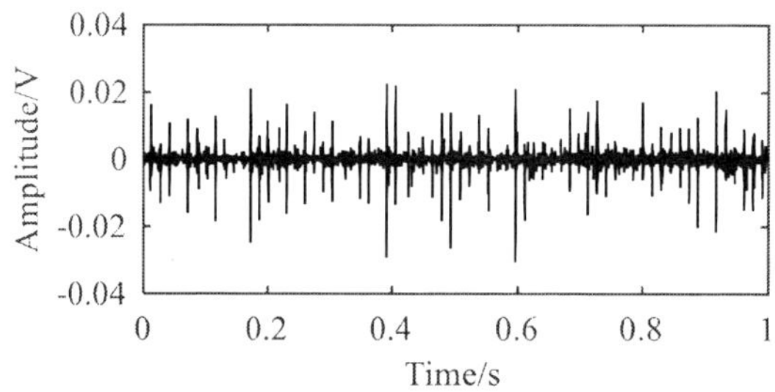

Fig. 4. De-nosing signal

Further to verify the effectiveness of the method, an envelope demodulation method based on Hilbert transform and fast Fourier transform is applied to the new signal. Figure 5 shows the envelope spectrum with different methods. From Figure 5, it is clear that the proposed method can extract more obvious fault feature than wavelet transform although both two methods are available. It is worthwhile to note that parameters in wavelet transform are difficult to choose, so sometimes it cannot get a good de-noising effect. And then calculate the SNR of de-noising signals. From Table 1, it can be seen that the proposed method has better de-noising effect.

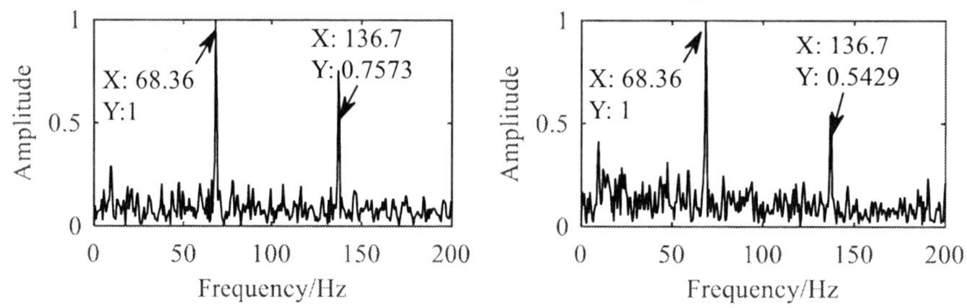

(a) Envelope spectrum with CEEMD and LTSA de-noising

(b) Envelope spectrum with wavelet transform

Fig. 5. Envelope spectrum of de-nosing signals with different methods

Table 1. SNR of de-nosing signals

Method	CEEMD and LTSA	Wavelet Transform
SNR(dB)	-0.6455	-5.4728

Then we also focused on outer ring fault signal. Similarly, processed by CEEMD and LTSA de-noising, the SNR of signal is -4.5037dB, higher than -12.1840dB solved by wavelet transform. So we conclude that the proposed method in this paper has better noise reduction performance than wavelet transform method.

5 Conclusions

This paper adopts IMFs set instead of high dimensional phase space to avoid huge computation and complex algorithm of phase space reconstruction. To overcome the limitation in selecting reduction target of LTSA, packing numbers is introduced to estimate the intrinsic dimension of IMFs set. In addition, the proposed method has better nonlinear de-noising effect by comparing to wavelet transform method. Experiments show the proposed method enhances SNR of signals and obviously extracts weak fault feature. So we can conclude that the proposed method does apply to early fault diagnosis of rolling bearings.

Acknowledgments
This work is supported by the National Natural Science Foundation of China (Grant no. 51375037) and PetroChina Innovation Fund (Grant no. 2015D-5006-0606).

References

1. Cong F, Chen J, Dong G, et al. Short-time matrix series based singular value decomposition for rolling bearing fault diagnosis. Mechanical Systems and Signal Processing, 2013, 34(1): 218-230.
2. Aguiar Conraria L, Soares M J. The continuous wavelet transform: moving beyond uni- and bivariate analysis. Journal of Economic Surveys, 2014, 28(2): 344-375.
3. Zhong M C, Zhang C L, Cheng L, et al. Vibration signal of recoil mechanism de-noising processing based on wavelet analysis. Applied Mechanics and Materials. 2013, 401: 1424-1427.
4. Takens F, Detecting sange attractors in turbulence. Lecture Notes in Math. New York: SpringeL 1981.
5. Small M, Tse C K. Optimal embedding parameters: a modelling paradigm. Physica D: Nonlinear Phenomena, 2004, 194(3): 283-296.
6. Kegl B. Intrinsic dimension estimation using packing numbers. In: Proe ot the Conference Information Processing Systems. Washington, USA, 2003, 864-871.
7. Shayegh F, Sadri S, Amirfattahi R, et al. A model-based method for computation of correlation dimension, Lyapunov exponents and synchronization from depth-EEG signals. Computer methods and programs in biomedicine, 2014, 113(1): 323-337.
8. Sadullaev A, Rakhimov K. Capacity dimension of the brjuno set. Indiana University Mathematics Journal, 2015, 64(6): 1829-1834.
9. Yeh J R, Shieh J S, Huang N E. Complementary ensemble empirical mode decomposition: A novel noise enhanced data analysis method. Advances in Adaptive Data Analysis, 2010, 2(02): 135-156.
10. Wang J. Local tangent space alignment. Geometric structure of high-dimensional data and dimensionality reduction. Springer Berlin Heidelberg, 2012: 221-234.
11. Su Z, Tang B, Deng L, et al. Fault diagnosis method using supervised extended local tangent space alignment for dimension reduction. Measurement, 2015, 62: 1-14.
12. Yongbin L, Lin Z, Qingbo H, et al. A fault feature extraction method for machine health diagnosis using manifold learning. Digital Manufacturing and Automation (ICDMA), 2012 Third International Conference on. IEEE, 2012: 46-49.

A maintenance system based on TBE monitoring for guaranteeing the manufacturing accuracy

Shengyu Shi[1, 2], Jing Lin[3, *], Ming Zhao[2], Xiufeng Wang[2]

[1] Laboratory of Operation and Application in Electromagnetic Domain
[2] Shaanxi Key Laboratory of Mechanical Product Quality Assurance and Diagnostics, School of Mechanical Engineering, Xi'an Jiaotong University, Xi'an, 710049, P.R. China
[3] State Key Laboratory for Manufacturing Systems Engineering, Xi'an Jiaotong University, Xi'an, 710049, P.R. China
[*] Corresponding author. Tel.: 86-29-83395183; Fax: 86-29-83395183; Email: jinglin@mail.xjtu.edu.cn(Lin), shi.shengyu@stu.xjtu.edu.cn(Shi)

Abstract. Backlash is one of the most commonly encountered nonlinear phenomena in feed drives, which degrades the manufacturing accuracy of machine tools. Application of the full-closed loops could significantly suppress the adverse effects of backlash. However, a transient backlash error (TBE) remains at the reversal points. This paper focused on developing a maintenance system based on TBE monitoring, which can help ensure the manufacturing accuracy as well as keep the life cycle cost minimal by eliminating the unnecessary maintenances. To achieve this aim, foremost, the backlash width was measured based on the signals from the built-in encoders. Then, the amplitude of TBE was estimated referring to the previously derived equation [3], which reveals the relationship of TBE with the backlash width and the machining parameters. The warning limit of TBE was specified according to the permissible machining error. Hence, the maintenance system was constructed.

Keywords: Transient backlash error (TBE), CNC machine tools, monitoring, maintenance, manufacturing error

1 Introduction

Backlash is a common but intractable problem in CNC machine tools, which degrades the machining accuracy severely. In semi-closed-loop CNC machine tools, the backlash results in ever-present positioning error that equals half of the backlash width [1-2]. In full-closed-loop CNC machine tools with higher precision, backlash brings in the transient backlash errors (TBE) at directional reversals, which is one of the most predominant manufacturing errors in these machines [3].

To ensure the quality of the manufactured components, TBE must be kept within the permissible machining error. There are mainly two ways to achieve this aim: compensation and maintenance.

In [4], Tsang studied the condition-based maintenance (CBM) system in detail. According to his opinion, a functional CBM system should own the following abilities:

(1) Sampling and recording the wanted signatures;
(2) Quantifying how serious the problem is based on the measured signatures;
(3) Identifying the specified components that are the root causes of the problem;
(4) Setting a reasonable warning limit (the trigger);
(5) Specify what maintenance action to take – how to restore the system to the desired state.

In our paper, a CBM system will be developed according to Trang's requirements to realize the goal that the TBE is always less than the permissible machining error. In detail, the built-in encoders will be employed to collect the position signals of the slide table and the servomotor; based on these signals, backlash width will be calculated. In [3], Shi et al. investigated the mechanism of the TBE, and meanwhile, proposed and validated the analytical equations that reveal the relationship of TBE with the backlash width and the machining parameters. Therefore, the TBE could be quantified once the backlash width and the machining parameters are known. The warning limit of TBE could be specified referring to the permissible machining error. Hence, the embryo idea of the maintenance system based on TBE monitoring is coming out. The rest of this paper will give a detailed introduction of this maintenance system.

2 Measuring the backlash width by using the built-in encoders

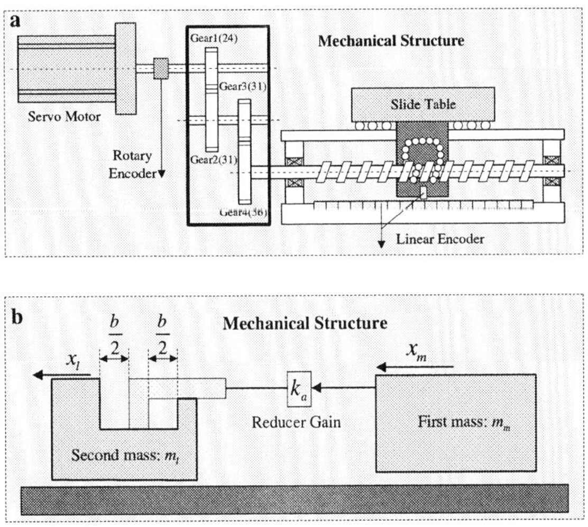

Fig. 1. Mechanical structure of the feed drives: (a) the physical diagram and (b) the simplified schematic diagram.

The mechanical structure of the feed drive under investigation is shown in Fig. 1(a). It can be seen that the feed drive is powered by a rotary servomotor via gear reducer and screw-nut assembly, which are the main sources of backlash. From the physical structure, a two-mass system combining with backlash element and a reducer gain could be derived, as shown in Fig. 1(b), where the first mass m_m is equivalent to the inertia of the rotating motor, the second mass m_l equals the mass of the slide table, and the transmission train between the motor and the table has been simplified as a backlash element and a reducer gain k_a.

In general, the model displayed in Fig. 1(b) could be further simplified (see Fig. 2(a)) and, thus, is termed the friction-driven hysteresis model [5-7]. The dynamic characteristics of this model are shown in Fig. 2(b). Here, b indicates the size of the total gap and is termed "backlash width".

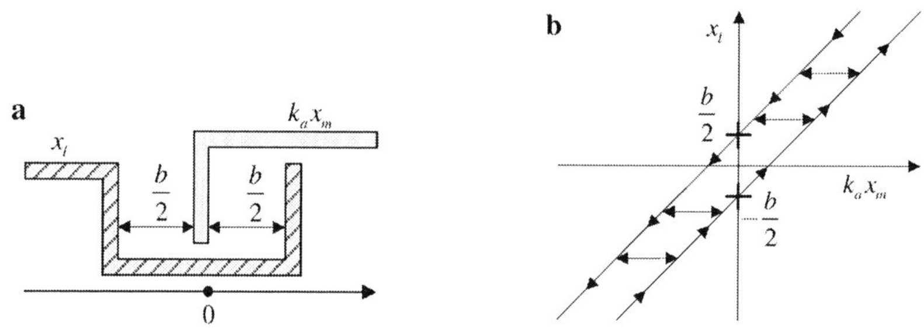

Fig. 2. Friction-driven hysteresis model: (a) the structure schematic diagram, and (b) the dynamic characteristics.

Referring to Fig. 2(b), the relationship between the table position x_l and the motor position x_m could be derived:

$$\begin{cases} x_{l+} = k_a x_m - b/2 & \text{feed in positive direction} \\ x_{l-} = k_a x_m + b/2 & \text{feed in negative direction} \end{cases} \tag{1}$$

Thus, the backlash width b can be easily measured by the following three steps.

First, program the axis to take a round trip along the measurement travel and record the motor position x_m and the table position x_l simultaneously from the built-in encoders.

Second, display the measured x_m vs. $k_a x_l$. The positions measured in *Y-axis* are shown in Fig.3, from which it can be found that two approximately parallel lines are formed. The drawing of partial enlargement exhibits that when the motor reached a certain position x_m respectively from the positive direction or the negative direction, the table positions were different: x_{l+}on the upward side and x_{l-} on the downward side.

Finally, the backlash width could be indicated by the difference between x_{l+} and x_{l-}.

The backlash widths of *Y*- and *Z-axis*, measured in May 2013, were exhibited in Fig. 4. It can be found that the backlash widths are non-negative position-dependent variables, and thereby, denoted as $b(x_m)$.

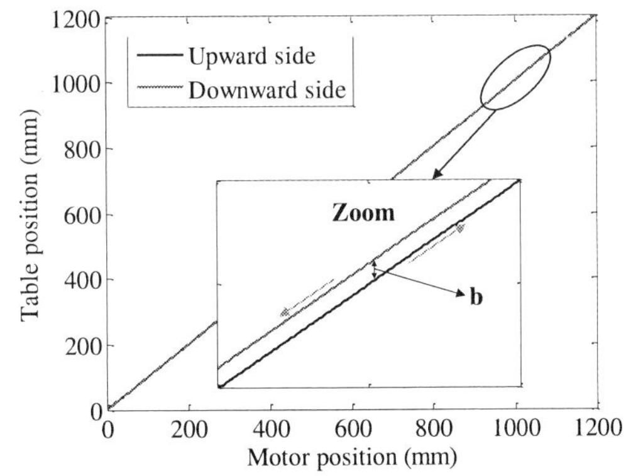

Fig. 3. The table position vs. the motor position measured in a round trip in *Y-axis*.

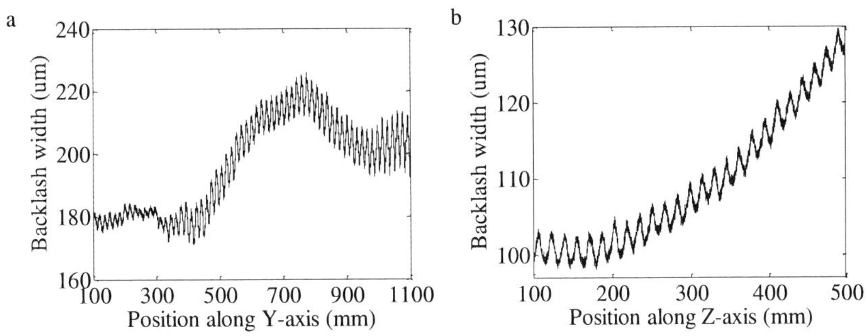

Fig. 4. The backlash widths in (a) *Y-axis* and (b) *Z-axis*, in May 2013.

3 Maintenance based on TBE monitoring

In general, once the manufacturing error could be quantified and monitored, the maintenance action could also be properly triggered to guarantee the quality as well as to keep the life cycle cost minimal by eliminating the unnecessary maintenances. In this maintenance system, the maintenance action is triggered when the estimated TBE exceeds the preset warning limit.

The precision of closed-loop CNC machine tools has been greatly improved, but TBEs still exist transiently at reversals [3]. In [3], the analytical equations are derived to reveal the relationship of TBE ε_b with the backlash width b and the machining parameters. Hence, ε_b could be indirectly quantified and monitored by frequent inspection of the backlash width b.

According to the equations, it can be found that when different profiles are machined or different parameters are employed, different TBEs would be estimated. Hence, certain machining parameters must be specified for estimation of ε_b. Here, the certain parameters are assumed to be sculpturing of a circular profile with radius

200mm, at the feed rate of 200mm/min. And the open-loop position gain is assumed 10. Under such assumption, the relationship between ε_b and b is derived and shown in Fig.5. Thus, once the backlash width b is obtained, ε_b could be uniquely obtained.

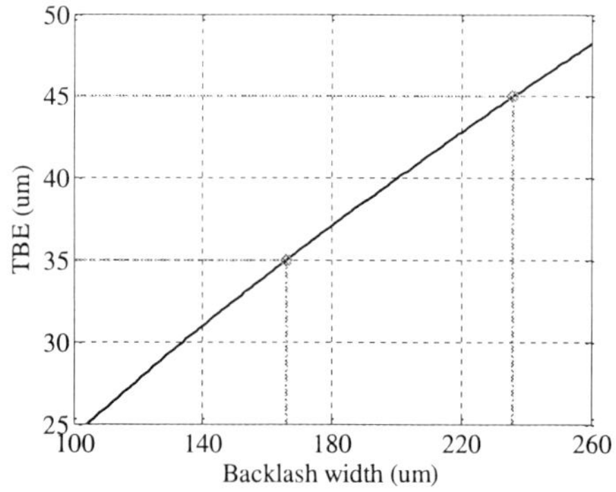

Fig. 5. The backlash width b vs. TBE ε_b encountered in circle machining with 200mm radius at 200mm/min.

The warning limit of manufacturing error could be achieved according to the permissible manufacturing error. Assuming that the permissible machining error is specified as that less than 35μm is tolerable and larger than 45μm is intolerable, the warning limit of TBE could be set: $\varepsilon_b < 35\mu m$ is permissible and $\varepsilon_b > 45\mu m$ is disallowed. Accordingly, it can be derived that the backlash width $b < 166\mu m$ is permissible and $b > 236\mu m$ is not allowed (as seen in Fig.5).

Hence, the maintenance decision could be made referring to Table 1.

Table 1. The maintenance decision

$\varepsilon_{b,r=200mm,v=200mm/min}$	backlash width b	maintenance decision
$\varepsilon_b < 35\mu m$	$b < 166\mu m$	periodic measurement
$35\mu m < \varepsilon_b < 45\mu m$	$166\mu m < b < 236\mu m$	frequent measurements or maintenance
$\varepsilon_b > 45\mu m$	$b > 236\mu m$	maintenance

In Fig.4 (a), it can be observed that in *Z-axis*, the backlash width doesn't exceed 166μm. Therefore, *Z-axis* didn't need maintenance. In *Y-axis*, the maximum of the backlash width is 226.1μm, larger than 166μm and smaller than 236μm. Therefore, *Y-axis* should be more frequently measured or maintained.

Hence, the maintenance system is developed based on monitoring of the TBE. With the aid of this system, the maintenance action could be properly triggered, neither too late nor too early, thus, ensuring the manufacturing accuracy of the machine tools, eliminating the unnecessary maintenance actions, and reducing scrap rates.

4 Conclusions

In this paper, a maintenance system based on quantitative monitoring of TBE was proposed, which is particularly good at ensuring the manufacturing accuracy. In detail, the system has the following advantages.

(1) The necessary position signals are collected by employing the built-in encoders. Therefore, external sensor as well as the associate cost is no longer required.

(2) Equations, which have been theoretically derived and experimentally validated in [3], were employed to map the measured backlash width onto the TBE. This implies that the manufacturing error, TBE, could be quantified precisely.

(3) The warning limit was reasonably specified in accordance with the permissible machining error.

(4) The transient backlash error (TBE), one of the most predominant machining errors in high-precision machine tools, is monitored. Hence, the manufacturing accuracy could be ensured and scrap rates could be reduced.

Acknowledgments

This research is financially supported by the National Science and Technology Major Project (2014ZX04001051), which are highly appreciated by the authors.

Reference

1. Kao, J.Y., Yeh, Z.M., Tarng, Y.S., Lin, Y.S.: A study of backlash on the motion accuracy of CNC lathes. Int. J. Mach. Tools Manuf. 36(5), 539--550 (1996)
2. Tarng, Y.S., Kao, J.Y., Lin, Y.S.: Identification of and compensation for backlash on the contouring accuracy of CNC machining centres. Int. J. Adv. Manuf. Technol. 13(2), 77--85 (1997)
3. Shi, S., Lin, J., Wang, X., Xu, X.: Analysis of the transient backlash error in CNC machine tools with closed loops. Int. J. Mach. Tools Manuf. 93, 49--60 (2015)
4. Tsang, A.H.C.: Condition-based maintenance: tools and decision making. J. Qual. Maint. Eng. 1(3), 3--17 (1995)
5. Nordin, M., Gutman, P.O.: Controlling mechanical systems with backlash—a survey. Automatica 38(10), 1633--1649 (2002)
6. Woo, K.T., Wang, L.X., Lewis, F.L., Li, Z.X.: A Fuzzy system compensator for backlash. In: 1998 IEEE International Conference on Robotics and Automation, pp.181--186. IEEE Press, Leuven (1998)
7. Barreiro, A., Baños, A.: Input–output stability of systems with backlash. Automatica 42(6), 1017--1024 (2006)

Research on Developing Blade Tip Timing System and Its Application

B Yang[1], C Shi[2], H Liu[3], L Yue[4]

College of Mechanical and Electrical Engineering, Nanjing University of Aeronautics and Astronautics, Nanjing 210016, China, yuelinme@nuaa.edu.cn

ABSTACT

Detection the vibration of blades is of great important to the compressor structural health. The paper focus on the research of rotating blade vibration non-contact measurement system based on blade tip-timing method. Thinking about the accurate and real time data acquisition of rotation speed and arrival time of blades, the core part of the system were the design of the data acquisition system using sensor technique, data acquisition (DAQ) and signal processing. The system functions were certified by 6 blades test rig. The blade vibration displacement related to the sweep rotation speed can be found. The blade vibration frequency and corresponding order were identified by means of blade vibration identification software. The result was consistent with the finite element method analyses result. With the seeded fault by changed thickness of blades then the transformation of the blade stiffness during the operating time can be discovered.

Keywords: Blade tip-timing, LabVIEW, vibration, applications, fault, system

I. Introduction

As turbomachinery is integral to production processes in many industries, the reliable operation of this machinery become critical to these processes. In one case, a low-pressure turbine blade failure not only led to the catastrophic failure of a turbo-generator set, but also a production time loss of about one year. It has been stated that about 30% of all turbine-forced outages are blade related[1] [2]. Hence, the ability to monitoring blade vibration is of the utmost importance.

High cycle fatigue occurring at low stress and high vibration frequency is a common reason for blade damage in turbomachinery. Contacting vibration measurement techniques, such as strain gauges[3] are well established, but they are costly to install, they have a limited operating life, they cannot monitor all blades and influence the properties of blade. Another technique is to measure blade vibration in a non-contact way. The blade tip timing technology consists in use of sensors which are mounted in the rotor casing. They measure the times of arrival of each blade. Then these timings are used to estimate the vibratory motions

Nowadays non-interference stress measurement system (NSMS)[4] has been widely used in foreign countries, for example NSMS by Hood Technology Corporation. However we cannot get the raw data and the system is very expensive. So this chapter describes design and development of blade tip-timing system (BTTS) based on LabVIEW.

II. Blade TIP-TIMING MEASURING SYSTEM

A. Introduction To Blade Tip-timing Method

Blade tip-timing technique is a non-contact measurement which uses several sensors to detect the blade passing signals. Sensors are mounted on rotor casing which detect the signal when a blade passes in front of them(Figure 1). They can be optical, capacitive or eddy-current sensor. Another sensor is mounted in front of the shaft and takes only one pulse per revolution. The data collected by this sensor are then used to estimate the rotor speed.

The time of arrival of one blade in per revolution could be predicated when blades do not vibrate. Once blade vibrates, it will affect the arrival times of blade. Then, vibration amplitudes of blades are estimated by subtracting measured times to theoretical ones.

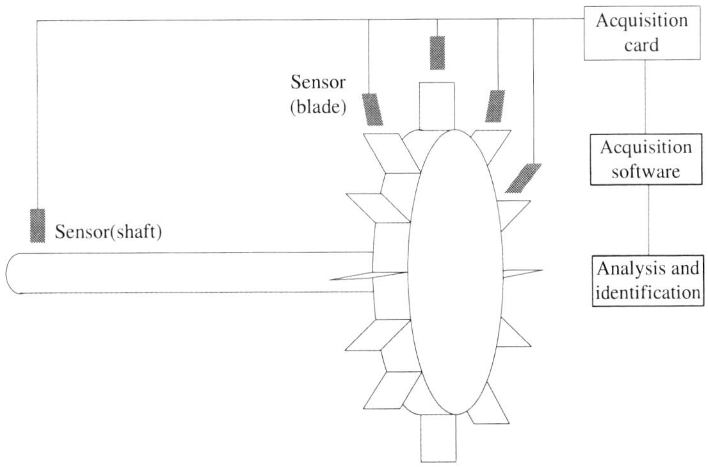

Figure 1 BLADE TIP-TIMING measuring system

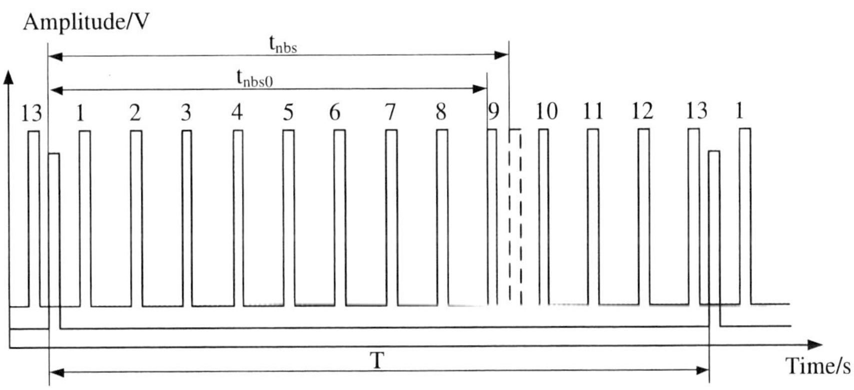

Figure 2 Pulse sequence

If there were no vibration, we could exactly determine the time of arrival of each blade in front of each sensor, such as t_{nbs}. n is the current cycle. b is the blade number. s is the sensor number .

But blades vibrate, and their vibratory displacements change these times, such as t_{nbs0} (figure 2). T is the meaning of speed pulse interval. When we know the radius R which is the distance between the blade tip and the center of shaft, we can get the displacement of vibration:

$$y_b = \frac{2\pi R}{T}(t_{nbs} - t_{nbs0}) \tag{1}$$

'Synchronous' or 'engine-ordered' blade vibration occurs when a blade vibration frequency is an integer multiple of the engine rotor speed. 'Asynchronous' blade vibration occurs when a blade vibration frequency is a non-integer multiple of the rotor speed. This chapter focuses on synchronous blade vibration.

B. BTT Experimental Test

The BTT rig was a fan test rig consisting of a motor, a tunnel, a rotor with 13 blades. The maximum rotation speed is 1500 revolutions-per-minute (RPM).Several blade modes are included in this period. Four eddy current sensors (Figure 3) were mounted in the rotor casing, as shown in (Figure 4) and can measure the time of arrival when every blade passes by which is related to once-per-revolution (OPR) sensor was mounted on the rotor shaft. A frequency converter was used to change the rotation speed.

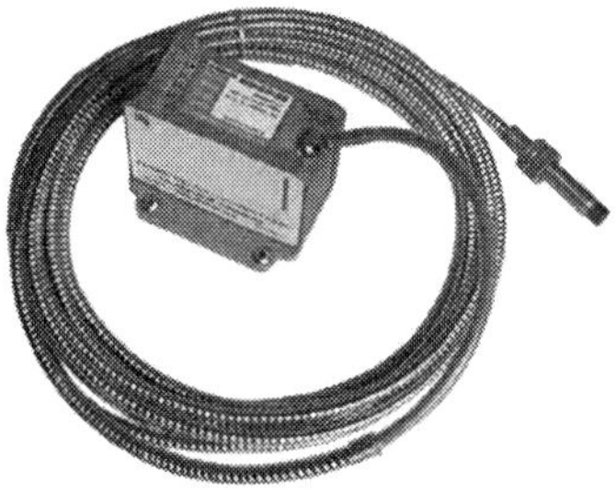

Figure 3 Eddy current sensor

Figure 4 The installation of the eddy current sensor

C. Acquisition Card

The signal of probes is passed into the data acquisition device which we adopt NI PCI-4472. The performance of PCI-4472 is as follows: 24-bit resolution ADCs with 110 dB dynamic range; 8 simultaneously sampled analog inputs at up to 102.4 kS/s; ±10 V input range or ±31 V with SMB-120 cable.

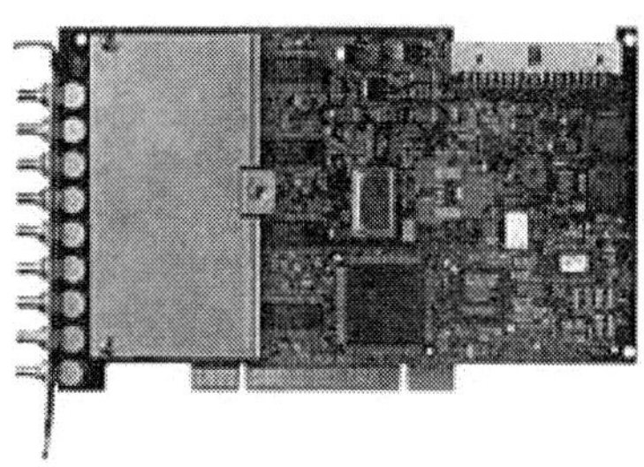

Figure 5 PCI-4472

D. Acquisition Software

Acquisition software was finished based on LabVIEW. The software structure of the system is shown as in Figure 6. The main function includes data acquisition system and BTT data postprocessing system. The acquisition module controls mainly the settings which can execute particular measurement through acquisition card. The BTT data postprocessing system analysis the data from DAQ and transform it into displacements of blades which will be discussed later. At the same time, the vibration displacements of blade can be shown.

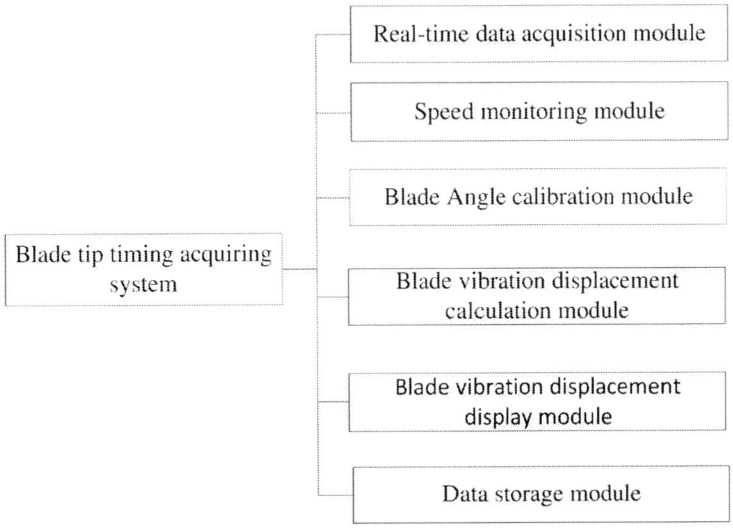

Figure 6 System modules

III. PROCESSING OF DATA COLLECTED FROM SENSORS

Since data acquisition system has been widely used in worldwide and became more and more efficient and stable, the DAQ are not discussed in this chapter. BTT data processing system are given as follows.

A. Measurements Of Pulse Peaks

The most important part of non-contact measurements of blade vibrations is the detect the arrival time of the blade when subsequent blades arrive at the measuring point of the sensor. For inductive sensors the presence of the blade arrival at the measuring point is associated with falling edge of the pulse via the voltage of 0V. But in order to achieve the experiment efficiently, we regard the measuring point as the pulse peak at the rising edge of the pulse. The steps of localization of peak point is as follows

1) Assure the angles of blades in once per-revolution.

2) Determine the intervals where the subsequent blade peaks will occur.

3)Detect the time of arrival of blades in time domain.

B. Calibration

The time of arrival in per revolution keeps stable when blades do not vibrate. So the blade vibration amplitudes are estimated by subtracting measured times to theoretical ones. The calibration of blades is very important to estimate the arrival time. As the rotation speed is higher and higher, we found that blades can lead to some deviations due to the force of inertia when the blades do not vibrate. In order to make the results more confident, we employ some new methods in the process of calibration which is ahead of the experiment. The tradition method calculates the angles in once per-revolution in low speed without considering the force of inertia. In the consideration of the deviation brought by inertia, we measure the angles of blades in 200rpm、

400rpm、600rpm、800rpm and 1000rpm where the blades do not vibrate according to the Campbell diagram. Then linear fit is done at coordinate graphs of angle and rotation speed so as to provide objective basement of angle for various speed. Vibration amplitudes are estimated by subtracting measured times to theoretical ones deserved by the linear fit.

IV. TEST

A. *Experiment Case*

Tests were made on our bladed disc. 4 sensors were disposed around the disc at locations: 7°, 22°, 37° and 55°. The rotate speed grows to 1500rpm and then stop. Synchronous vibration appear at N = 628(Figure 7), N=725(Figure 8), N=758 (Figure 9) during the reduction of speed. Red line represents the change of rotor speed, and other different color curves represent a blade vibration displacement measured by four different sensors. The whole test process is 53s. We adopted three identification methods: Two parameter analysis[5], Autoregressive analysis[6] and sine fitting analysis[7]. Then we get vibration engine order and vibration frequency under three different rotating speed (Table.1). The results of three methods are consistent and can be consistent with the simulation prediction results which come from Campbell diagram.

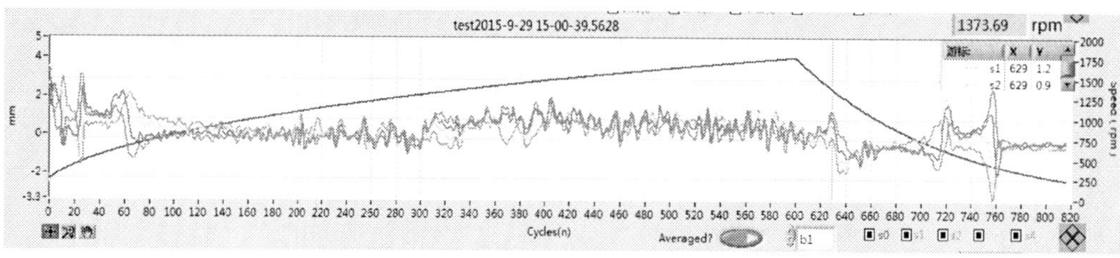

Figure 7 Blade vibration displacement curve, N=628, Speed=1373rpm

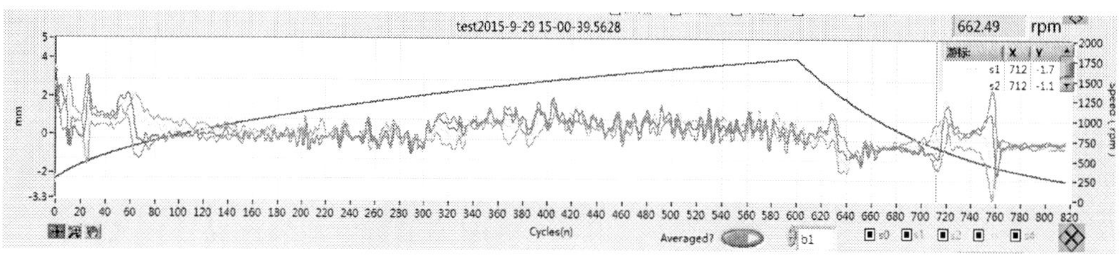

Figure 8 Blade vibration displacement curve, N=725, Speed=662rpm

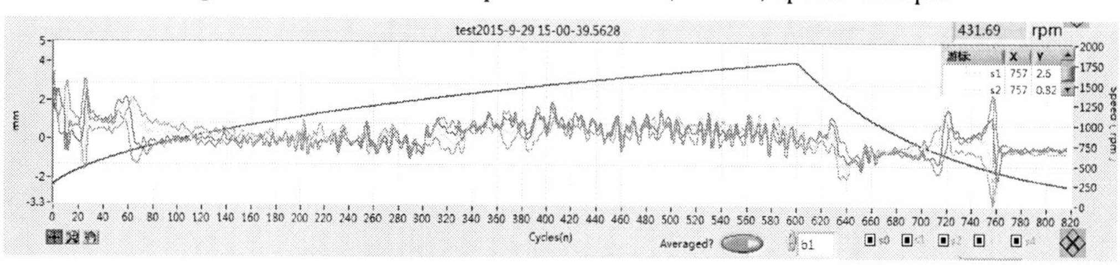

Figure 9 Blade vibration displacement curve, N=758, Speed =431rpm

Table 1 The vibration parameter results under three different rotating speed

Analysis methods	Speed=1373rpm		Speed=662rpm		Speed=431rpm	
	order	Frequency	order	frequency	order	frequency
Two parameter	2.03	45.16Hz	3.05	32.35Hz	3.9	28.92Hz
Autoregressive	1.91	40.41Hz	2.92	30.55Hz	3.79	26.9Hz
sine fitting	2	44.27Hz	3	31.3Hz	4	28.46Hz
Campbell diagram	2	55Hz	3	37Hz	4	32Hz

B. Conclusion

In this paper we build blade rotor vibration test system according to blade tip timing test technique to detection the vibration of blades. The system mainly includes two aspects of hardware and software. The hardware includes four eddy current sensors and acquisition card. In software we finished the acquisition of blade passing signals and calculated vibration displacement. Finally, we get the second, third and fourth order synchronous vibration of blades.

REFERENCES

[1] Luo J B, Tan S S, Yuan L P. The accident cause analysis of large turbine blade[J]. Electric power security technology, 2002, 4（8）:11-12.

[2] Carter T J. Common failures in gas turbine blades[J]. Engineering Failure Analysis, 2005, 12(2): 237-247.

[3] Heath S, Imregun M. A survey of blade tip-timing measurement techniques for turbomachinery vibration[J]. Journal of Engineering for Gas Turbines and Power, 1998, 120(4): 784-791.

[4] Fang Z Q. The technology research of the principle and application of turbine blade vibration non-contact detection [D]. Tianjin University, 2007.

[5] Heath S. A new technique for identifying synchronous resonance using tip timing[J]. Journal of Engineering for Gas Turbine and Power, 2000, 122(10): 219~225.

[6] Carrington I B, Wright J R., Cooper J E, et al. A comparison of blade tip timing data analysis methods[J] .Journal of Aerospace Engineering, 2001, 215(5):301~312.

[7] Joung K K, Kang S C, Paeng K S, et al. Analysis of vibration of the turbine blades using non-intrusive stress measurement system[C]//ASME 2006 Power Conference. American Society of Mechanical Engineers, 2006: 391-397.

An Improved Detection Method for Induction Motors with Stator Inter-turn Short Circuit

Yucheng Zou, Ling Chen, Niaoqing Hu, Yitian Wang and Ming Gao

Laboratory of Science and Technology on Integrated Logistics Support,
National University of Defense Technology, Changsha, China.
E-mail:1575609178@qq.com

Abstract. As affected by many factors such as the unbalanced supply power and the asymmetric structure, Park's Vector Approach is usually subject to certain limitations in the practical application. As the relationship between the long axis and the short axis of the ellipse was utilized, an improved method based on the Park transform and Hilbert transform was put forward in this paper. The finite element simulation and experiment results show that the faults were easier to detect.

Keywords: Induction motor; stator; Park transform; Hilbert transform

1 Introduction

Stator winding inter-turn short-circuit fault is one of the common faults of induction motors, which accounts for about 30%-40% [1]. Due to heavy localized heating, insulation protection layer will soon be ineffective [2]. To some extent, it may resulting in fire, casualties and economic losses [3]. Therefore, it seems particularly important to detect the inter-turn short-circuit faults at an early stage.

The Park's Vector Approach and extended Park's Vector Approach was introduced to detect the inter-turn stator winding fault [4-5]. In [6], a new method combining between the Hilbert transform and the Extended Park's Vector Approach was proposed to get the fault frequency. The Park's Vector Approach was used to diagnose the faults by transforming the signal into an approximate circle. But as a matter of fact, the unbalanced power supply and asymmetric structure could also cause the change of the figure.

Moreover, the figure was difficult to identify and the degree of the fault was hard to judge. In order to solve this problem, an improved method was proposed in this paper. First of all, Park transform and Hilbert transform were applied respectively. And then, in order to describe the degree of the fault, a symbol K, represented as the ratio of long axis and short axis, was defined. Finally, the aim of stator fault diagnose was achieved by using Park transform repeatedly to detect the faults on different phases. Simulation and experiment are carried out to verify the validity of the method.

2 METHODOLOGY

Equation (1) is the Park transformation. θ is set to be zero at first. The three-phase currents of an ideal motor are transformed from *abc* coordinate system to *dq0* coordinate system.

$$\begin{bmatrix} I_d \\ I_q \\ I_0 \end{bmatrix} = \frac{2}{3}\begin{bmatrix} \cos\theta & \cos(\theta-\frac{2\pi}{3}) & \cos(\theta+\frac{2\pi}{3}) \\ -\sin\theta & -\sin(\theta-\frac{2\pi}{3}) & -\sin(\theta+\frac{2\pi}{3}) \\ \frac{1}{2} & \frac{1}{2} & \frac{1}{2} \end{bmatrix}\begin{bmatrix} i_a \\ i_b \\ i_c \end{bmatrix} \quad (1)$$

And for an ideal motor, *Id* and *Iq* can be expressed as equation (2).

$$\begin{cases} I_d = \frac{\sqrt{6}}{2} i_M \sin(wt) \\ I_q = \frac{\sqrt{6}}{2} i_M \sin(wt-\frac{\pi}{2}) \end{cases} \quad (2)$$

Where i_M is the maximum value of the supply phase current (A) [4], w is the angular supply frequency (rad/s), t is the time (t).

$$\begin{cases} I_d = a_{10} + \sum_{i=1}^{n}(a_{1i}\cos w_{1i}t + b_{1i}\sin w_{1i}t) \\ I_q = a_{20} + \sum_{i=1}^{n}(a_{2i}\sin w_{2i}t + b_{2i}\cos w_{2i}t) \end{cases} \quad (3)$$

For actual motors or motors with stator inter-turn short-circuit faults, equation (2) would be expressed as equation (3). Apparently, equation (2) is the particular form of equation (3). For an ideal motor, a circle can be gotten with I_d and I_q. But for actual motors, as shown in Fig. 1, the circle can be changed and it is difficult to judge whether the faults exist.

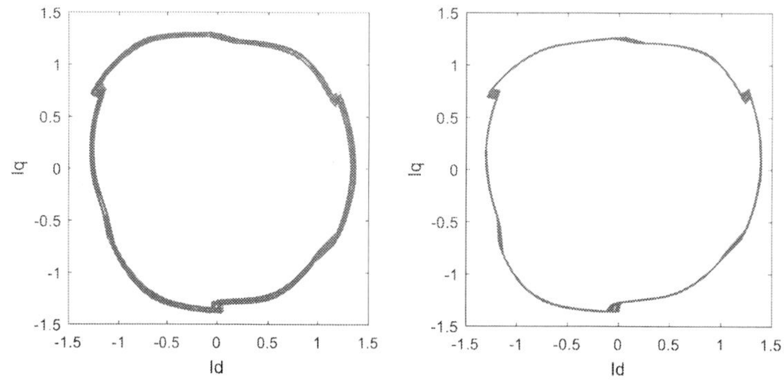

Fig. 1. The *id* and *iq* of a healthy experimental motor and a faulty experimental motor. (15*Hz*)

For ideal motors or normal motors, amplitude of each frequency should be almost the same. But as far as the inter-turn short-circuit faults occurred, coefficients of I_d and I_q will be no longer equal to each other. As a result, it might be a fault feature to determine the occurrence of inter-turn short circuit.

Using the Hilbert transform, equation (4) and equation (5) could be obtained:

$$\begin{cases} |HI_d| = \sqrt{I_d^2 + \hat{I}_d^2} = \sqrt{\sum_{i=0}^{n}(a_{1i}^2 + b_{1i}^2) + e_1} \\ |HI_q| = \sqrt{I_q^2 + \hat{I}_q^2} = \sqrt{\sum_{i=0}^{n}(a_{2i}^2 + b_{2i}^2) + e_2} \end{cases} \tag{4}$$

Where e_1, e_2 are functions containing sine/cosine.

$$\begin{cases} h_{id} = \sqrt{\sum_{i=0}^{n}(a_{1i}^2 + b_{1i}^2)} \\ h_{iq} = \sqrt{\sum_{i=0}^{n}(a_{2i}^2 + b_{2i}^2)} \end{cases} \tag{5}$$

In order to describe the degree of the fault, a symbol K, represented as the ratio of long axis to the short axis, was defined as equation (6). In order to make the effect better, θ should be set to $2\pi/3$ and $4\pi/3$ respectively and compute again.

$$K = \frac{h_{id}}{h_{iq}} - 1 \tag{6}$$

For ideal motors, K should be zero. For actual motors with no faults, K might be larger because of the slight unbalanced power supply and the asymmetric structure. For actual motors with inter-turn short circuit faults, K should be much larger.

3 Simulation Verification

In this paper, the FEM (Finite Element Method) model of Y160M-4 under normal condition was established firstly [7]. Each step was 0.001s and about 1500 points were acquired. Part of the parameters were changed in order to simulate the motor with stator winding inter-turn short-circuit faults or the supply voltage unbalance. The proposed method was verified with the simulation data and the results were listed in Fig. 2. 'o' represents 'θ= 0'. '*' represents 'θ = 2π/3'. '□' represents 'θ= 4π/3'.

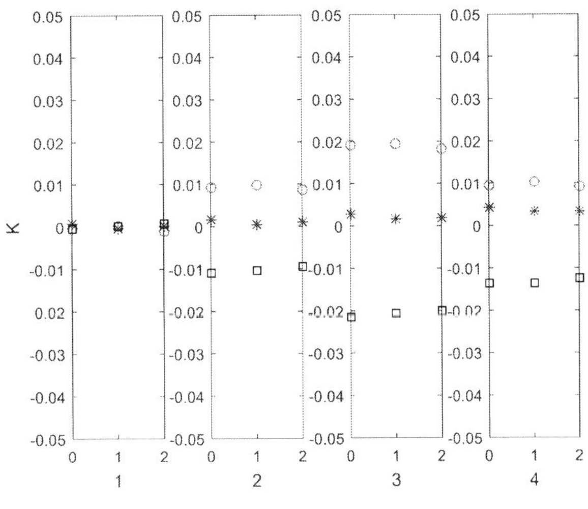

Fig. 2. Simulation results under different working conditions (1, 2, 3 and 4)

According to the Fig.2, when the simulation motor was normal, the K were almost zero. When the A-Phase was faulty (one turn), the K were from about -0.01 to 0.01. When the A-Phase was faulty (two turns), the K were from about -0.02 to 0.02. When the voltage of phase A is 101% times of the other two phases, the K were from about -0.013 to 0.01. Obviously, the K were larger with the number of faulty turn increasing, but the method was also affected by the supply power.

4 Validation by experimental data

An experiment was carried out on an experimental platform to validate the validity of the improved method. A healthy induction motor and an induction motor with fault in stator were detected sequentially. The motor power supply was obtained by the frequency converter. The motors were connected to the load through the coupling. The sampling frequency was 1000 *Hz* for each motor.

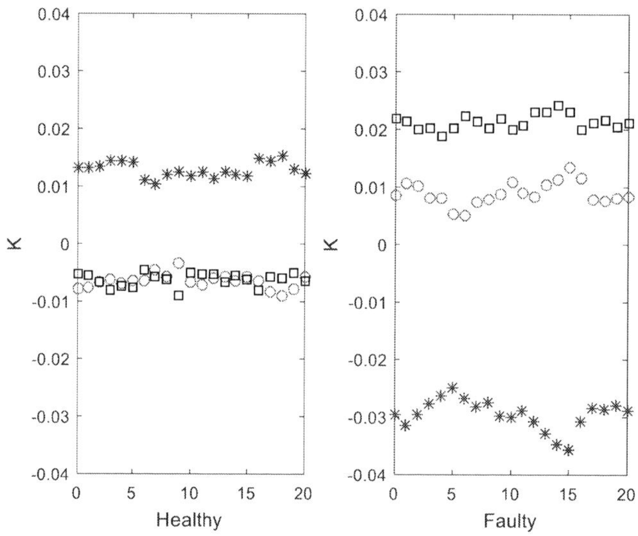

Fig. 3. The results of a healthy and a faulty motor. (15Hz)

About 40960 points were calculated and the data were divided into 20 sections. In practical computation, in order to avoid errors in both ends of the data, only points from 256 to 1792 of each section were calculated to obtain the h_{id} and h_{iq}. From the Fig.3, affected by the slight unbalanced power supply and the asymmetric structure, the *K* of the healthy motor ranged from -0.01 to -0.01 approximately. Comparatively speaking, *K* of the faulty motor ranged from -0.03 to -0.02 approximately and the fault might occur because the sequence changed.

5 Conclusion

In summary, this paper proposed an improved method to detect the stator inter-turn short-circuit of induction motors. And the validity of the method was confirmed by the simulation and the experiment results. Comparing the Park's Vector Approach, the improved method can detect the stator short-circuit fault more effectively. But problem still exists that the method was influenced by the supply power. As a result, more efforts will be taken to study the relationship between supply power and inter-turn short circuit.

6 Acknowledgement

This work is supported by National Science Foundation of China (51475463). Our deepest gratitude goes to the anonymous referees for their helpful comments. My thanks also go to my family for their love and support.

References

1. Nandi, S., Toliyat, H.A.: Novel frequency-domain-based technique to detect stator interturn faults in induction machines using stator-induced voltages after switch-off. IEEE T IND APPL, 38(1), 101-109 (2002)
2. Eftekhari, M., Moallem, M., Sadri, S., Shojaei, A.: Review of induction motor testing and monitoring methods for inter-turn stator winding faults. Electrical Engineering, 1-6. (2013)
3. S Gazzana, D.D., Pereira, L.A., Fernandes, D.: A system for incipient fault detection and fault diagnosis based on MCSA. Transmission and Distribution Conference and Exposition, 2010 IEEE PES, 1-6. (2010)
4. Cardoso, A.J.M., Cruz, S.M.A., Fonseca, D.S.B.: Inter-turn stator winding fault diagnosis in three-phase induction motors, by Park's Vector approach. Electric Machines and Drives Conference Record, . IEEE International, B1-B5. (1997)
5. Cruz, S.M.A., Cardoso, A.J.M.: Stator winding fault diagnosis in three-phase synchronous and asynchronous motors, by the extended Park's vector approach. Industry Applications IEEE Transactions on, 37(5), 1227-1233 (2001)
6. Sahraoui, M., Zouzou, S.E., Ghoggal, A., Guedidi, S.: A New Method To Detect Inter-Turn Short-Circuit In Induction Motors. Electrical Machines (ICEM), 2010 XIX International Conference on, 1-6. (2010)
7. Bo, Z.: The application of Ansoft 12 in engineering electromagnetic field. China Water Power Press. Beijing (2010).(In Chinese)

Compressed Vibration Signal Reconstruction via Sparsity-promotion with Segmentation Threshold Denoising

Yanliang Ke, Huaqing Wang, Lingyang Li and Gang Tang*

College of Mechanical and Electrical Engineering, Beijing University of Chemical Technology, No. 15 in East Road of North Third Ring, Chaoyang District, Beijing, 100029, China
Gang Tang is the corresponding author
E-mail: tanggang@mail.buct.edu.cn

Abstract. Most vibration-based approaches for rotating machinery fault diagnosis are achieved based on the Shannon sampling theorem, through which a large amount of vibration signals will be measured. That leads to a big-data problem, which brings great pressure on data storage and transmission. To deal with the big-data problem, compressed sensing was proposed, the main idea of which is to reconstruct the target signals from a limited amount of measurements. The proposal of the compressed sensing theory brings a novel insight that we can store and transmit small samples on the sending-end, then reconstructing the limited observations with reconstructing algorithm on the receiving-end. However, there is an obstacle that the vibration signals are inadequately sparse. Therefore, a simple and effective method was developed to increase the sparsity of the vibration signals based on the segmentation threshold denoising. When the sparsity of the target signals is promoted, it is much easier to reconstruct the original signals from limited samples using compressed sensing reconstruction methods. Applications to roller bearing vibration signals validate the effectiveness of the proposed method.

Keywords: Rotating machinery, compressed sensing, sparsity, segmentation threshold denoising

1 Introduction

Vibration-based analysis for rotating machinery condition monitoring is always achieved by sampling vibration signals based on the Shannon principle [1]. Under this sampling framework, a large amount of redundant signals will be measured, leading to a big-data problem, which brings great pressure on the data storage and transmission. Thus, it is indispensable to compress the vibration signals while preserving adequate information.

An alternative sampling scheme termed compressed sensing theory has been developed in recent years based on sparse representation of the analyzed signals [1]. With a well-designed measurement strategy, the original signals can be reconstructed from dimension-reduced observations with a desired accuracy [2]. Since its proposal, the compressed sensing theory has been widely used in image processing, remote sensing, et al. Various studies indicate that it is possible to apply the compressed sensing theory to fault diagnosis. Wang et al. investigated several approaches for fault detection of roller bearing based on the compressed sensing [3]. Tang et al. developed a fault classification strategy for rotating machinery faults based on compressed observations [4]. Chen et al. presented a learning dictionary for extracting impulse components under the sparse model of compressed sensing [5]. Although these literatures focused on the fault feature extraction as well as fault pattern recognition based on the compressed sensing, few investigations have been devoted to the reconstruction of faulty vibration signals. Compressed sensing is a promising type of data compression method based on the sparse representation of the analyzed signals. Unfortunately, it is extremely difficult to find a suitable method to achieve the sparse representation of vibration signals. Thus, a sparsity-promoted method based on segmentation threshold denoising is proposed in this study. The original signals can be divided into several segmentations for the reason that a suitable dimension of the constructed measurement matrix might be achieved. Then, the threshold can be determined according to the detected impulse, through which the interference can be eliminated while preserving the valuable information. Assisted by the matching pursuit algorithm, the original signals can be recovered successfully.

The rest of this paper are organized as follows. The proposed method is illustrated in Section 2. Then, the comparative and experiment results are presented in Section 3. Section 4 shows the conclusion and discussion of the proposed method.

2 Compressed vibration signal reconstruction

Compressed sensing has attracted more and more attentions, which beats the Shannon sampling theorem by reconstructing the original signals from a limited amount of measurements. Assuming that $x \in \mathrm{R}^{N\times 1}$ is the original signal and it can be represented by a dictionary $D \in R^{N\times N}$ in a sparse way via $x = D\alpha$ and $\alpha \in R^{N\times 1}$ is the representation coefficients of the signal x. The signal can be considered to be K-sparse, when the representation coefficients α contain only K non-zeros ($K<<N$). Next, a measurement matrix $A \in R^{M\times N}$ is utilized to compress a high-dimensional signal to a low-dimensional one through the operation $y = Ax = AD\alpha$.

The solution can be achieved by solving the sparsity-promoted optimization equation: $\alpha^* = \arg\min \|\alpha\|_0 \quad s.t.\ y = AD\alpha$

Then, the reconstructed signals $x^* = D\alpha^*$ can be obtained.

However, when applying the compressed sensing theory to the reconstruction of vibration signals, there is a major obstacle that it is overwhelmingly difficult to select a suitable transform to represent the vibration signals in a sparse way. Thus, an unsatisfactory reconstruction results will be achieved, as illustrated in Fig. 1(a).

To overcome the problem induced by the undetermined sparsity of vibration signals, a sparsity-enabled approach via the segmentation threshold denoising is developed in this study as presented in Fig. 1(b). The original signals should be divided into several segments according to the location of impact, through which measurement matrix can be achieved with desired dimensions. In each data segments, a threshold can be determined, through which the peaks in each segments can be preserved with other components discarded. Then, the sparsity of the analyzed signals can be largely increased. Since the signals in time-domain are adequately sparse, identity matrix is selected as sparse dictionary. Finally, the matching pursuit algorithm is used to reconstruct the signals.

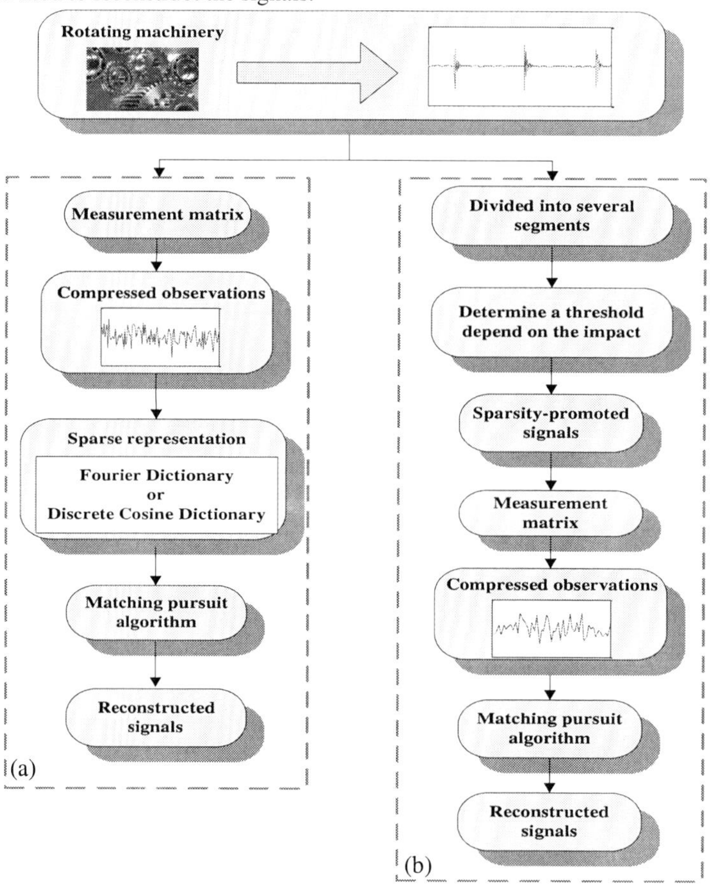

Fig. 1. Two different signal reconstruction scheme based on the compressed sensing: (a) Signal reconstruction based on Fourier or Discrete Cosine Dictionary (b) Signal reconstruction based on the proposed method

3 Experimental results

To validate the effectiveness of the proposed method, vibration signals measured from a roller bearing with a defect in the outer race are employed. The roller bearing is running at a speed of 1300 rpm and the sampling frequency is 100 kHz. The normalized vibration signal with outer-race defect is shown in Fig. 2. The Fourier coefficients of vibration signal acquired by FFT are presented in Fig. 3(a), which indicate that the vibration signals cannot be represented sparsely via FFT. The reconstructed signal by orthogonal matching pursuit in Fig. 3(b) shows that it cannot acquire desired results using the compressed sensing based on the Fourier basis.

To increase the sparsity of the vibration signals and preserve adequate information, the proposed sparsity-promoted strategy is employed. The original signals are divided into several segments according to the location of peaks, as shown in Fig. 4(a1)-(a3). Then, a threshold can be determined to increase the sparsity of each segment and the results in Fig. 4(b1)-(b3) demonstrate that the sparsity of each segment is promoted largely compared to the signals in Fig. 4(a1)-(a3). Then, the Gaussian random matrix is used to compress the signal and the compressed signal is acquired in Fig. 5. With the promotion of vibration signal sparsity, the reconstructed result can be achieved, as presented in Fig. 6(a). To evaluate the performance of the proposed method, the mean square error is used. The mean square error of the reconstruction signal in Fig. 3(b) is 2.346, while the error of the signal in Fig. 6 (a) is 1.032, which indicates that the proposed reconstructing strategy outperforms the FFT-based compressed sensing. Then, the conclusion can be draw that there is a fault form the frequency spectrum of the reconstructed signal in Fig. 6(b).

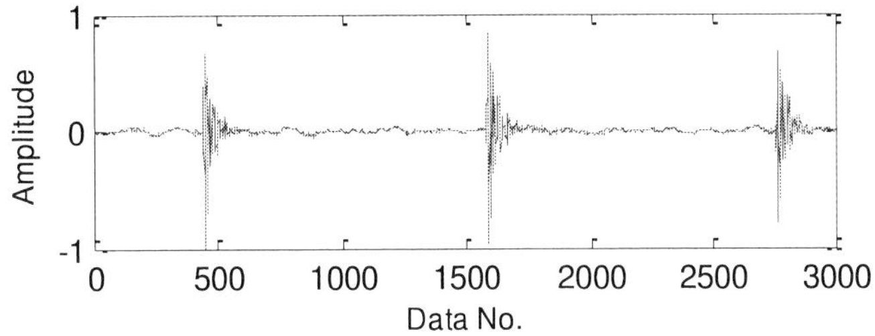

Fig. 2. The normalized vibration signal with outer-race fault

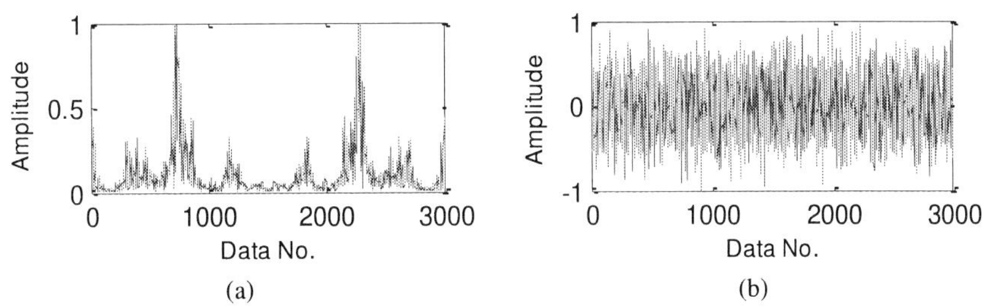

Fig. 3. (a) Fourier coefficients of vibration signal (b) Reconstructed signal based on Fourier basis

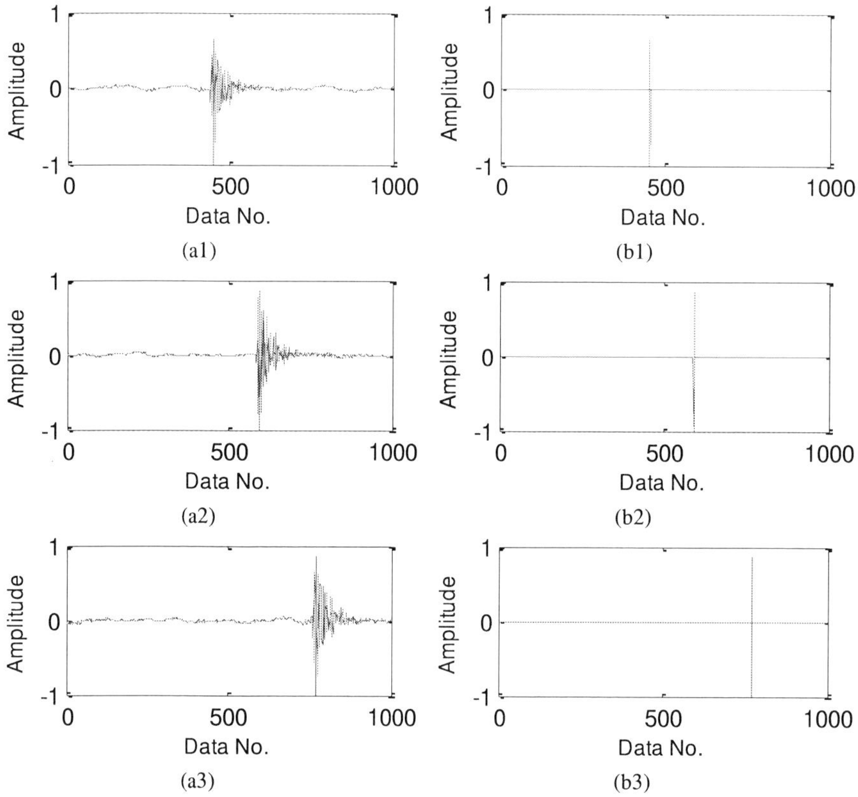

Fig. 4. (a1)-(a3) The divided data segments (b1)-(b3) Reconstructed signal in each segments

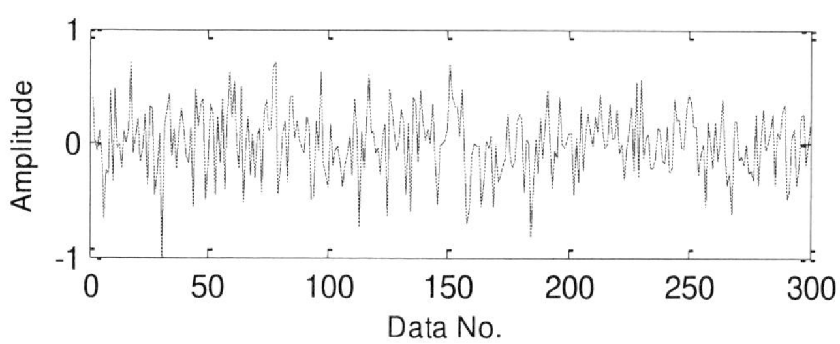

Fig. 5. Compressed signal by Gaussian random matrix

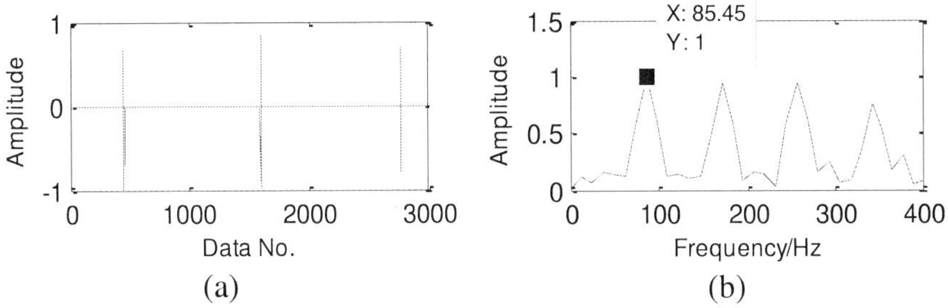

Fig. 6. (a) Reconstructed signal by the proposed method (b) its spectrum

4 Conclusion and discussion

To overcome the difficulty in compressed-sensing-based signal reconstruction strategy generated by the non-sparse property of vibration signals in Fourier or Discrete Cosine dictionary, a sparsity-promoted method based on segmentation threshold denoising is proposed. The comparative and experimental results indicate that the sparsity can be largely promoted by the proposed method, which outperforms the representation based on Fourier or Discrete Cosine dictionary. The reconstruction results show the effectiveness of the proposed method. However, there are still some difficulty should be overcome in the future research, such as how to enhance the impact when the peaks of the signals are buried by the strong interference.

Acknowledgments

This work is partially supported by the National Science Foundation of China (Grant no. 51375037, 51405012) and PetroChina Innovation Foundation (2015D-5006-0606, 2014D-5006-0104).

References

1. X. P. Zhan, N. Q. Hu, L. Hu, L. Chen and Z. Cheng. A bear fault diagnosis method based on the low-dimensional compressed vibration signal. Advances in Mechanical Engineering, 2015, 7(7): 1-12.
2. D. L. Donoho. Compressed sensing. Information Theory, IEEE Transactions on, 2006, 52(4): 1289-1306.
3. H. Q. Wang, Y. L. Ke, G. G. Luo and G. Tang. Compressed sensing of roller bearing fault based on multiple down-sampling strategy [J]. Measurement Science and Technology, 2015, 27(2): 025009.
4. G. Tang, Q. Yang, H. Q. Wang, G. G. Luo and J. W. Ma. Sparse classification of rotating machinery faults based on compressive sensing strategy. Mechatronics, 2015, 31: 60-67.
5. X. F. Chen, Z. H. Du, J. M. Li, X. Li and H. Zhang. Compressed sensing based on dictionary learning for extracting impulse components. Signal processing, 2014, 96: 94-109.

Aero-engine Intermediate Bearing Fault Diagnosis Based on Rough Set Theory and NNs

Bei Wang[1] Fengtao Wang [2] Tao Zhang [3] Mingli Cui [4] Hongkun Li [5]

[1]School of Mechanical Engineering, Dalian University of Technology, Dalian, China, 1286075488@qq.com

[2]School of Mechanical Engineering, Dalian University of Technology, Dalian, China, wangft@dlut.edu.cn

Abstract. Novel fault identification method is proposed based on the Rough Set Theory and improved NNs to identify the aero-engine intermediate bearing fault type under the weakly fault characteristic signal and the complex transmission path. Firstly wavelet packet is used to noise reduction. Secondly the data of the RMS values of the reconstruction signal are calculated and attribute reduction based on Rough Set Theory. Finally the fault classification results will be getting through the improved NNs (neural networks). Study shows that the data processing effect is more obvious after Rough set theory and has obvious advantages in choosing bearing test position in the complex condition of the aero-engine intermediate bearing. The improved NNs have the advantage of small error, fast learning. This method applies not only to the aero-engine intermediate bearing fault classification and diagnosis but also applied to other areas.

Keywords: Aero-engine, intermediate bearing, rough set theory, fault diagnosis, improved neural networks

1 Introduction

Aero-engine is the "heart" of the airplane [1]. The intermediate bearings as the key components of the aircraft-engine, the inner and outer rings rotate at the same time, the outer ring is connected with the high pressure end of the aero-engine, and the inner ring is connected with the low pressure end rotor, which plays the role of intermediary support, the intermediate bearing as a supporting part, the intermediary of inner and outer ring of bearing support with high pressure and low rotation of the rotor, so at the time of the experiment should consider the speed difference and the transmission path. Hu Yaobin, Zhao Yuanxi et al applied neural network to the rolling bearing fault identification [2-3]. Neural networks are a kind of intelligent method which has the advantage of strong nonlinear mapping ability and self-learning, self-organization and self-adaptation. Due to the slow convergence rate of standard neural network algorithm and oscillation often occurs during the training process, the author improves the neural network algorithm by combining with the increase of momentum term and adaptive learning rate. In combination with the actual aero-engine and aero engine test rig, the vibration signal transmission path is complex [4-5], the background noise is more serious, and the SNR is very low. Due to complex transmission paths, more measure points, on the choice of the model input variables, the most by correlation to artificially set the threshold analysis method is selected [6], the method inevitably bring personal subjective factors, cause information redundancy or loss; moreover, the factors that affect the vibration signal is various, if they are included in the model input, will aggravate the burden for the training of the neural network and reduce the prediction accuracy of the model [7]. Firstly, the wavelet packets to signal noise reduction [8], to solve the problem of noise interference signal source, then by applying rough set theory to attribute reduction [9], by using the rough set theory under the premise of not reducing information, attribute reduction, reduce the dimension of the input vector [10], solves the model input redundancy and information loss. Finally, the reduced data is used as input, and the improved neural network is used for fault recognition. The experimental results show that the method is feasible and effective.

2 Theoretical background

2.1 Wavelet packet

Adopt the method of noise reduction based on wavelet packet, the SNR can be improved, the choice of response frequency band can be adapted according to the characteristic of the analyzed signal, and make it match the signal spectrum.

Wavelet packet decomposition;

$$d_j^{2n}(t) = \sum_{l \in Z} h_{l-2k} d_{j-1}(l) \tag{1}$$

$$d_j^{2n+1}(t) = \sum_{l \in Z} g_{l-2k} d_{j-1}(l) \tag{2}$$

Wavelet packet reconstruction;

$$d_{j-1}^{n}(t) = \sum_{l \in Z} h_{l-2k} d_{j-1}(l) + \sum_{l \in Z} g_{l-2k} d_{j-1}(l) \tag{3}$$

The basic process of wavelet packet decomposition is as follows: First the noise signal by wavelet packet decomposition, due to noise is usually in the high frequency coefficients, then the wavelet packet decomposition of the high frequency coefficients in the form of threshold quantization. Finally, the signal reconstruction can be reached the target of reducing the noise.

2.2 Rough set theory

Rough set theory is a kind of data analysis theory proposed by Poland scientist Z.Pawlak in 1982. It is a kind of effective dealing with imprecise, uncertain and incomplete data tools, its main idea is reducing knowledge, exporting problem of decision-making or classification rules on the premise of the ability of classification unchanged [11]; Specific steps are as follows:

1. Sample data discretization. Multiple attributes discretization method is: equidistance method, equal frequency method, maximum entropy method, etc.
2. Attribute reduction of sample data. Attribute reduction requires the classification and decision ability not changed of the knowledge, delete irrelevant or unimportant attributes [12].

2.3 Improved BP neural network

The BP neural network is mainly composed of input layer, hidden layer and output layer. Neural network is error back propagation learning algorithm. After offered neural network learning samples, the network will start learning as follows:

The output of the input layer neurons is equal to its input. The output of the hidden layer and output layer neurons calculated according to (4)

$$o_{pj}^{(l)}(n) = f_i[net_{pj}^{(l)}(n)] = f_j[\sum\nolimits_i w_{ji}^{(l)}(n) o_i^{l-1}(n) - \theta_j^{l-1}(n)] \tag{4}$$

Where $w_{ji}^{(l)}(n)$ is the connection weight of n time from the l-1 layer neuron I to the l layer neuron j, $o_{pj}^{(l)}(n)$ is the output in l layer neuron j of sample p at the time n. f_j is the transformation function for the j neuron of the l layer.

Computational training error

Input layer:

$$\delta_{pj}^{(l)} = o_{pj}^{(l)}\{[1-o_{pj}^{(l)}][t_{pj} - o_{pj}^{(l)}]\} \tag{5}$$

Hidden layer:

$$\delta_{pj}^{(l)} = o_{pj}^{(l)}[1-o_{pj}^{(l)}]\sum\nolimits_k \delta_{pk}^{(l+1)}(n) w_{kj}^{(l+!)}(n) \tag{6}$$

Modified weight

$$w_{kj}^{(l+1)}(n+1) = w_{kj}^{(l+!)}(n) + \eta \delta_{pj}^{(l)}(n) o_{pj}^{(l-1)}(n) \tag{7}$$

η is the learning rate.

Repeating the above steps until satisfied $\frac{1}{2}\sum\nolimits_j [t_{pj} - o_{pj}^{(l)}(n)]^2 < \varepsilon$

3 Methodology

Measuring the vibration signal of aero-engine intermediate bearing by using acceleration sensor in the two sides of the intermediate bearing housing, the following process of :

1. Multi-scale decomposition of wavelet packet, wavelet packet decomposition of quantized treatment to the high frequency coefficients in the form of threshold value, finally, the signal reconstruction X(t) is obtained;
2. The RMS of different measurement points of the signal X (t) are calculated;
3. Discretization of the RMS of the training samples;
4. The discrete RMS input to the Rough set model to attribute reduction;
5. The training samples that has been reduced put into the improved neural network, until the model training error reaches the set value;
6. As the input of the corresponding minimum condition attribute test sample input model has been trained to identify the fault. The proposed method as follows:

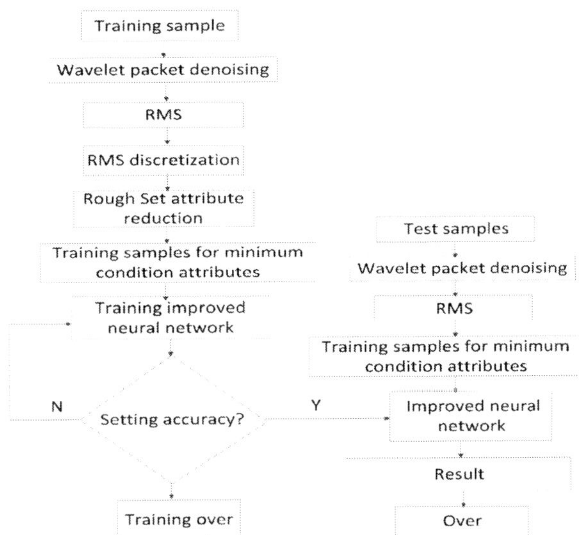

Fig. 1. Schematic diagram of proposed method

4 Application and results

Intermediate bearing vibration signal acquisition from the aero-engine intermediary bearing double rotor test rig used the acceleration sensor in the bearing housing on both sides of the intermediate bearing. The arrangement of the intermediate bearing and the acceleration sensor on the aero-engine intermediary bearing double rotor test rig is shown in Figure 2:

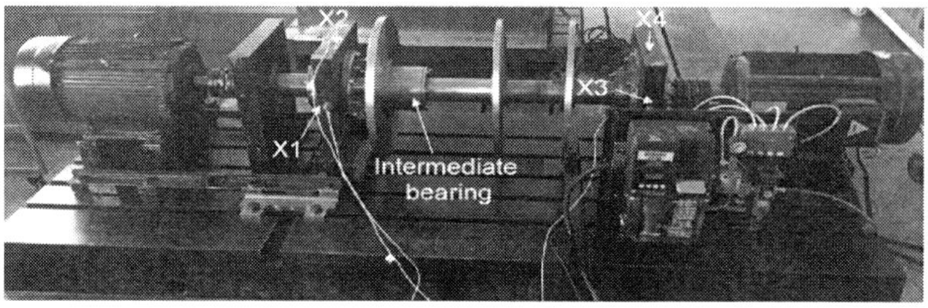

Fig. 2. Aero-engine intermediary bearing double rotor test rig

The experiment selected the type of NU1013 bearing as intermediate bearing. There were a total of 40 bearing samples, which included normally bearing 10, the rolling element fault 10, the outer fault 10, the inner ring fault 10. The bearing samples were divided into 10 groups, each group of four bearings including normal bearing, rolling element fault, outer ring fault and inner fault. Fig 3 is the intermediary bearing.

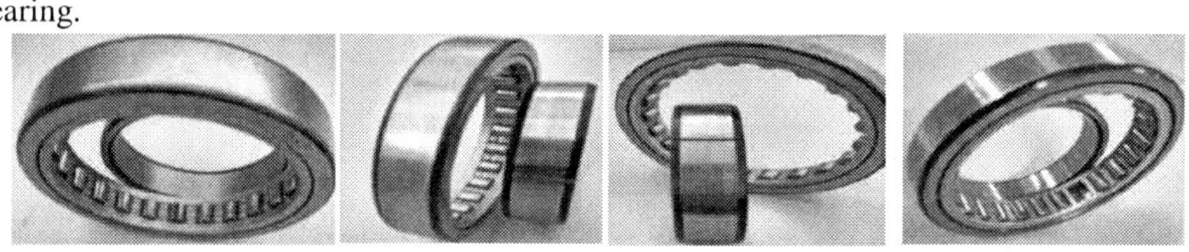

Fig. 3. Intermediary bearing

With acceleration sensors to collect the vibration signal, in the process of sampling, the high pressure rotor speed was 750 r/min, the low pressure rotor rotating speed was 750 r/min, and the rotational speed difference was 1500 r/min. First, the signals were reduced the noise by wavelet packet, Fig 4 is the intermediate bearing signal:

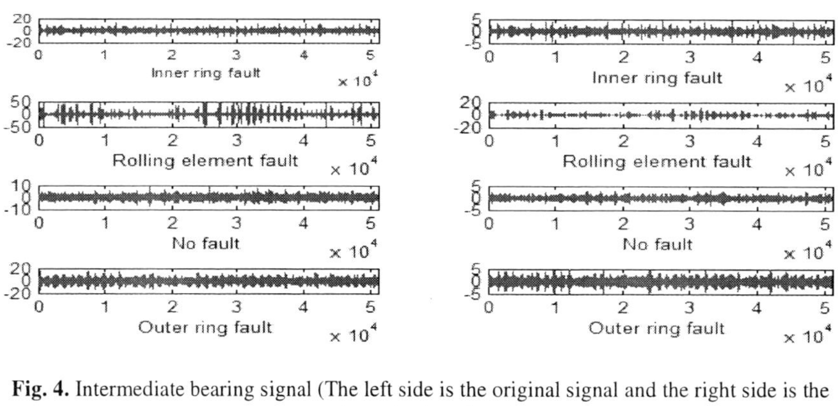

Fig. 4. Intermediate bearing signal (The left side is the original signal and the right side is the noise reduction signal)

According to the signal had been reduced, calculated the RMS of the training samples and test samples respectively on the four measuring points X1, X2, X3 and X4 under different fault conditions and recorded as condition attributes. Decision attribute values are depending on the type of fault. And equal discrete frequency method is simple and practical, the effect is better [13], therefore, the article adopted it to discrete the training sample set.

The attribute discretization decision was reduced by rough sets, the core attributes of decision is X2, X3. With improved BP neural network as classifier of fault identification of bearing test, first of all, the neural networks improvement, initial weights of neural value token (1, 1) random number; setting minimum expected error as 0.01; the maximum number of cycles as 1000; learning rate as 0.01; momentum factor mc as 0.95, the hidden layer transfer function as Sigmoid type. Results are shown in Fig 5, we can see from Fig 5, the training time of the algorithm is reduced to 1/9 of the original method.

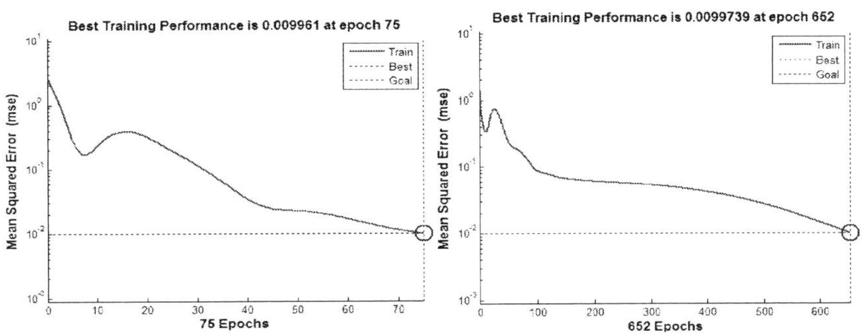

a Improved neural network b BP neural network

Fig. 5. The iterative process of fault identification for the intermediate bearing

Experiment with 20 bearings test samples, which are the inner fault 5, outer ring fault 5, rolling body fault 5, and normal bearing 5. The bearing fault identification results shown in Table 1, BP neural network four kinds of bearing fault identification number is: the inner ring fault 5, the outer ring fault 4, rolling fault 3 normal bearings 5, the average recognition accuracy rate is 85%; while the wavelet packet and Rough set theory combined with the improved neural network four kinds of bearing fault identification number is: the inner ring fault 5, the outer ring fault 5, rolling failure 5 and normal bearing 5, the average correct identification rate of 100%.

Table 1. Bearing fault recognition results

Recognize patterns	Inner fault	Outer fault	Rolling element fault	Normal bearings	Recognize rate
	5	5	5	5	
BP neural network	5	4	3	5	85%
Algorithm in this paper	5	5	5	5	100%

Bearing fault diagnosis experiment results show that using wavelet packet denoising improves the signal-to-noise ratio of the original signal, the influence of the neural network on the recognition time and the influence of the noise signal in the process of fault identification are solved. Rough set theory reduce the fault signal measuring points from 4 to 2, the average correct recognition rate from 85% to 100%. The improved neural network training time is decreased significantly. Therefore, the wavelet packet and Rough set theory combined with the improved neural network approach to fault identification ability is strong, shortly training time.

5. Conclusions

In this study, an effective method is proposed for the fault diagnosis of the aero-engine intermediate bearing, solved the problem of fault identification of the medium bearing in the strong noise and complex path. The average correct recognition rate of the medium bearing reached 100%. The high accuracy of fault recognition indicates that the proposed could be used as a reliable tool to fault identification.

Reference

1. Sun C, He ZJ, Zhang ZS, et al: Reliability evaluation of aero-engine based on state information. J. Journal of Mechanical Engineering. 06, 30-37 (2013).
2. Hu YB, Li SY, Hu LB: Research on fault diagnosis method of rolling bearing based on Neural Network. J. Mechanical design and manufacturing. 02, 187-189 (2012).
3. Zhao YX, Xu YG, Gao LX, et al: Fault pattern recognition of rolling bearing based on harmonic wavelet packet and BP neural network. J. Journal vibration and shock. 10, 162-165 (2010).
4. Chen YN, Zhang HB: Research progress and Prospect of the whole dynamics of the air engine. J. Acta Aeronautica et Astronautica Sinica. 08, 1371-1391 (2011).
5. Liao MF, Ma ZG, Liu YQ: Fault characteristics and diagnosis methods of aero engine intermediate bearing. J. Journal of Aerospace Power. (12), 2752-2758 (2013).
6. Li G, Shi J: On comparing three artificial neural networks for wind speed forecasting. J. Applied Energy. 87, 2313-2320 (2010).
7. Wang ZY, Guo CX, Cao YJ: Short term load forecasting method based on fuzzy rough set and neural network. J. Proceedings of the CSEE. 25, 7-11 (2005).
8. Sun W, Xiong BS, Wang JP, et al: Rolling bearing fault diagnosis method based on wavelet packet denoising and LMD. J. Journal vibration and shock. 31, 153-156(2012).
9. Huang XM, Yi JK: Development status and application prospect of Rough set theory. J. Journal of Beijing University of Technology. 31, 29-34 (2012).
10. Li XF: Establishment and application of Rough set -ANN model of enterprise financial crisis early warning. J. System engineering theory and Practice. 10, 8-14(2004).
11. Ryszkiewicz M: Rough set approach to incomplete information systems. J. Information Sciences. 112:39-49 (1998).
12. Jelonek J, Krawiec K, Slowinski R.: Rough set reduction of attributes and their domains for neural networks. J. Computational Intelligence. 11, 339-347 (1995).
13. Chan CC: A rough set approach to attribute generalization in data mining. J. Information Sciences. 107: 169-176 (1998).

Fault diagnosis of rolling bearings based on class information incorporated kernel entropy component analysis

Hongdi Zhou[1], Tielin Shi[1 *], Jianping Xuan[1], Guanglan Liao[1], Wuxing Lai[1], Jie Duan[1], Lei Su[2], Ke Li[2]

[1] School of Mechanical Science and Engineering, Huazhong University of Science and Technology, Wuhan 430074, Peoples R China.
{ zh_hongdi, tlshi, jpxuan, guanglan.liao, wxlai, jduan }@hust.edu.cn

[2] Jiangsu Key Laboratory of Advanced Food Manufacturing Equipment & Technology, Jiangnan University, Wuxi 214122, China
{ lei_su2015, like }@jiangnan.edu.cn

Abstract. Rolling bearings are present ubiquitously in industrial fields. Vibration-based fault diagnosis has received extensive studies to improve existing techniques toward to be more accurate in dealing with various problems, such as varying load effect and noise contamination. Especially, the sensitivity of diagnostic features from the vibration signals may vary with load conditions due to nonlinear effect and non-stationary noise. Kernel entropy component analysis (KECA) is a novel unsupervised manifold learning method for feature extraction. In order to sufficiently utilize the class information of given data in feature extraction, a class-information-incorporated KECA (CIKECA) is proposed for fault diagnosis of rolling bearings. The experimental results demonstrated the feasibility and effectiveness of the proposed method.

Keywords: fault diagnosis, kernel entropy component analysis, class information incorporated KECA

---

* Corresponding author (T. L. Shi). Tel/fax: +86 2787556835.

1 Introduction

Rolling element bearings, one of the most prevalent components, have been widely used in rotating machines of modern industry. There is no doubt that the bearing failure is one of the most frequent reasons for machine breakdown, unexpected failures may cause huge economic losses and even lead to casualties [1-2]. Therefore, it is necessary to study a kind of technology which can rapidly diagnose the incipient bearing faults in rotating machines. The major challenge for bearing condition monitoring lies on how to extract reliable and sensitive features from the original signals. The vibration signals are usually complicated, including some useful information and noises. To obtained more fault information to reflect the conditions of bearings, a high dimensional features sets involving time domain, frequency domain and time frequency domain are often used for diagnosis[3]. Obviously, more features bring in not only more information, but also the increasing of computation burden and lower recognition accuracy. Thus, it is essential to develop effective dimension and feature extraction techniques.

As an effective dimension reduction method, manifold learning aims at discovering the intrinsic structure of nonlinear data. Principal component analysis (PCA)[4] is the most widely used method for its simplicity and effectiveness. It is a linear and unsupervised method that maximizes second order statistics of original features. However, since it is linear, the performance may degenerate for nonlinear data [5]. Kernel methods are a promising approach, since they can constitute a framework to formulate nonlinear versions from linear algorithms [6]. An influential method is kernel principal component analysis (KPCA) [7], which is an extension of traditional linear PCA. However, these methods still choose top eigenvalues and corresponding eigenvectors ignoring the nature of original feature sets.

Kernel entropy component analysis (KECA) is a novel nonlinear feature extraction method, which was firstly proposed and employed in pattern recognition by Jenssen[8]. Unlike the methods depending on the second order statistics of the data set, KECA is developed based on information theory and preserves the maximum Renyi entropy of the data set[8]. It is a promising way to feature extraction and recognition. However, KECA is unsupervised without considering label information, which may discard some useful classification information. To fully utilize tagged information, we proposed a supervised method called class-information-incorporated KECA (CIKECA) based on KECA inspired by CIPCA[9]. CIKECA is presented in two ways: one is to make full use of tagged information of the given data, the other is to still follow the same simple mathematical formulation as KECA. Compared with KECA, CIKECA improves the performance in pattern recognition.

2 Principle of the CIKECA

2.1 Kernel entropy component analysis

For a given sample $\boldsymbol{X}=x_1, \ldots, x_N$, its Renyi quadratic entropy[10] is $H(p)=-\lg\int p^2(x)dx$, where $p(x)$ is the probability density function. Since the logarithm is monotonous, only the integral function $V(p)=\int p^2(x)dx$ needs to be considered. Parzen window density $\overline{p}(x)=\frac{1}{N}\sum_{x_i\subset D}K_\sigma(x,x_i)$ is applied to estimate $V(p)$, and hence $H(p)$. Here, $\boldsymbol{K_\sigma}(x, x_i)$ is the so-called Parzen window or kernel function [11], centered at x_i with a width dominated by the parameter σ. Using the sample mean approximation of the expectation operator, we have

$$\overline{V}(p)=\frac{1}{N}\sum_{x_i\in D}\overline{p}(x_i)=\frac{1}{N}\sum_{i=1}^{N}\sum_{j=1}^{N}K_\sigma(x_i,x_j)=\frac{1}{N^2}1^T K1 \tag{1}$$

where $\boldsymbol{K}$ is a $N\times N$ kernel matrix, the element (i, j) of $\boldsymbol{K}$ is $K_\sigma(x_i, x_j)$, and $\boldsymbol{1}$ is the $N\times 1$ vector (all elements are one).

The Renyi entropy could be expressed by the eigenvalues and eigenvectors of the kernel matrix which could be eigen-decomposed as $\boldsymbol{K=ED_\lambda E^T}$, where $\boldsymbol{D_\lambda}$ is a diagonal matrix storing eigenvalues $\lambda_1,\ldots, \lambda_N$ and $\boldsymbol{E}$ is a matrix with corresponding eigenvectors $e_1, \ldots, e_N$ as columns. Thus, the formula (1) can be expressed as follows:

$$\overline{V}(p)=\frac{1}{N^2}\sum_{i=1}^{N}\left(\sqrt{\lambda_i}e_i^T 1\right)^2 \tag{2}$$

Each term in this equation will contribute to the entropy estimate. The eigenvalues and eigenvectors of KECA are selected based on the first d largest contributions to the entropy estimate.

2.2 Class-information-incorporated kernel entropy component analysis

Given a set of c-class training sample patterns $x_j^i \in R^n$ (i=1, 2, ...,C; j=1,2, ..., n_i), then constructed the corresponding class labels y_i. Where x_j^i is the jth training sample of ith class, y_i is encoded in 1-of-c ways, i.e., only the ith component is set to 1, 0 otherwise. Now connecting each sample x_j^i and its label y_i to develop an augmented column vector $V_j^i=[(x_j^i)^T, y_i^T]^T$, $V_j^i \in R^{(n+c)\times d}$. Constructed the kernel matrix $K_t = \frac{1}{N}\sum_{i=1}^{c}\sum_{j=1}^{n_i}(V_j^i-\mu)(V_j^i-\mu)^T$. Here, $\mu = \frac{1}{N}\sum_{i=1}^{c}\sum_{j=1}^{n_i}V_j^i$. Eigen-decompose $\boldsymbol{K}_t$ and compute $V(p)$, then the projection matrix $U \in R^{(n+c)\times d}$ consisting of a set orthogonal unit vectors { $u_k \in R^{(n+c)}$, k=1, ..., d} can be obtained. Hence, the $\boldsymbol{U}$ is formed by the first d eigenvectors, corresponding to the first d largest entropy estimate, in a decreasing order. The number, d, of the projection vectors to be kept is determined in terms of $\sum_{j=1}^{d}\lambda_j / \sum_{j=1}^{n+c}\lambda_j \geq \alpha$ (set to 0.95 here both for CIKECA and KECA).

For a pattern to be classified, we can't directly use $\boldsymbol{U}$ to extract so-needed features of testing samples due to the unknown label. To perform an effective extraction task, decomposed the projection matrix $\boldsymbol{U}$ into two parts $\boldsymbol{U_x}$ and $\boldsymbol{U_y}$ as follows:

$$U = (u_1, u_2, ..., u_d) = \begin{pmatrix} U_x \\ U_y \end{pmatrix}, u_i = \begin{pmatrix} u_{xi} \\ u_{yi} \end{pmatrix} \quad (3)$$

$\boldsymbol{U_x}=(u_{x1}, u_{x2}, ..., u_{xd}) \in R^n x^d$, $\boldsymbol{U_y}=(u_{y1}, u_{y2}, ..., u_{yd}) \in R^c x^d$, i=1, 2, ..., d. To avoid the redundancy of the features and improve the recognition accuracy after the projection, the matrix $\boldsymbol{U_x}$ and $\boldsymbol{U_y}$ should be orthogonalized by Schmidt.

For the testing sample x, assuming its unknown class label is y, the corresponding augmented vector $\boldsymbol{v}$ can be approximately represented by a set of the found principal components. Now we can not only get so-needed features directly but also estimate its corresponding class label indirectly from its extracted features as follows:

$$V = \begin{pmatrix} X \\ Y \end{pmatrix} \approx \sum_{i=1}^{d} a_{vi} u_i = \begin{pmatrix} \sum_{i=1}^{d} a_{vi} u_{xi} \\ \sum_{i=1}^{d} a_{vi} u_y \end{pmatrix} = \begin{pmatrix} U_x a_v \\ U_y a_v \end{pmatrix} \quad (4)$$

where $\boldsymbol{a}_v=(a_{v1}, a_{v2}, \ldots, a_{vd})^T$ is the vector with the data projected in the new vector basis. We can obtain $X \approx \boldsymbol{U}_x \boldsymbol{a}_v$ from Eq. (4) and then can get its least-squared approximate solution to $\boldsymbol{a}_v$, i. e., $\boldsymbol{a}_v=(\boldsymbol{U}_x)^+X$. Now substituting it into $Y \approx \boldsymbol{U}_y \boldsymbol{a}_v$, the approximate class label $\bar{y}= \boldsymbol{U}_y(\boldsymbol{U}_x)^+\boldsymbol{a}_v$ for the testing samples could be estimated. Here, $(\boldsymbol{U}_x)^+$ is the pseudo-inverse matrix of $\boldsymbol{U}_x$. As seen, CIKECA provides class estimation for each testing samples distinctly different from KECA where we can't conduct the estimate.

3 Fault diagnosis of rolling bearings based on CIKECA

3.1 Data description

The raw vibration signals of fault bearings were obtained from the bearing data center[12]. The data has been validated in many researches and become a standard reference in the bearing diagnostics filed[2]. The test rig consists of a 2 hp Reliance Electric motor driving a shaft on which a torque transducer and encoder are mounted. The sampling frequency is 12 kHz, and the motor speed is 1797rpm under 0 hp. The faults of testing bearings were seeded on the drive-end bearings (SKF 6205-2RS) of the motor by electro-discharge machining (EDM). Four operating conditions (normal, rolling elements fault, inner and outer races fault) were introduced in the experiments. The fault diameter of ball fault, inner and outer race fault were 0.18 mm, 0.71 mm and 0.36 mm, respectively. 100 samples for each state were acquired, each of which contained 1024 data points. 50 samples were randomly selected for training and the remaining were used for testing.

3.2 Qualitative comparison

In order to examine whether such an introduction of class label to KECA is helpful for recognition, we took the qualitative comparison for the KECA and CIKECA. For intuitively observed, the results of the first two principal components were plotted after feature extraction with these methods , as shown in Figure 1-2.

Obviously, KECA cannot separate the ball fault and outer race fault absolutely in Figure 1, where a small part of those classes are overlapped. Figure 2 shows that CIKECA obtains a fairly good and clear separation of the clustering results for those four different operating conditions.

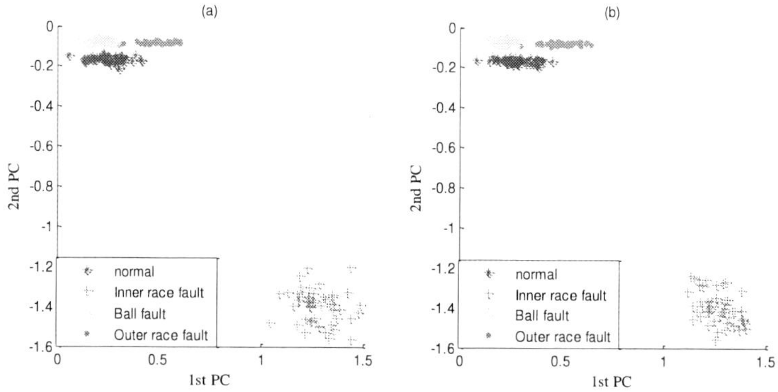

Fig. 1. the clustering results with KECA: (a) training samples, (b) testing samples

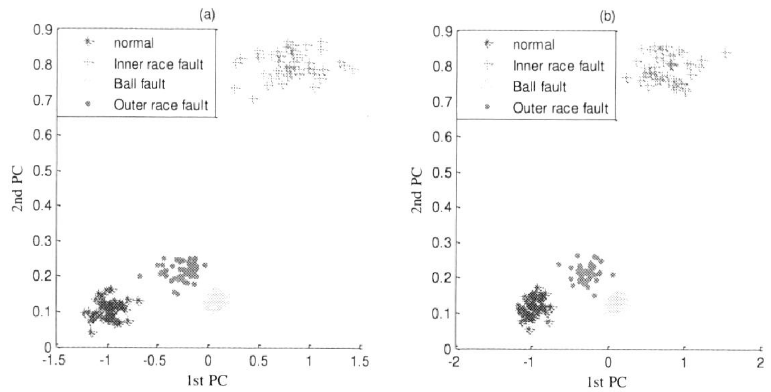

Fig. 2. the clustering results with CIKECA: (a) training samples, (b) testing samples

Though having obtained the class label estimate and extracted features, we still need to determine the true class of the testing samples. K-nearest neighbor (KNN) classifier based on Euclidean distance was used as final evaluation criteria. In KECA, only extracted features were directly fed into KNN, while we combined the extracted feature a_y and the estimated label y to decide the category in CIKECA. The classification results are shown in table 1. The accuracies of original data, KECA and CIKECA are 85%, 99.5% and 100%, respectively. It can be seen that both CIKECA and KECA are effective on feature extraction and can improve the classification accuracy, while CIKECA has better classification performance than KECA.

Table 1. classification results (%) of each method with KNN classifier

Fault type	Normal	Inner race fault	Ball fault	outer race fault	accuracy
origin	82	100	96	62	85
KECA	100	100	100	98	99.5
CIKECA	100	100	100	100	100

In order to evaluate the effect of different training sample sizes upon the performance based on CIKECA, the experiments were conducted with PCA, KECA and LDA for comparison. Table 2 reveals that the classification accuracy increased with the raising of training sample size. When the training sample size is less than 20, CIKECA has the highest accuracy than other methods. The above analysis proves that CIKECA can capture the intrinsic geometric structure embedded in the data and achieve efficient performance in feature extraction and classification.

Table 2. classification results (%) of different method with different training samples

Training sample size	10	20	40	50	70
origin feature	80.28	77.19	83.75	85	85.83
PCA	30.83	65.93	96.25	97.5	99.17
LDA	67.22	99.69	100	100	100
KECA	56.39	96.87	98.75	99.5	100
CIKECA	100	100	100	100	100

4 Conclusions

In this study, a feature extraction method called class-information-incorporated kernel entropy component analysis (CIKECA) is proposed for fault diagnosis of rolling bearings. CIKECA is presented in two objectives: one is to make full use of tagged information of the given data, the other is to still follow the same simple mathematical formulation as KECA. First, the multi-domain features involving time domain, frequency domain and time frequency domain are constructed. Then CIKECA is used to dimension reduction which can extract the intrinsic geometric features and remove the superfluous information. Finally, the extracted features of the feature space derived from the original vibration signals are fed into the k-nearest neighbour (KNN) classifier to recognize different operating conditions of bearings. The results demonstrated the feasibility and effectiveness of the proposed method for fault diagnosis of rolling bearings.

Acknowledgements

This research is supported by the National Natural Science Foundation of China (Grant No 51575202). The authors also thank the free download of the rolling element bearing fault datasets provided by Case Western Reserve University, Rockwell Science Office of Naval Research and CVX.

Reference:

1. Rai, A., Upadhyay, S.H.: A review on signal processing techniques utilized in the fault diagnosis of rolling element bearings. Tribol. Int. 96, 289-306 (2016)

2. Smith, W.A., Randall, R.B.: Rolling element bearing diagnostics using the Case Western Reserve University data: A benchmark study. Mech. Syst. Signal PR. 64-65, 100-131 (2015)

3. Lei, Y., He, Z., Zi, Y., Hu, Q.: Fault diagnosis of rotating machinery based on multiple ANFIS combination with GAs. Mech. Syst. Signal PR. 21(5), 2280-2294 (2007)

4. Jollife, I.T., Principal Component Analysis. John Wiley & Sons, Ltd (2002)

5. Li, B.W., Zhang, Y.: Supervised locally linear embedding projection (SLLEP) for machinery fault diagnosis. Mech. Syst. Signal PR. 25(8), 3125-3134 (2011)

6. Gomez-Chova, L., Jenssen, R., Camps-Valls, G.: Kernel Entropy Component Analysis for Remote Sensing Image Clustering. IEEE Geosci. Remote S. 9(2), 312-316 (2012)

7. Schölkopf, B., Smola, A., Müller, K. R.: Nonlinear component analysis as a kernel eigenvalue problem. Neural Comput. 10(5), 1299-1319 (1998)

8. Jenssen, R.: Kernel Entropy Component Analysis. IEEE T. Pattern Anal. 32(5), 847-860 (2010)

9. Chen, S., Sun, T.: Class-information-incorporated principal component analysis. Neurocomputing. 69(1-3), 216-223 (2005)

10. RRNYI, A.: On Measures of Entropy and Information. Fourth Berkeley symposium on mathematical statistics and probability, pp. 565-580. (1961)

11. Parzen, E.: On estimation of a probability density function and mode. The annals of mathematical statistics. 33(3), 1065-1076 (1962)

12. Case Western Reserve University Bearings Data Center Website, http://csegroups.case.edu/bearingdatacenter/home

Gearbox Fault Diagnosis Based on Resonance-Based Sparse Signal Decomposition and Information Entropy

Xiao Wang[1,2], Yuangeng Guo[1], Meng Chen[1], Ping Xie[1*], Guoqian Jiang[1], Jimeng Li[1]

[1] Yanshan University, Qinhuangdao
[2] The Sixth branch of Qinhuangdao Port Co.,Ltd., Qinhuangdao
*Corresponding author: Telephone: +86 0335-8057041; E-mail: pingx@ysu.edu.cn;
Address: No. 438, West Part of Hebei Avenue, Qinhuagndao 066004, PR China

Abstract. This paper aims to address coupling faults diagnosis issue of the gearbox system. When coupling faults occur in the gearbox system, it will present the coupling and non-stationary complex characteristics. This paper proposes a new fault diagnosis method based on the resonance-based sparse signal decomposition (RB-SSD) and permutation entropy(PE). In this method, RB-SSD method is first used to decompose the processed signal into nonlinear independent components according to its different oscillating behaviors. Thus, the optimal signal sparse representation is formed with the high- and low-resonance components. Then the permutation entropy is applied to analyze quantitatively the low-resonance components. The real measured vibration data of those coupling faults signals is analyzed by this method and the results demonstrate that the proposed method can identify the coupling fault features more accurately than simply using the method of the permutation entropy.

Keywords: Resonance-based sparse signal decomposition; Permutation entropy; Coupling faults, Gearbox fault diagnosis

1 Introduction

The gearbox system is an elastic system, especially when the fault occurs, the vibration signal is non-stationary and nonlinear, even chaotic[1,2]. It is a complex signal that contains the continuous oscillation component and the transient impulse component, which is not suitable for the Fourier analysis based on the stationary assumption. Wavelet transform is a kind of effective time-frequency analysis method, which can be used to characterize the local characteristics of the signal in both time and frequency domain. But the wavelet decomposition is lack of adaptability, and the choice of the wavelet function is difficult to determine. EMD is a kind of adaptive signal decomposition method, which can reflect the characteristics of the signal by decomposing signal into several IMFs. EMD method has been successfully applied to fault diagnosis of bearing and achieved good results[3]. However, EMD is essentially a binary filter bank, When the fault frequency band overlaps with the natural frequency of the system, EMD can't separate the signal well[4].

Selesnick proposed the resonance-based sparse signal decomposition(RB-SSD)[5]. The method can be used to separate the sustained oscillation and transient impulse from the high- and low-resonance components. Therefore, according to the difference of Q-factor, the signal can be separated. Firstly, on the basis of the signal to be analyzed to select two different Q-factors. By means of TQWT, the signal is expressed as high- and low-resonance components. Then use morphological component analysis(MCA) to establish the sparse decomposition objective function. Finally, by splitting the augmented Lagrange contraction algorithm to optimize the decomposition, we can get the high- and low-resonance components, and realize the separation of the sustained oscillations and the transient impulse.

In this paper, we first use the RB-SSD to separate the target signal. After this treatment, the periodic component of the gear meshing frequency, quasi periodic components and periodic pulse component of failure can be separated. Subsequently, the low-resonance component is processed by the information entropy. Permutation entropy(PE), as a kind of information entropy, is suitable for the detection of the signal mutation and we use it to analyze the low-resonance component. The difference between the different fault types are obtained by entropy curves. The correctness and validity of the method are verified by the engineering application.

2 Resonance-based Sparse Signal Decomposition

2.1 Tunable Q-factor Wavelet Transformation

The resonance of signals can be quantified by its Q-factor, $Q=f_c/B$, with f_c is the frequency of the signal and B is the signal bandwidth. When the signal's center frequency are similar and bandwidths are overlapped the linear time-invariant filtering will be unable to yield the separation. However the signal resonance attribute may be different. At this point, we can separate the signal components of the central frequency band overlapping each other but with different Q-factor from the resonance property of the signal. Thus the he problem of linear filter is overcome effectively.

In this paper, the TQWT method can overcome the shortcoming of the conventional constant-Q wavelet transforms, which can tune Q-factor according to the oscillatory behavior of the signal[6]. Based on the radix-2 FFT, the computation is more efficient, it can easily quantify the redundancy and Q-factor, and has the characteristics of complete discrete, perfect reconstruction, and moderately complete. When the signal to be analyzed has high- and low-resonance component, the basis function libraries of high and low Q-factor transforms are obtained respectively, and the corresponding transform coefficients are calculated. Meanwhile, the signal is decomposed by the wavelet basis functions with different quality factor. $\beta=2/(Q+1)$, with β is high-pass scale factor. $\alpha=1-\beta/r$, with α is low-pass scale factor. r is redundant. αf_s is the sampling frequency of v_0. βf_s is the sampling frequency of v_1. f_s is the sampling frequency of x. In the RB-SSD process, the max layers L_{max} of TQWT can be adjusted by α, β.

2.2 Separation of high- and low-resonance Components

RB-SSD can be used to separate the signal nonlinearly by MCA based on the oscillation behaviors[7,8]. Then the optimal sparse representation of high- and low-resonance component is established. The signal x is composed of high resonance attribute signal x_1 and low resonance attribute signal x_2, $x = x_1 + x_2$. x must be a nonlinear signal, and this kind of signal cannot be analyzed by the traditional linear filtering. The purpose of MCA is to estimate the source signal x_1, x_2 from the signal x. The assumption that the signal x_1, x_2 can be expressed respectively with low correlation basis functions libraries S_1 and S_2. An objective function of MCA can be expressed as

$$J(w_1, w_1) = \|x - S_1 W_1 - S_2 W_2\|_2^2 + \lambda_1 \|W_1\|_1 + \lambda_2 \|W_2\|_1 \quad (1)$$

where W_1, W_2 represent the coefficients of the basis function libraries S_1 and S_2. λ_1, λ_2 are regularization parameters, their values have an effect on the energy of the high- and low-resonance components of the decomposition.

Noted that it is difficult to minimized the objective function (1) due to the large number of variables and the non-differentiability of the l_1-norm[9,10]. The split augmented Lagrangian shrinkage algorithm(SALSA) particularly is chosen to solves the problems which for MCA can be solved easily. Then the estimates are provided as

$$x_1 = S_1 W_1^*, x_2 = S_2 W_2^* \quad (2)$$

3 Permutation Entropy

The permutation entropy(PE) reflects the complexity of one dimensional time series, and detect the dynamic mutation of complex system. When the system equipment is faulted, the influence of the nonlinear factors is different, the performance of the signal complexity will not be the same[11,12]. Therefore, it is feasible to use the PE for fault diagnosis.

Given time series that data length is $n, \{x(i), i = 1, 2, ..., n\}$, the phase space reconstruction is first carried out and the result is $X(t) = \left[x_t, x_{t+\tau}, ..., x_{t+(m-1)\tau}\right]$, where m is embedding dimension and τ is delay time. Then ranking the m elements of $X(t)$ ordered arrangement. Supposing that there are equal numbers, sort by the original time order, the resulting sequence is one of $m!$ kinds of ordering patterns.

Ordering pattern passes for a symbol, and the reconstructed sequence can be viewed as a sequence of symbols. For symbolic sequence π, $f(\pi)$ indicates the number of times it appears in time series analysis, then its relative frequency is $p_i(\pi) = f(\pi)/\left(N-(m-1)\tau\right)$, and PE is defined as $H_p(m) = -\sum_{j=1}^{N-(m-1)\tau} p_i(\pi)\ln p_i(\pi)$. After normalization, $PE = H_p(m)/\ln(m!)$.

4 Fault analysis of PE based on RB-SSD

When the gear box failure occurs, vibration signal is mainly composed of the transient shock component and the harmonic component. RB-SSD method can nonlinearly separate fault signal into high- and low-resonance components and the residual. Due to the traditional signal processing method based on signal stationarity assumption is not suitable for processing such non-stationary signals, the PE is chosen to analyze the low-resonance components. In this method the result can well reflect the change of the parameters of nonlinear dynamic system. Through above analysis, the procedures of the proposed method are as follows:

Step1: According to the fault signal x, the Q-factors(Q_1,Q_2) and the redundance (r_1,r_2) of RB-SSD can be determined. After repeated experiments the TQWT parameters are chosen to be Q_1=1, Q_2=4, r_1=r_2=3 in this paper.

Step2: Based on the parameters in step1, we can respectively obtain the basis function libraries S_1,S_2 and get transformation coefficients W_1,W_2 of the TQWT.

Step3: To determine the regularization parameter λ_1, λ_2 and establish the objective function as function(1). Then, the estimates $\hat{x}_1, \hat{x}_2$ will be provided by SALSA. The signal is separate into high- and low- resonance components by RB-SSD.

Step4: Due to the fault information is mostly concentrated in the low-resonance component, we analyze this component through the method of PE. Then the PE curves can be used to identify faults.

5 Engineering Application

In this section, the proposed method is applied to vibration signals analysis of a gear box as shown in the Fig.1(a). The vibration signals are measured from acceleration sensors set in the high-speed shaft under different rotational speed. The signals were sampled at 66666 Hz. We need to divide the vibration time series into several sub sequences. According to the sampling frequency and other related information of the signal, data length is determined as 8192, resampling is carried out at 324 intervals.

When there are a variety of complex fault, the coupling of the signal are large, and failure frequency aliasing. Due to the ambiguity and overlap between fault features, the result of using the PE is not effective. Table 1 shows the three kinds of coupling fault types, the results of PE are shown in Fig.2(a). Fault 2 and 3 can be clearly distinguished, but fault 1 and other faults have cross in entropy curve.

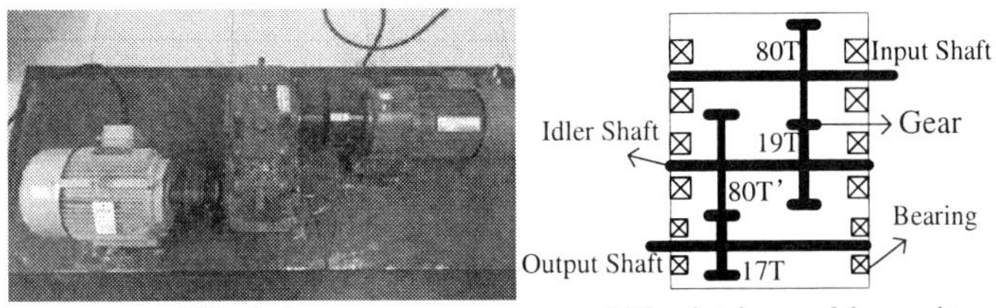

(a)Testbed (b)The sketch map of the gear box

Fig. 1. Testbed

Table 1. The coupling fault type

	Gear:80T	Gear:19T	Gear:80T'	Gear:17T	IS Bearing
Fault type 1	Chipped	Good	Good	Eccentric	Good
Fault type 2	Good	Good	Eccentric	Chipped	Good
Fault type 3	Good	Good	Broken	Broken	Ball

we first decompose the signal into high and low-resonance components by the RB-SSD based on the different resonance attribute of signal. The results are shown in Fig.3. It can highlight the low-resonance component of fault with the form of a pulse signal.Then PE method is used to deal with the low-resonance component and get the corresponding curves. The curves correspond to different fault types. Fig2(b) shows curves that are obviously different. This method is suitable to distinguish the different coupling faults. We use this method to analyze another 10 sets of data, which can effectively separate the faults, it is an effective way to diagnose the fault types.

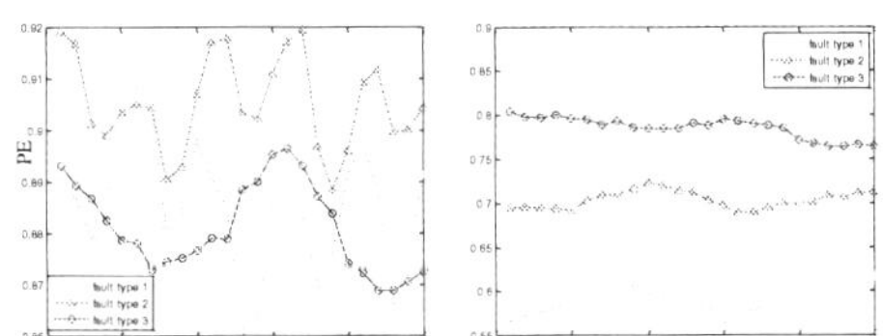

(a) (b)

Fig. 2. The curves of (a) PE (b)PE of RB-SSD under different

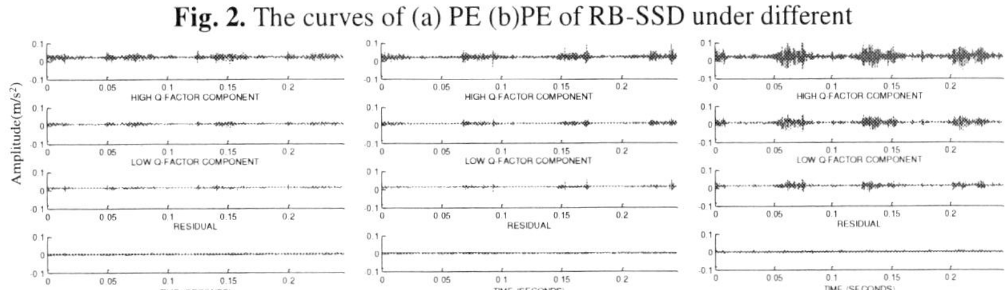

(a)fault type 1 (b)fault type 2 (c)fault type 3

Fig. 3. Processed results by using the RB-SSD

6 Conclusion

The RS-SSD can be used to separate the vibration signals with similar center frequencies and overlapped bandwidths into high- and low-resonance components according to the corresponding Q-factors. The low-resonance component effectively highlight the fault with the form of a pulse signal. Moreover, PE curves were used in low-resonance component to quantificationally evaluate the faults and the fault types. The result of engineering application shows that the proposed method can act as a promising tool for diagnosing the fault types of rotating machinery.

Acknowledgements

This work was supported by the Hebei Province Natural Science Fund (Grant No. F2016203421), Key Project of Science and Technology Research of Higher Education in Hebei Province (Grant No. ZD20131080)

References

1. Lin, J.S., Chen, Q.: Fault feature extraction of gearboxes based on multifractal detrended fluctuation analysis. Journal of Vibration and Sound 32, 97-101 (2013)
2. Cui, L., Song X.G., Zheng J.R.: Nonlinear dynamic analysis for a geared flexible rotor system considering multi-clearance. Journal of Vibration and Sound 32, 171-178 (2013)
3. Yu, D.J., Cheng, J.S., Yang Y.: Fault diagnosis approach for roller bearings based in empirical mode decomposition method and Hilbert transform. Chinese Journal of Mechanical Engineering 18, 267-270 (2005)
4. Flandrin, P., Rilling G., Goncalves. P.: Empirical mode decomposition as a filter bank. IEEE Signal Processing Letters 11,112-114 (2004)
5. Selesnick, I.W. : Resonance-Based Signal Decomposition: A New Sparsity-Enabled Signal Analysis Method. Signal Processing 91, 2793-2809. (2011)
6. Starck, J.L., Moudden, Y.B.J, et al.: Morphological component analysis: an adaptive thresholding strategy. Image Proceedings of IEEE, 16, 2675-2681 (2007)
7. Wright, S.J., Nowak, R.D., Figueiredo, M.A.T.: Sparse reconstruction by separable approximation. IEEE Trans. On Signal Processing 57, 2476-2493 (2009)
8. Selesnick, I.W., Wavelet transform with tunable Q-factor. IEEE Transactions in Signal Processing 59, 3560-3575 (2011)
9. Afonso, M.V., Bioucas, J.M., Figueiredo, M.A.T.: Fast image recovery using variable splitting and constrained optim ization. IEEE Transactions on Image Processing 19, 2345-2356 (2010)
10. Afonso, M.V., Bioucas, J.M., Figueiredo, M.A.T.: An augmented lagrangian approach to the constrained optim ization formulation of imaging inverse problems. IEEE Transactions on Image Processing 20, 681-695 (2011)
11. YanR, L.Y., Gao, R.X.: Permutation entropy: a nonlinear statistical measure for status characterization of rotary machines. Mechanical Systems and Signal Processing 29, 474-484 (2012)
12. Li, X., Ouyang, G., Richards, D.A.: Predict ability analysis of absence seizures with permutation entropy. Epilepsy Research 77, 70-74 (2007)

Improved Constrained ICA with Wavelet and EEMD Decomposition for Fault Diagnosis of Roller Bearings

Gang Tang, Guozheng Li, Ganggang Luo, Huaqing Wang

College of Mechanical and Electrical Engineering, Beijing University of Chemical Technology, Beijing 100029, China
E-mail: tanggang@mail.buct.edu.cn

Abstract. As an improvement for the traditional ICA algorithm, a so-called constrained independent component analysis (CICA) incorporates prior information of measured mixed signals as reference signals into ICA, and then extracts the desired independent component directly. In the traditional CICA method, square pulse signal is utilized to establish the reference signals. However, the selections of duty ratio and initial phase are relatively complex, which will obviously affect the fault feature extraction. To solve this problem, this paper proposes an improved CICA algorithm based on discrete wavelet transform and ensemble empirical mode decomposition (EEMD). The wavelet method is firstly utilized to decompose the single channel vibration signal into several wavelet coefficients. Then reference signals are designed using EEMD method. The source signal is decomposed into several intrinsic mode functions (IMFs), and then the fault characteristic frequency energy ratio is calculated in different IMFs. The reference signals are determined according to the criterion of maximum energy ratio. Experimental results show that the improved CICA method can successfully extract the fault signal of rotating machinery.

Keywords: fault diagnosis; constrained ICA; discrete wavelet transform; EEMD; reference signal

1 Introduction

Independent component analysis (ICA) has been widely used in mechanical vibration signal processing. However, the traditional ICA method usually needs to calculate all the independent components firstly, and then pick out the independent components what we need. So the algorithm is complex and the accuracy is often not high enough.

In order to solve this problem, constrained independent component analysis (CICA) was proposed [1], which needs to incorporate the prior information as a reference signal into ICA algorithm so that the algorithm converges to only interested signals [2]. The CICA method can significantly improve the effectiveness and accuracy of fault diagnosis, thereby saving a lot of time and space costs. In the traditional CICA approach, square pulse signal is utilized to establish the reference signal. However, the selections of duty ratio and initial phase are relatively complex and affected the accuracy of the extraction result largely.

This paper proposes an improved CICA algorithm based on discrete wavelet transform (DWT) and ensemble empirical mode decomposition (EEMD). With the approach, a vibration signal is firstly decomposed into multichannel signals by DWT method to transform underdetermined ICA model into positive definite ICA model. Secondly, an approach to design reference signals based on EEMD method is mainly introduced. Finally, experimental results show that the improved CICA method can effectively separate the fault signal of roller bearings.

2 Principle of the CICA Algorithm

2.1 Principle of constrained independent component analysis

Constrained independent component analysis (CICA) incorporates the prior information of measured mixed signals as reference signals into the traditional ICA algorithm so that the algorithm converges to only interested signals, and then extract the desired independent component directly. The model can be written as [3]:

$$\max J(y) \approx \rho\{E[G(y)] - E[G(v)]\}^2$$
$$s.\,t. \begin{cases} g(w) = \varepsilon(y,r) - \xi \le 0 \\ h(w) = E(y^2) - 1 = 0 \end{cases} \quad (1)$$

Where y is the algorithm output, ρ is a positive constant, v is a Gaussian variable with zero mean and unit variance, and G is a non-quadratic function. Eq. (2) is essentially a constrained optimization problem, which can use the Lagrange multiplier method to calculate.

$$L(w,u,\lambda) = J(y) - \frac{1}{2\gamma}[\max{}^2\{\mu + \gamma g(w), 0\} - \mu^2] - \lambda h(w) - \frac{1}{2}\gamma\|h(w)\}\|^2 \quad (2)$$

Where μ and λ are Lagrange multipliers, updated based on a gradient-ascent method.

2.2 The shortages of traditional CICA

Constructing a pulse signal is a traditional way to design reference signal [4]. The pulse signal contains in the periodic characteristics of vibration signals of rotating machinery system which we interest in. It can be written as

$$r = square(2\pi f/fs + \theta, \omega) \quad (3)$$

Where θ is the phase angle and ω is the duty ratio. The phase angle and the duty ratio of reference signal will largely affect the extracting effect of fault features. The actual signal may be collected from any point-in-time of the fault period, which leads to the actual initial phase of fault signals being collected indefinitely. If we run CICA algorithm repeatedly to get the best phase angle, it will increase the calculation greatly and reduce the operation rate. On the other hand, the duty ratio of a signal is also an important indicator, which affects the extraction results greatly. A wrong duty ratio will make the algorithm non-convergent. Therefore, the traditional CICA algorithm is very sensitive to the selections of parameters. As a matter of fact, there is no good method to determine these parameters.

3 Improved CICA algorithm based on DWT and EEMD

3.1 Multiplexing single signal channel based on DWT

In engineering practice, the number of source signals is usually unknown. A small quantity of observed signal from sensor is often less than the number of source signals. This type of ICA model is called underdetermined ICA model. To solve the problem of underdetermined model of ICA algorithm, single channel signal multiplexing method is widely used. Discrete Wavelet Transform (DWT) [5] is a kind transformation analysis method, which develops the localization of the short-time Fourier transform (STFT). At the same time, it can provide a changed time-frequency window following its frequency. It can be written as

$$WT(a,b) = \frac{1}{\sqrt{a}} \int x(t) \psi^{*}(\frac{t-b}{a}) dt \tag{4}$$

where the parameter a and b are the change parameters, a is a scale parameter (in some sense means the frequency of concept), b is the time parameter or translation parameter.

3.2 The selection of reference signal based on EEMD

Huang [6] presented ensemble empirical mode decomposition (EEMD) method by adding auxiliary noise signal, which had good performance in signal decomposition. This chapter proposes an approach to design reference signals based on EEMD method. Firstly, the source signal is decomposed into several IMFs, and there will be pseudo components in the process of the EEMD decomposition. The IMFs with large correlation coefficient will be retained for next step. In order to measure the fault characteristic frequency of the IMFs, calculate the fault characteristic frequency energy ratio in different IMFs [7]. The energy ratio can be calculated by

$$ER = \frac{E_f}{E} \times 100\% \tag{5}$$

where, E_f is the energy value of the characteristic frequency of the desired fault signal, E is the total energy of the IMF. The IMF signal contains more fault information if the value of ER is larger. The reference signals were determined according to the criterion of maximum energy ratio.

4 Application of improved CICA in bearing compound fault diagnosis

A vibration signal is gathered from acceleration sensors of a rotary machinery shown in Fig.5. For the fault diagnosis test, an outer-race defect (0.3 mm (width) × 0.05 mm (depth)) and a roller defect (0.3 mm (width) × 0.05 mm (depth)) are artificially created using a wire-cutting machine. The rotational speed of the bearing is 1300 r/min and a large sampling frequency (100 KHz) is chosen. The sampling process continues for 10 seconds. The passing frequency of the outer-race defect and the roller defect is 86.3 Hz and 102.3 Hz, respectively.

As mentioned above, the single channel signal should be multiplexed in order to solve the problem of undetermined ICA model. The DWT method is firstly utilized to decompose the single channel vibration signal into three-layer wavelet coefficients [8]. The detailed signals named d1, d2, d3 and the approximate signals named a1, a2, a3 can be reconstructed by using wavelet coefficients. These reconstructed signals will be the input matrix of the CICA algorithm. Then the collected source signal is decomposed into several IMFs. Cross correlation coefficients of each IMF and source signal are calculated, as listed in Table 1.

Table 1. Cross correlation coefficient of each IMF and source signal

IMF1	IMF2	IMF3	IMF4	IMF5	IMF6
1.0000	0.3263	0.5055	0.6716	0.5394	0.2756
IMF7	IMF8	IMF9	IMF10	IMF11	IMF12
0.2683	0.1643	0.0728	0.0369	0.0216	0.0066
IMF13	IMF14	IMF15	IMF16	r_n	
0.0073	0.0058	0.0040	0.0072	-0.0003	

The IMFs with large correlation coefficient will be retained for calculating the ER of the outer-race fault frequency and the roller flaw frequency according to Eq. (5), as listed in Table 2.

Table 2. ER of outer-race and roller of each IMF

TYPE	ER(%)							
	IMF1	IMF2	IMF3	IMF4	IMF5	IMF6	IMF7	IMF8
OUTER-RACE	32.34	10.23	10.64	17.56	20.27	0.81	1.65	1.25
ROLLER	0.35	0.28	3.49	1.04	0.14	13.34	4.85	5.25

Table 2 shows that IMF1 has the maximum ER of the out-race fault frequency and IMF6 has the maximum ER of the roller fault frequency. As a consequence, IMF1 and IMF6 are selected as reference signals to extract the outer-race fault signal and roller fault signal respectively. The reference signals and extracted signals are shown in Fig.1 and Fig.2 respectively.

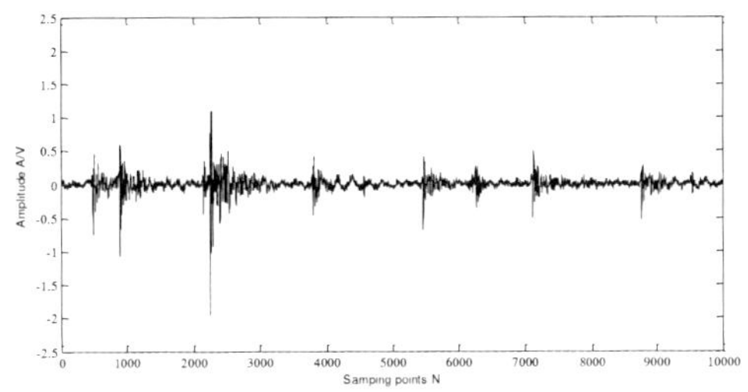

(a) the reference signal

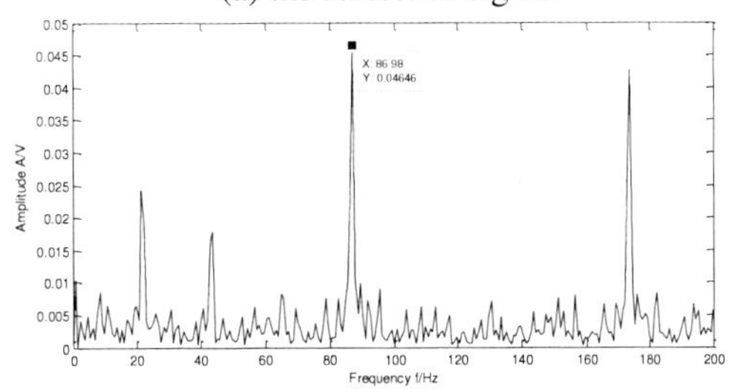

(b) the envelope spectrum

Fig. 1. Reference signal and envelope spectrum of an outer-race defect

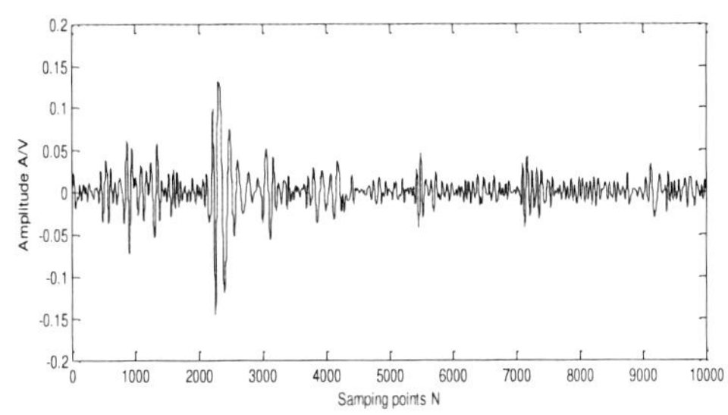

(a) the reference signal

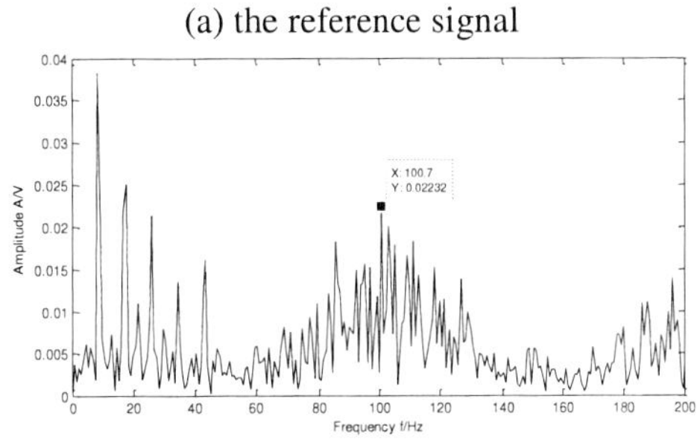

(b) the envelope spectrum

Fig. 2. Reference signal and envelope spectrum of a roller defect

We can see that the desired fault signals are extracted successfully.

5 Conclusions

CICA method can only extract the desired fault signal by incorporating the prior information into the ICA algorithm. The traditional way to construct pause signal as a reference signal is very sensitive to the selections of parameters. In order to solve the problem of undetermined ICA model, single channel signal is multiplexed by DWT. The fault frequency energy ratio of IMFs decomposed by EEMD is calculated in order to select reference signal. Experimental results show that the improved CICA method could effectively extracted the fault signal of rotating machinery. Thus it can optimize the selection of reference signals.

Acknowledgments. This work is partially supported by the National Natural Science Foundation of China (Grant no.51405012, 51375037) and The Open Fund of State Key Laboratory, Southwest Jiaotong University (TPL1603). The first author would like to thank the Public Hatching Platform for Recruited Talents of Beijing University of Chemical Technology.

References

1. Lu W, Rajapakse J C. ICA with reference[J]. Neurocomputing, 2006, 69(16): 2244-2257.
2. Mi J X. A novel algorithm for independent component analysis with reference and methods for its applications[J]. PloS one, 2014, 9(5): e93984..
3. Hyvärinen A, Karhunen J, Oja E. Independent component analysis[M]. John Wiley & Sons, 2004.
4. Wang Z, Chen J, Dong G, et al. Constrained independent component analysis and its application to machine fault diagnosis[J]. Mechanical Systems and Signal Processing, 2011, 25(7): 2501-2512.
5. Daubechies I. Ten lectures on wavelets[M]. Philadelphia: Society for industrial and applied mathematics, 1992.
6. Huang N E, Shen Z, Long S R, et al. The empirical mode decomposition and the Hilbert spectrum for nonlinear and non-stationary time series analysis[C]//Proceedings of the Royal Society of London A: Mathematical, Physical and Engineering Sciences. The Royal Society, 1998, 454(1971): 903-995.
7. Wang H, Li R, Tang G, et al. A compound fault diagnosis for rolling bearings method based on blind source separation and ensemble empirical mode decomposition[J]. PloS one, 2014, 9(10): e109166.
8. Chen G, Qie L, Zhang A, et al. Improved CICA algorithm used for single channel compound fault diagnosis of rolling bearings[J]. Chinese Journal of Mechanical Engineering, 2016, 29(1): 204-211.

Research of Machining Center Spindles Condition Monitoring System based on Wavelet Theory

Guo-fa Li [1,2] , Jing-hua Dong[1,2] ,Bin-bin Xu[1,2] ,Quan-pu Li[3] ,ChaoHou[1,2]

and Sheng-xu Wang[1,2]

[1]College of Mechanical Science and Engineering, Jilin University, Changchun 130025, China;
[2]Key Laboratory of CNC Equipment Reliability Technique of Machinery Industry, Jilin University, Changchun 130025, China
[3]Dalian Machine Tools Group Co., LTD, Dalian 116000,China

Abstract. In order to achieve the condition monitoring of machining center spindles, The paper established a spindle condition monitoring system based on wavelet theory. Through the analysis of the spindle machining state and failure information, spindle vibration, spindle motor current , spindle motor temperature and spindle speed were determined as monitoring parameters .The system monitored the parameters by the hardware system made use of wireless vibration sensor, current transformer, temperature transmitter, frequency transducer. And it transmitted the data to industrial computer through PCI data acquisition card. The software system in industrial computer was programmed by LABVIEW and MATLAB, and it could realize the signal processing and analysis. The cutting experiments were carried out by taking some processing technologies for example. And it showed some conclusions: the system could realize signal filtering. The constancy of spindle temperature could indicate the normal operation condition of spindle motor. The current changes could reflect the change of cutting power and tool breaking fault. Auto-correlation analysis of vibration signal could judge the signal's periodicity. Spectrum analysis could determine the vibration signal frequency components. And the wavelet analysis could further determine the time-frequency relationship of the vibration signal. All of the analyses could be employed to distinguish normal condition and failure condition. The experiments showed that the system have the functions of spindles condition and failure monitoring.

Keywords: machining center spindles, condition monitoring, hardware system, software system, cutting experiments, signal analysis

1. Research status both at home and abroad

At present, the condition monitoring and failure diagnosis of machine tools has

Fund project: China's National Science and Technology Major Project（2013ZX04011-012） "High-end CNC Machine Tools and Basic Manufacturing Equipment-The Reliability Promoting Project for 1000 Domestic Machining Centers"
Author introduction: Guo-fa Li (1970 -), male, professor, doctoral supervisor, research direction: CNC equipment reliability technology and theory, E - mail: ligf@jlu.edu.cn
Corresponding author: Bin-bin Xu (1982 -), female, lecturer, research direction: the theory of CNC equipment reliability, maintainability. E-mail: xubinbin@jlu.edu.cn

made a series of achievements at home and abroad research. Yangang X et al. [1]took the radial displacement of the spindle, spindle speed, vibration, temperature as the condition characteristics of motorized spindle, and developed the corresponding program of the signal analysis and display in the Monitoring and Control General System (MCGS) software platform. Chang etal.[2] developed the spindle vibration monitoring system based on spectrum analysis on the basis of the analysis of vibration signal spectrum. Yuqing Zhou et al.[3] researched the influence of the load ,the position and the acceleration for the measured signals, and put forward to a machine tool condition monitoring system which could trace fault source. The literature [4] used wavelet noise reduction for the spindle non-stationary signal, and used Short Time Fourier transform to obtain the spindle vibration amplitude. The literature [5, 6] acquired cutting signals of tools through a large number of experiments, and used multi-resolution wavelet analysis method to extract the features of tools wear state. And it also made use of the advantages of neural network technology. Finally it effectively realized the function of tools wear state recognition. All of the research provides a number of research ideas and methods for this article. The spindle is one of the key function parts of CNC machine tools. Its basic status and reliability largely determines the working status and reliability of CNC machine tools. But in view of the commonly used mechanical spindle, related research on the overall state monitoring is still less.

More modern factories use the method of production line, a machine tool of production line is generally responsible for the specific process. The same machining parameters make the normal state stable, so it provides the convenience for distinguishing between normal and fault condition. So the article, combining previous research results, has developed a kind of spindle condition monitoring system based on wavelet theory to meet the needs of the actual industrial production. The system had low cost, high practicability, real-time monitoring, and many other advantages, and it could be independent of the numerical control system. The system based on signal acquisition and analysis of spindle vibration, current, temperature, speed and other parameters of the actual processing condition, realized the condition monitoring for the spindle processing. The article selected the VDL - 1000 machining center as a machine tool used in the system development.

2. The selection of monitoring parameters

In order to determine the monitoring parameters of spindle, firstly we needed to understand the relationship between the spindle running status and each monitoring parameter. The spindle speed as one of the major parameters of spindle should be monitored as a parameter to reflect the spindle basic processing information. In practical machining the spindle tends to run for a long time, so it needed to monitor the temperature of spindle motor to prevent motor temperature becoming too high and cause the phenomenon of downtime. The spindle motor current could reflect the change of spindle processing power and tool condition (such as tool wear and tool broken, etc.), therefore the spindle motor current needed to be monitored. The spindle vibration is different under different processing condition, and the abnormal vibration of the spindle often contains the failure information. In addition, it was also needed to understand the relationship between the common failures spindle system and each monitoring parameter. With early field test track of a batch of machining center, a large number of failure data are obtained. In the field test track there were 138 failures data, the spindle system failure had 25 of them. This showed, the spindle system failure was one of the main failures of machine tools, so monitoring for the spindle system was meaningful. There was Spindle alarm because the temperature is too high, so it was needed to monitor temperature of spindle motor. The spindle motor current could reflect tools crashing, tools falling and tool stuck. The vibration signal could reflect the upper cutter holder of Spindle box being loose and the spindle tool vibration. By the analyses for spindle machining status and the common failures, spindle speed, the spindle vibration, spindle motor temperature, spindle motor current were taken as the four monitoring parameters of the system.

3. The construction of the monitoring system

The monitoring system consisted of hardware system and software system. It had the function of Signal acquisition and processing of monitoring parameters, the real-time signal display and the data storage.

3.1 The construction of the hardware system

The hardware system consisted of the sensor part and the data acquisition part. The connection figure was shown in Fig.1.

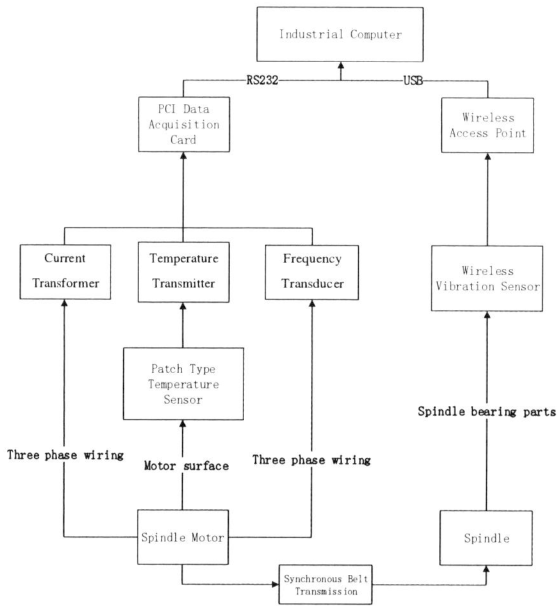

Fig. 1. The hardware system connection figure

In the choice of the sensors, the main principles were the principle of meeting the measurement requirements and the principle of simple and feasible. The spindle speed measurement usually used the speed sensor, But the speed sensor installation was easy to affect the machining process and even damage the spindle structure. So the system adopted a indirect measurement method of spindle speed. Based on the characteristic of the spindle motor speed control by the change of current frequency ,it used the formula "n = 60 f/p " (n as the spindle speed, p as the spindle motor pole logarithm, f as the spindle motor current frequency).So it could indirectly obtain the spindle speed "n" by current frequency of frequency transducer. The frequency transducer could be installed in the machine tool's electric cabinet, so the installation was simple and didn't affect the normal processing. This system adopted the current transformer to collect spindle motor current and the current could be measured by putting the spindle motor's wiring through current transformer. Because the frequency range of VDL - 1000 machining center spindle motor current is 10-300 Hz, the frequency transducer and the current transformer used in this system should be the sensors of wide frequency range. Considering the motor's characteristics of not easy to tear open outfit, the system used a patch type temperature sensor when it was monitoring the temperature of spindle motor. The probe parts of the temperature sensor should be put on the surface of the motor, so it was easy to directly measure the motor's surface temperature. Then it transmitted the signals to the PCI acquisition card through the temperature transmitter. As a result of the spindle machining in high speed and the installation of sensors which couldn't destroy the spindle structure, the system chose the magnetic suction wireless vibration sensor. And the system chose the way of indirect measurement to measure the spindle vibration. The vibration sensor measured the vibration by being adsorbed on the spindle bearing parts. Then it transmitted signals to the industrial computer through the wireless gateway.

The system's data acquisition part adopted the PCI data acquisition card of high speed and high precision. The sensors' signal output terminals were connected to the PCI acquisition card, so the PCI acquisition card could collect multi-channel signals of the required measurements at the same time. The PCI acquisition card could be connected to the industrial computer through the RS - 232 port, so it could transmit the data to the industrial computer for processing and analysis. And the industrial computer also could be used as a display module for signal display and data storage by software system.

3.2 The software system development

The software system adopted LABVIEW programming language, and it also combined with signal processing and analysis of MATLAB. The software system mainly included the basic information management module, the data storage module, the condition monitoring module.

The condition monitoring module was the main module of software system. Its main function was to realize the sensor signal acquisition, storage, filter and analysis. The signal acquisition is a process of transforming the analog signals collected by the sensors into the digital signals through a series of signal transformation, then transmitting the digital signals to the PC. According to the Nyquist theorem [7], when the sampling frequency is more than twice the highest frequency of the signal, the sampling digital signal obtains the intact information of the original analog signal relatively.

Due to the original signals have much noise and interference, so the collected signals need filtering processing. Considering the complex compositions and the non-stationary situation of the vibration signal, the effect of the traditional filtering method couldn't meet the requirements. The system adopted the method of wavelet d-noising, which could effectively eliminate clutter interference to obtain useful signal. The basic principles of wavelet d-noising are: firstly setting signal's threshold, after the wavelet decomposition the coefficients' module of each layer which are more than and less than the threshold been respectively for processing, then to finish the inverse transformation of the wavelet coefficients been processing and finally obtaining d-noising signal after the signal reconstruction.

In order to obtain characteristics of collected signals, signal analysis was also required. The system adopted the method of signal characteristics to identify and distinguish with auto-correlation analysis, spectrum analysis and wavelet analysis.

The auto-correlation function could distinguish the signal types, and can judge whether the analyzed signal is cyclical. Spectrum analysis can transform the time domain signal to the frequency domain signal, but its window function is fixed in the whole process of analysis. So it is failed to reflect the unsteady feature of signal, the transient characteristics and the time-frequency local features. In view of the unsteady process of the monitoring signal, the system also adopted wavelet analysis for signal analysis. Wavelet analysis [8,9] kind of fine transient signal analysis method. Its window has the characteristics of area fixed but shape variable, namely, the wavelet analysis is a local signal analysis method whose resolution value is variable on the time and the frequency.

4. The experiments of monitoring system

To test and verify the function of the monitoring system, the article presented the monitoring results by the simulation of machining process based on the specific cutting experiments. The system was mainly used in a machine tool of the production line. Due to the machine tool's machining process wouldn't change, so it could reflect the running state of the machine tool through the change of the system monitoring parameters. On the basis of the preliminary acquisition of the processing load data of a certain production line in a factory, the typical machining method and parameters were chosen to simulate machining process (as shown in Table1).

Experiment Number	Tool Diameter d (mm)	Work piece Material	Tool Material	Tool type	Spindle Speed (r/min)	Feed Speed(mm/min)	Back Cutting Depth ap (mm)	Cutting Width aw (mm)	Tool Teeth z(number)
1	7	Gray cast iron	Alloyed tool steel	Twist drill	950	95	1	6.7	1
2	30	Gray cast iron	Alloyed tool steel	Twist drill	200	10	1	30	1
3	16	45 # steel	Alloyed tool steel	Ball nose end milling tool	700	107	0.3	8	4
4	16	45 # steel	Alloyed tool steel	End mill	3000	1400	0.3	4	3

Table 1. Experiments conditions graph

4.1 State monitoring function experiments

Fig.2 wasthe spindle motor current situation of the spindle idling and of the drilling test. Comparingthem, the drilling test'scurrent was increased, so it reflectedthat the cutting process neededbigger cutting torqueand higher power.

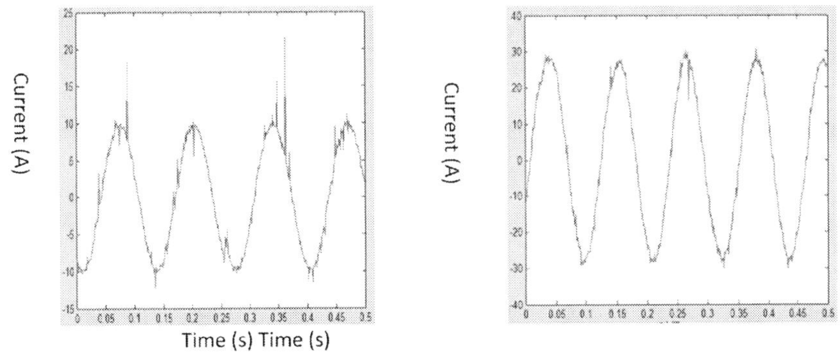

Fig. 2

As shown in Fig.3, in the process of the Experiment 1, it appeared the phenomenon of tool broken. And the system detected a sharp rise in the spindle motor current and then back to normal soon. The current change could reflect the phenomenon of tool broken, and at this time the machine tools should be stopped to replace the tool. In addition, if it appeared the phenomenon such as tool stuck and tool crashing, the motor current would increase rapidly. At this time the machine tool should stop processing in a timely manner to avoid damage to the motor.

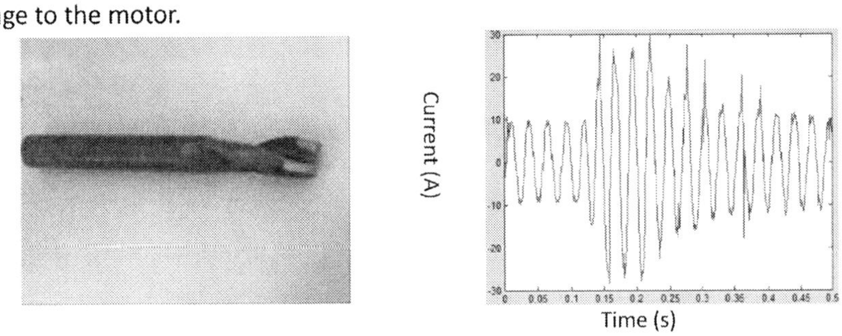

Fig. 3

4.2 Signal filtering and analysis function experiments

To test and verify the filtering function of the system, the waveform comparison of spindle vibration signals before and after filtering were shown in Fig.4.It could be seen that the filtering effect was obvious. The high frequency interference was effectively removed. After filtering the current signal had better smoothness and no distortion of signal curve.

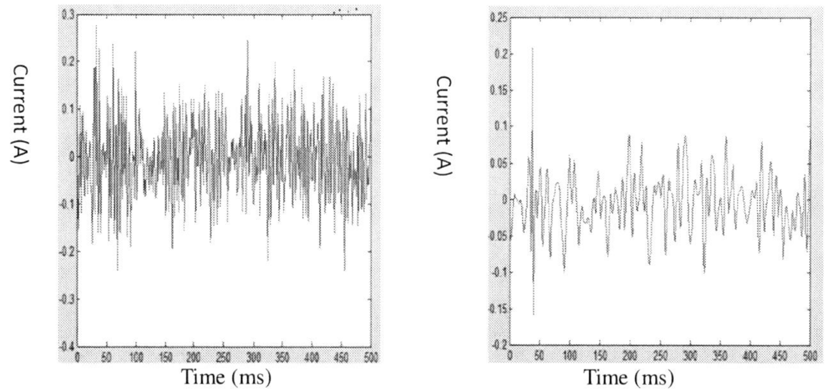

Fig. 4

According to the theory of auto-correlation analysis [10], as shown in Fig.8, the vibration signal and the current signal had certain periodicity, and they consisted of a periodic signal and some noise.

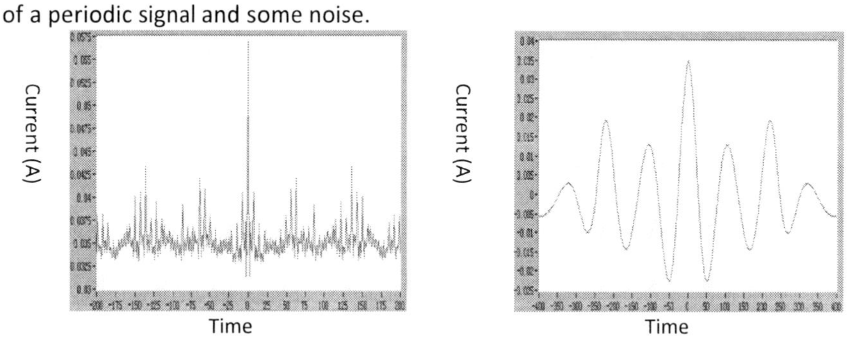

Vibration signal auto-correlation analysis   Current signal auto-correlation analysis

Fig. 5

As shown in Fig.6, by the spectrum analysis of the Experiment 4 (spindle speed 3000 r/min), the vibration signal maximum amplitude points was near the spindle rotation frequency, so at this time the spindle running was normal. In addition, there was some second harmonic frequency composition. By rotating machinery vibration theory [11], the spindle may have phenomenon of misalignment. And third harmonic frequency signal could reflect that the spindle may have internal friction. The spectrum analysis couldn't reflect the relationship between the amplitude and the time, therefore the wavelet analysis to signal was needed.

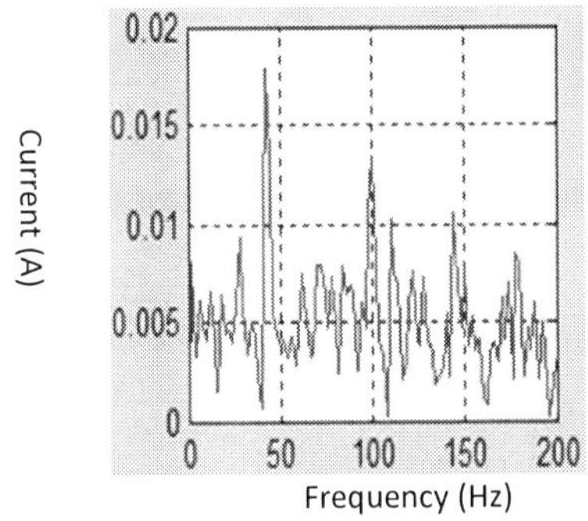

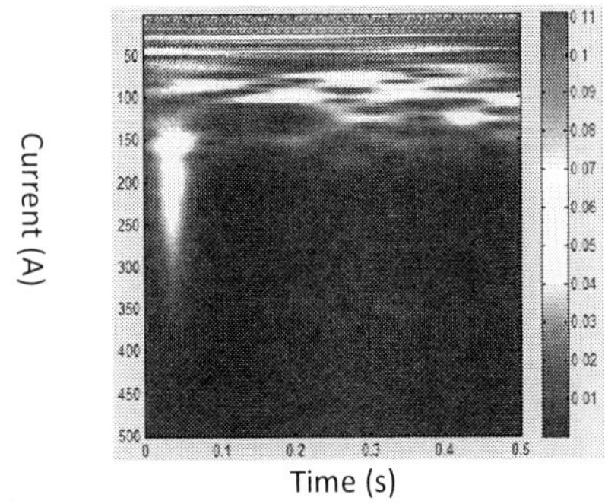

Fig. 6. Thevibration spectrum analysis figure **Fig.7.**The wavelet time-frequency figure

As shown in Fig.7, it was the vibration signal wavelet time-frequency figure which was corresponding toFig.6. The sampling frequency is 1000 Hz, and the color change showed the size of the wavelet coefficients. By the analysis of the figure it showed that the basic frequency components corresponded to spectrum analysis results. The spindle rotation frequency was the main frequency components. Its energy would change over time, so it had a certain non-stationary. The second harmonic frequency components changed significantly over time, so it didn't always exist. That is to say, the phenomenon of misalignment was slight. And the third harmonic frequency appeared only in a relatively short period of time. It indicted that there was only the mild local friction of the spindle, so it had little influence on normal processing. Through the above analysis it was concluded that the current state of the spindle run properly. Although there were parts of the second harmonic frequency, the third harmonic frequency and even more times frequency components, only the frequency components continued to occur and their amplitudes became larger, the spindle would fail.

In conclusion, the system through the spindle machining status monitoring not only could reflect the failure such as tool broken etc, but also could distinguish between normal and fault condition by analyzing the frequency components of the rotated spindle.

5. Conclusion

The article developed a machining center spindle condition monitoring system based on the monitoring objects –the domestic VDL - 1000 machining center. By use of each sensor, data acquisition card and so on to build the hardware system it could monitor the spindle status information. Using wavelet d-noising and wavelet analysis, through LABVIEW and MATLAB programming software system could realize the signal acquisition, storage, filter, analysis, and other functions. At last, through the cutting experiments of the actual machining process simulation main functions of the system was verified. The system not only could implement the spindle running state monitoring and analysis, but also could be used for failure condition monitoring. And the system could be applied to other kinds of CNC machine tools' spindles. So it had a good adaptability and high feasibility.

Reference

1. Yangang X, Han W, Xingqi L, et al. Monitor System Design for Machine Electric Spindle based on MCGS [J]. Information Engineering (ICIE), 2010 WASE International Conference on, 2010, 03:248 - 252.
2. Chang C F, Chen J J. Vibration monitoring of motorized spindles using spectral analysis techniques [J]. Mechatronics, 2009, 19(5):726-734
3. Yu-qing Zhou, Xue-song Mei, Ge-dong Jiang, et al. Technology on Large Scale Numerical Control Machine Tool Condition Monitoring Based on Built-in Sensors. Journal of Mechanical Engineering, 2009, 45(4) : 125-130
4. Zhen-bang Hu, Mu-xun Xu, et al. Analysis of non-stationary signal of a sudden unbalanced spindle based on wavelet noise reduction and short-time Fourier transformation [J]. Journal of Vibration and Shock, 2014, 33(5):20-23.
5. Dimla E, Dimla. Sensor signals for tool-wear monitoring in metal cuttingoperations—a review of methods. International Journal of Machined Tool & Manufacture, 2000,40(8); 1073-1098.
6. Xiao-li Li. A brief review acoustic emission method for tool wear monitoring during turning, International Journal of Machined Tool & Manufacture, 2002,42(2); 157-165.
7. Chuan-ben Lu. Introduction to Modern Signal Processing [M]. Beijing: Beijing University of Posts and Telecommunications Publishing House, 2003.
8. Liao T W, Ting C F, Qu J, et al. A wavelet-based methodology for grinding wheel condition monitoring [J] . International Journal of Advanced Manufacturing Technology, 2007, 47: 580 -592.
9. Xiao-chu Tang. The Wavelet Analysis and Its Application [M]. Chongqing: Chongqing University Publishing House, 2006.
10. Shi-bo Xiong, Chang-yi Huang. Mechanical Engineering Testing Technology [M]. Beijing: Mechanical industry Publishing House, 2009.
11. Bing Lv. The Research on Vibration Detection and Evaluation Technology of Main Spindle of Machining Center [D]. Lanzhou, Lanzhou University of Technology, 2008.

COMADEM
International Journal of Condition Monitoring and Diagnostic Engineering Management

Condition monitoring of closed metal structure based on ultrasonic technology

Yunzhe Sun 1[a], Dingxin Yang 2[b]*, Zheng Hu 3[c], and Baojian Hou 4[d]

[a]*National University of Defense Technology, Laboratory of science and technology on integrated logistics Support, Changsha 410073, China*
[b]*National University of Defense Technology, Changsha 410073, China*
[c]*National University of Defense Technology, Changsha 410073, China*
[d]*National University of Defense Technology, Changsha 410073, China*
**Corresponding author. Tel.:+86-15581600458; e-mail:sunyunze@126.com*

ABSTRACT

A new condition monitoring technology of through closed metal structure based on the ultrasonic is introduced in this paper. By using this promising technique, data communication could be achieved without physical penetrations. This paper focused on the key technology of high reliability communication method through a metal wall in theory and experiment based on ultrasonic. And a bi-directional reliability data transmission method is presented, they are digital amplitude modulation when the data transmits from the external to the inside and data transmission based on impedance modulation when the data transmits from internal to external. Meanwhile, weak signal-enhanced transmission methods based on stochastic resonance theory is studied, which improved the reliability of digital signal under noise background. Then an experimental plat form to monitoring the state within a metal structure is build which indicates a high engineering value.

Keywords: *Condition monitoring technology; Wireless communication; Energy harvesting; Piezoelectric material; Impedance modulate.*

1. Introduction

In modern aeronautics and aerospace industry there is a great demand for sensing the internal information of a sealed metal structure [1-2]. To obtain the internal state information of a closed metal structure, a through-wall communication system which permits to complete the bidirectional transmission of data is needed [3-4]. Unfortunately, traditional data transmission technology has strong disadvantage of limiting or potentially compromising. Firstly of all, traditional electromagnetic wave through a metal wall cannot transfer without attenuation because of the electromagnetic shielding, which will greatly reduce the reliability of the data transmission [5]. Secondly, wired communications through the metal wall will undermine the structural integrity and environmental isolations, for instance a sealed vessel which containing toxic gases. Meanwhile, holes in the metal wall which allow wires to pass through will undesirable because they reduce the integrity of the metal wall [6]. Then an alternative technique for transmitting data through metal barriers which can be used to condition monitoring and diagnostic engineering is ultrasonic communication system. Because of its perfect performance in transmission capacity and power transmission efficiency, ultrasonic transmission soon got the attention of the public [7].

Piezoelectric materials are now widely used for actuator applications because of its excellent performance. A through-wall ultrasonic energy transmission system which permits acoustic energy to be transformed into electric energy without the need for extra power to supply is introduced in this paper. Piezoelectric energy harvesting from ultrasonic through a metal wall is one of the promising alternatives method [8]. A deformation of polarized material results in an electric potential and an external electric potential will induce a deformation of the piezoelectric material, which are two progresses of the piezoelectric effect and inverse piezoelectric effect respective. On the theoretical basis of above we can model and simulate the progress of energy transmission through the metal wall.

2. system structure

An ultrasonic through-wall communication system is established by attaching a set of two piezoelectric transducers to a metal barrier as a structure of sandwich, and the diagram of the system is illustrated in Figure.1.

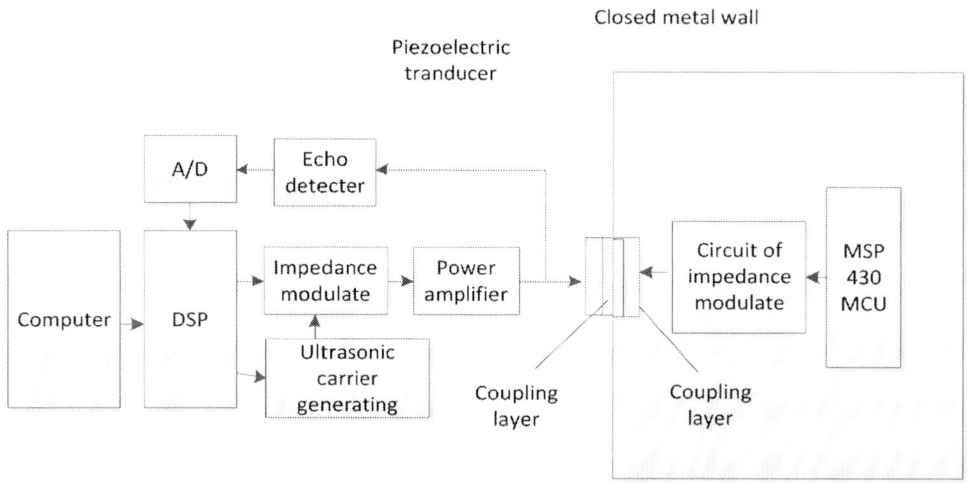

Figure 1. Diagram of the wireless communication system

The ultrasonic-electric channel is constructed to test the capabilities and limitation of a data transmission channel which is establishes by attaching a set of two piezoelectric transducers to a sealed metal wall as depicted in fig.1. The outside transmit and inside receive piezoelectric transducer are axially aligned on the block of wall. the piezoelectric transducer with a diameter of 16.5mm are made of APC850 material from American Piezoelectric Corp and the thickness of these transducers are selected to give them a nominal thickness-mode resonance at 1MHz. Ultrasonic frequency signal is generated by DDS signal source of AD9850. Power amplifier is HP1171 which can amplifier maximum output voltage 30V, used to amplify the ultrasonic signal generated by DDS signal source.

Data transmission process from outside of a metal wall to internal is as following: high frequency carrier signal generated from DDS signal source which controlled by a PC and microprocessor will connect to a power amplifier. The signals after amplification sent to a piezoelectric transducer. Analog information obtained from an inside sensor is converted into digital bit.

3. Digital pulse signal enhancement based on bistable system transmission model

The simplest method of transmitting data through a metal barrier is by pulse amplitude modulation, which will cause a series of echoes. The primary received pulse corresponds to the transmitted source data; meanwhile the echoes may cause interference with the subsequent data signal. The response of the transmitting data through a metal wall barrier is shown in Figure 2.

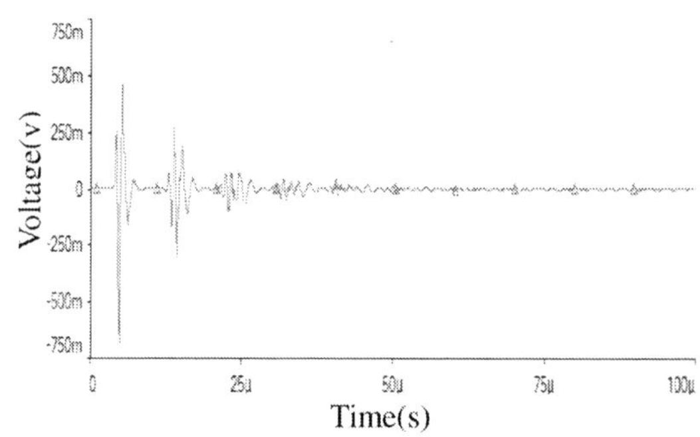

Figure 2. Response of transmitting data through a metal wall barrier

For these reasons, the balance method used to overcome the effect of echo is needed to ensure a reliable and effectively transmitted in the system of ultrasonic communication through metal wall.

For periodic and non-periodic digital signal transmission is different with sinusoidal signal detection in strong noise background which main problem is to judge there has signal or not.

The signals after processing of stochastic resonance can be as possible to restore the original signal from the output of digital pulse signal, in order to guarantee the whole system operate normally.

This section mainly discusses the problem of binary bipolar digital signal pulse transmission. The duration time T_c of each bit of the binary pulse signal is fixed. In general case, periodic array pulse symbols have a regular cycle change such as "0,1,0,1...". While for non-periodic array there is not a regularity to follow. The emphasis of this paper is to survey the duration of a code element. The relation of the bitable nonlinear dynamical systems is given by

$$\tau_0 \dot{x}(t) = x(t) - x^3(t) + s(t) + \Gamma(t) \qquad (1)$$

Where $s(t)$ is the digital pulse signal sequence, and $\Gamma(t)$

is the Gaussian white noise with the means of 0,and RMS is σ.The different of this model with normalized bitable system model is the former is introduction the parameter of τ_0.the system time parameters τ_0 is proportional to the time needed to reach steady state which can reflect the reaction speed. The smaller τ_0 is the more quickly system responds which means the more input signal can be handled.

Original binary digital pulse signal $s(t)$ can be written as follows:

$$s(t) = A\sum_{i=-\infty}^{+\infty} S_i P(t - iT_c) \tag{2}$$

Where S_i is an independently distributed random binary sequence, and $P(t)$ is a rectangular pulse.

$$P(t) = \begin{cases} 1 \rightarrow t \in [0, T_c] \\ 0 \rightarrow else \end{cases} \tag{3}$$

Pulse signal $s(t)$ can be seen converted from the digital sequence of "0" and "1". the binary digital signal "0" is mapped into $-A$, correspondingly the binary digital signal "1"is mapped into $+A$. Each one symbol at the start time is $t = iT_c$ and last time is T_c.

Numerical simulation methods are used to study the digital pulse signal transmission. The input signal is sampled can be expressed as follows:

$$s(nh) + \Gamma(nh) \tag{4}$$

Where n=1,2 ⋯ N. Because there are only two possible states of the input signal in a symbol interval, substituting (4) into (1) the model can be expressed as shown in

$$\tau_0 \dot{x}(t) = x(t) - x^3(t) \pm A + \Gamma(nh) \tag{5}$$

Due to the output of the system is analog signal, it is necessary to restore it as a digital pulse sequence. The equations can be expressed as following:

$$y(t) = sign[x(t)] \tag{6}$$

Bit error rate in a digital pulse signal can be indicated the accuracy of signal transmission, which is an important indicate to reflect the quality of digital signal transmission. In addition, the nonlinear system itself can be seen as a data transmission channel, according to information theory, channel capacity can be used to describe the data transmission capabilities of the system.

4. Bidirectional digital communications through a metal wall

4.1 theory of data transmission in the metal wall

The pressure field distributed in the metal wall is presented in Fig. 5. For the resonant frequency of the piezoelectric transducer is simulated and the simulation result shows the maximum pressure appear near the resonance frequency.

Point to be noted that the physical parameters of the piezoelectric transducers used in this paper vary depending on piezoelectric transducer's statements: hanging freely in air or attached to the metal wall with epoxy or a cyanoacrylate adhesive. Piezoelectric transducer coupled with different metal wall will get the different results.

Table 1. The material properties used in the experiment

Property	Value
Density of piezoelectric[g/cm^3]	7800
Area of piezoelectric transducer on wall[m^2]	0.01
Wall thickness[m]	0.022
Piezoelectric transducer thickness[m]	0.002
Piezoelectric transducer coefficient e_{33}[C/m^2]	23.2
Elastic stiffness at constant displacement c^D_{33} [N/m^2]	15.87
Mechanical quality factor Q	80

4.1 Working principle of data transmission based on ultrasonic waves

When ultrasonic waves which made from the outside of piezoelectric transducer reach the metal, part of the energy will be reflected back, and the remaining ultrasonic energy go through the metal wall. An ultrasonic carrier is applied to the outside piezoelectric transducer and travels to the inside piezoelectric transducer, where a fraction of the acoustic vibration power is reflected back towards the outside. Changing the electrical load on the terminals of the transducer (for instance, setting opening or shorting the terminals) alters its electric impedance which, in turn, modulates the size of the reflected signal. The changes in the reflected signal are seen by the outside transducer. The changes in the outside transducer electrical impedance that are produced by the variations in the reflected signal result in amplitude variations in the applied voltage [9]. These amplitude variations are related to the data being sent from the inside and are easily detected to recover the binary data. And the communications modulation circuit is shown in Figure 3 as following

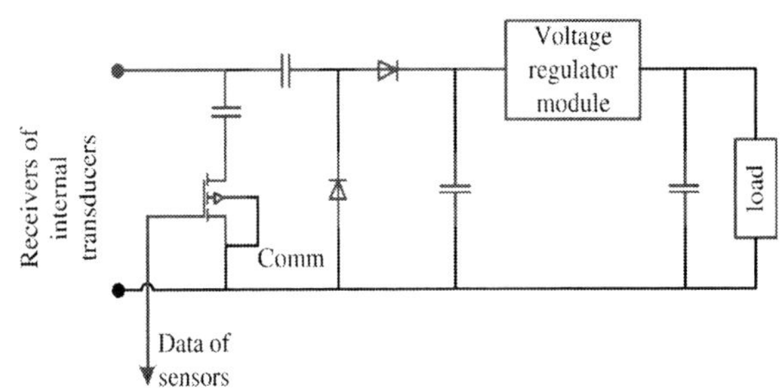

Figure 3. Communications modulation circuit

4.2 Data transmission system

The system is designed to operate with transducers having a center frequency of approximately 1MHz, though other frequencies can be used.

Rather than have a resonant circuit on board whose output frequency would be temperature dependent, the clock is derived from the carrier signal using a simple zero-crossing detector comparator circuit, the third direct connection to the inside transducer. This circuit generates a clock at the exact carrier frequency (approximately 1 MHz) to drive the logic electric components. Knowing the data rate exactly makes it relatively easy for the outside of the metal wall to synchronize to the incoming data stream. Finally, a metal-oxide semiconductor field-effect transistor (MOSFET) is used to open or short the transducer terminals in accordance with the data being sent. The internal MCU control the modulation circuit to generate a modulation rate and receiving transducer on other side will detect the received signal. After passing through the filter circuit we can get the data of information which is shown in Figure 4 .

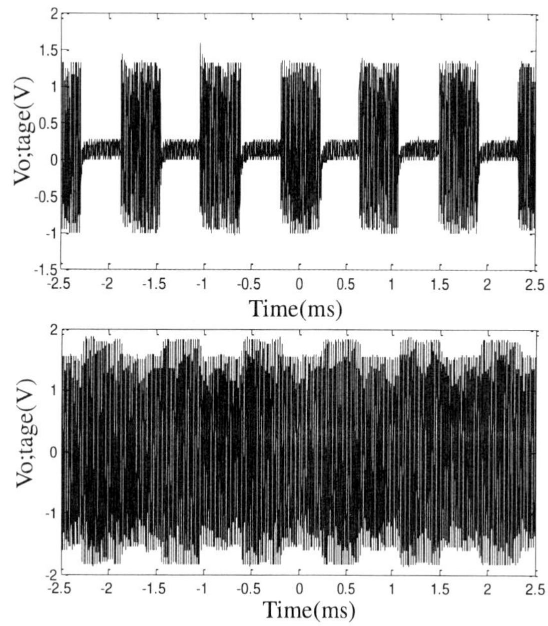

Figure 4. Data transmission waveforms

The true communication need to send the regular data of zeros and ones. The waveform of the temperature sensor data is shown in Figure 5.

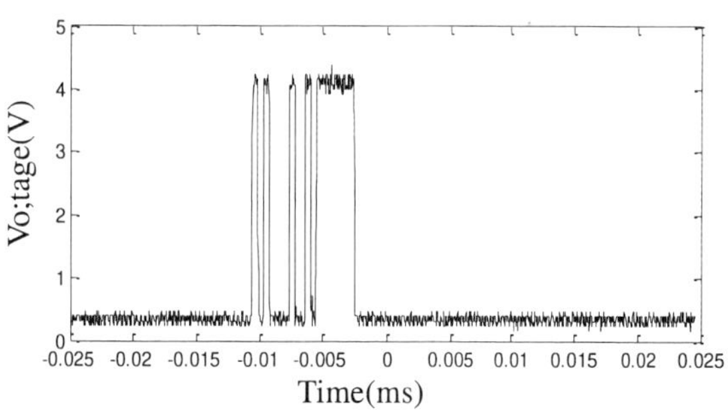

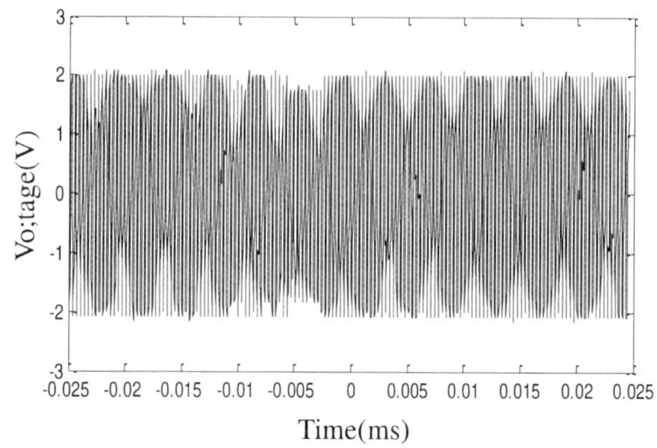

Figure 5. Waveform of the temperature sensor data

5. Energy harvesting consideration

As a closed structure, it is import without the external electric power supply. So the energy harvesting based on the piezoelectric material is significantly useful.

A piezoelectric transducer can be used either to transform an electric current to an acoustic pressure field, or the opposite to produce an electric current from an acoustic field [11]. The absorbed portion of the ultrasonic wave is converted into electrical energy through piezoelectric transducers and harvest by the internal electric circuitry [8].

6. Discussion

This paper has presented the approach used for condition monitoring and fault diagnosis through the closed metal wall. Technology for communication digital sensor data through a based on the ultrasonic is introduced. The results show that the system performs well at low data rates. However, the degradation due to multiple reflections within the metal barrier become significant and will adversely affect performance.

Acknowledgements

The authors would greatly appreciate the support provided by National Nature Science Foundation of China No.51375485 and No.51305449 for this work.

Reference

1. Lei, Y., Lin, J., Zuo, M. J ,He, Z. (2014). Condition monitoring and fault diagnosis of planetary gearboxes: a review. Measurement, Volume: 48 Issue:1. 292 – 305.
2. Duan, Y., Toliyat, H. (2012). A review of condition monitoring and fault diagnosis for permanent magnet machines.Power and Energy Society General Meeting IEEE, Volume: 59 Issue:5. 1 – 4.
3. Primerano, R. A. (2009). High bit-rate digital communication through metal channels. Information Sciences and Systems, 2009. CISS 2009. 43rd Annual Conference on IEEE. 902 - 906.
4. Roa-Prada, S., Scarton, H. A., Saulnier, G. J., Shoudy, D. A., Ashdown, J. D., and Das, P. K., et al. (2013). An ultrasonic through-wall communication (utwc) system model. Journal of Vibration & Acoustics, Vol.135,issue 1, 48-65.
5. Lawry, T. J., Wilt, K. R., Ashdown, J. D., Scarton, H. A., and Saulnier,

G. J. (2013). A high-performance ultrasonic system for the simultaneous transmission of data and power through solid metal barriers.. IEEE Transactions on Ultrasonics Ferroelectrics & Frequency Control, Vol.60, issue 1, 194-203.

6. Ozeri, S., Shmilovitz, D., Singer, S., and Wang, C. C. (2010). Ultrasonic transcutaneous energy transfer using a continuous wave 650kHz Gaussian shaded transmitter. Ultrasonics, Vol. 50，issue 7, 666-674.

7. Yang, Z., Yang, J., and Hu, Y. (2008). Nonlinear behavior of electric power transmission through an elastic wall by acoustic waves and piezoelectric transducers. Ultrasonics, Ferroelectrics, and Frequency Control, IEEE Transactions on, Vol.55, issue 11, 2527-2531.

8. Hu, Y., Zhang, X., Yang, J., and Jiang, Q. (2003). Transmitting electric energy through a metal wall by acoustic waves using piezoelectric transducers. Ultrasonics, Ferroelectrics, and Frequency Control, IEEE Transactions on, Vol.50，issue 7, 773-781.

9. Primerano, Richard A.(2009). High bit-rate digital communication through metal channels. Information Sciences and Systems. CISS 2009. 43rd Annual Conference on IEEE, 2009:902 - 906.

10. Proakis, J. G., Sozer, E. M., Rice, J. A., & Stojanovic, M. (2001). Shallow water acoustic networks. Communications Magazine, IEEE, Vol.39，issue 11, 114-119.

11. Yang D X, Hu Z, Zhao H, et al. (2015). Through-Metal-Wall Power Delivery and Data Transmission for Enclosed Sensors: A Review, Sensors, Vol. 15, issue 12, 31581-31605.

A FAULT DIAGNOSIS METHOD BASED ON WPD AND t-SNE

A. Gu Yuhai[1] B. He Linfeng[2] and C.Ma chao[3]

[1] Beijing Information Science and Technology University, The ministry of education key laboratory of modern measurement and control technology, No.12 Xiaoying east road, Qinghe, Haidian district, Beijing, 100192,China.
guyuhai@bistu.edu.cn

[2] Beijing Information Science and Technology University, The ministry of education key laboratory of modern measurement and control technology, No.12 Xiaoying east road, Qinghe, Haidian district, Beijing, 100192,China.
494185471@qq.com

[3] Beijing Information Science and Technology University, The ministry of education key laboratory of modern measurement and control technology, No.12 Xiaoying east road, Qinghe, Haidian district, Beijing, 100192,China.
machao@bistu.edu.cn

Abstract. A fault diagnosis method based on 2D manifold map by using vibration signal of rotating machinery is presented, which combines the method of time-frequency analysis and manifold learning. The sampled vibration signal is decomposed to multilayer information with wavelet packet decomposition (WPD) method to construct high dimensional data vectors as matrix. The matrix structure is different with different fault conditions. The every level data of wavelet packet decomposition is processed by envelope demodulation of Hilbert, FIR filter, and resampling to extract the modulation signal contained in each level and reducing the data length. Further, every level data vector is deal with autocorrelation computation and normalization and constructed as high dimensional data vectors. Finally, the manifold learning methods of t distributed stochastic neighbor embedding (t-SNE) is applied to do dimension reduction with the vectors to generate 2D visualization manifold map. The generated map of vibration signal is applied to fault diagnosis and classification of rotary. With the experiment test, the feasibility and effectiveness of this method is verified.

Keywords: WPD, Hilbert demodulation, t-SNE, Visualization fault diagnosis method

1 Introduction

Fault diagnosis for rotating machinery is now principally realized by acquiring vibration signals from key parts of rotating machinery and extracting and analyzing fault-related sensitive characteristics from vibration signal.[1] Since not a little char-

Foundation-supported program: National Natural Science Foundation-supported Program (51275052); Key Program Funded by Beijing Municipal Natural Science Foundation (3131002).

acteristic information in vibration signal is associated with the operating state of machine, it is almost impossible to effectively identify the operating state only by separately extracting one kind of characteristic information from vibration signal. Although fault identification for operating state of equipment could be performed more accurately by attaining a few features by using feature extraction method and making integrated use of the characteristic information, the increase of characteristic information may lead to the remarkable increase of information dimensions, thereby resulting in difficulty in engineering application. The fault state may vary depending on structure of high-dimensional feature vector composed of various feature parameters; manifold learning helps to express the internal structure of high-dimensional data visually so that the fault state could be identified in an intuitive manner.

Manifold Learning Algorithm is a non-linear machine learning algorithm which maps high-dimensional space data to low-dimensional space based on the inherent law or spatial geometry of high-dimensional data, and visually expresses the degree of similarity or distance between data in high dimensional space in the form of two or three-dimensional manifold distribution. At present, the frequently studied manifold learning algorithms principally include principal component analysis (PCA) algorithm, locally linear embedding (LLE) algorithm, isometric mapping (Isomap) algorithm and local tangent space alignment (LTSA) algorithm, etc. [2]-[5]

Despite their extensive application for machinery fault diagnosis, these manifold learning algorithms normally bring about such problems as nonlinear data congestion and inadequately clear low dimensional manifold expression[6]-[7]. To make up for the shortcomings of above-noted manifold learning algorithms, the following procedure could be followed based on t-distributed stochastic neighborhood embedding (t-SNE) algorithm[8]: Decompose the acquired vibration signal into multi-layer signal data through wavelet packet decomposition; demodulate each layer of data respectively through Hilbert transform to obtain fault-induced modulation signal components; use a FIR filter to remove high-frequency noise, retain low-frequency signal through low frequency resampling, reduce the data length; make the processed layers of data into high-dimensional data vector, and subject high-dimensional vector to dimension reduction through manifold learning algorithm to generate two or three-dimension graphics[9]; since the manifold result varies depending on fault state, the fault state of equipment could be identified based on manifold shapes.

2 Principle of Wavelet Packet Decomposition-based Manifold Learning-aided Fault Diagnosis

Acquire the vibration data from specified parts of machinery equipment, decompose one-dimension data into multi-layer detail data through wavelet packet decomposition, and make multiple layers of data into high-dimensional data vector.[10] The characteristic information from low frequency to high frequency varies depending on the layer of data packet, while the frequency content distribution varies at different layers depending on fault characteristics. Hence, the internal structure generated between high-dimensional data varies depending on fault state. Ultimately, the mapping relation comes into being between two-dimensional visualized graphic shape and fault state after manifold learning-based dimension reduction.[11] From the spatial perspective after wavelet packet decomposition, the original one-dimension data is decomposed into high-dimensional data that represents different frequency components; the high-dimensional data is subjected to feature extraction; finally, the visual graphics vary after manifold learning-based dimension reduction. The procedure of manifold learning-based fault diagnosis method is shown in Figure 1.

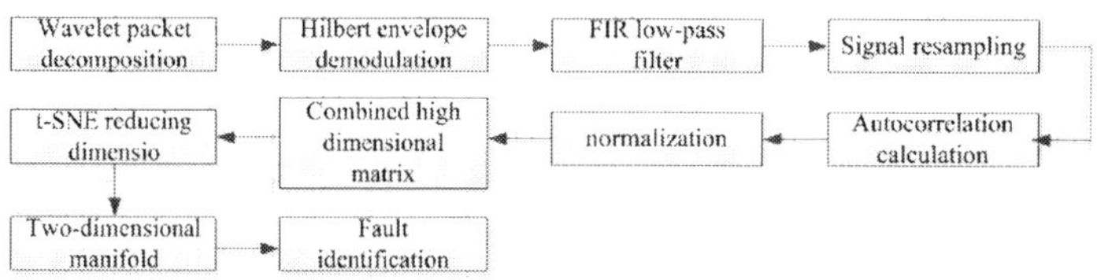

Fig. 1. The generation procedure of high-dimensional data based on the time domain features of wavelet packet decomposition

Subject the acquired vibration data to N-layer wavelet packet decomposition to generate 2N component time-domain signals; subject the component signals to Hilbert envelope demodulation,[12] and extract the fault-induced modulation signals from each component; subject demodulation signals to FIR filtering so as to eliminate the high-frequency noise from each component; resample each component signal at low sampling frequency so as to reduce data length, thereby accelerating the calculation; subject resampling signals to autocorrelation calculation, and normalize to autocorrelation coefficient; finally, combine the normalized components into L×2N-dimension matrix, where L stands for the data length after resampling of component signal. Subject the high-dimensional matrix of the constructed time domain signals to dimension reduction through t-SNE manifold learning algorithm to generate two - dimension graphics. Since the manifold result varies depending on fault state, the fault state of equipment could be identified based on manifold shapes.

3 Test Data Acquisition and Test Data Analysis

Bentley RK4 micro-rotor experiment table was used to acquire actual vibration signals in order to verify the effectiveness of wavelet packet decomposition-based manifold learning-aided fault diagnosis method. Data acquisition unit INV3062 manufactured by Beijing Coinv was used for the experiment; INV3062 is designed with 4-channel synchronous acquisition, and each channel is designed for 24-bit data acquisition at a sampling frequency of up to 51.2KHz; this data acquisition unit could be connected to ICP acceleration sensor or voltage input signal. Rotor experiment table bearing block vibration data was acquired with three-axis acceleration sensor INV9832A at a sampling frequency of 20.48KHz.

The data of normal state, rotor imbalance, rub impact and base looseness acquired at Bentley RK4 rotor experiment table was processed into high-dimensional feature data through wavelet packet decomposition, and the fault was identified through manifold learning. The length of data acquired for each fault state was 40960 points; decompose the vibration data of each of the 4 states into 16 component data through 4-layer wavelet packet decomposition using "db1" wavelet; subject each component data to Hilbert envelope demodulation, and subject the demodulated data to FIR digital low-pass filter at a frequency that is equal to 1/8 of original sampling frequency; resample the filtered data at a frequency that is equal to 1/4 of original sampling frequency; calculate the autocorrelation function of resampled data and normalize the data; combine the normalized data into 640×16-dimension characteristic data. Figure 3 shows the two-dimensional visual graphics generated through dimension reduction of 640×16-dimension characteristic data using PCA algorithm.

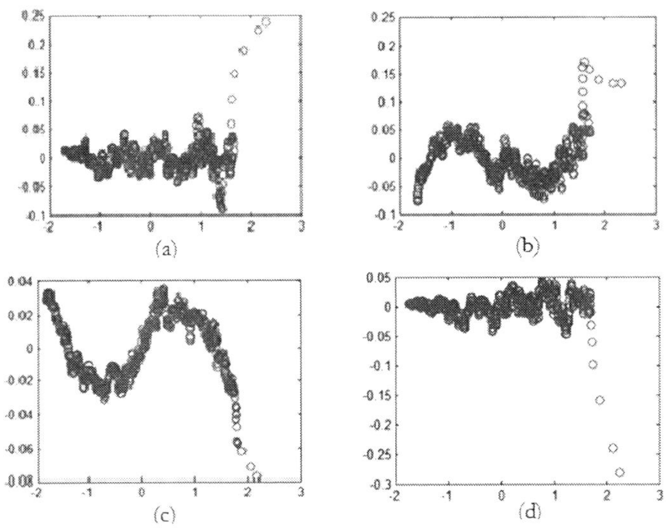

Fig. 2.(a) (b) (c) (d) : The two-dimensional manifold figure of imbalanced fault , rub-impact fault and foundation looseness fault and the normal state through dimension reduction based on the PCA.

Figure 2(a), (b), (c)and (d) show the result of PCA manifold learning-based dimension reduction after preliminary treatment performed with Wavelet packet decomposition signal; as shown in those figures, different fault states exhibit different two-dimensional manifold forms, but the structure of two-dimensional manifold exhibits poor and inadequately clear regularity.

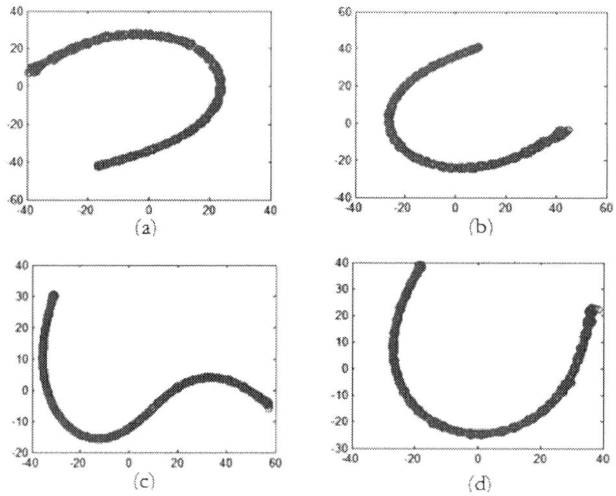

Fig. 3. (a) : The two-dimensional manifold figure of the imbalanced fault **(b)** : The two-dimensional manifold figure of the rub-impact fault **(c):** The two-dimensional manifold figure of the foundation looseness fault **(d):** The normal two-dimensional manifold figure

Figure 3 (a), (b), (c) and (d) shows the two-dimensional Visual manifold generated through dimension reduction of the same set of 640×16-dimension characteristic data using t-SNE algorithm. As shown in those figures, the two-dimensional manifolds of different fault states after t-SNE manifold learning-based dimension reduction are more regular and clear as compared with PCA algorithm; t-SNE algorithm offers better post-dimension reduction manifold structure when compared with PCA.

4 Conclusion

Use wavelet packet decomposition, Hilbert transform envelope demodulation, FIR filtering and low frequency resampling to build high-dimensional data vector that represents the fault state structure characteristics, generate two-dimensional low-dimensional manifold by performing dimension reduction of high-dimensional data through t-SNE manifold learning algorithm, identify the fault state of machinery equipment through different manifold structural forms, and create the fault identification of machinery equipment. This method enables wavelet packet decomposition for vibration signals from a number of sensitive characteristic areas of machine respectively, and generates high-dimensional data in a unified manner; this signal processing and dimension reduction method is used to to obtain low dimensional manifold for the identification of the fault state of machinery equipment.

References

1. Wang Hongjun, Wan Peng. Sensitive features extraction of early fault based on eemd and wpt. Transactions of Beijing Institute of Technology[J].2013.33(9): 945-950.

2. Renping Shao, Wentao Hu, Yayun Wang, Xiankun Qi. The fault feature extraction and classification of gear using principal component analysis and kernel principal component analysis based on the wavelet packet transform. Measurement[J], 2014, 54(6):118-132.

3. Roweis S.T , SaulL.K. Nonlinear dimensionality reduction by locally linear embedding [J].Science , 2000 , 290:2323-2326.

4. He Qing, Xie Fangfang, Li Hong, Lan Lan.Feature extraction of vibration of turbine unit based on manifold learning method. Vibration test and diagnosis | Vibration test diagnosis[J].2014,34(4):705-708.

5. Balasubramanian M, Schwartz E, Tenenbaum J, de Silva V, Langford J, The isomap algorithm and topological stability[J]. Science 295:7, 2002.

6. Hinton G,Roweis S.Stochastic Neighbor Embedding. Advances in Neural Information Processing Systems . 2003.

7. HINTON G,ROWEIS S.Stochastic neighbor embedding. Neural Information Proceeding Systems:Natural and Synthetic . 2002.

8. L. vander Maaten and G. Hinton, "Visualizing data using t-SNE," Journal of Machine Learning Research.2008. 9 (2605):2579-2605.

9. Tenenbaum J.B, Silva V.D, Langfold J.C. A global geometric framework for nonlinear dimensionality reduction [J]. Science, 2000, 290:2319-2323.

10. Wan Peng, Wang Hongjun, Xu Xiaoli. Fault diagnosis model based on local tangent space alignment and support vector machine. Chinese Journal of Scientific Instrument[J].2012.33(12):2699-2795.

11. Xiang Dan, Ge Shuang. A model of fault feature extraction based on wavelet packet sample entropy and manifold learning. Journal of vibration and shock[J],2014,33(11):1-5.

12. HU Yi Ding, R EN Weixin, Yang Dong, LI Miao. Demodulation of non- stationary amplitude modulated signal based on Hilbert transform. Journal of vibration and shock[J].2013,32(10):181-183.

Remote Knowledge Base Building Method on Locomotive Traction System

Bin Ren[1], Shaopu Yang[1], Xiaoli Xu[2], Junwu Zhao[1]

[1] Shijiazhuang TieDao University, 17 Northeast, Second Inner Ring , Shijiazhuang, Hebei.
renbin@stdu.edu.cn

[2] Beijing Information Science&Technology University, No.12 Qinghe Xiaoying-donglu, HaiDian Qu,Beijing

Abstract. The safety and efficiency of railway transportation are affected directly by the stability and reliability of locomotive traction system. In order to effectively guarantee the reliable, stability and safety of railway transportation, research on the method of remote knowledge base building on the key parts of locomotive traction system. Facing the heavy load locomotive, the state detection test platform taking traction system as core was constructed, knowledge acquisition method based on the variant precision rough was proposed, construction method of knowledge base model was studied, knowledge acquisition test platform based on remote network was established, data inquiry of locomotive status was provided for users, which will provide the key test technologies to reveal dynamic performance and fault mechanism of locomotive and the internal relation between characteristics parameters and various faults.

Keywords: locomotive traction system; variant precision rough sets; remote knowledge base

1 Introduction

Railway transportation in China is developing in the direction of safety, high speed, heavy load and intelligent. Locomotive traction device is used as the key equipment in the modem development of railway transportation, its running stability, reliability and safety of a direct impact on the development of the economy. Because traction system integrates mechanical, electronic and information technology into one intelligent system, and mechanical structure is complicated, which leads to various hidden failures inevitably in running process, thus leading to the descend of equipment dynamic, decline of precision, increase of accident rate and so on[1]. In recent years, researchers applied artificial intelligence such as expert system, pattern recognition and rough analysis to fault diagnosis of locomotive key parts. Oliveviler Ondel et al. used Kalman filter estimation method to build coupling mode recognition model of motor, and determine the fault type accurately in different conditions [2]. Stator current is used as the researched object, 5 HMM were established by Hidden Markov Model, and the locomotive fault state recognition was realized by calculating the largest probability[3]. Combine wavelet packet characteristic entropy and rough set theory to realize fault diagnosis of electric locomotive traction motor [4]. Fault diagnosis model based on information temporal fusion is applied to fault diagnosis analysis of electric locomotive traction motor to detect the fault position and reason accurately [5]. Therefore, in order to study fault predicting technology of locomotive traction system deeply, the multi-sensor fusion technology will be used to establish online detecting experimental platform with the goal of constructing rolling bearing knowledge base, combine the special law of rolling bearings during operating state and adopt the variable precision rough set as theoretical basis to research knowledge acquisition and method of the database construction by modern measurement technology and information processing technology. Reliable knowledge rules base will be provided for fault diagnosis of locomotive traction system and alert research. Reliable sample data can be provided for signal analysis and the research of unit detection technology by constructing remote online detecting experimental platform, providing corresponding knowledge acquisition and the method of the database building, which is helpful for us to understand the effect of the key parts on locomotive's operation process. It's significant for improving railway transport efficiency and maintaining good operation of locomotive [6].

2 Dynamic Monitoring Experimental Platform

The operation process of locomotive traction system is variable, how to acquire the data of operating status of equipment on polytropic conditions is a foundation of predicting and diagnosis. Taking the rolling bearing of traction system as test object, the detection equipment required to evaluate the system dynamic performance is equipped to implement the fault simulation of rolling bearing, which provides fundamental experimental condition for acquisition of knowledge data, typical method of fault monitoring and simulation verification. The construction plan of dynamic monitoring experimental platform is shown in fig 1.

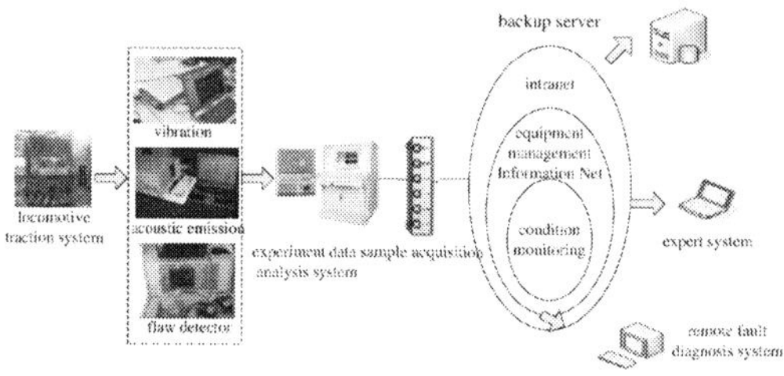

Figure1 Dynamic performance fault monitoring test platform

3 Knowledge Mining Based On Variable Precision Rough Sets

3.1 Summary of knowledge mining theory

The theory of Rough Sets, presented in 1982 by Polish mathematician Pawlak Z, which is a new mathematical tool for researching in uncertain, fuzzy knowledge of the knowledge representation and data processing. Its principle can be concluded that the setting of universe---feature attribute discretization---knowledge reduction---logical rules [7]. Thus knowledge rule can be achieved. Because of the classifications of the Pawlak's rough sets model must be precise, and the objects analyzed must be known, the practical applications will be limited to a great extent. The Variable precision rough set theory is the expansion according to the Rough Set Theory by Ziarko, which introduces the threshold β（$0 \le \beta \le 0.5$）, in order to solve the data classification problems when there is no function or uncertainty relation between attributes caused by noise.

the setting of universe

Let X and Y expressed nonempty subset of the finit case U. If for every e∈X, and e∈Y, then Y includes X, denoted by Y⊇X. Letbe

$$c(X,Y)=\begin{cases}\dfrac{|X\cap Y|}{|X|}, & |X|>0,\\ 0, & |X|=0,\end{cases} \qquad (1)$$

Called c(X， Y) is the relative classification correct rate of set X for set Y.

$$\underline{R}_{\beta}X = \cup\left\{[x]_R \mid \frac{|[x]_R \cap X|}{|[x]_R|} \geq \beta\right\} \qquad (2)$$

For a certain X⊒U and a indiscernibility relation R⊒A in U , threshold 0.5<β≤1 is given, then there is, which is β approximate set under the R of X, orβ positive region of X, denoted by, and namely get objects of U divided into object set of X with classification accuracy that is not less than β.

feature attribute discretization

Attribute discretization is to make the choice of interpolation in the objective observations of some attribute for all objects in the universe, and the purpose is to divide continuous characteristic space into groups effectively, then the size of the attribute range sets will be reduced.

The knowledge reduction

On the premise of maintaining origin information system or classification ability of decision tables, the knowledge reduction is to delete redundant knowledge of knowledge representation system for getting the smallest knowledge covering of decision tables. Effective information with more condition attributes can be rapidly obtained by information content and importance degree, which conduct the knowledge reduction based on variable precision rough set. The reduction method as following:

Step1 According to finding Core algorithm, core attribute is obtained

Step2 According to equation （2） and （3） respectively calculates information content and importance degree of core attribute set.

$$I(c) = 1 - \frac{1}{|U|^2}\sum_{i=1}^{n}|X_i|^2 \qquad (3)$$

$$E(c) = \frac{\sum_{i=1}^{n}|pos_c(D_i)|}{|U|} \qquad (4)$$

$|X_i|$represents the number of objects included some equivalence class that is divided by some equivalence relationship. D_i represents some equivalence class of decision class, and $|pos_c(D_i)|$ represents rate of object that condition attribute set can distinguish.

Step 3 In the equation of c\core, let

$$red = core \bigcup \max sig(c_i) \quad (5)$$

Step 4 I(core)=I(c) and E(core)=E(c) will be judged . If equations are tenable, the "red" can be regarded as a reduction, or calculate $sig(c_i)$, which $c_i \in C - red$.

The logical rules

According to all kinds of simplified form of reduction, the knowledge of nuclear is determined, and the knowledge rules set for all the decision classes will be obtained.

3.2 knowledge base construction

Taking the traction system of the HXD3B type electric locomotive as research object, the experimental of sample acquisition is researched by the reduction rule of variable precision rough set, a preliminary classification is conducted to attribute conditions which can represent equipment running condition, then the initial attribute subsets is got. Attributes reduction is optimized based on the heuristic method and the rules coverage and confidence degree is used to evaluate rules, then samples knowledge acquisition models of traction system will be established [8].

According to the complex features of electric locomotive running condition, the fault diagnosis knowledge database of traction system will be built, and the method as shown in fig 2:

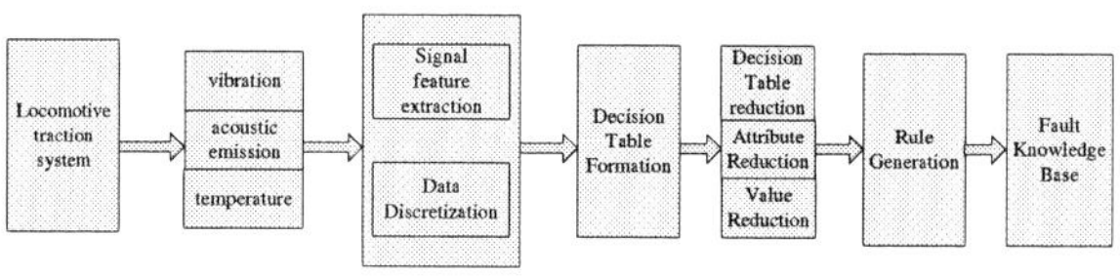

Figure 2 Traction system sample fault case database

The method of knowledge acquisition and database building based on the variable precision rough set as follows:

(1)The information of normal and wear condition on traction system is got by the test platform, the main information includes vibration (amplitude, frequency), sound emission, temperature and precision, which the vibration signal can be collected by two directions on vertical and transverse of traction motor. Acoustic emission signal can be collected by direction on vertical of traction motor. Then the acquisition signal will be made to time domain analysis, and extract features which can reflect fault characteristic information and operating status of equipment.

(2)Based on multiple symptom characteristics information, the different condition information collected is made to signal processing, and data discretization after fault feature extraction. Taking feature information attributes as condition attribute, and different equipment fault state as decision attribute, then fault diagnosis decision table will be built.

(3)Based on the established knowledge decision table, the characteristic attribute reduction is conducted by the information quantity and importance degree. which is an object set, that is the finite nonempty set of universe; D is a decision attribute set , is a condition attribute set , AT is an attribute set; V is attribute value domain, and is an information function, he method of knowledge reduction mentioned above are used to obtain effective knowledge.

(4)The rule confidence is used as the evaluation index to measure and evaluate on the decision rules reduced [9]. Because of the inconsistent fault examples included in the fault diagnosis decision table, and the different performance characteristics of different diagnosis rules, let $\{rule_1, rule_2, \cdots, rule_n\}$ as the generalized decision rules set, and every sequence such as $c_1(rule_i), c_2(rule_i), \cdots, c_n(rule_i), d_1(rule_i), d_2(rule_i), \cdots d_m(rule_i)$ is determined by every rule $rule_i$.

$C' = \{c_1, c_2, \cdots, c_n\}$ is the subset of the conditional attribute set of the decision table,

$D = \{d\}$ is decision attribute;

The rules confidence α

$\alpha(rule_i) = \frac{card(C'(rule_i) \cap D(rule_i)))}{card(C'(rule_i))}$,can be expressed by the value of rough membership function. Which $C'(rule_i)$ is the condition attribute, $D(rule_i)$ is the decision attribute; $card(C'(rule_i))$ is the number of examples that meets the condition attribute ($C'(rule_i)$) of $rule_i$.

When $card(C'(rule_i)) \neq 0$, $card(C'(rule_i) \cap D(rule_i))$ is the number of examples that meets the condition attribute ($C'(rule_i)$) and the decision attribute ($D(rule_i)$) of $rule_i$, which reflect the rules reliability.

4 Based on the Remote Network Knowledge Base Building Technology

In order to forecast locomotive traction system and verify the methods of fault warning diagnosis effectively, the samples and knowledge base with an open style and reconfigurable environment is established, which can be used to provide data and knowledge for the evaluation of the running conditions of system performance and the fault early warning. Functions such as obtaining、modifying and updating are realized by the establishment of the system based on remote network knowledge base. The modular design is used to meet the function of data administration. According to the practical needs, relevant data and information are obtained automatically by the various methods. The analysis process of data and logical rules is realized by the network data analysis and fault diagnosis system, response to service request of users, a bridge between user service and data service will be built [10].

The fault analog signal acquisition can be conducted by several kinds of equipped test instruments on the knowledge acquisition test platform, that is the experimental study, which is aimed to obtain the accuracy、vibration、temperature and acoustic emission of different key parts. Then the text forming sample data obtained are guided into the data base, which can be used to the analysis and management of data acquisition.

Taking the rolling bearing as the study object, the INV9824 type acceleration sensor is used to acquisition acceleration signal samples. There are three kinds of information types: bearing crack、bearing peeling and gear wear. Gathered signals will be input with the type of text format specified by the system, and data sharing will be realized by remote network. The system is shown in figure 3.

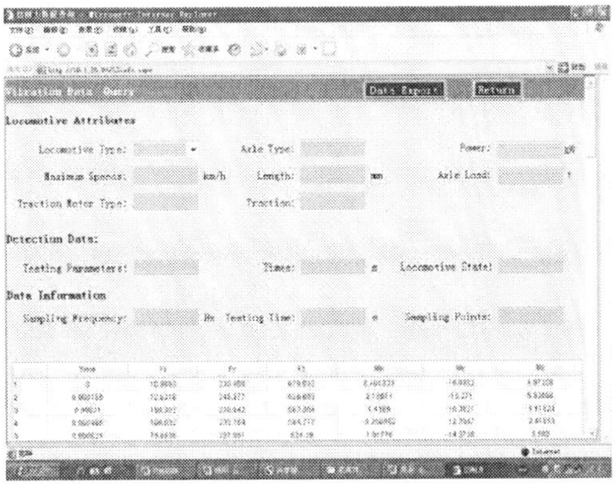

Figure 3 Remote network knowledge base input system

The knowledge acquisition system is designed by the Microsoft Visual Studio 2008 as developing environment, and is developed in the way of WebForm based on the Framework3.0、IIS6.0, which such as C#、ASP.NET、JavaScript、SQL2008 and Matlab are used as develop language and technology. Functions such as entering、saving、searching and trend display are realized on the developed system. The system can be divided into three parts mainly: data input、data analysis and knowledge base. Different analysis can be completed by C# calling Matlab program.

Through the features extract, the energy spectrum of vibration signal is used as characteristic parameter to compose the initial decision chart. At the same time, information rule base is obtained by attribute reduction based on the variable precision rough algorithm, and confidence is used as the evaluation index. Knowledge base system is shown as Figure 4.

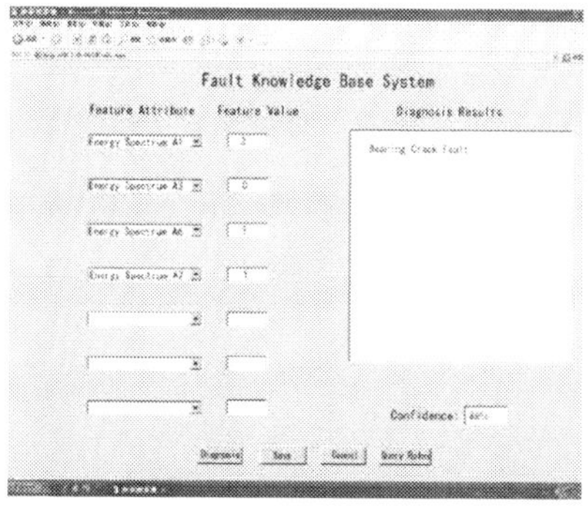

Figure 4 Fault knowledge base input system

According to the diagnosis rules obtained, which can satisfy the actual working conditions by experimental verification. The acquisition of final rule can reveal some redundant information in the characteristic data after data analysis. The briefest rules table can be obtained by rough reasoning, which can provide the reliable data sources for locomotive fault prediction.

5 Conclusions

Research on remote network Knowledge Base of Locomotive Traction System, can assure the key parts of locomotive safety, stability and reliability, and can prevent the fault, decrease the downtime and increase the efficiency of railway transportation. Which have very important significance to improve the economic efficiency for extend the time of operating without failure.

Remote knowledge base of locomotive traction system was built, which can realize data acquisition detection of locomotive dynamic performance. The test platform with core of sample acquisition and database experiment, equipped with the detection system of precision measurement, vibration signal, AE signal and noise signals. Multiple data sample acquisition test system has been built, and the method of sample acquisition test is studies. The method of data acquisition based on variable precision rough sets theory has been studied, which provide the method of database construction based on knowledge. The data acquisition platform is built by remote network, which provides the experimental equipment and key technology for the analysis of locomotive dynamic performance, the forming mechanisms of fault and the internal relation between characteristics parameters and various faults.

ACKNOWLEDGMENT

This work is partially supported by the National Natural Science Foundation of China (Grant No. 51405313, 11227201, 11202141).

REFERENCES

1. Wei Liu, Haibo Qiu. Research on the technology of data fusion in locomotive traction motor fault diagnosis system[J]. Silicon Valley,2010(02):139.
2. Olivier Ondel, Emmanuel Boutleux, Eric Blanco. Coupling pattern recognition with state estimation using kalman filiter for fault diagnosis[J]. IEEE Trans on Industion Elect,2012,59(11):4293-4230.
3. H.Nakamura. Diagnosis of short circuit fault of induction motor based on hidden markov model[C].2007 Annual Report Conference on Electrical Insulation and Dielectric Phenomena, 61-65.
4. Guojun Ding,Lide Wang, Juan Song, Zhui Lin. Neural network based on wavelet packet-characteristic entropy and rough set theory for fault diagnosis[J]. 2010 International Conference on Computer Engineering and Technology, Proceedings, 2010, 1:1560-1564
5. Xiaoxuan Chen, Dayong Luo, Yongteng Shan.Fault diagnosis of locomotive traction motor based on neural network information fusion[J]. Computer Measurement & Control,2007,15(5): 563-573.
6. Jie Lian, Zhanhong Li, Xiaoqun Li. Remote information acquisition and fault diagnosis on the DF7G Type locomotive[J]. Railway Quality Control,2013,41(6):25-27.
7. PAWLAK Z. Rough Sets theoretical aspects of reasoning about data[M]. Kluwer Academic Publishers,Dordrecht,1991,39(1):110-113.
8. Yunge Jing. The research based on attribute reduction of Rough set[M]. Southwest Jiaotong University Press.2013,7:45-48.
9. Miao D Q, Li D G. Rough Sets Theory Algorithms and Applications[M]. TsingHua University Press. Beijing , 2008.4.
10. Xu X L, Wen J P,Xu Y M, et al. Remote monitor & measurement instrument system faced to internet interoperable service of manufacturing. ISIST'2004, Xian,2004,3:1112-1116.

DIAGNOSE AND CONTROL THE OPERATION OF THE PUMP UNIT

Author: Zheng xin The second production plant in Da gang oilfield In Hua hang of Hebei province Tel:15931716915
Author: Zheng bochen The sixth production plant in Da gang oilfield In Hua hang of Hebei province

Abstract: for the same equipment, the more accurate methods you adopt, the more outstanding effect has. This paper briefly describes that how to monitor convinently the condition of the pump unit and quickly find the way of the major failure. The method of diagnosis is to apply vibration **spectrum technology、analyze vibration spectrum information and collect relevant information which can reduce the errors of judgment。The method includes Vibration spectrum information recognition、identification the running state of bearings with the pulse impact value and temperature gradient value of the method of fault localization. If pump unit appears vibration phenomenon, dynamic load would enlarge, in this case will increases power consumption and it's component will wear out. Concerning this issue, after years of research I have summed up some successful experiences and practices. With the apply of these methods, pump unit will be greatly reduced energy consumption and extended service life. In order to ensure the normal operation of pump unit, we need to control satisfaction value of the pump unit, observe the comprehensive information, directly judge the location of the vibration source without pumping off and whether to add lubricant by identification of the pulse impact value. The article shows the measured image of each gauge and the specific value of all kinds of pumps of data sheet。**

keywords: pump unit diagnosis in control of graphic solution Implementation efficiency

Technical person need to make the right judgment to the equipment operation, otherwise will lead to the low efficiency and cost raising. There are many ways for examining pump unit malfunction, but vibration recognition is necessary. By collecting vibration velocity , acceleration , vibration displacement, apply off-line monitoring to judge the state of vibration is very effective, According to the international standard, ISO2372 vibration velocity root mean square value sheet and ISO2373 absolute value of vibration limit. When vibration velocity of the pump gets close to the dissatisfaction value, vibration velocity of the electrical machinery approach to the limiting value, acceleration reachs $30mm/s^2$ (experience points), types of failure are hard to judge, so we need to collect Indirect parameter. TO judge the major failure of the equipment, we can apply vibration graph to judge failure, analyze information about fundamental frequency、octave、frequency division and looseness in vibration velocity graph, analyze information on shock and friction in acceleration graph, also observe the fundamental frequency scope information as a reference in vibration displacement graph, also include information of the other instrument as a reference.

Here are my methods about rapid identification、less misjudgement and judgment procedure

1.How to control rapid identification and reduce misjudgement

The concatenation of motor and pump、close coupling、mutual clamp、all kinds of the impact of unknown energy and vibration source will produce different signs. When the physical components of the transported media change, it will lead to the vibration. When a part malfunctions, will also lead to the change of the related components, besides, collected vibration values are instantaneous, respective value changes over time which has no connection. Due to many factors, it is very difficult to examine fault and easy to make mistakes. Which ways should we take to achieve good result depend on collecting multiple information analyze comprehensively, which can ensure we are able to get right methods. In the operation of the pump unit, it will bring about vibration as well as the signs of temperature and sound, so it is essential to gather these information for analysis.

1)Disadvantages and responses of vibration frequency spectrum analysis

Disadvantages and supplement of vibration frequency spectrum information

Vibration signals contain lots of information of the abnormity or fault which are characterized by vibration phenomenon, vibration amplitude is an important symbol of equipment operation. Mastering vibration information is crucial. Vibration frequency spectrum which is the most effective judgement for the revolving equipments reveals many fault frequency. Due to many factors vibration frequency spectrum information may distort and hard to recognize, there is not a one to one correspondence between vibration fault and diagnosis information, one abnormal information probably corresponds to multiple different faults, one fault is characterized by many different information. This relationship has close links with operation condition、environmental factors、process system、ancient fault and maintenance、etc. Sometimes vibration fault has nothing to do with velocity and frequency. Vibration spectum presents a variety of mixed frequency peak values、waveform、atypical harmonic oscillation signal, which is hard to identify by nonprofessional, especially in rolling bearing damage and lubricant quantity. When outer race and inner race of bearing、rolling elements、cage have tiny depressions、flakings、scratches and cracks, in rotation rolling element is experiencing periodic and skipping motion which can produce shock fault. If other factors produce high vibration, in the view of vibration total energy, vibration reflects continuous and energy-released process. Vibration causes large energy, impact is short, so accumulated is very little, the impacted information is so weak that usually would be submerged, therefore, is hard to identify in vibration spectrum, so do the other signals , it is very necessary to collect multiple signals to examine.

information supplement one Apply shock pulse to identify rolling bearing fault. The value displayed by impulse meter is an important parameter of bearing operation, which means impulse information is major, spectrum is minor. The reason why we adopt shock pulse identification is that impulse measurement technology is designed to identify the abrasion and damage of the bearing. The impulse method is able to directly judge the operation and service life of the bearing without professional's analysis. Through the analysis of the impulse value, we can not only discover early initial fatigue point and corrosion spot of the bearing, but also detect poor assembly and poor lubrication with other reference information.

information supplement two: collecting temperature to identify. High temperature value is used to judge fault. Components are sensitive to temperature, and temperature change is easy to grasp. This is because equipment operation is closely related to temperature which an important factor which is an important factor in causing mechanical failure. Temperature anomaly is the thermal signal of mechanical failure, it can be used to examine some defects and take control of changes caused by various stress. Two examples, example one: High temperature judge equipment failure. According to the temperature value, analyze whether deep- groove ball is suffered the great axial force. The temperature of the bearing only suffered by radial force will be rising under the action of axial force, at the same time, the shock pulse value will go up. abnormal information displayed by vibration spectrum meter is hard to confirm, therefore, it is the best way to judge fault through the surface temperature of components, there are many reasons for high temperature, what need to do is to take control of much more information to clear u possible malfunction. Example two : make use of high temperature of the bearing to deduce correlative faults. Through high temperature, analyze components which are closely related to the bearing. We had monitored motor which had bearing N317 replaced for a while. Before the replacement of bearing, vibration spectrum showed motor had slightly one time frequency malfunction , after the replacement, spectrum showed that motor did not have one time frequency malfunction with the collected information. But the temperature of bearing room got higher ,when it rised to 80℃, after disassembly we found bearing clearance was so small that constraint centrifugal force of the motor rotor. rotor showed no sign of imbalance, which is hard to identify in the spectrum.

information supplement three, intuitive judgment to malfunction. The intuitive judgment method is the best. Why do we adopt intuitive judgment? Simply put, different structures of the machines and information distortion will lead to the errors of judgment, through intuitive judgment and the identification of the endometriosis or remnant of the surface, are able to be familiar with pump operation and quickly eliminate the looseness point in a short time. Example one judge vibration with feeling. Slight looseness of component connected with the motor shell can be confirmed by hand feeling, will be eliminated rapidly. the looseness of the motor shell are able to feel by hand, just like looseness of fan shroud, base plate and bearing room ,etc. subtle movement of two groups of components can be felt by hand and the two frequencies are different. After confirmation, screw the loose nuts, vibration will be eliminated. Example two, hand feeling judgement and wear trace confirmation method. This method is common in judgement. If the two half-couplings of the pump unit have great loosen, this will lead to the mutual collision between two half-couplings. You can feel continuous collision with hand, the vibration of front part of the motor and pump is bigger than that of back part. You can see whether there are some of debris of the rubber and nylon from gear coupling on the base of the coupling, if any, inner soft gasket of the coupling is damaged, if the coupling does not loosen, rotor unit are loose. According to the size of vibration value of the check points, confirm the location of the vibration source. Example three, hearing voices to confirm malfunction source. This way can identify the fault of the small parts, there is a small cover installed on the bearing outer, the interference fit between the inner hole of small cover and axis had emerged a gap which would lead to that the rotation of small cover and axis out of sync are out of sync. The rotation of the sync will produce grating, this grating is so small that can not be showed by the vibration spectrum meter, but person can tell this information.

2.measuring instruments of implementing rapid diagnosis

In conclusion, we can make a judgement rapidly in spot by the use of the four meters which are portable vibration spectrum analyzer(handheld computers, analyzer quickly show the spectrum as soon as collect information)，shock pulse bearing analyzer, mobile infrared temperature measuring instrument, mechanical failure stethoscope(this meter can acquire information which can not be acquired by man)

Sample Pictures

portable vibration spectrum analyzer

shock pulse bearing analyzer

mobile infrared temperature measuring instrument

mechanical failure stethoscope

3.The examples of the identification of common failures

1) Identification to the rotor of static balance difference

There are four spots which need to be monitored in single shaft multi-stage pump bearing room, if vibration velocity of the two horizontal radial observation points are bigger than that of two Vertical radial, spectrum meter shows that vibration velocity will reach one time frequency, the other frequency amplitude are less than that of one time frequency, the vibration velocity of axis observation points are no larger than that of 1/3 radial, the above information shows that impeller of multistage pumps rotor tend to weight after the impeller and bearing have been assembled, and judge whether have the other disturb factors, acceleration peak value is less than 20,shock pulse meter shows that high frequency weak pulse value is below 15dBc, low frequency strong pulse value is below 25dBm, the surface temperature of bearing room is below 60℃,which account for the bearing is in good condition.

2)Identification to the operation of bearing and pitting corrosion

We need to collect much more information to identify the operation of bearing, if the bearing has no slight pitting corrosion, the interference fit of bearing assemblies is bigger, so that shock pulse will be slightly higher, but time- domain waveform、amplitude spectrum have no clear information, if the bearing has pitting corrosion, time- domain waveform、 amplitude spectrum have clear information., to avoid mistakes, the best result is to combine with information of two meters. In the course of monitoring, we found that shock strong pulse value reach 30dBm, acceleration spectrum peak value reachs 30mm/s^2,we can hear slight noise of bearing(the noise of gear damage is different from that of bearing), to judge the bearing has whether pitting corrosion, these information need to be mastered. as a little hint, the biggest obstacle to identify shock pulse is intense external shock which probably lead to measure distortion, therefore, we should be more adept at removing disturbances, besides, we should refer to the kurtosis index of acceleration spectrum and bearing operation. If kurtosis value reachs 2~3, bearing motion pair local damage increases, kurtosis index will increase. If margin index value is bigger than 4, bearing motion pair is damaged and gap increases.

There are some problems which include looseness and crack in bearing cage, fault characteristic frequency can be calculated by the formulas, if the size of bearing is unknown, we can adopt approximate algorithm formula to work out fault characteristic frequency, according to the vibration velocity parameter table, in the range of 15~25Hz, see whether have isolated non integer high amplitude crests or not, with the mechanical failure stethoscope, hear whether have the sound of high and low frequency and grating, this two information mean that cage has the problems. It is hard to judge the failure with shock pulse.

3) identification to accuracy between rolling bearing and related fittings grouping

To identify operational state of rolling bearing and related fittings grouping is to collect pulse value and temperature value. The collected shock pulse dBm is just below 25 , dBc is just below 15, we can confirm that motion pair has not serious problems, the condition of related fittings grouping is judged by temperature value, we can use infrared thermometer to check temperature gradient of assembly, if the radial temperature of chassis of bearing outer ring is high, bearing inner diameter is the highest(higher than bearing inner ring temperature). The farther the distance, the lower the temperature of bearing. The temperature of seal cover is less than that of bearing outer ring chassis and bearing, which means the bearing and cap have no friction. because of gap, rotation between rolling bearing inner diameter and bearing is out of sync. You can see acceleration、 vibration velocity frequency range chart. The spectrum probable show the triple frequency， if gaps are left between rolling bearing outer ring and chassis, bearing outer ring will be rotating, shock pulse dBm、 dBc are negative. If dBc is just negative, it means that slight gap left between rolling bearing outer ring and chassis, because of two values are negative, we can infer that collected vibration values are damped.

4) identification to friction between the different parts of the system

Correct identification to friction between rotator and stator is to need to collect more information. if acceleration spectrum shows that waveforms have on-going cycle of up and down and burr, wave crests may have clipping, which mean that friction will emerge. frequency division is apparent, if local friction is slight, frequency division is small and vice versa. The location of friction source is identified by the portable infrared temperature instrument. The high temperature spots of concatenated components are the friction source. The current can also real the condition of friction.

The implementation of low-powered and longstanding operation of pump unit

Pros and cons of running state of horizontal rolling pump reference table

			18
	unqualified		11
			7
	unsatisfactory		5
			3.5
	satisfactory		2
			1.5
			0.7
	good		0.5
below17Kw	22-100kw	110-900kw	mm/s

Practice of low-powered and longstanding operation of pump unit

1.ensure satisfaction scales operation, according to the international standards ISO2372、2373 vibration intensity maximum unsatisfactory value, control velocity below 70%, as a little hint, when maximum unsatisfactory value of large-scale horizontal rolling pump is 7mm/s, we should control under 5 mm/s, when motor bearing height is between 225 and 400,velocity is 4.5 mm/s, control velocity under 3 mm/s, temperature of working bearing keeps below 60℃. These measures can effectively reduce energy loss.

How to determine the satisfaction value of pump unit depends on two factors, one is to internal and overseas standard, the other is to gather experience, with the repair 、debugging and assembly precision.

2.dynamic diagnosis to remove faulty

use visual vibration spectrum to identify at least once a month. Simply put, whether the equipment has problem depends on amplitude、 frequency、 dimensionless value、 dimension value。 If so, determine the source of trouble and remove. Through the shock pulse measurement instrument, we can identify the condition of bearing and wear degree, add lubricant as required and ensure economic operation. If lack of lubrication, when conditions are the same, monitor at different time, the bearing analyzer shows shock pulse are more than 3 times which are more than 5dBc of previous measurements. In this case, grease needs to be added.

We need to toughen the inspection to eliminate the looseness produced by vibration, Specific practice is that check the condition of vibration with hand, through the hand, we can find vibration sources, tighten the loose parts with no downtime.

Productive pump unit picture

Example: there are four 250kw, 3000r / min of crude oil transportation pump units in The second production plant in Dagang oilfield, the effect is evident with the above methods. This four pump units have been operating from February,2010 up to now, cumulative time is more than 40, 000 hours, after fault removal, following a period of test operation, the equipments have been keeping in normal running. we aim to achieve the best result from the least investment to increase production efficiency without 2nd Level & 3rd Level maintenance.

The smaller vibration of pump unit is ,the smaller mechanical damage is, energy consumption as well. We have lots of experiments on low vibration operational parameters of pump unit, as a result, service life increases and cost reduces. This is the best result we want. I believe this implementation will come into widespread use, it might be worth learning、 referring、 copying and sharing.

Design of Test Platform for Power Transmission and Hydraulic System of Tracked Vehicle

Sun tao[1], Zhan shaomin[1], Yang zhida[2], Meng xiangliang[2]

1 Changchun Engineering & Technology College, department of mechanical engineering, Changchun 130117 China,

2 Beijing Information Science and Technology University, The ministry of education key laboratory of modern measurement and control technology, Beijing100192,China

ABSTRACT: The maneuverability, stability and safety of high-power tracked vehicle were directly affected by Hydro-mechanical transmission, which was the key part of vehicle. In order to improve the technological level of tracked vehicle power transmission system's reliability and condition monitoring, this paper designed comprehensive transmission and hydraulic system test platform, test hardware system and software system. The test platform can control the engine start-stop and speed , monitor different types of integrated transmission shifting, achieve pressure buffering, transmission efficiency, no-load loss and torque detection, also accomplish the hydraulic system feature detection . After the completion of the test, the platform condition monitoring report can be printed automatically. Experimental and performance study shows the test platform is stable and reliable, and the cycle of power transmission detection, the support capacity of equipment maintenance can be shortened.[1]

KEYWORDS: Tracked Vehicle; Hydro-mechanical transmission; power cabin; hydraulic system; test platform.

---

[1] Sun tao, (1978—), male,Changchun Engineering & Technology College, lecturer. research direction :Transmission Measurement and Control.

INTRODUTION

The performance of hydro-mechanical transmission, as the principal transmission for high-power vehicle, could immediately affect the vehicle maneuverability, stability and safety; as the core component that expands the limited speed and torque range of engine so as to get vehicle adapted to complex road and operating conditions, hydro-mechanical transmission mainly consists of hydraulic torque converter, multi-shift gear, wet shift clutch, and hydraulic steering mechanism etc. Its electronic control and sensing devices are now normally tested through off-line disassembly or external cable connection, which is subject to such shortcomings as magnetic interference, difficult operation, high misdiagnosis rate, high maintenance cost and low efficiency [1]. On the other hand, the late participation of China in research and development of hydro-mechanical transmission led to poor research capacity for test techniques and equipment fault test, which severely restricts the design of new integrated transmission, as well as the development of equipment quality reliability test and fault diagnosis technologies [2-3]. On this account, it's of crucial importance to develop an operationally reliable and easy-to-handle indoor testing system for comprehensive test on integrated transmission, test data acquisition and automatic test report printing.

The integrated test platform for power transmission and hydraulic system developed to address the need for repair of integrated transmission and the transmission system performance test with full consideration given to the manual operation of test through the integration with computer technology, testing technology and sensor technology is characterized by intellectualization, multi-functional integration, and excellent stability.

1. Test Platform Structure Design

The integrated test platform for power transmission and hydraulic system is an integrated monitoring & testing system [4] incorporating gearbox, engine, hydraulic pump station, test condition control and the acquisition, analysis and processing of various signals; the architecture is as shown in Fig. 1; this platform is designed for test and fault diagnosis of various performance parameters for power cabin of various types of vehicles. The system is principally composed of hardware and software. The hardware comprises operating console, sensor, signal conditioning cabinet, data acquisition card, and various instruments and meters etc., while the software consists of measurement & control software at upper computer.

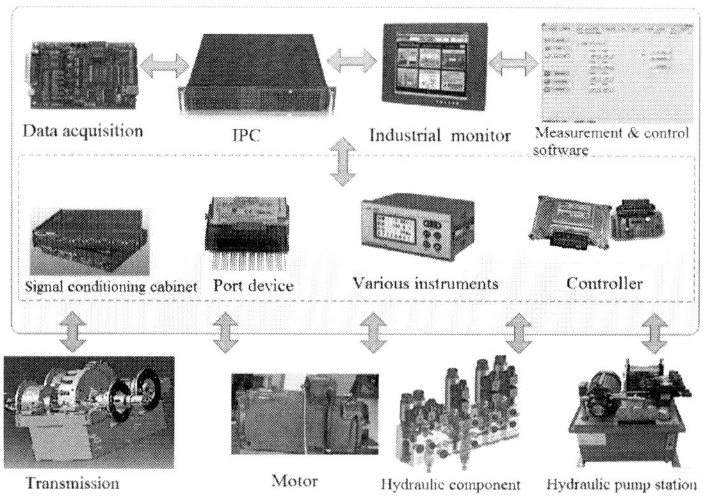

Fig. 1 Architecture of measurement & control system

This platform is designed for the control of tested power cabin and the acquisition, display and processing of data required. The power cabin test-related parameters like temperature, pressure, flow rate, speed and torque are collected and delivered in a form of analog quantity or digital quantity to measurement & control software for analysis and processing [5]. . The principal structural configuration of operating console is shown in Fig. 2 and mainly composed of data display section, manual adjustment section, electric control section and electronics section. Data display section comprises monitor and data display instrument. The monitor displays all parameters collected through measurement & control software and all the control signals (including input/output speed, line pressure, and gear pressure etc.) transmitted; data display instrument indicates the current parameters of each section of the system, e.g. total system current, loading valve current, and fan pump current etc. Manual adjustment section is mainly designed for the control of gearbox, engine, and motor. The electric section is used for mains power supply and current limiting protection for system. The auxiliary function control section principally constitutes a cooling system. The meters are used for the conditioning, acquisition, and transfer etc. of signals..

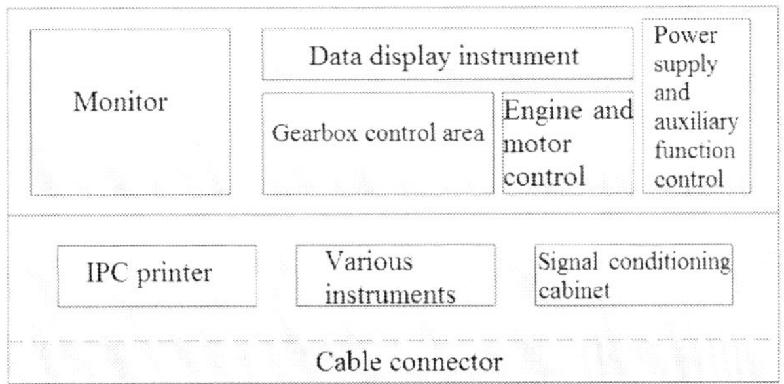

Fig. 2 Structural configuration of operating console

2. Hardware System Design for Test Platform.

The block diagram of signal acquisition and processing module is shown in Fig. 3.

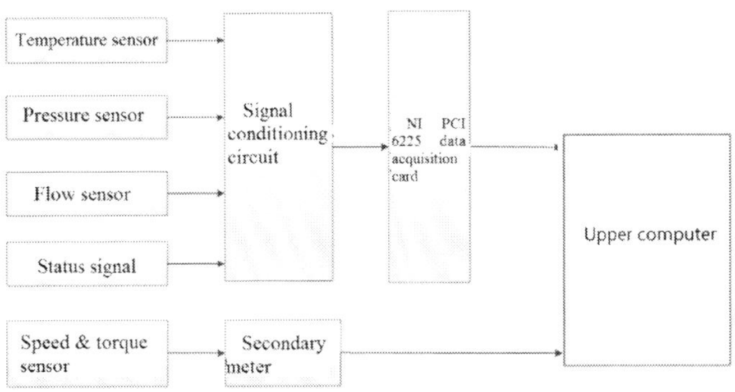

Fig. 3 Block diagram of signal acquisition and processing module

this module is designed for measurement and data acquisition of such parameters as gearbox speed, torque, oil temperature, pressure and flow rate. This section is principally composed of sensor set, signal conditioning module, secondary meter, data acquisition card, and measurement & control software. The sensor set acquires various physical parameters of engine and gearbox; the signal conditioning module converts different types of signals from sensors into standard electric signal; the data acquisition card enables the A/D conversion of signal, and delivers the signal to measurement & control software. The signals from temperature, pressure and flow sensors are converted into standard electric signal through signal conditioning module, and put into the data acquisition card, while the signal from speed & torque sensor is changed through secondary meter into digital quantity, which is transmitted to measurement & control software via RS232 bus.

3. Test Platform Software Developing

The LabVIEW, as a powerful and smart instrumental analysis software and application development tool developed by US-based NI, is one of the most widely used virtual instrument development environments all over the world, and is principally designed for instrument control, data acquisition, data analysis, and data display etc. [6] The functional block diagram of software system developed in a modular and hierarchical manner by dint of LabVIEW platform is shown in Fig. 4.

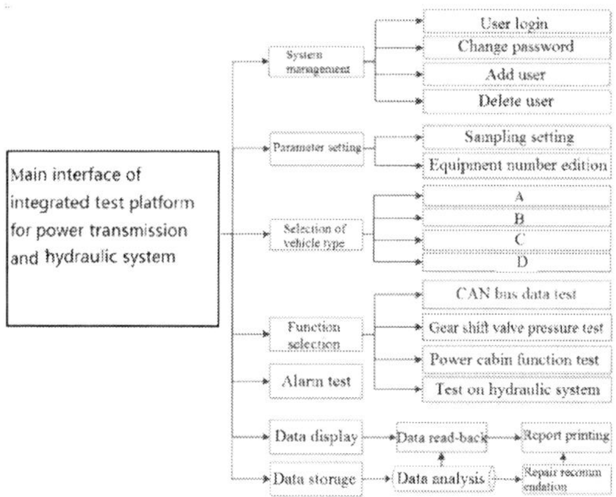

Fig. 4 Block diagram of software function module

The integrated test platform for power transmission and hydraulic system as developed on a generalized, integrated, standardized, modular and intellectualized basis incorporates such techniques as test condition control, signal conditioning & acquisition, data analysis & processing, and diagnostic information output & printing, and enables the test on power cabin CAN bus data, gear shift valve pressure test, power cabin function test and experimental test on hydraulic system for a variety of vehicle types.

The platform is provided with a user login system accessible only to authorized users. After the log-in and parameter setup by user, the operator could choose the vehicle type to be tested and perform test analysis for various features.

By offering a bidirectional serial multi-node digital communication system, the controller area network (CAN) serves as a serial communication network that could effectively support distributed control and real-time control [7]. CAN bus has been extensively used for data communication of military vehicles; where the overall communication for power cabin is realized with CAN bus, all sensor data enters vehicle terminal via the bus; the test stand captures bus data of vehicle terminal and delivers it to the upper computer for CAN bus data test.

4. Platform Test-based Testing

4.1 Gear Shift Valve Pressure Test.

The operating pressure waveform experiment-based testing software for gear shift valve principally comprises pressure value display area, pressure waveform display area and gear display area. The waveform display area indicates the clutch pressure waveform, the test on hydraulic buffer time of clutch and the test on variation law, and enables the storage, printing and playback of pressure waveform. To facilitate the observation, the software is designed with a cursor location module, where the test personnel could move the cursor to oil filling waveform and set the positions before and after oil filling so that it's possible to display the time difference in between, i.e. the oil filling time.Oil filled time measurement is shown in Fig. 5.

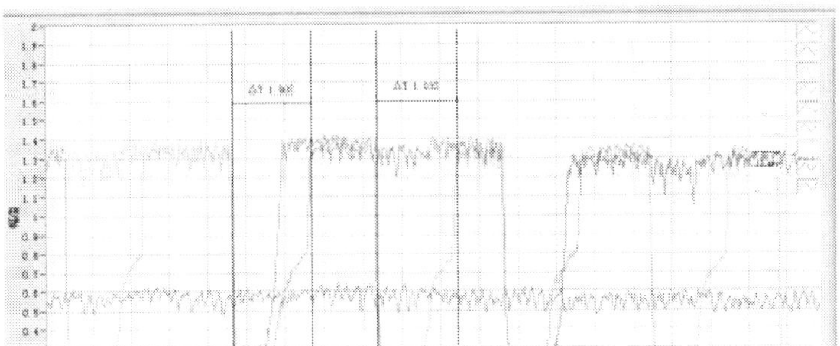

Fig. 5 Oil filled time measurement

4.2 Power Cabin Function Test

The power cabin function test enables the no-load loss test on power cabin, the loaded test on fan pump, the loaded test on gear pump and the parameter test on sensor for 4 types of vehicles; the software could display multiple speed & torque data, no-load loss value, gear pump pressure, gear pump load pressure, pump discharge, torque converter flow rate, oil temperature and other physical parameters, and could save and print reports through operation.

4.3 Experimental Test on Hydraulic System.

Hydraulic system test module enables hydraulic system-related control and testing, the on-line data acquisition for 4 pressures, 3 flow rates, and 2 temperatures, the storage and loading of files, and the printing of reports.

5. Conclusion.

The platform incorporates data acquisition technology, virtual instrument and office automation technology for status monitoring and test control of integrated transmission. According to the practical operation of integrated test platform for power transmission and hydraulic system, the test stand functions solidly and reliably, and the parameter measurement is accurate; the platform could remarkably shorten the state test cycle, and is of far reaching importance for improvement in integrated transmission performance.

Acknowledgements

This work was completed with the support of the Beijing finance found of science and technology planning project.(KZ201611232032) （KM201611232004）, and young prominent talent projects. (No.2014000026833ZK25) (No. 2014000020124G096) （CIT&TCD201404112）

References

[1] Lin Hongjun. Tracked vehicle transmission test system research [D]. Jilin university, 2009.

[2] Ji Jiang. The design of the tracked vehicle integrated transmission system performance test rig [D]. Shandong university, 2011.

[3] Liu Yang. Vehicle integrated transmission test rig measurement and control system research [D]. Henan university of science and technology, 2013.

[4] Feng Fuzhou, Cao Yukun, Wang Zhanjun, An Gang, An Hongwei. General equipment mechanical hydraulic system design and development of test platform is [J]. Journal of armored force engineering institute, 2008.

[5] Zhao Ding,Fang Guo-chang . Wheeled armored vehicles gravity tank vibration testing system research [J]. Science and technology innovation and productivity, 2015.

[6] Yu Tieyue, Lin Jianhuan, Huang yi-jian . Introduction of and the LabVIEW virtual instrument [J]. Journal of fujian computer, 2004.

[7] Huang Huan. Research and development based on CAN bus network monitor system [D]. Chengdu university of technology, 2009.

Feature Extraction of Vibration Signals Based on Fusion Sparse Coding

Gang Tang, Yao Zhang and Huaqing Wang

College of Mechanical and Electrical Engineering, Beijing University of Chemical Technology, Beijing 100029, China
Huaqing Wang: the Corresponding Author
E-mail: wanghq_buct@hotmail.com

Abstract: Feature extraction of vibration signals plays an important role in the machine fault diagnosis. Sparse coding (SC) provides an effective way to make it. However, SC cannot accurately extract impulse components and identify fault character frequencies. In addition, it may remove some useful information of the signal. To improve the performance of SC, this paper uses fusion sparse coding to extract the fault features of vibration signals, which fuses vibration signals from different sensors to acquire better sparse coefficients. Firstly, a vibration signal is processed by several kinds of SC in parallel. Then, the fusion scheme is developed to get new sparse coefficients. The vibration signals generated by a roller bearing with an inner-race fault are employed to validate the effectiveness of the proposed method and the results show that the fusion sparse coding outperforms single sparse coding.

Keywords: Vibration signals; feature extraction; sparse coding; information fusion

1 Introduction

Rotary machinery is commonly used at present. It is important to guarantee the safety of machines to avoid disastrous accidents. So fault diagnosis and condition monitoring method are significant to be studied. Fault characteristics analysis is usually used for fault diagnosis, which extracts fault features by processing signals with spectral analysis, coherence analysis and so on. To extract impulse features in vibration signal, there are many kinds of methods e.g., Fourier transform and wavelet transform. However, as there are always many interferences in the vibration signals, it is necessary to seek a better method to deal with signal flexibly. Sparse coding provides an effective way to make it. Mallat and Zhang put forward a signal decomposition method by over-completed dictionary with fewer coefficients [1], which can represent a signal much more flexibly and briefly.

In the process of feature extraction by sparse representation, an effective sparse coding algorithm may greatly benefit it. There are several common sparse coding (SC) algorithms reported in literatures, e.g., orthogonal matching pursuit (OMP), basic pursuit (BP), particle swarm optimization (PSO). Guo et al. exploit the PSO to search atoms from the dictionary that optimally match with the vibration signals of rolling bearings [2]. Chen et al. proposes a sparse decomposition method for resonance signal sparse decomposition which decomposes vibration signals into high resonance components, low resonance components and residual components. Then the fault features can be identified from the envelope spectrum of low resonance components [3].

However, there are still many shortcomings with SC because of sensor errors, environment interferences and so on. These interferences will generate kinds of noises embedded in vibration signals, which will make some useful components submerged. Therefore, it is essential to improve the performance of sparse coding.

According to the theory of information fusion [4], the information from different sensors can be fused to be more reliable ones. Therefore, there may be a better performance if introducing the fusion theory into SC [5]. In this paper, a kind of fusion sparse coding is developed. The main steps can be described as follows: Firstly, several kinds of SC are used to deal with vibration signals in parallel. Then, the fusion scheme is executed to get new accurate sparse coefficients to reduce the error. Any kind of SC can be fused as the participating algorithm with no limitation. Meanwhile, the redundant dictionary is learned from the vibration signals with K-SVD [6] realizing sparse representation of signal by constructing the dictionary adapted to the structure of signal.

The paper is organized as follows: A brief description of sparse coding will be provided in Section 2. The fusion sparse coding (FSC) will be presented in details in Section 3. An actual experiment with FSC will be shown in Section 4. The conclusion is presented in Section 5.

2 Sparse coding for Vibration Signals

It is crucial to reduce the noise embedded in signals to make the fault characteristic frequency prominent. Sparse coding is a good choice to achieve such an object.

The main advantage of signal decomposition with redundant dictionary is that, the atoms, which are adaptive to the structure of vibration signals, can be chosen efficiently by linear combination. As for a noised-signals $s(n)$, it can be expressed by

$$s(n) = x(n) + v(n) \quad 0 \le n \le N-1 \tag{1}$$

Where $x(n)$ denotes the noiseless signal and $v(n)$ denotes the noise. With the transform matrix $D \in R^{N \times L}, N < L$, the noiseless signal can be expressed by

$$x = Du \tag{2}$$

Where u denotes sparse coefficients. With Eq. (1) and Eq. (2), it can be expressed by

$$s = Du + v \tag{3}$$

Where $D \in R^{N \times L}$ denotes redundant dictionary.

The main purpose of SC is to acquire the sparse coefficients with redundant dictionary. But there are infinite solutions to Eq. (3) because of the redundancy of dictionary. In order to get the sparsest solution, with the known dictionary D and signal s, it can be solved by

$$\min_u \|u\|_0 \ s.t. \ \|s - Du\|_2 \le \gamma \tag{4}$$

Where $\| \|_0$ and $\| \|_2$ denote l_0 - and l_2 - norm respectively, parameter γ denotes the reconstruction error. It is a NP hard problem because it is a non-convex optimization problem [7]. To ensure that the coefficient is sparse enough, there are many methods proposed, such as Orthogonal Matching Pursuit (OMP), Basic Pursuit (BP), The Least Absolute Shrinkage and Selectionator operator (LASSO) and so on.

OMP is based on Matching Pursuit (MP), which is a kind of pure greedy iterative algorithm. The atoms chosen by MP may be not orthogonal with each other, which cannot get optimal results. However, OMP can make sure that signals can be decomposed orthogonally because of the constant adjustment of the chosen atoms in each process of iteration. So OMP can get the sparsest solution approximately with constant iterative effectively [8].

BP makes use of the equivalency of l_0 – and l_2 – norm under a certain condition, so the Eq. (4) can be transformed into

$$\hat{u} = \arg\min_u \|u\|_1 + \lambda \|s - Du\|_2^2 \tag{5}$$

Therefore, the NP-hard problem becomes a convex optimization problem, which can be solved by linear programming. In order to acquire better result, it is crucial to choose an appropriate value for λ. While the energy of noise is large, the value should be small. Otherwise, it should be large [9].

3 Fusion Sparse Coding for Vibration Signals

Although single sparse coding, such as OMP and BP, plays an important role in the sparse representation of signals, there are still some disadvantages for any sparse coding, such as reconstruction error and missing of useful impulse. Especially, while the prior information of vibration signals is unknown, it will be difficult to choose an appropriate sparse coding.

According to the information fusion theory, different data from different sensors can be fused to acquire more reliable information than that from single sensor. Since it is difficult to determine an appropriate sparse coding, several kinds of sparse coding can be fused to improve the performance of sparse reconstruction. This paper designs fusion sparse coding to extract the fault features of vibration signal. The main steps can be described as follows:

1. $k (k \ge 2)$ kinds of sparse coding algorithms are executed independently to get the sparse coefficients of input signal with K sparse level.

2. Record the data sets which are defined as the positions of K-dominate elements in each sparse coding algorithm, and acquire the union set of these data sets.

3. Leave those columns in dictionary which is corresponding to the union set, and remove others. So a higher dimensional dictionary can be converted into a lower dimensional one.

4. Compute the pseudo-inverse matrix of the new lower dimensional dictionary $D_N \in R^{N \times P}, (N > P)$ to get the estimation of sparse coefficients, which can be described as:

$$\hat{u} = D_N^+ s \tag{6}$$

Then select the K-maximum value of $\hat{u}$ as the estimation of sparse coefficients.

5. Place the estimation of sparse coefficients from step 4 in the position, according to the data set from step 2, to constitute a new sparse matrix, namely fusion sparse matrix.

In this paper, OMP, BP and Lasso will be fused as the joint algorithms of FSC. The redundant learning dictionary is trained by K-SVD.

4 Experimental Results

To verify the effectiveness of FSC, the proposed method is utilized to extract the impulse component from the vibration signals induced by a roller bearing with a defect in the inner race. The joint algorithms include OMP, BP and Lasso with the redundant learning dictionary trained by K-SVD.

The fault characteristic frequency of the vibration signals generated by a roller bearing with an inner-race fault is 145.84Hz calculated by theoretical equation. The time-domain waveform of the vibration signal is shown in Fig. 1, and the corresponding envelope spectrum is presented in Fig. 2, where the fault features are buried by the interferences.

As shown in Fig .1, there are obvious impulses, which means there is a defect in the rolling bearing. However, the fault characteristic frequency cannot be identified clearly because of some interferences. In order to acquire the fault characteristic frequency, the FSC is applied to make it. Furthermore, the FSC is compared with basic pursuit (BP) to evaluate the performance effectively.

The result is shown in Fig.3. The impulse features extracted by BP and FSC is shown in Fig.3a-b and the corresponding envelope spectrum of signal can be seen in Fig.3c-d. As shown in Fig.3, the method of BP has weak performances and it removes some effective components in the signal, which may cause some impulse features ignored. However, the performance is much more accurate while the signal is reconstructed with the FSC. The fault character frequency can be identified more clearly than that of BP.

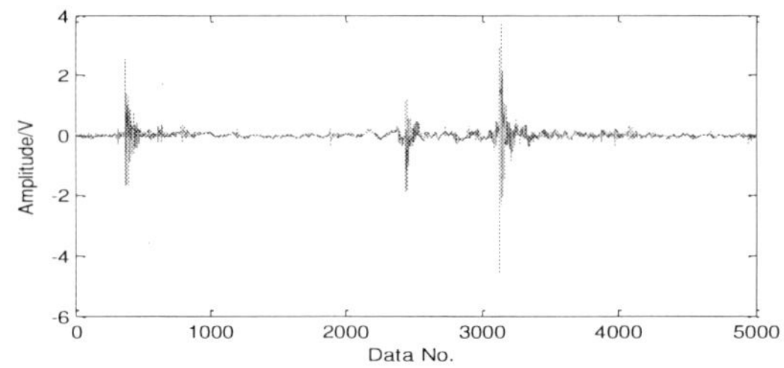

Fig. 1. A vibration signal in time-domain

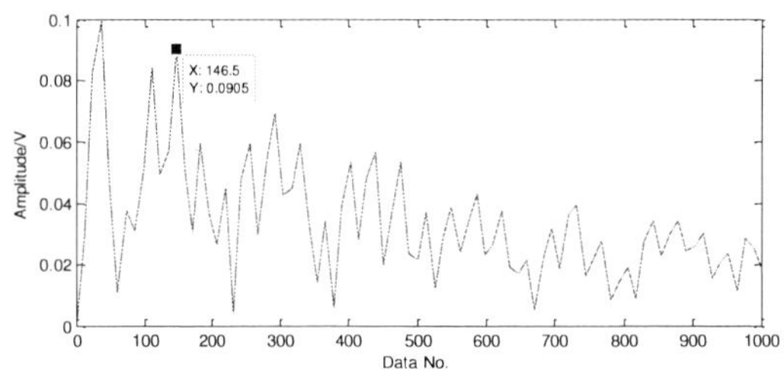

Fig. 2. Envelope spectrum of the vibration signal

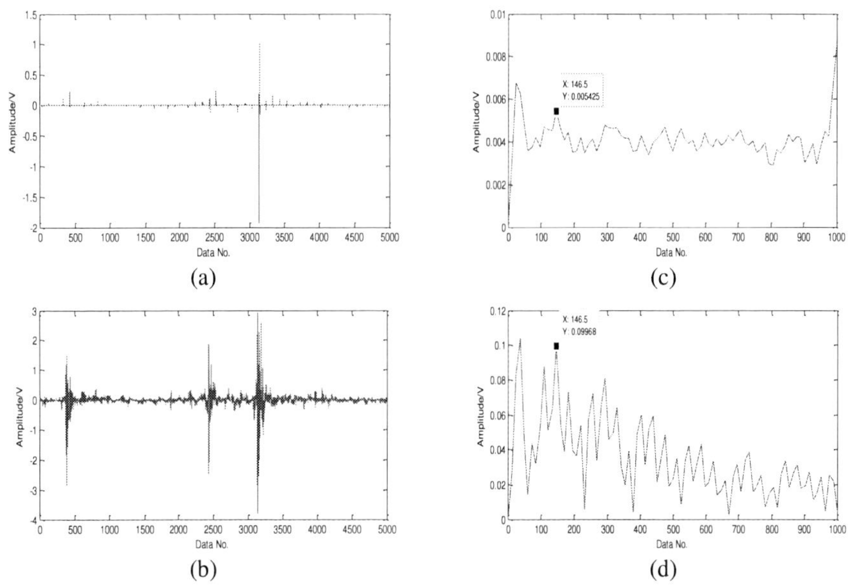

Fig. 3. Waveform of impulse features in time-domain extracted by (a) BP, (b) FSC; and corresponding envelope spectrum of de-noised signals: (c) BP, (d) FSC.

5 Conclusion

In this paper, a kind of fusion sparse coding is developed with OMP, BP and Lasso being the joint SC algorithms. Firstly, several single SC algorithms are operated independently to get sparse coefficients. Secondly, the fusion of different sparse coding is presented by the union of the sparse coefficient to get new sparse coefficients. Lastly, vibration signals of a rolling bearing are used to verify the validity of the proposed method. As shown in the results, the performance of FSC is better than that of BP. However, there are still some interferences not removed by FSC. So the fusion strategy is still to be further studied to make a better performance.

Acknowledgments

This work is partially supported by China Fundamental Research Funds for the Central Universities (BUCTRC201509), PetroChina Innovation Foundation (2014D-5006-0104, 2015D-5006-0606) and the Open Fund of State Key Laboratory, Southwest Jiaotong University (TPL1603). The first author would like to thank the Public Hatching Platform for Recruited Talents of Beijing University of Chemical Technology.

References

1. Mallat S, Zhang Z: Matching pursuits with time-frequency dictionaries. J. IEEE Trans Signal Process. 41(12), 3397-3415(1993)
2. GUO Jun feng, Zheng Xiaohu, Wei Xin chun: Research of the rolling bearings vibration signal's sparse representation. J. Journal of Gansu Science. 26(3), 91-94(2014)
3. Chen Xiangmin, Yu Dejie, Luo Jiesi: Envelope demodulation method based on resonance-based sparse signal decompos it ion and its appl ication in roller bearing fault diagnosis. J. Journal of Vibration Engineering. 25(6), 628-626(2012)
4. L. J. Xu, J. Q. Zhang, Y. Yan: A Wavelet-Based Multisensor Data Fusion Algorithm. J. IEEE Transactions on Instrument and Measurement. 53(6), 1539-1545(2004)
5. Deng Sen, Jing Bo, Sheng Sheng, Huang Yifeng, Zhou Hongliang: Impulse feature extraction method for machinery fault detection using fusion sparse coding and online dictionary learning. J. Chinese Journal of Aeronautics. 28(2), 488-498(2015)
6. Michal Aharon, Michael Elad, and Alfred Bruckstein: K-SVD: An Algorithm for Designing Over-complete Dictionaries for Sparse Representation. J. IEEE Transactions on signal processing. 54(11), 4311-4322(2006)
7. Davis G, Mallat S, Avellaneda M: Adaptive greedy approximation. J. Constr Approx. 13(1), 957-989(1997)
8. K. Kreutz-Delgado, J. F. Murray, B. D. Rao, Kjersti Engan, Te-Won Lee, T. J. Sejnowski: Dictionary learning algorithms for sparse representation. J. Neural Computation, 15(2), 349-396(2003)
9. Yi Qin, Yongfang Mao, Baoping Tang: Vibration signal component separation by iteratively using basis pursuit and its application in mechanical fault detection. J. Journal of Sound and Vibration, 332, 5217-5235(2013)

Signal Denoise Method of rolling bearing vibration signal Based on EMD-Sample Entropy

Zhang Tao[1] Wang Fengtao[2] Wang Bei[3] Li Hongkun[4] HanQingkai[5] Wang Lei[6]

1School of Mechanical Engineering, Dalian University of Technology, Dalian, China,
ztaodlut@163.com

2 School of Mechanical Engineering, Dalian University of Technology, Dalian, China,
wangft@dlut.edu.cn

Abstract. Aiming at the low signal noise ratio characteristic of rolling bearing early fault signal, one rolling bearing fault signal denoise method based on EMD-sample entropy is proposed. Firstly, the vibration signal of a rolling bearing is decomposed into a number of IMF components using EMD method. IMF highlights the various components of the local characteristics of the signal, comprising the main bearing fault features. Then, the optimal reconstructed signal can be determined by calculating the cross-correlation coefficient between the reconstructed signal and the original signal. Finally, combined with the spectral kurtosis algorithm and resonance demodulation method, the optimal reconstructed signal is utilized to diagnose the fault status of rolling bearing. The effectiveness of the proposed method is verified through the practical vibration signal with early fault.

Keywords: EMD; Sample entropy; Spectral kurtosis; Fault diagnosis; Rolling bearing

1 Introduction

Rolling bearing is an universal, precision and key part and is one of the most easily damaged part, the running status of which will directly influence the whole machine performance[1], so it is of great significance to research the condition monitoring and fault diagnosis methods of the rolling bearing.

The early weak fault vibration signals of rolling bearing usually exhibit nonlinear characteristics, non-stationary characteristicsand large noise. Empirical mode decomposition[2] is first proposed by the American scholar Huang in 1998, whose main idea is to decompose the signal of a time series into the sum of several intrinsic mode function(IMF). The procedure of EMD is as follows: decompose the signal of a time series into a set of IMF, then combine some of the IMF to constitute a high-pass, low-pass, band-pass filter. Each IMF highlights the local characteristics of the signal, so

EMD is appropriate to deal with nonlinear, non-stationary signals and has been successfully applied in roller bearing fault diagnosis.

Demodulated resonance method, which is called as high-frequency resonance method or the envelopment analysis, is one of the most commonly used methods in rolling bearing fault diagnosis currently[3,4]. The selection of bandpass filter parameter is the key issue for the resonance demodulation. Spectral kurtosis[5] was first proposed by Dwyer, whose essence is to calculate kurtosis values of each spectrum to reflect the spectral kurtosis of original signal at a frequency component.Therefor, it is particularly sensitive to non-stationary signals and it can pinpoint frequency band containg the fault information even the signal contains noise. Literature[6] had combined spectral kurtosis and resonance demodulation.

'Entropy' is a thermodynamic concept to quantitatively describe the system's disorder status. Richman J S and Moorman R J[7] proposed the approximate entropy algorithm in 1991, and makes the approximate entropy become feature extraction parameters. Sample entropy[8] is an algorithm which is different from approximate entropy, and it overcomes the bias of the approximate entropy. Sample entropy shows probability of occurrence of new model of the nonlinear dynamical system.

Based on the research and the analysis of a large number of sample entropy value, this article proposes a new method that based on EMD-sample entropy aiming at rolling bearings early fault signal noise reduction.

2 Theory basis

2.1 EMD method

EMD adaptively decomposes a non-stationary signal into a plurality of the IMF components. Each IMF meets the following two requirements[9]: first, in the entire sequence, the number of extreme points(including maxima and minima) and the number of over zero must be equal, or the difference between the number must be under one; Second, to any point, the mean between the upper envelope forming from local maxima and the under envelope forming from local minima is zero. EMD decomposition method is based on multi-scale filtering process and has the following characteristics: adaptive, orthogonality and completeness.

2.2 Resonance demodulation and spectral kurtosis

The resonance demodulated method is as follows: use the bearing or the detection system as a resonant body to amplify the high-frequency resonance response generated by the fault impact, get the low-frequency waveforms containing fault feature information by envelope detection method, then use the spectral analysis to find fault characteristic frequency so as to determine the type of failure[6]. To any Non-stationary random process with zero mean $x(n), n \in Z$,

$$x(n)=\int_{-1/2}^{+1/2} H(n,f)e^{j2\pi fn} dZ_x(f) \tag{1}$$

Where, $dZ_x(f)$ is quadrature spectrum incremental, and H(n,f) is the complex envelope of x(n) where the frequency is f.

Spectrum Kurtosis can be defined as the fourth-order normalized cumulant:

$$K_x(f)=\frac{\langle |H(n,f)|^4 \rangle}{\langle |H(n,f)|^2 \rangle^4}-2 \tag{2}$$

$\langle f(n) \rangle = \lim_{N\to\infty} \sum_N f(n)$ shows the time mean of the function f(n), and H(n,f) is plural. The constant '-2' replaces '-3' in the classics kurtosis.

The calculation of spectral kurtosis[10] is essentially the selection of $f/\Delta f$. The question is converted into the following questions: how to choose the best frequency resolution of the spectrum kurtosis to max the spectral kurtosis value at the given frequency for any signal[11]. The scope of the filter can be determined by this method in the signal processing.

3 EMD-sample entropy denoise and diagnostic methods

There is objectivity and broad practicality using sample entropy as EMD noise criteria, at the same time, spectral kurtosis algorithm itself also has a strong diagnostic capability, and it will be applied to the resonance demodulation. It is expected to achieve better results through combining EMD-sample entropy-spectral kurtosis. Therefore, this article proposes rolling bearing fault denoise and new diagnosis method based on the combination of the three above. The detailed diagnosis steps are shown in Figure 1.

1. Decompose the collected signals using EMD algorithm;
2. Calculate sample entropy value of each IMF;
3. Cluster reconfiguration, caculate the crosscorrelation between the reconstructed signal and the original signal to determine the optimal reconstructed signal;
4. Analyze the optimal reconstructed signal: get the fast kurtosis chart, select the center frequency and bandwidth where the kurtosis is largest at the figure;
5. Do bandpass filtered aimed at the optimal reconstructed signal using the parameters-the center frequency and bandwidth get from the step 4;
6. Do squaring envelope on the filtered signal, and obtain the envelope spectrum through fourier transform;
7. Compare the rolling bearing fault characteristic frequency with the frequency in the biggest envelope spectrum peak then determine the fault condition.

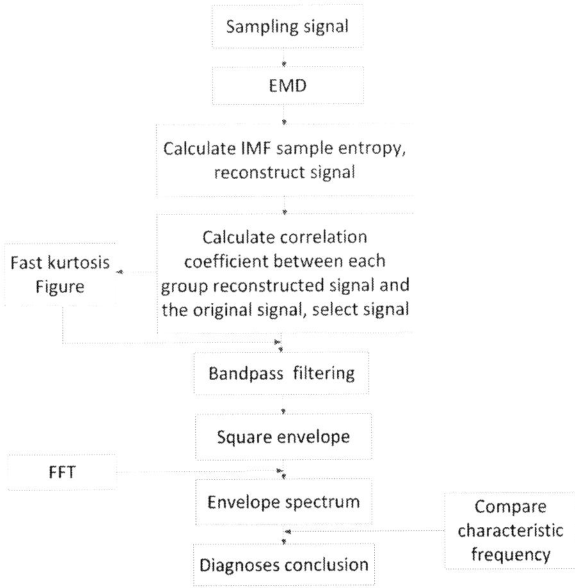

Fig. 1. Rolling bearing diagnostic process

4 Project example

Analyse rolling bearing fault acceleration signal to verify the method's practical performance, the signal is collected from bearing at the petrochemical plant gearbox. The device structure diagram is shown in Figure 2.

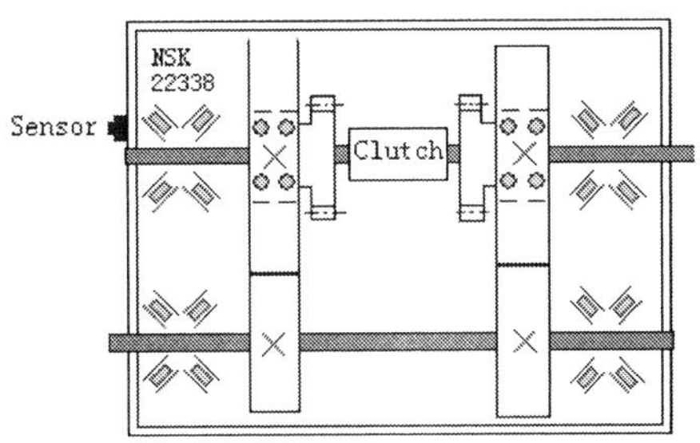

Fig. 2. Schematic diagram of the device

For more information about the signal is shown in Table 1.

Table 1. The Fault characteristic frequency of rolling bearing

The inner ring	The outer ring	Rolling elements	Cage
68.75Hz	48.2Hz	21.58Hz	3.44Hz

The signal is decomposed into 12 IMF components with EMD, and the IMF components aren't showed all in the article because of the space limitations. Calculate the correlation coefficients about the first seven components, and show them in Table 2.

Table 2. Each IMF component of the original cross-correlation coefficient

IMF	1	2	3	4	5	6	7
coefficients	0.72	0.43	0.29	0.23	0.24	0.12	0.13

It is difficult to choose the compents because the cross-correlation coefficients are very similar among IMF3, IMF4 and IMF5. First, calculate the former seven components's sample entropy using the sample entropy clustering criterion proposed above, the result is showed in Table 3.

Table 3. The IMF signal sample entropy

IMF	1	2	3	4	5	6	7
sample entropy	1.85	1.77	1.37	0.83	0.41	0.21	0.10

It is clear that the sample entropy of IMF1, IMF2 and IMF3 are bigger than 1 and the amplitude changes little. The sample entropy of IMF4, IMF5, IMF6 and IMF7 are at the same level. So IMF1-IMF3 components can be reconstructed as Sig1, IMF4-IMF7 components can be reconstructed as Sig2. The cross-correlation coefficients between the original signal and reconstructed signal are 0.8904 and 0.4107. The cross-correlation coefficient of the reconstruction Sig1 is improved compared to single IMF component and cross-correlation coefficient is much higher than Sig2. What's more, it retains most of the impact characteristic information of the original signal. So, choose Sig1 signal as the optimal reconstructed signal to do spectral kurtosis analysis.

The fast spectral kurtosis of Sig1 is shown in Figure 3. The maximum spectral kurtosis is found where the $f/\Delta f$ is 5400Hz/800Hz, then the range of the filter is determined. Do envelope demodulation after the filtering of the original signal, the envelope spectrum is get finally, which is showed in Figure 4.

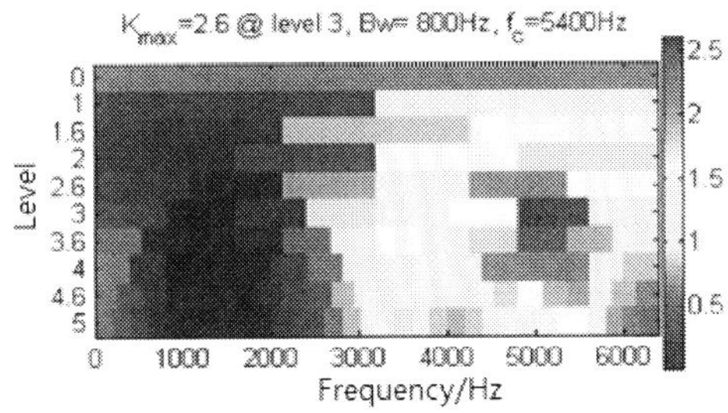

Fig. 3. Optimal fast spectral kurtosis reconstructed signal diagram

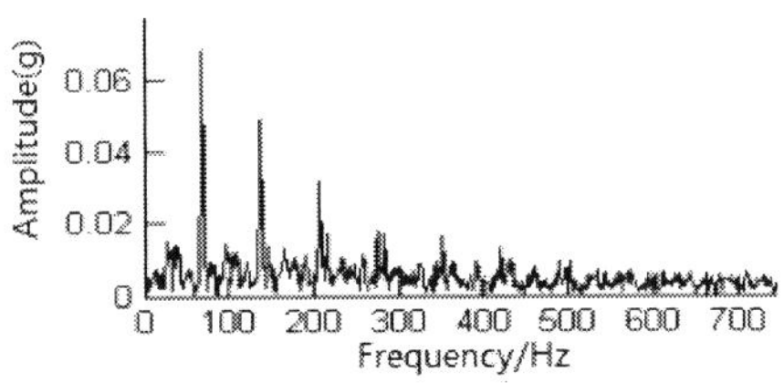

Fig. 4. Optimal *reconstructed signal envelope spectrum*

The frequency of 68.75Hz and multiplication frequency can be clearly seen in the optimal reconstructed signal envelope spectrum. This frequency is very consistent to the inner ring fault characteristic frequency. Then the fault can be determined for the inner rolling bearing faults.

5 The conclusion

This article proposes one denoise criterion based on the sample entropy clustering reconstruction, the SNR of the signal is largely improved. Early fault signal denoise and fault diagnosis of rolling bearings is achieved by spectral kurtosis algorithm in demodulated resonance. The objectivity of sample entropy clustering reconstruction criterion used in EMD denoise and the efficiency of the fault diagnosis method based on EMD-sample entropy-spectral kurtosis are verified through analysing the practical early fault signal of the roll bearing. While, this diagnostic method needs human experience, automation diagnosis method should be studied in the future research.

Reference

1. Mao K, Wu Y: Fault Diagnosis of Rolling Element Bearing Based on Vibration Frequency Analysis.Measuring Technology and Mechatronics Automation (ICMTMA), 2011 Third International Conference on. IEEE, 2011:198-201(2011)

2. Huang NE, Shen Z, Long SR: The empirical mode decomposition and the Hilbert spectrum for nonlinear and nonstationarytime series analysis.J.Proc.R.Soc 454:903-905(1998)

3. Wang P, Ding W Q, Mei Q, et al: Adaptive demodulated resonance diagnosis for the gear faults. Prognostics and System Health Management Conference. IEEE, 2014:208 – 212(2014)

4. HoD, RandallRB: Optimisation of bearing diagnostic techniques using simulated and actual bearing fault signals.J.Mechanical Systems and Signal Processing. 14(5):763—788(2000)

5. J.Antoni, R.B.Randall: The spectral kurtosis: Application to the vibratory surveillance and diagnostics of rotating machines.J.Mechanical Systems and Signal Processing. (20):308-331(2006)

6. Su Wensheng, Wang Fengtao, et al: Application of EMD denoising and spectral kurtosis in early fault diagnosis of rolling element bearings.J.Journal of Vibration and Shock .29(3):18-21(2010)

7. Richman J S, Moorman R J: Physiological time-series analysis using approximate entropy and sample entropy.J.Am J Physiol Heart Circ Physiol .278(6):H2039～2049(2000)

8. Costa M, Goldberger AL, Peng CK: Multiscale entropy analysis of biological signals.J. Physical Review E. 71:021906(2005)

9. Wang Lei, Wang Fengtao, et al: Empirical mode decomposition based on kernel principal component analysis with application.J.Journal of Vibration and Shock. 29(2):39-42(2010)

10. Liu Huasheng: The research of fault diagnosis of rolling bearing based on EMD.D.Dalian: School of Mechanical Engineering, Dalian University of Technology (2007)

11. Li Changyou, Xu Minqiang, et al: Fault diagnosis for rolling element bearings based on independent component analysis.J.Journal of Harbin Institite of Technology.40(9):1363-1365 (2008)

Fault Detection of Gear using Periodicity-induced Overlapping Group Shrinkage

Zhiwei Ye[1], Binqiang Chen[1], Wangpeng He[2], Yanyang Zi[2]

[1] Xiamen University, Xiamen 361005, P.R. China.
[2] Xi'an Jiaotong University, Xi'an 710049, P. R. China.

Abstract: The paper addresses the weak fault periodic oscillatory features extraction from vibration signals using periodicity-induced overlapping group shrinkage (POGS). Incipient signals are considered as periodic group sparse signals submerged in noise. Based on the minimization of a convex cost function involving a group sparsity promoting penalty function and the principle of majorization-minimization, a fast iterative algorithm is engaged for its solution. The method is applied to analyze gearbox vibration signals which acquired from a machinery fault simulator. The results demonstrate that the method can satisfactorily reveal periodic impulses.

Keywords: Fault detection, periodicity-induced overlapping group shrinkage, majorization-minimization, convex optimization

1 Introduction

Rotating machinery is one of the most representative common types of mechanical equipment which plays a prominent role in high running speed application. Gear system is an important component of rotating machinery which widely used in aerospace, automotive and heavy industry application[1]. Incipient fault detection of gear system not only offers an effective prevent method on factory accident but also avoids insufficient maintenance or excess maintenance [2]. Therefore, gear diagnosis technology is significant to reduce the economic losses[3].

Recently, overlapping group shrinkage (OGS) was established for estimating group-sparse signals in noise which initially formulated as a convex optimization by a convex regularization [4]. After the development of OGS, the improved version of OGS which utilizes a non-convex regularization is also developed [5]. Based on periodic transient pulses, He [6] proposed an approach for rotating machinery fault diagnosis and acquired an extraordinary feature extraction.

This paper addresses the weak fault features extraction from vibration signals using periodicity-induced overlapping group shrinkage (POGS). The fundamental theory of overlapping group shrinkage (OGS) is given in Section 2. Periodicity-induced OGS (POGS) is introduced in Section 3. Section 4 carried on an experimental to validate the method. Conclusion is presented in Section 5.

2 2. Overlapping group shrinkage (OGS)

2.1 Signal Modeling

Assuming the measured discrete-time series y_n as:

$$y_n = x_n + \omega_n, \quad n = 0, \ldots, N-1 \tag{1}$$

where signal x_n is the signal which has a group sparse property and ωn is the white Gaussian noise. A group-sparse signal is one where large magnitude signal values tend not to be isolated. Instead, these large magnitude values tend to form groups. The purpose of de-noising is to estimate the pure signal from the white Gaussian noise.

2.2 Convex Optimization

A common used model to estimate sparse vectors from noisy signals is:

$$x^* = \arg\min_x \left\{ F(x) = \frac{1}{2}\|y - x\|_2^2 + \lambda\Phi(x) \right\} \tag{2}$$

where λ is the regularization parameter, $\phi(x)$ is the sparsity-promoting penalty function. Convex optimization is a reliable and efficiency method to estimate sparse vectors from noisy signals. Using non-convex regularization, Chen [4] proposed an improved method to ensure convexity of the objective function. The problem can be solved by an iterative algorithm based on majorization-minimization (MM) [6]. With a group size of K and convex or non-convex regularization, the model is denoted as:

$$x^* = \arg\min_{x \in \mathbf{R}^N} \left\{ P_0(x) = \frac{1}{2}\|y - x\|_2^2 + \lambda \sum_n \phi\left([(\sum_{k\in\kappa} x_{n+k}^2)]^{1/2}; a\right) \right\} \tag{3}$$

where $\kappa = \{0, 1, \ldots, K-1\}$, $y \in R^N$ is the noisy observation, ϕ is the penalty function, the parameter 'a' can be chosen so that is convex even if ϕ is not. Ref. [5] reported an method to prove the objective function P_0 in (3) is strictly convex if

$$0 \leqslant a < 1/(\lambda K) \tag{4}$$

3 3. Periodicity-induced OGS (POGS)

3.1 3.1 Binary Weights

To facilitate the following derivation, we define a binary sequence $b = [b_0, b_1, \ldots, b_{K-1}]$, with $b_k \in \{0,1\}$, and sets

$$\kappa := \{0, 1, \ldots, K-1\}; \kappa_0 := \{k \in \kappa : b_k = 0\}; \kappa_1 := \{k \in \kappa : b_k = 1\}. \tag{5}$$

Since b is a binary vector, we have $\kappa = \kappa_0 \bigcup \kappa_1$ and $\kappa_0 \bigcap \kappa_1 = \varnothing$ is the empty set. We denote the cardinality (size) of the sets κ, κ_1 and κ_0 as K, K_1, and K_0, respectively, so that $K = K_0 + K_1$, and

$$\sum_{k \in \kappa} b_k = \sum_{k \in \kappa_1} b_k = K_1 \tag{6}$$

Therefore, the objective function P_1 can be written as:

$$x^* = \arg\min_{x \in \mathbf{R}^N} \left\{ P_1(x) = \frac{1}{2}\|y - x\|_2^2 + \lambda \sum_n \phi\big(\theta(x,b,n);a\big) \right\} \tag{7}$$

where the binary-weight grouping function θ is defined as:

$$\theta(x,b,n) := \left[\sum_{k=0}^{K-1} b_k x_{n+k}^2 \right]^{1/2} \tag{8}$$

which is the Euclidean norm of a binary weighted block. The objective function can be reorganized as:

$$\begin{aligned} P_1(x) &= \sum_{n=0}^{N-K} \left[\frac{1}{2K_1} \sum_k b_k x_{n+k}^2 + \lambda \phi\left(\left[\sum_{k \in \kappa} b_k x_{n+k}^2 \right]^{1/2}; a \right) \right] + L(x) \\ &= \sum_{n=0}^{N-K} P(B_n(x)) + L(x) \end{aligned} \tag{9}$$

where $L(x)$ is linear in x, $B_n(x)$ is defined as:

$$B_n(x) := \left[x_n, x_{n+1}, \ldots x_{n+K-1} \right] \tag{10}$$

Therefore, if P is strictly convex, then P1 is strictly convex. The condition for convexity of P is shown as:

$$\phi''(x;a) > -1/(K_1\lambda),\ x \neq 0 \tag{11}$$

The time interval between two consecutive faults can be considered identical within the support of a group. When the period T of a potential fault is known or predictable from the knowledge of the machinery, we can select the group with a length K and its zero and non-zero entries by

$$\begin{aligned} N_0 + N_1 &\approx f_s T, \\ N_0 + N_1 &= K / M. \end{aligned} \tag{12}$$

where N1 and N0 are the estimated length (in samples) of impulsive transients and the time interval between them, and integer $M \geqslant 2$ is the number of periods contained in one group, and f_s is the sampling rate. Thus, in one group, the numbers of zero and non-zero entries are $K_0 = MN_0$ and $K_1 = MN_1$, respectively. Moreover, when the transient sequence is periodical, the binary weight b_k are chosen according to a periodic group structure. Specifically, binary vector b can be represented as:

$$b = [\underbrace{1,1,\ldots,1}_{N_1},\underbrace{0,0,\ldots,0}_{N_0},\ldots,\underbrace{1,1,\ldots,1}_{N_1},\underbrace{0,0,\ldots,0}_{N_0}] \tag{13}$$

where in each period, there are N_1 non-zero entries grouped according to the impulsive signal, and the entire group comprises M periods. In this case, the last N_0 zeros has no effect in Equ.13. In practice, selected N_1 and M, the pattern of b can be determined.

3.2 3.2Parameter Selection

Iit is feasible to estimate the period of the transients based by component geometry and rolling speed, but it might be difficult to estimate the duration of the transient. Since the binary structure b is acting as a sliding window capturing the global periodicity structure, it is not necessary to match the length of a transient exactly. In order to explore the correlation of the regularization parameter λ to the binary pattern b, we show the optimal λ as a function of the standard deviation σ of noise using different binary patterns in Figure. We define the optimal as the value minimizing the RMSE for each fixed binary pattern b. Moreover, all the 16 curves in Fig. 6 show that the optimal λ varies approximately linearly with noise level. In practice, we suggest to chose λ in proportional to the noise level as $\lambda = r\sigma$.

4 Application

4.1 4.1 Gearbox Experiment

Data used to verify the proposed denoising algorithm are collected on a custom-built gearbox test bed, as shown in Fig.1. The platform is constructed of speed control box, AC servo motor, gearbox, magnetic powder brake, rotor, dynamic balance quality panel, loading device and other auxiliary mechanism. As can be seen in Fig.1(a), the measuring point position is located in the box lateral wall of fault gear. The test gear had a simulated crack at the root of one tooth 2mm×0.1mm across the whole width (shown in Fig. 1(b)).

4.2 Features Extraction

We apply it to the simulated data from the fault diagnosis comprehensive experimental platform. The signal is a 0.5-second signal with sampling rate fs =12800Hz (shown in Fig. 2), and is composed of a periodic sequence of transients occurring with 13Hz. The current rotating speed is approximately 780rpm and the test gear rotating speed is 572rpm (9.53Hz). As can be seen in Fig.2, the fault diagnosis time-domain vibration signal is mixed and disorderly.

Fig. 1. (a) Fault diagnosis comprehensive experimental platform ; (b) Weak tooth crack fault

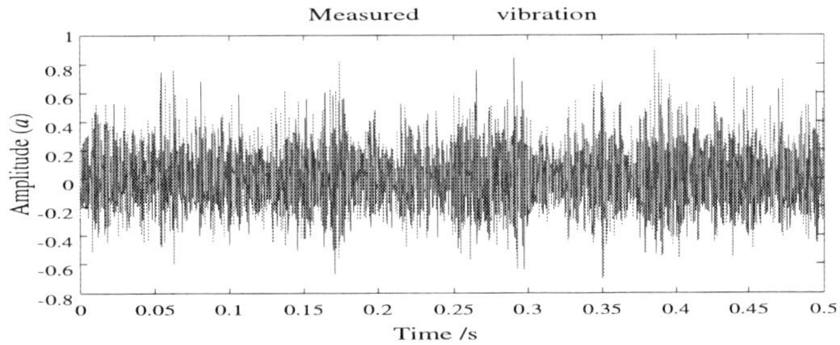

Fig. 2. Measured vibration signal of test gear

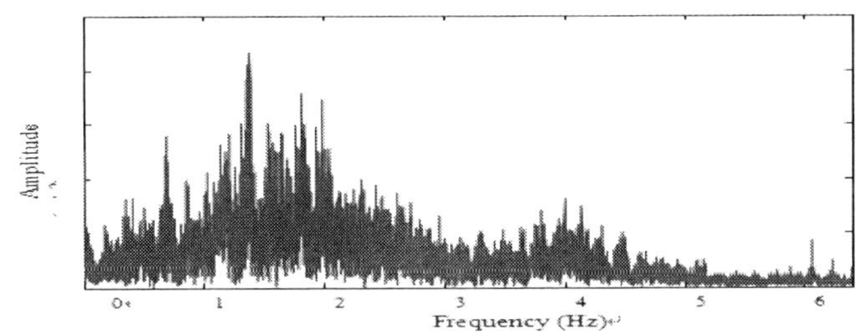

Fig. 3. Fourier spectrum of measured signal

Periodic group sparse signals are buried in strong background noise and irrelevant interference. The fourier spectrum is shown in Fig.3. It can be observed from the figure that the energy of the signal distributed along the whole frequency range. The constituent frequency component is too complex to extract the characteristic frequency component.

POGS approach is utilized to process the vibration signal. The atan penalty function is engaged to ensure the convexity of the objective function. The value of standard deviation σ of noise is calculated approximately 0.1985. Setting M=4 & N1=4, the multiplier r is chosen to be 0.325 according to Table 2. Therefore, the regularization parameter λ can be calculated as $\lambda = r\sigma = 0.0645$ .The result is shown in Fig.4.

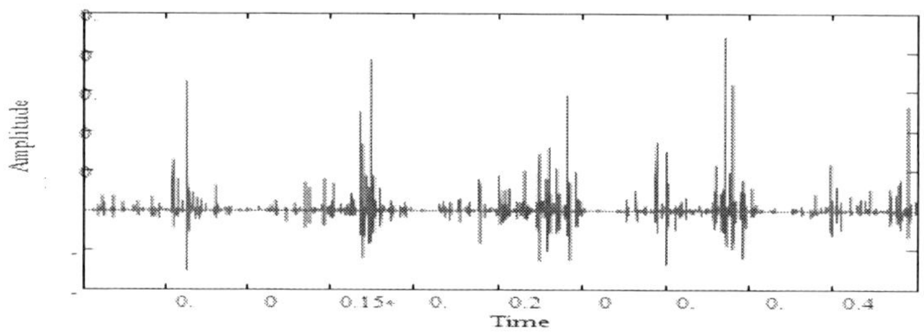

Fig. 4. Denoised fault signal

As can be seen in Fig.4, the shock signal which contains the fault characteristics is extracted successfully. The denoised fault signal shows a distinct shock signal generating time and position. The fault shock signal is periodic occurred every 0.105s (9.53Hz). The frequency of periodic shock signal is equal with the test gear rotating frequency. Therefore, the presented experiment proved that POGS approach is an efficiency method to the compound faults detecting.

5 Conclusion

In this paper, we address the weak fault periodic oscillatory features extraction from vibration signals using POGS approach. A faults detection experiment is carried on to verify the proposed denoising algorithm. The experiment result shows that the method can satisfactorily reveal the fault periodic impulses from gearbox vibration signals.

6 References

1. Lei Y, Lin J, Zuo M J, et al. Condition monitoring and fault diagnosis of planetary gearboxes: A review. Measurement, 2014, 48(1):292-305.
2. Heng A, Zhang S, Tan A C C, et al. Rotating machinery prognostics: State of the art, challenges and opportunities. Mechanical Systems & Signal Processing, 2009, 23(3):724-739.
3. Hou T J, Shi C J, Sun J J. Gear Fault Diagnosis Based on CepstrumAnalysis[J]. Coal Mine Machinery, 2011.
4. Po-Yu Chen, Ivan W. Selesnick. Translation-invariant shrinkage/thresholding of group sparse signals . Signal Processing, 2014, 94(1):476-489.
5. Chen P Y, Selesnick I W. Group-Sparse Signal Denoising: Non-Convex Regularization, Convex Optimization[J]. IEEE Transactions on Signal Processing, 2013, 62(13):3464-3478.
6. Sriperumbudur B K, Torres D A, Lanckriet G R G. A majorization-minimization approach to the sparse generalized eigenvalue problem[J]. Machine Learning, 2011, 85(1-2):3-39.
7. He W, Ding Y, Zi Y, et al. Sparsity-based Algorithm for Detecting Faults in Rotating Machines[J]. Mechanical Systems & Signal Processing, 2015.

Mechanical Parameters Identification for Laminated Composites Based on the Impulse Excitation Technique

Wenlei Song [1] ,Yongteng Zhong [1] , Jiawei Xiang [1]

[1] Wenzhou University, 325035 Wenzhou, China

songwenlei1991@163.com, zhongyongteng@wzu.edu.cn, jwxiang@wzu.edu.cn

Abstract. Beams, shafts, plates and shells are the basic structures in mechanical systems and the corresponding mechanical parameters (dynamic Young's modulus, shear modulus and Poisson's ratio) are extremely important for the mechanical property and fault analysis. The impulse excitation technique (IET) is a nondestructive and very convenient testing method to obtain the mechanical parameters of the materials. It could be applied at different circumstances such as high temperatures, low temperatures and different engineering materials. The IET is based on the analysis of the vibration of a test sample after it was impulse excited. The mechanical properties of a test sample are related to its mechanical resonance frequency. Dynamic Young's modulus is determined by the resonant frequency in the flexural mode, whereas the dynamic shear modulus can be calculated by torsional resonant frequency. In the present, the IET is employed to identify the mechanical parameters of laminated composite materials. Numerical simulation of the rectangular and cylinder bars and the laminated composites are given to testify the performance of the present technique. Moreover, a series of actual experiments were also done to verify correctness of the simulation results.

Keywords: impulse excitation technology; mechanical parameters; laminated composites; temperature loadings.

1 Introduction

There are several techniques can be used to study the mechanical parameters [1-2], such as the Single Edge Notched Beam (SENB) and the Impulse Excitation Technique (IET). By comparison, the method using impulse excitation technology (IET) has received much attention in recent years due to its simple and fast properties for the measurement of mechanical parameters. As we know well, the mechanical parameters are the key to simulate the relative true dynamic response of mechanical systems with faults. In practical engineering application, the working temperature of components is often variable. However, the dynamic response has close relationship with the temperature variations [3-5]. The IET allows the measurement from room temperature to high temperature (up to 1750 $^{\circ}C$) [6], and can be used widely.

Previous researchers did many investigations using IET to determine the mechanical parameters (dynamic Young's modulus, shear modulus and Poisson's ratio) for different materials such as ZnO ceramics [7], particulate epoxy composites [8], concrete [9] and ceramic particle reinforced aluminum [10].

When using IET measure the Young's modulus (*E*) and shear modulus (*G*), the flexural resonant frequency f_f and the torsional resonant frequency f_t must be obtained, respectively. American Society for Testing and Materials (ASTM) provide several more authoritative test method and procedure [11-14]. After we get the values of f_f and f_t, the Young's modulus *E* and shear modulus *G* can be calculated accordingly. And the Poisson's ratio μ will be also calculated by the relationship among μ , *E* and *G*. However, few researchers focused on the mechanical properties of composites. In the present, we extend IET to determine the mechanical parameters of laminated composites. Using finite element method (FEM), the rectangular and cylinder bars and the composites plate under different temperature loadings are given.

2 Technique method

We put the structures on the two supports located at the fundamental nodal point. The locations of fundamental nodal point are 0.224L from each end for flexural modes and it should be the center point of bars for torsional mode, as shown in Fig.1.

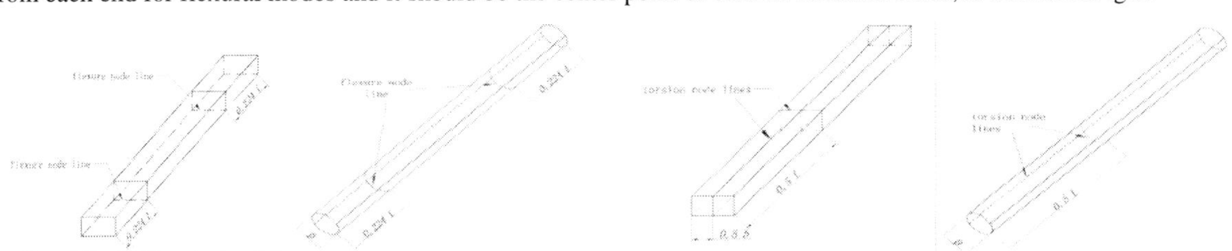

(1) (2) (1) (2)

(a) The flexural modes (b) The torsional modes

Fig.1. The diagrammatic sketch of structures

By experimental modal analysis (EMA), the fundamental flexural resonant frequency and fundamental torsional resonant frequency will be obtained accordingly. After we get the fundamental flexural resonant frequency and fundamental torsional resonant frequency, the Young's modulus and shear modulus will finally be calculated form the formula, as shown in the following two sub-sections.

2.1 The calculation of dynamic Young's modulus

(1) If the structures are rectangular bars, the fundamental flexure frequency can be calculated by [12]

$$E = 0.9465(mf_f^2 / b)(L^3 / t^3)T_1 \tag{1}$$

where E is Young's modulus (Pa), m is mass (g), b is width (mm), L is length (mm), t is thickness (mm), f_f is the fundamental flexure frequency (Hz), T_1 is correction factor for fundamental flexural mode to account for the thickness of the bar, which is calculated by

$$T_1 = 1 + 6.585(1 + 0.0752\mu + 0.8109\mu^2)(t/L)^2 - 0.868(t/L)^4 - \left[\frac{8.340(1+0.2023\mu+2.173\mu^2)(L/t)^4}{1.000+6.338(1+1.1408\mu+1.536\mu^2)(L/t)^2}\right] \tag{2}$$

where μ is Poisson's ratio. To simple the calculation, if $L/t \geq 20$, T_1 can be simplified as

$$T_1 = 1.000 + 6.585(L/t)^2 \tag{3}$$

and E can be directly obtained using Eq.(1).

If $L/t \leq 20$ and μ is known, T_1 can be calculated using Eq.(2). Then E can also be directly obtained using Eq.(1).

If $L/t \leq 20$ and μ is not known, then we must assume an initial Poisson's ratio and use an iterative process to determine a value of Poisson's ratio. The iterative process is shown in Fig.2.

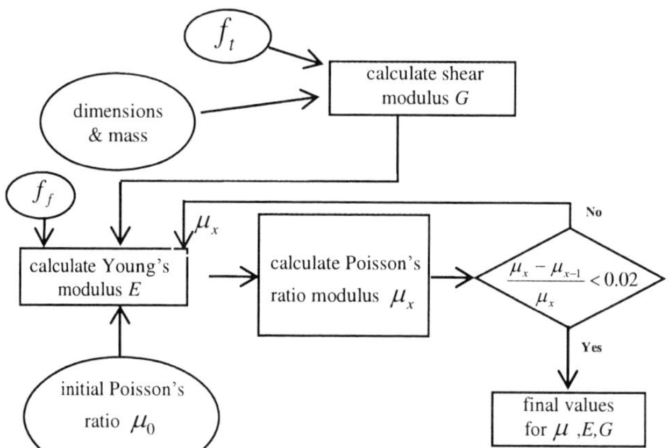

Fig.2. The flow diagram of iterative process

The procedure of iterative process can be described as follow, firstly, using the Eqs mentioned above, calculate the dynamic Young's modulus E from the fundamental flexure frequency f_f and the initial Poisson's ratio μ_0, calculate the dynamic shear modulus G from the fundamental torsional frequency f_t. Secondly, substitute E and G into the Eq $\mu = \frac{E}{2G} - 1$ and calculate a new value of Poisson's ratio for anther iteration beginning. Repeat the iterative process until no significant difference (0.02 or less) is observe between μ_x and μ_{x-1}.

(2) When the structures are rods of circular cross section, if we have got the fundamental flexure frequency f_f , E can be calculated as follow

$$E = 1.6067(L^3 / D^4)(mf_f^2)T_1 \quad (4)$$

where D is the diameter of rod (mm), $T_1^{'}$ is correction factor for fundamental flexural mode to account for the diameter of structures, Poisson's ratio, and so forth.

$$T_1^{'} = 1 + 4.939(1 + 0.0752\mu + 0.8109\mu^2)(D/L)^2 - 0.4088(D/L)^4 - \left[\frac{4.691(1 + 0.2023\mu + 2.173\mu^2)(D/L)^4}{1.000 + 4.754(1 + 0.1408\mu + 1.536\mu^2)(D/L)^2}\right] \quad (5)$$

If $L/t \geq 20$, $T_1^{'}$ can be simplified as

$$T_1^{'} = 1.000 + 4.939(D/L)^2 \quad (6)$$

and E can be calculated easily.

If $L/t \leq 20$ and μ is known, $T_1^{'}$ can be calculated using Eq.(5) directly and then E can be also calculated.

If $L/t \leq 20$ and μ is not known, the iterative process shown in Fig.2 must be done to start the computation.

2.2 The calculation of dynamic shear modulus

(1) If the structures are rectangular bars and we have got the fundamental torsional frequency f_t, the shear modulus G can be calculated as follow

$$G = \frac{4Lmf_t^2}{bt}\left[B/(1 + A)\right] \quad (7)$$

where G is dynamic shear modulus (Pa), f_t is fundamental resonant frequency of the rod (Hz). A and B are the empirical correction factor dependent on the width-to-thickness.

$$B = \left[\frac{b/t + t/b}{4(t/b) - 2.52(t/b)^2 + 0.21(t/b)^6}\right] \quad (8)$$

and

$$A = \left[\frac{\left[0.5062 - 0.8776(b/t) + 0.3504(b/t)^2 - 0.0078(b/t)^3\right]}{\left[12.03(b/t) + 9.892(b/t)^2\right]}\right] \quad (9)$$

(2) If the structures are cylindrical rods and we have got the fundamental torsional frequency, the shear modulus G can be calculated as follow

$$G = 16mf_t^2(L/\pi D^2) \quad (10)$$

3 Simulation procedure

It is known from the formulas mentioned above. The mechanical parameters of solid structures are mainly related to their mass and dimensions. The structures' dimensions and mass can be measured by measuring instruments .Meanwhile, the fundamental flexure resonant frequency and fundamental resonant torsional frequency can be obtained using impulse excitation technology respectively. After the several variable quantities are measured or known, we can calculate the structures' dynamic Young's modulus (E) and dynamic shear modulus (G) using Eqs mentioned above.

In this section, there will list four different types of examples and implement the simulation using commercial FE software ANSYS.

3.1 Traditional single materials

Suppose that the structures are rectangular bars with different dimensions at room temperature.

In this section, it will simulate five types of structures in different structural dimension combinations as specified in Table 1 using software. Modal analysis is performed to get the resonant frequency in flexural modes and torsional modes. The structures' support locations and displacement locations are shown in Fig.1-a-1 and Fig.1-b-1.

Table 1. The dimension of structures

No.	Length (m)	Width (m)	Thickness (m)
S_{11}	0.15	0.01	0.005
S_{12}	0.20	0.01	0.005
S_{13}	0.15	0.02	0.005
S_{14}	0.15	0.01	0.006
S_{15}	0.20	0.02	0.006

During every simulation procedure, the measured results of f_f and f_t will be recorded in Table 2 and then they will be used to calculate the values of E and G through Eqs.(1,3,7,8,9), can be calculated by $\mu = \left(\frac{E}{2G}\right) - 1$. We can analyze the results and absolute relative errors by do comparisons between theoretical values and measured values.

The comparisons clearly reflect the quality of simulations, and the size of absolute relative error is an important standard to estimate the reliability of the simulation.

Table 2. The comparison between theoretical value and measured value for rectangular bar

NO.		S_{11}	S_{12}	S_{13}	S_{14}	S_{15}	No.		S_{11}	S_{12}	S_{13}	S_{14}	S_{15}	No.		S_{11}	S_{12}	S_{13}	S_{14}	S_{15}
The theoretical value	f_{ft} (Hz)	1176.2	662.7	1176.2	1409.1	794.4	The measured value	f_f (Hz)	1176.4	662.8	1179.0	1409.1	795.4	The absolute relative error ε (%)	f_f	0.017	0.016	0.26	0	0.12
	f_{tt} (Hz)	7925.2	5923.9	4781.9	8698.3	4164.3		f_t (Hz)	7918.4	5938.2	4771.3	8699.7	4153.7		f_t	0.085	0.096	0.22	0.016	0.25
	E_t (GPa)	210						E (GPa)	210.08	210.07	211.00	210.00	210.51		E	0.038	0.033	0.48	0	0.24
	G_t (GPa)	80.77						G (GPa)	80.632	80.615	80.413	80.797	80.358		G	0.17	0.19	0.44	0.033	0.51
	μ_t	0.3						μ	0.3027	0.3029	0.3120	0.2996	0.3098		μ	0.9	0.97	4	0.13	3.27

ε is the absolute relative errors between the measured and theoretical values.

The above results demonstrate that the absolute relative errors between theoretical values and measured values are logical and acceptable. Therefore the simulation's results are reasonable and credible.

3.2 The composite materials

In this section, a cylindrical rod is chosen as analysis object, its material is laminated composites, its dimension is determined (L=1m, D=0.04m) and structure of layers is shown in Fig.3.and its structure's directional angles are 0◦, 90◦, 0◦ We build the models followed Fig.1-a-2 and Fig.1-b-2. Four different temperatures applied in structures separately. And then flexure frequency and torsional frequency will be recorded into table 4. E, G, μ would be calculated through the Eqs above. The results of measurement and calculation are listed in Table 3. And then compare and analyze the final results of simulation.

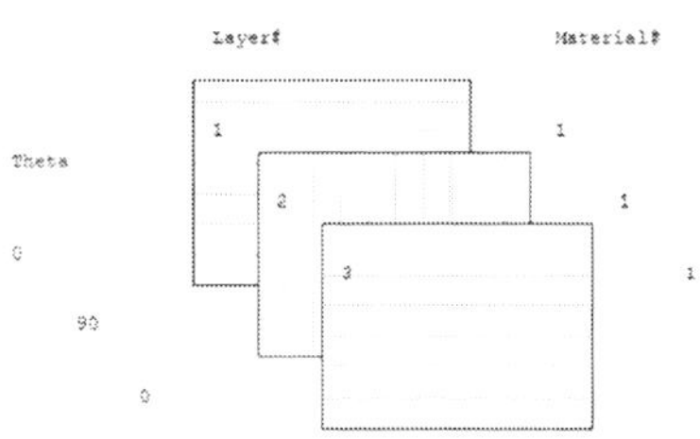

Fig.3. The structure of composites

Table 3. The results of simulations

temperature	f_f (Hz)	f_t (Hz)	E (GPa)	G (GPa)	μ
0°C	2194	18075	210	80.77	0.3
10°C	1897.3	9303.9	157.03	59.112	0.33
20°C	1537.9	7404.9	103.16	37.44	0.38
30°C	1055	4803.3	48.541	15.758	0.54

From the results of simulations, we can see the change regulation of dynamic Young's modulus and dynamic shear modulus when the temperature is elevated. The resonant frequency f_f and f_t will reduce and the value of E and G will become smaller after calculation with Eqs mentioned above. And then the Poisson's radio will elevated.

4 Experimental investigation

The IET measure the fundamental resonant frequency of test sample by exciting them mechanically by a single elastic strike with an impulse tool and measuring the vibrational response with an accelerometer. The signal was analyzed with a commercial instrument evaluating the fundamental frequency.

We did a series of experiments at room temperature. Laminated composites epoxy board (3240) was selected as samples for use in experiments, a small impact hammer was used to excite resonant vibrations in a nodally supported sample, an accelerometer was used to feed the samples' vibration signal and analyzed with a commercial instrument AVANT MI-7016 evaluating the fundamental frequency. The figures of the original frequency signal and frequency spectrum were displayed by a computer. The specimens were put on pieces of thin steel sheets (1*mm*) for flexural mode and supported on traditional cotton threads for torsional mode at the positions of the vibrational node to ensure free vibrating. The flexural nodal point correspond to 0.224 of the length of the board from each end and the torsional nodal point correspond to 0.5 of the length of the board from each end, show as Fig.1. Small deviations from the exact support locations have a minimal effect on the measure vibration frequency. It is applicable to a variety of regular-shaped sample geometries such as bar, board, disks and cylinders. Here, we restrict our analysis to rectangular board samples.

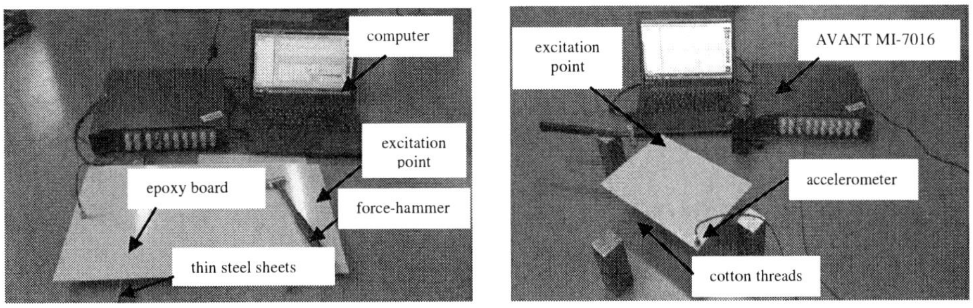

(a) flexural mode (b) torsional mode

Fig.4. Experimental apparatus

The experiments set-up is shown in Fig.4.The sample was put on the supports at the position of the vibrational nodes to ensure free vibration. An accelerometer was positioned at either end of board to measure the vibration. To excite the board in the desired flexural mode and torsional mode the simple was hit by a small force-hammer in the center of the board or at end of the board show as Fig.4.a-b. The signal was transmitted to the AVANT MI-7016 instrument, which could estimate the frequency and displayed the figures of the original frequency signal and frequency spectrum through a computer.

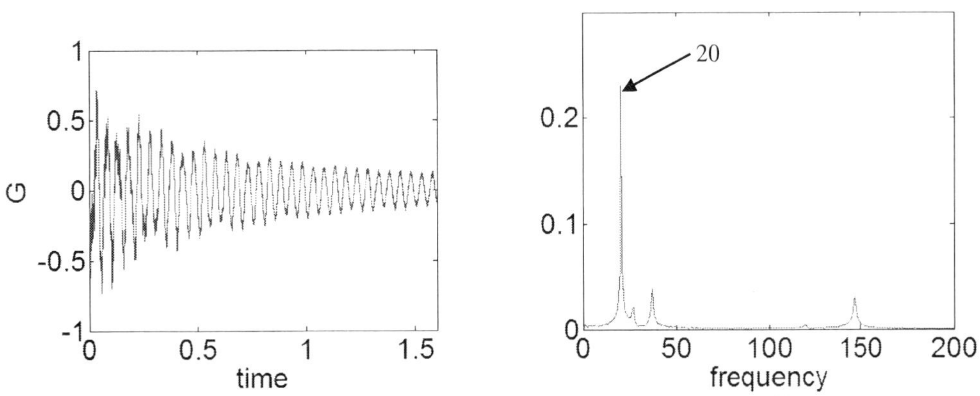

(a) (b)

Fig.5. the figures of the original frequency signal and frequency spectrum for flexural mode

From the Fig.5-b, we could obtain the fundamental flexural resonant frequency is 20 *Hz*, so we would get the value of the Young's modulus E^*=2.396e10 *Pa* using equations form (1) to Eq. (3).

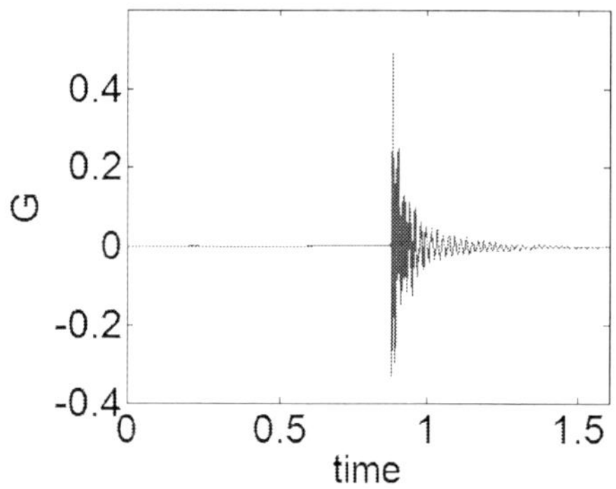

(a)

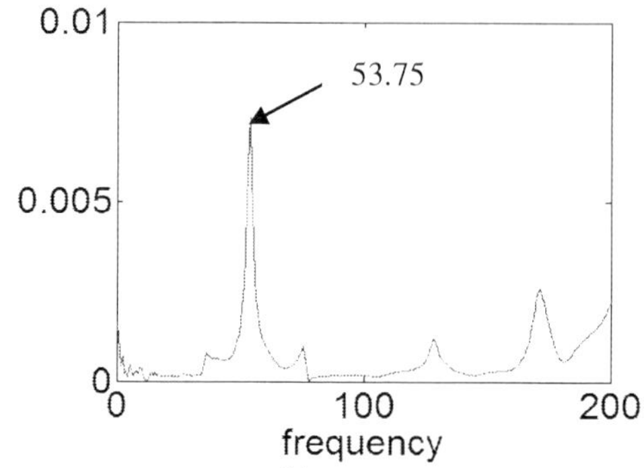

(b)

Fig.6. the figures of the original frequency signal and frequency spectrum for torsional mode

From the Fig.6-b, we could obtain the fundamental torsional resonant frequency is 53.75 *Hz*, so we would get the value of the shear modulus G^*=0.818e10 *Pa* using equations form (7) to Eq. (9).

Finally, the Poisson's ratio of laminated epoxy plate can be calculated by $\mu = \left(\frac{E}{2G}\right) - 1$ as $\mu^* = 0.464$.

5 Conclusions

The method of using the impulse excitation technology to measure structures' mechanical parameters is feasible through the analysis and simulations mentioned above. There are also many advantages of IET under the structures' actual working conditions.

From simulations, we will find the change regulation of the dynamic Young's modulus and dynamic shear modulus of composites when the temperature is varied. When the temperature of structures' is elevated, the values of resonant frequency (f_f and f_t) will be reduced and the value of Young's modulus (*E*) and shear modulus (*G*) will also become smaller ,and then the Poisson's ratio become elevated for composites.

6 Acknowledgments

The authors are grateful to the support from the National Science Foundation of China (Nos. 51575400, 51505339, 51405346), the Zhejiang Provincial Natural Science Foundation of China (Nos.LR13E050002, LQ16105005), the Zhejiang Technologies R&D Program of China (No. 2014C31103).

References

1. Sousa, F.J.P. , M.Bo Dal, Guglielmi, P.O. , et al.: Characterization of Young's modulus and fracture toughness of albite glass by different techniques. Ceram. Int. 40, 10893--10899 (2014)
2. M. Alfano, Pagnotta, L.: A non-destructive technique for the elastic characterization of thin isotropic plates. NDT&E. Int. 40, 112--120 (2007)
3. de Oliveira, A.P.N., Vilches, E.S., Soler, V.C. Villegas, and F.A.G.: Relationship between Young's modulus and temperature in porcelain tiles. J. Eur. Ceram. Soc. 32, 2853--2858 (2012).
4. Dickson, R. W., Wachtman, J. B.: An Alumina Standard Reference Material for Resonance Frequency and Dynamic Elastic Moduli Measurement I. For Use at 25°C. Phys. Chem. 75A, 155--162 (1971)
5. R.W. Dickson, R.W., Schreiber, E.: An Alumina Standard Reference Material for Resonance Frequency and Dynamic Elastic Moduli Measurement II. Characteristic Values for Use from 25 °C to 10000 °C .Phys. Chem. 77A, 391--394 (1971)
6. Roebben, G., Bollen, B., Brebels, A., Van Humbeeck, J., Van der Biest, O.: Impulse excitation apparatus to measure resonant frequencies, elastic moduli, andinternal friction at room and high temperature. Rev. Sci. Instrum. 68, 4511--4515 (1997)
7. Swarnakar, A.K. , Donzel, L., Vleugels, J., Van der Biest, O.: High temperature properties of ZnO ceramics studied by the impulse excitation technique. J. Eur. Cera. Soc. 29, 2991—2998 (2009)
8. Tognana, S., Salgueiro, W., Somoza, A., Marzocca, A.: Measurement of the Young's modulus in particulate epoxy composites using the impulse excitation technique. Mater. Sci. Eng. A.527, 4619--4623(2010)
9. Bahr, O., Schaumann, P., Bollen, B., Bracke, J.: Young's modulus and Poisson's ratio of concrete at high temperatures: Experimental investigations. Mater. Des. 45, 421—429 (2013)
10. Hauert, A., Rossoll, A., Mortensen, A.: Young's modulus of ceramic particle reinforced aluminium: Measurement by the Impulse Excitation Technique and confrontation with analytical models. Compos. A.40, 524—529 (2009)
11. ASTM Standard E1875-08,: Standard Test Method for Dynamic Young's Modulus, Shear Modulus, and Poisson's Ratio by Sonic Resonance. ASTM International, West Conshohocken, PA.
12. ASTM Standard E1548-02,: Standard Test Method for Dynamic Young's Modulus, Shear Modulus, and Poisson's Ratio of Refractory Materials by Impulse Excitation of Vibration. ASTM International, West Conshohocken, PA.
13. ASTM Standard E1876-01,: Standard Test Method for Dynamic Young's Modulus, Shear Modulus, and Poisson's Ratio by Impulse Excitation of Vibration. ASTM International, West Conshohocken, PA.
14. ASTM Standard C1198-09,: Standard Test Method for Dynamic Young's Modulus, Shear Modulus, and Poisson's Ratio for Advanced Ceramics by Sonic Resonance. ASTM International, West Conshohocken, PA.

Mechanical Fault Detection Using The Skewness Maximization Based Recovering Periodic Impulsive Signals

Shuhui Wang[1], Yuqing Zhou[1] Jiawei Xiang[1]

Wenzhou University, 325035 Wenzhou, China

caomaowsh@163.com, zhouyq@wzu.edu.cn, jwxiang@wzu.edu.cn

Abstract. Skewness index is the third moment of the probability density function (PDF) which reflects the asymmetry of the center of the signal probability distribution. With the larger value of the skewness, the asymmetry becomes more serious. The proposed approach can reveal hidden periodic impulses via maximizing the skewness of a measured signal by an adaptive filter. Skewness is employed as an objective function to train the adaptive filter on periodic impulsive source in a blind fashion. Incorporating with the adaptive finite impulse response (FIR) filter, the skewness tends to the maximum. Skewness maximization (SM) realizes to suppress the undesired components, while enhancing the desired periodical impulsiveness. At last, the numerical simulation of the faults and the experimental investigations are given to validate the effectiveness of the approach.

Keywords: REBs; fault detection; MED; skewness; adaptive filtering.

1 Introduction

The rotating machinery is one of the most widely used operating equipment in almost all the modern industry sections. Therefore, condition monitoring and fault diagnosis of the operating machine accurately has been of great importance [1-3]. Rolling element bearings (REBs) are crucial components in mechanical transmission apparatus which means any defects that occurred to the REBs may lead to serious damages to the whole system. Vibration analysis is the most commonly method for detecting failures within the REBs. Once there are any damages occupied to certain surface of REBs, the contact surface between rolling elements and the bearing race creates periodically impacts each time when they strike to each other. It is the periodic impulsive vibration signals that provide information about the current status of the REBs [4]. Such periodical impacts generate harmonically related sinusoids [5] in the frequency spectrum which can be used to extract the characteristic frequency corresponding to the damages [6]. However, the unavoidable ambient random noise and disturbance (i.e. low frequency sinusoids) makes the measured signal deviated from the original vibration impulses. To some extent, the vibration signals measured from the transducer is a convolution of the various signals and the linear filter [7]. While there usually requires a high signal quality, a priori appropriate signal pre-processing technique is exploited before further analysis. The wavelet based method [8], empirical mode decomposition (EMD) [9] and local mean decomposition (LMD) [10] etc., have been performed in dealing with the raw signals which give the superior reliability but suffer from computational demanding when there is no excellent power computing performance and sufficient processing capacity.

Blind deconvolution as a well-known technique is conditioned on some statistical characteristics (i.e. probability density function) of the interested signal. Minimum entropy deconvolution (MED) has a majority share of applications in the engineering fields. MED method was first proposed by Wiggins [11] and successfully applied to geographical surveying to deconvolute the seismic waves. In the literature, the author choose kurtosis (the fourth order statistics) as the objective function of the impulse train (the desired signals). Donoho [12] summarized principal ideology of the MED method and gave the general formula. R.B.Randall [13] and his colleagues exploited the MED method for damage detection of the REBs and gear boxes and the results shows the excellent performance of MED method.

In recent studies, Pääjärvi and LeBlanc [14] presented a new method by choosing skewness instead of the kurtosis as the objective function for impulsive source in MED and manifest its admirable behavior than that of the kurtosis. Ovacıklı et al. [15] implement the approach recovering periodic impulsive signals. In this paper, skewness is performed as the objective function in MED dealing with the fault detection in RBEs. The outline of this paper is as follows. The fundamental theories of MED method are briefly summarized in Section 2. In Section 3, the skewness based MED method validated by simulation analysis and experimental results. Finally, conclusion remarks are drawn in Section 4.

2 Method

2.1 Minimum entropy deconvolution

MED is a method that using adaptive filter deconvolute the mixed signals obtained by the transducer. Entropy is an indicator to reflect the status of disorder. It is well known that random noise inclines to have a normal distribution (i.e. more Gaussian) which means a high value of the entropy. Minimizing the entropy will enhance the structured components of the vibration signals while attenuating the Gaussian distributed ones. Vibration signals de-noising of the mixture can be achieved in this way

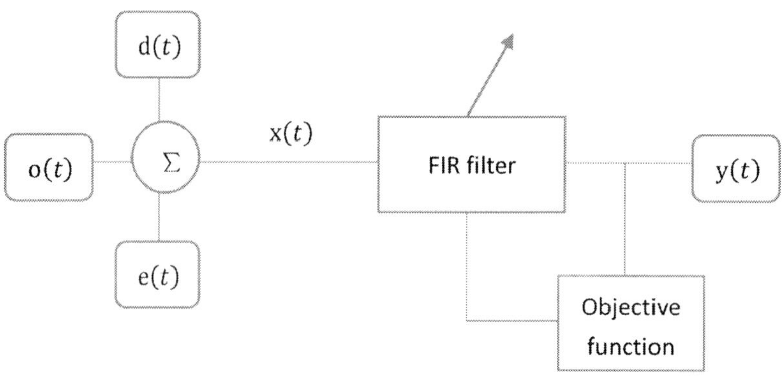

Fig.1. The general MED model system

where $o(t)$ is the desired fault impulse, $d(t)$ is the low frequency sinusoids disturbances and $e(t)$ is the random noise; $h_o(t)$, $h_d(t)$ and $h_e(t)$ are the corresponding linear time invariant (LIT) transfer function respectively; f is the finite impulsive response (FIR) filter of length L. $x(t)$ is the MED system input and $y(t)$ is the output.

As is shown in figure 1, the obtained vibration signals from the transducer are modeled by three participations and the signals are used as inputs to the adaptive filter. The weight of the filter is adjusted iteratively by search the optimum value of the objective function. It should be noted that the objective function is a function of the output signals. Accompanied by the filter weight adaption, the outputs are gradually asymptotic to the fault-impulse. The model for the inputs and the outputs are given as

$$x(t) = h_o(t) * o(t) + h_d(t) * d(t) + h_e(t) * e(t) \tag{1}$$

$$y(t) = f * x(t) \tag{2}$$

The output is a function of the FIR filter weight. The objective function is based on higher-order statistics of $y(t)$. For any output sequence $y(n)$ of length N, the objective function is generalized by [donoho],

$$o_s^r(y(n)) = \frac{\frac{1}{N}\sum_{n=1}^{N}|y_n|^r}{\left(\frac{1}{N}\sum_{n=1}^{N}|y_n|^s\right)^{r/s}} \tag{3}$$

where r denotes the rth-order and generally $s = 2$.

2.2 Skewness maximization

Skewness represents the statistical asymmetry of the vibration signals and it can be used to describe the statistical characteristics of impulses, while the fault impulses have the asymmetry characteristic which different from noise and low frequency sinusoids [10]. The skewness is employed as the objective function in the MED scheme and defined as

$$S(y) = \frac{\sum_{n=1}^{N}(y_n)^3}{\left[\sum_{n=1}^{N}(y_n)^2\right]^{3/2}} \tag{2}$$

With FIR filter weight vector $F = [f_1 \quad f_2 \quad \cdots \quad f_L]^T$ of length L and input sequence $x(n) = [x_n \quad x_{n-1} \quad \cdots \quad x_{n-l+1} \quad \cdots \quad x_{n-L+1}]^T$ at time n where the corresponding elements are padded with zero while the subscript is less than one, the output at time n can be expressed as

$$y_n = \sum_{l=1}^{L} f_l \cdot x_{n-l+1} \tag{3}$$

y_n is a function of FIR filter weight f and the derivative is

$$\frac{\partial y_n}{\partial f_l} = x_{n-l+1}$$

Noting that $S(y)$ is a function of FIR filter weight f, maximizing skewness (i.e. searching for the optimum) is equivalent to take the derivative with respect to weight vector f of the objective function $\nabla_{S(y)} = 0$, we have

$$\nabla_{S(y)} = \left[\frac{\partial S}{\partial f_1} \quad \frac{\partial S}{\partial f_2} \quad \cdots \quad \frac{\partial S}{\partial f_l} \quad \cdots \quad \frac{\partial S}{\partial f_L}\right]^T = 0 \tag{4}$$

Consider Eq.(2), we have

$$\frac{\partial S}{\partial f_l} = \frac{\partial}{\partial f_l}\left[\frac{\sum_{n=1}^{N}(y_n)^3}{\left[\sum_{n=1}^{N}(y_n)^2\right]^{3/2}}\right]$$

$$= 3\left[\sum_{n=1}^{N} {y_n}^2 x_{n-l+1}\right]\left[\sum_{n=1}^{N} {y_n}^2\right]^{-3/2} - 3\left[\sum_{n=1}^{N} y_n x_{n-l+1}\right]\left[\sum_{n=1}^{N} {y_n}^3\right]\left[\sum_{n=1}^{N} {y_n}^2\right]^{-5/2} = 0 \tag{5}$$

Rearrange Eq.(5), we obtain

$$\sum_{n=1}^{N} y_n x_{n-l+1} = \frac{\sum_{n=1}^{N} {y_n}^2}{\sum_{n=1}^{N} {y_n}^3}\left[\sum_{n=1}^{N} {y_n}^2 x_{n-l+1}\right] \tag{6}$$

Denoting $A = \frac{\sum_{n=1}^{N} {y_n}^2}{\sum_{n=1}^{N} {y_n}^3}$. Using (3) in (6), the left hand side

$$\sum_{n=1}^{N} y_n x_{n-l+1} = \sum_{n=1}^{N}\left\{\left[\sum_{k=1}^{L} f_k x_{n-k+1}\right] x_{n-l+1}\right\}$$

$$= [x_{2-l} \quad x_{3-l} \quad \cdots \quad x_{N-l+1}]\begin{bmatrix} x_1 & 0 & 0 & & 0 \\ x_2 & x_1 & 0 & \cdots & 0 \\ x_3 & x_2 & x_1 & & 0 \\ & \vdots & & \ddots & \vdots \\ x_N & x_{N-1} & x_{N-2} & \cdots & x_{N-L+1} \end{bmatrix}\begin{bmatrix} f_1 \\ f_2 \\ f_3 \\ \vdots \\ f_L \end{bmatrix}$$

And the right hand side

$$\sum_{n=1}^{N} {y_n}^2 x_{n-l+1} = [x_{2-l} \quad x_{3-l} \quad \cdots \quad x_{N-l+1}]\begin{bmatrix} {y_1}^2 \\ {y_2}^2 \\ {y_3}^2 \\ \vdots \\ {y_N}^2 \end{bmatrix}$$

Therefore, $\nabla_{S(y)} = 0$

$$\begin{bmatrix} x_1 & x_2 & x_3 & & x_N \\ 0 & x_1 & x_2 & \cdots & x_{N-1} \\ 0 & 0 & x_1 & & x_{N-2} \\ & \vdots & & \ddots & \vdots \\ 0 & 0 & 0 & \cdots & x_{N-L+1} \end{bmatrix} \begin{bmatrix} x_1 & 0 & 0 & & 0 \\ x_2 & x_1 & 0 & \cdots & 0 \\ x_3 & x_2 & x_1 & & 0 \\ & \vdots & & \ddots & \vdots \\ x_N & x_{N-1} & x_{N-2} & \cdots & x_{N-L+1} \end{bmatrix} \begin{bmatrix} f_1 \\ f_2 \\ f_3 \\ \vdots \\ f_L \end{bmatrix}$$

$$= \begin{bmatrix} x_1 & x_2 & x_3 & & x_N \\ 0 & x_1 & x_2 & \cdots & x_{N-1} \\ 0 & 0 & x_1 & & x_{N-2} \\ & \vdots & & \ddots & \vdots \\ 0 & 0 & 0 & \cdots & x_{N-L+1} \end{bmatrix} \begin{bmatrix} {y_1}^2 \\ {y_2}^2 \\ {y_3}^2 \\ \vdots \\ {y_N}^2 \end{bmatrix} A$$

Denoting $X_0 = \begin{bmatrix} x_1 & x_2 & x_3 & & x_N \\ 0 & x_1 & x_2 & \cdots & x_{N-1} \\ 0 & 0 & x_1 & & x_{N-2} \\ & \vdots & & \ddots & \vdots \\ 0 & 0 & 0 & \cdots & x_{N-L+1} \end{bmatrix}$ and $Y = [{y_1}^2 \quad {y_2}^2 \quad \cdots \quad {y_N}^2]^T$

Eventually

$$\mathrm{F} = \left(X_0 {X_0}^T\right)^{-1} (X_0 Y) A$$

Where $X_0 {X_0}^T$ is the autocorrelation matrix and $X_0 Y$ is the cross-correlation matrix. The iteration procedure is concluded as followings:

1.Compute the autocorrelation matrix $X_0 {X_0}^T$ and its inverse.

2.Initiate the FIR filter weight vector F.

3.Compute the output sequence $y(m) = [{y_1}^{(m)} \quad {y_2}^{(m)} \quad \cdots \quad {y_N}^{(m)}]^T$. m is the iteration index.

4.Compute the cross-correlation matrix $X_0 Y$ and A.

5.Update the value of F.

6.Compute the error criterion to determine the iteration process.

3 Simulation ananlysis and experimental investigation

A simulation signal is built to extract the impulsive features, suppress the sinusoids and the white noise feature. The fault vibration signal model in (1) is adopted and the skewness maximization is verified respectively.

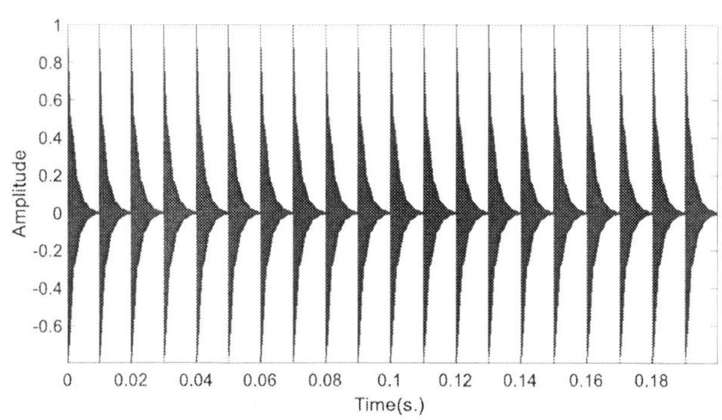

Fig.2. The original fault impulse

Assuming an impulsive train with a period $\mathrm{T} = 0.01$(i.e. frequency $\mathrm{f} = 100$) and the time domain waveform of the original impulsive signal is shown in Fig. 2.

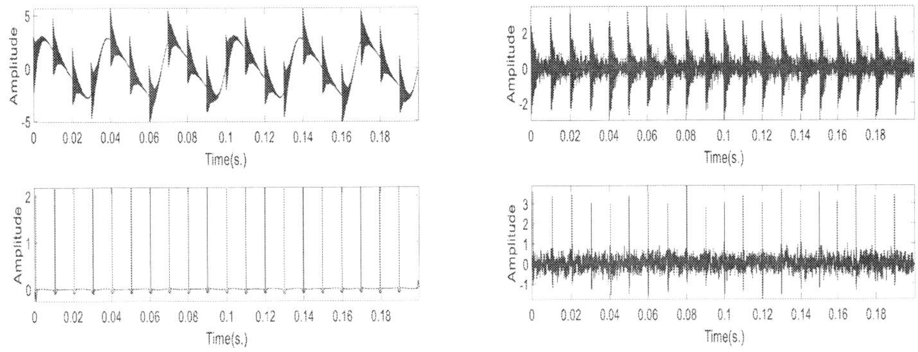

Fig.3. The smeared signal and the processed signal after SM

Fig. 3 shows the sinusoid disturbance and the noise interference to the impulse respectively. The left side skewness increased from 0.02 to 13.3 while the value of the right side rise from 0.13 to 2.56 via SM, which manifest the skewness maximization suppression ability to sinusoid and noise.

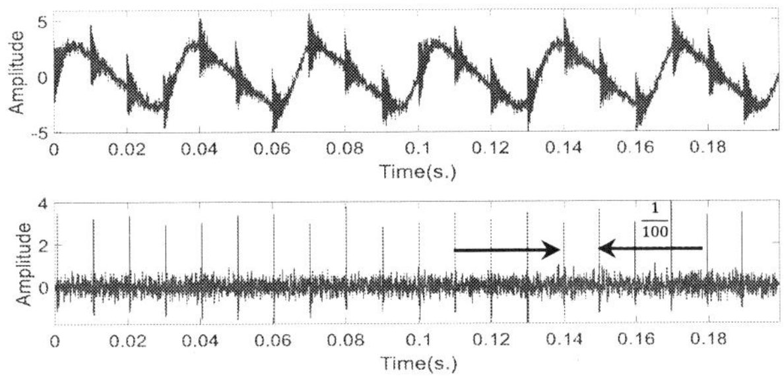

Fig.4. the mixture interference to impulse and the processed signal

Fig. 4 demonstrates the skewness maximization capacity dealing with the mixture of the two kinds of interferences. The mixture skewness is 0.03 while the value of the processed signal is 2.24.

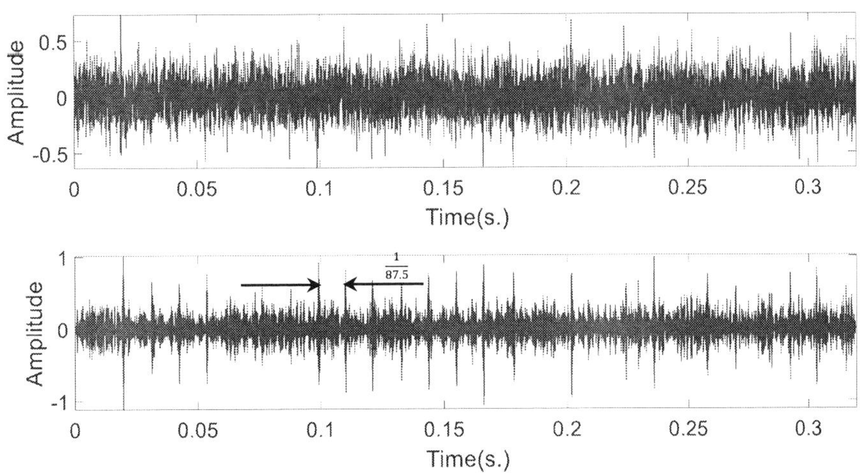

Fig.5. vibrations from outer race and its processd signal

The experimental data was acquired from a deep groove bearing (ER-12K), the shaft rotating frequency is estimated to be within 28.4 and 29.2 Hz. The outer race defect frequency lies within 86.6 and 89.1 Hz. The processed signal are visible at a frequency of 87.5 Hz The results seen in Figure 5 obviously manifest the availability of the method.

4 Conclusions

Skewness maximization based recovering periodic impulsive signals can be achieved by adapting a FIR filter to update the filter weight iteratively to restore the periodical impulse with a low computational cost and high performance. The interferences such as noise and sinusoid disturbance effect were studied and discussed respectively. From the simulation analysis and experimental results, it can be obviously draw out that maximizing the third order statistics is effectively attenuating the undesired components and preserve the periodic impulsive peaks.

Acknowledgments

The authors are grateful to the support from the National Science Foundation of China (Nos. 51575400, 51505339, 51405346), the Zhejiang Provincial Natural Science Foundation of China (Nos.LR13E050002, LQ16105005), the Zhejiang Technologies R&D Program of China (No. 2014C31103).

References

1. Meng, L. J., Xiang, J. W., Wang, Y. X., Jiang, Y. Y., Gao, H.F.: A Hybrid Fault Diagnosis Method using Morphological Filter-translation Invariant Wavelet and Improved Ensemble Empirical Mode Decomposition. Mech. Syst. Signal Pr., 50-51, 101--115(2015).
2. Shi, M, Wang, D, Zhang, J.: Nonlinear Dynamic Analysis of a Vertical Rotor-Bearing System. J. Mech. Sci. Technol., vol. 27, pp. 9--19 (2013).
3. Lim, G.M., Bae, D.M., Kim, J.H..: Fault Diagnosis of Rotating Machine by Thermography Method on Support Vector Machine. J. Mech. Sci. Technol., vol. 28, pp. 2947--2952 (2014).
4. McFadden, P.D., Smith , J.D.: Model for the Vibration Produced by a Single Point Defect in a Rolling Element Bearing. J. Sound. Vib. vol. 96, pp. 69--82 (1984)
5. Ovacıklı, A.K., Pääjärvi, P., LeBlanc, J. P., Carlson, J. E..: Uncovering Harmonic Content via Skewness Maximization - A Fourier Analysis. In: 22nd European Signal Processing Conference, pp. 481--485. Lisbon (2014).
6. Tandon, N., Choudhury, A.: A Review of Vibration and Acoustic Measurement Method for the Detection of Defects in Rolling Element Bearings. Tribol. Int., vol. 32, pp. 469--480 (1999)
7. Pääjärvi, P., LeBlanc, J. P..: Online adaptive blind deconvolution based on third-order moments. IEEE Signal Proc. Let, vol. 12, pp. 863--866(2005)
8. Hu, Q., He, Z.J., Zhang, Z., et al..: Fault Diagnosis of Rotating Machinery based on Improved Wavelet Package Transform and SVMs Ensemble. Mech. Syst. Signal Pr., vol. 21, pp. 688-705 (2007).
9. Huang, N.E., Shen, Z., Long, S.R..: The Empirical mode decomposition and the Hilbert spectrum for nonlinear and non-stationary time se-ries analysis . P. Roy. Soc. A-Math. Phy., vol. 454, pp. 903--995 (1998).
10. Smith, J.S..: The local mean decomposition and its application to EEG perception data. J. R. Soc. Interface., vol. 2, pp. 443--454 (2005)
11. Wiggins, R.A..: Minimum Entropy Deconvolution. Geoexploretaion, vol. 16, pp. 21--35, (1977).
12. Donoho, D. L..: On Minimum Entropy Deconvolution. Applied Time Series Analysis, Findley, D. F., Ed. New York (1981)
13. Endo, H., Randall, R.B..: Enhancement of Autoregressive Model based Gear Tooth Fault Detection Technique by the Use of Minimum Entropy Deconvolution Filter. Mech Syst Signal Pr, vol. 21, pp. 906--919 (2007).
14. Pääjärvi, P., LeBlanc, J. P..: Skewness Maximization for Impulsive Sources in Blind Deconvolution. In: 6th Nordic Signal Processing Symposium, pp. 304--307. Espoo,Finland (2004)..
15. Ovacıklı, A.K., Pääjärvi, P., LeBlanc, J. P., Carlson, J. E..: Recovering Periodic Impulsive Signals through Skewness Maximization. IEEE T. Signal Proces., vol. 64, pp. 1586--1596 (2016).

Rolling Element Bearing Fault Detection Based on Bayesian Residual Transform

Kai Lou [1], Jiawei Xiang [1], Yongteng Zhong [1]
[1]Wenzhou University,325035 Wenzhou,China
loukailk1992@163.com, jwxiang@wzu.edu.cn, zhongyongteng@wzu.edu.cn

Abstract: The raw vibration signal carrying a great deal of information in rotating machines. However, in the working condition, faulty rolling bearing response is often characterized by the presence of different kinds of impulses. Therefore, signal denoising problem is one of the most important tasks in the fault detection for rolling bearing.In the present, Bayesian residual transform (BRT) method is proposed in this paper. The BRT could decomposes a signal into a set of residual signals, these kinds of signals at different scales could be decoupled for the purpose of signal analysis and noise suppression. Moreover, to illustrate the feasibility of utilizing the BRT method for processing faulty signals of rolling bearings, the numerical simulation and experimental investigation are performed to verify the performance of the BRT method. Results in this paper indicate that it is effective to utilize the proposed method for the fault detection of rolling bearings.

Keywords:denoising problem, signal processing, bayesian residual transform, noise suppression

1 Introduction

Bearing faults are among the most common causes of failure in rotating machines. The root cause of such failures include spalling, pitting, wear, and so on. In order to obtain the fault feature, the vibration signal is one of the most important sources of information for monitoring conditions of rolling element bearing. However, such vibration signals are often contaminated by various noises, so it is usually difficult to obtain high-fidelity signals. Therefore, a lot of advanced approaches have been proposed, but such methods also have some disadvantages.

The wavelet denoising method has been most commonly used. In wavelet transforms [1-3],it has a number of advantages for the purpose of faulty signal processing, particularly pertaining to noise suppression. A signal $f(t)$ is decomposed into a set of wavelet coefficient, at each scale using the noise threshold θ,finally get the reconstructed signals. It is a good choice to deal with the non-stationary signals.However, one limitation worth noting that the usage of wavelet transforms to suppress noise will often lead to oscillation effects when dealing with the low signal-to-noise ratio (SNR) scenarios.

Empirical mode decomposition (EMD) has been recently developed in faulty diagnosis of rotating machinery [4], [5]. EMD deteriorates the signal into a number of intrinsic mode functions (IMFs) followed by the Hilbert spectrum analysis to reveal the frequency information hidden in the signal. It is a self-adaptive signal processing method that can be applied to non-linear and non-stationary process perfectly.However, one limitation is the distortion of the faulty impulses. In this condition, ensemble empirical mode decomposition (EEMD) is developed by Wu and Huang [6]. EEMD can apply to the spectral kurtosis optimized band pass filtered signal and obtain a number of IMFs. However, the reducing of the narrow band impulses and random noises embedded in original signal are still need to be solved.

For the above reasons, Bayesian method is proposed in this paper. In this perspective, a fault signal is made up of a number of signals including noise. According to multi-scale signal decomposition, each characterizing signal information at a different scale.So using this perspective is a great ideal to signal processing problem. First, as the wavelet transform, a signal decomposing a number of scales using Bayesian-based method which is called Bayesian Residual Transform (BRT). At each scale, information does not interfere with each other scale. Second, using the noise level estimation method proposed by Donoho [2] at each scale. And then through the inverse BRT of each scale signal, finally the new fault signal of noise suppression is obtained.

The paper is organized as follows.We briefly describe the fundamental theory of Bayesian Residual Transform (BRT) in Section 2. Then a simulation analysis is carried out to validate the effectiveness of the selection method and the experimental evaluation is further tested using BRT method in Section 3. The conclusion are drawn in Section 4.

2 Theoretical description of Bayesian Residual Transform and denoising method

2.1 Signal decomposition based on Bayesian Residual Transform

Briefly, the proposed Bayesian Residual Transform (BRT) can be summarized as follows. In the BRT, a signal $f(t)$ is represented by the summation of n residual signals, which is defined as follows:

$$f(t)=\sum_{i=1}^{n} r_i(t)=f_{\Sigma,1}(t) \tag{1}$$

where $r_i(t)$ is a residual signal characterizing the signal information at the ith scale. In order to decompose the signal $f(t)$ into a number of residual signals, a deep framework can be rewrited as follows:

$$f_{\Sigma,1}(t)=f_{\Sigma,2}(t)+r_1(t) \tag{2}$$

where $f_{\Sigma,2}(t)$ denote the summation of all residual signals at scale [2,n].In this step,estimating $\hat{f}_{\Sigma,2}(t)$ by given $f_{\Sigma,1}(t)$ is the key to solve the problem. Therefore, a kernel regression strategy used in conditional expectation. One can express as:

$$\hat{f}_{\Sigma,2}(t)=E\left(f_{\Sigma,2}(t)\middle|\hat{f}_{\Sigma,1}(t)\right) \tag{3}$$

where $E\left(f_{\Sigma,2}(t)\middle|\hat{f}_{\Sigma,1}(t)\right)$ [7] based on nonparametric Nadaraya-Waston Kernel regression [8],[9], using a kernel function K_j. Consider Eq.(3), $\hat{f}_{\Sigma,2}(t)$ can be obtained by:

$$\hat{f}_{\Sigma,2}(t)=K_j\left(\hat{f}_{\Sigma,2}(t)-\hat{f}_{\Sigma,2}(t_i)\right)=e^{-\frac{1}{\lambda_2^2}\left(\hat{f}_{\Sigma,2}(t)-\hat{f}_{\Sigma,2}(t_i)\right)^2} \tag{4}$$

A deep cascading framework when using Eq.(4) can be expressed as follows:

$$r_j(t)=\hat{f}_{\Sigma,j}(t)-\hat{f}_{\Sigma,j+1}(t) \tag{5}$$

where

$$\hat{f}_{\Sigma,j+1}(t) = K_j\left(\hat{f}_{\Sigma,j+1}(t) - \hat{f}_{\Sigma,j+1}(t_i)\right) = E\left(f_{\Sigma,j+1}(t) \middle| \hat{f}_{\Sigma,j}(t)\right) \tag{6}$$

Finally, the signal decomposed realization of BRT via kernel regression at each scale is completed,and the inverse BRT is simply the summation of all residual signals $r_1(t), r_2(t),..., r_n(t)$.Besides, the residual signals at the lower scales contain great characteristics of the signal,while the signal contain coarser signal characteristics.

2.2 Noise suppression

In this paper, the work is to illustrate the feasibility of utilizing the BRT for processing fault signal. However, at each scale, there are some random noise and useless information. Motivated by Donoho [2], a noise thresholding strategy proposed to have greater content at each scale.

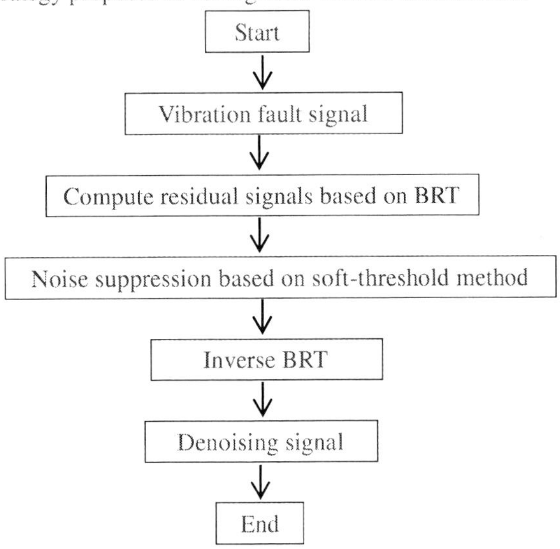

Fig.1 Flow chart of the proposed BRT method

The noise thresholding strategy can be achieved as follows.Since residual signals at different scales are obtained. The problem is to estimate the noise threshold θ_j at scale j, using adaptive noise estimate, which is defined as:

$$\theta_j = {MAD(r_j)}/{0.6745} \tag{7}$$

where MAD is the median absolute deviation. Based on soft-thresholding, $r'_j(t)$ can be obtained by:

$$r'_j(t) = \begin{cases} 0 & if\,|r_j(t)| < \theta_j \\ r_j(t) & otherelse \end{cases} \tag{8}$$

Using Eq.(8) and Eq.(5), the final residual signal is obtained. Finally, using the inverse BRT to obtain the new denoising fault signal. According to the above method mentioned, the complete process of the proposed method is shown in Fig.1

3 Simulation analysis and Experimental evaluation

3.1 Simulation analysis

In this section, to validate the effectiveness of the proposed method, a simulated signal which contains random noise is constructed. The model of the signal is expressed as follow:

$$f(t) = x(t) + R(t) \tag{9}$$

where $x(t)$ is the periodic impulse signal without noise, $R(t)$ is the random impulse noise. Suppose the sampling frequency is 20K and the sampling points are 8192. The impulse signal and the noisy input signal are shown in Fig.2(a) and Fig.2(b).It is clearly to see that the signal is immersed in heavy noises.

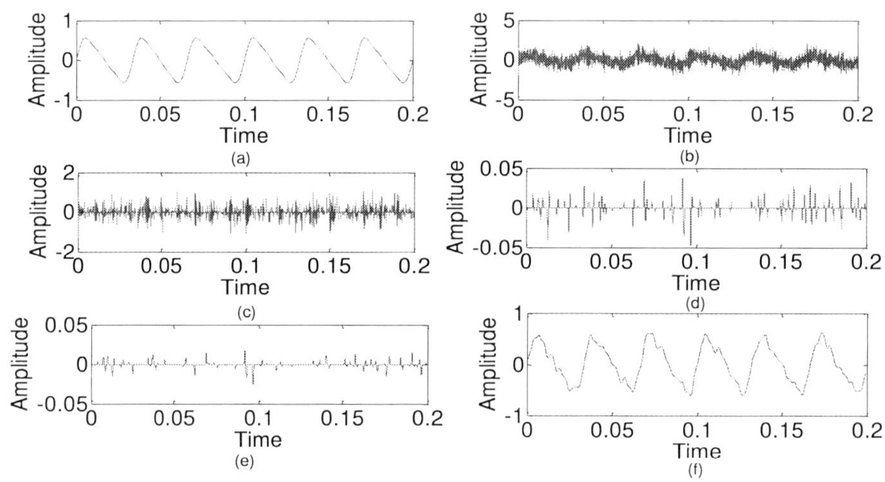

Fig.2 The simulated signal (a) Test signal (b) Noisy input signal (c)-(h) Signal decomposition using the BRT at different scales

The denoising reasults of using the above methods are shown in Fig2.(c)-(f). As shown in Fig2.(c)-(f), the BRT method can effectively eliminate the narrow band impulses almost completely. It can be observed that the noise process is well characterized at the low scales and the structural characteristics of the test signal is well processed.Also it can be observed that the characterize of residual signals become coarser and coarser with the scale j increases. The proposed method can be regarded as a promising method for denoising.

3.2 Experimental evaluation

In this section, the proposed approach is evaluated by defective a rolling element bearing with inner race.A deep groove bearing (ER-12K) is used in our experiment and the bearing parameters are: the number of rolling element $N_b = 8$, ball diameter $B_d = 0.3125$ inch, pitch diameter $P_d = 1.318$ inch, the contact angle $\alpha = 0°$. In the experiment processing, the sampling frequency f_s is 25.6 kHz.For the bearing with inner race fault, the total collected data is 12288 points, the shaft rotating frequency is $f_{shaft} = 29.87Hz$ and the ball pass frequency of the outer race (BPFO) is:

$$BPFO = \frac{N_b}{2} f_{shaft}\left(1 - \frac{B_d}{P_d}\right)\cos\alpha = 3.0516 f_{shaft} = 91.15Hz \tag{10}$$

From Fig.3(a), the raw signal features are always immersed in heavy noise. And the results of BRT method processing is shown in Fig.3(b) , the corresponding envelope spectrum of (a) is shown in Fig.3(c) and corresponding envelope spectrum of (b) is shown in Fig.3(d).

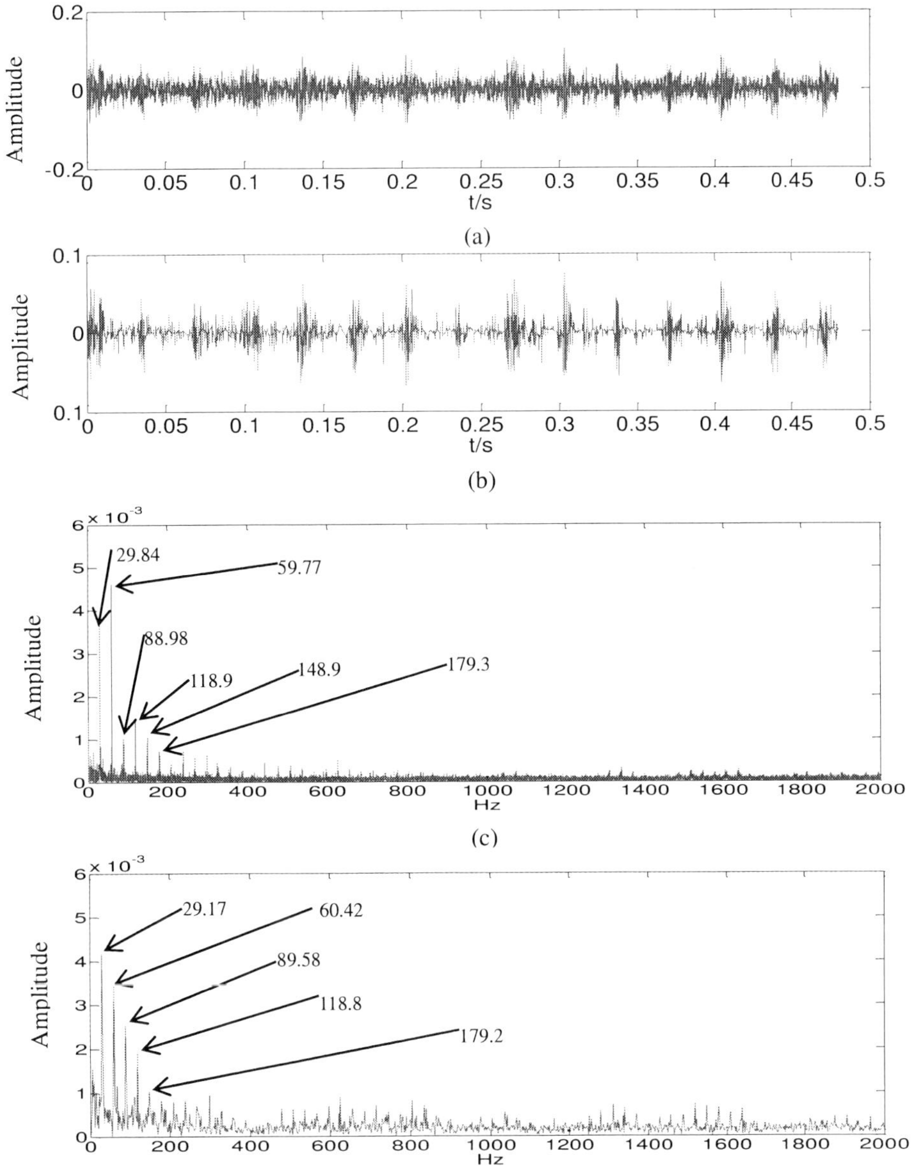

Fig.3 The raw signal, denoise signal and the corresponding envelope spectrum for the outer race: (a)The raw signal; (b) The denoise signal; (c) The envelope spectrum of (a), (d) The envelope spectrum of (b)

In Fig.3(c), the shafting rotating frequency (29.84 Hz) and its second harmonic (59.77 Hz). The feature frequency (88.98 Hz) and its harmonics are revealed.However,it is quite volatility in Fig.3(c).Therefore, in the Fig.3(d),the shafting rotating frequency (29.17 Hz) and its second harmonic (60.42 Hz). And the feature frequency (89.58Hz) and its harmonic,such as 179.2 Hz are absolutely obvious.The comparison shows the effectiveness of the BRT method. Therefore, the results in this paper is helpful to use the BRT method for processing the inner race fault signals for noise suppression.

Conclusion

In this paper, aiming at the shortcoming of the conventional denoising method for bearing fault signal under variation conditions, Bayesian Residual Transform based approach is proposed. The BRT method decomposes a signal into a number of residual signals, at different scales signals containing characteristic of the information and the useless information. It allows noise suppression method to get rid of the noise information while preserving the important signal characteristics. The simulation and experimental validation demonstrated the feasible of using the BRT method. It can be pointed out that the BRT method in this study will lead to good results of the noise suppression.

Acknowledgments

The authors are grateful to the support from the National Science Foundation of China (Nos. 51575400, 51505339, 51405346), the Zhejiang Provincial Natural Science Foundation of China (Nos.LR13E050002, LQ16105005), the Zhejiang Technologies R&D Program of China (No. 2014C31103).

References

1. Donoho,D. L., Johnstone, I. M.: Ideal Spatial Adaptation Via Wavelet Shrinkag. Biometrika.41,425--455(1994)
2. Donoho, D. L.: De-noising by Soft-thresholding. IEEE Trans. Inf. Theory. vol.57, 613--627(1995).
3. Harsha,S.P., Kankar,P.K., Sharma,S.C.: Rolling Element Bearing Fault Fiagnosis using Wavelet Transform, Neurocomputing.74,1638--1645(2011)
4. Huang,N.E., Shen,Z.,. Long, S.R., et al.: The Empirical Mode Decomposition and The Hilbert Spectrum for Nonlinear and Non-stationary Time Series Analysis, Proc. R. Soc. Lond. Ser. A. 454,903--995(1998)
5. Lei, Y., Lin, J., He, Z., Zuo MJ.: A Review on Empirical Mode Decomposition in Fault Diagnosis of Rotating Machinery. Mech Syst Signal Process. 35,108--126(2013)
6. Wu,Z. Huang,N.E.: Ensemble Empirical Mode Decomposition: A Noise-assisted Data Analysis Method, Adv. Adapt. Data Anal. 01,1(2009).
7. Fieguth,P.: Statistical Image Processing and Multidimensional Modeling. New York(2010).
8. Nadaraya,E. A.: On Estimating Regression, Theory Probab. 9, 141--142(1964)
9. Watson,G. S.: Smooth Regression Analysi, Indian J. Statist., 26, 359--372(1964)
10. Daubechies,I.: Ten Lectures on Wavelets. Philadelphia, PA, USA: SIAM(1992)
11. Wong,A., Mishra,A. Bizheva, K. and. Claus, D. A.i: General Bayesian Estimation for Speckle Noise Reduction in Optical Coherence Tomography Retinal Imagery, Optics Express, 8, 8338--8352(2010)
12. Wong,A., Xiao Yu ,Wang.: A Bayesian Residual Transform for Signal Processing,Access, IEEE, 3, 709--717(2015)
13. Donoho,D L., Johnstone,I.: Adapting to Unknown Smoothness via Wavelet Shrinkage. Journal of the American Statistical Association, 90, 1200--1224(1995)
14. Meng, L. J., Xiang, J. W., Wang, Y. X., Jiang, Y. Y., Gao, H.F.: A Hybrid Fault Diagnosis Method using Morphological Filter-translation Invariant Wavelet and Improved Ensemble Empirical Mode Decomposition. Mech. Syst. Signal Pr., 50-51, 101--115(2015).

Multivariable support vector machine bearing life prediction method based on principal component analysis

Baojian Wang [1], Xiaoli Zhang[2], Zhibin Zhao[1], Xuefeng Chen[1]

[1] State Key Laboratory for Manufacturing and Systems Engineering, Xi'an Jiaotong University,Xi'an 710049, China; E-Mails: wangbaojian@mail.xjtu.edu.cn ;

[2] Key Laboratory of Road Construction Technology and Equipment, Ministry of Education,Chang'an University, Xi'an 710064, China; E-Mails: lilyzhang@chd.edu.cn

Abstract: As one of the important components of aero-engine, the main bearing is one of the main causes of engine's fatal faults, so it is especially important to predict the residual life of the main bearing. Support vector machine is an efficient machine learning method which can solve practical engineering problem in small sample cases well and providee conditions for life prediction. This paper proposes a method of building samples based on principal component analysis combined multivariable prediction with support vector machine regression, which can realize the multi-feature information input and fully digs the information content. Finally the prediction effect is better with the proposed method. Applying principal component analysis to train multivariable sample structure can effectively reduce the dimension of training samples and data redundancy. Employing contribution value to select main component and building multivariate sample matrix, the cut-and-trial method obtains the approximate optimal value range and the particle swarm optimization algorithm searches global optimal value. The proposed method avoids the problem that directly global search is easy to fall into local optimum and inherits global search efficiency. Based on the above, the bearing's fatigue life experiment is carried out. The effectiveness and advantage of the method is verified.

Keywords: Aero-engine, Life prediction, Principal component analysis, Support vector machine, Particle swarm optimization

1 Introduction

In modern times, data becomes an important driving force so the prediction method by using data to train and mine potential change rule is an important content worth of studying. Statistical learning theory is proposed by doctor V. Vapnik et al. in AT&T bell laboratory in 1960s, which aims at solving the problem of machine learning under small sample[1]. Under this new theory, a new general method-Support Vector Machine (SVM) was developed based on structural risk minimization principle, which is different from the traditional learning method based on empirical risk minimization principle. Thus SVM gains many unique advantages in solving the problem of small sample, nonlinear and high dimensional[2].

2 Multivariate SVM prediction method

2.1 SVM's multivariate construction based on principle component analysis

Thesis[3],[4] introduces some indexes to evaluate the wearing degree of bearing. Thesis [5] analyzes some statistical indicators of time domain and frequency domain, finding that none of them has the sensitivity and consistency of the damage of bearing. So single index can not reflect the degradation process of the bearing and multi-index is necessary. But when the multivariate training sample of time domain is built, the variables have correlation with each other. As information is redundant when we use them directly, we present a multivariate construction method based on principle component analysis (PCA).

The purposed PCA is to find a set of optimal orthogonal unit vector machine (principle component) through linear transformation, then reconstruct the sample by them and insure the sample has the minimum mean square error with the original sample after reconstruction.

Assume $\{x_i | i=1,2,\cdots,N\}$ is the given sample data set, $X=[x_1,x_2,\cdots,x_N]$ is the matrix form, then the mean vector of the sample is:

$$\mu=\frac{1}{N}\sum_{i=1}^{N}x_i \quad (1)$$

The covariance matrix S_t of the sample X is:

$$S_t=\frac{1}{N}(X-\overline{X})(X-\overline{X})^T=\frac{1}{N}\sum_{i=1}^{N}(x_i-\mu)(x_i-\mu)^T \quad (2)$$

Assume the rank of S_t is m, $\lambda_1,\lambda_2,\cdots,\lambda_m$ ($\lambda_1 \geq \lambda_2 \geq \cdots \geq \lambda_m$) is the eigenvalue of S_t, $w_i, i=1,2,\cdots,m$ is the eigenvector of S_t.

$W=[w_1,w_2,\cdots,w_m]$ is the component matrix of sample X, w_i is the principle component.

In this article, we use some time domain index to construct matrix X, then use the principle component analysis to process X and get the component matrix of it.

Then we use the first l rows of the component matrix to construct the training sample matrix, demand that the first l rows should express the 95% information of the original data.

2.2 SVM parameter selection based on Particle Swarm Optimization

Vapnik et al. has said that the selection of SVM parameters has a significant influence on its learning ability and generalization ability[6].The three main parameters are kernel function parameter, penalty factor and insensitive loss factor. If these three parameters are not selected properly, the accuracy and generalization ability of SVM will be influenced.

Now there are four main methods that are used to optimize the parameters of SVM: test errors, cross validation, the generalization error estimation and gradient descent, artificial intelligence and evolutionary algorithm[7]. For the high speed and efficiency of the fourth method, we introduce the particle swarm optimization. It is first introduced by Kennedy and Eberhart in 1995 inspired by the bird flock looking for food[8]. So ciobiologist E.owllson believe the member in biotic community can learn from other members when they are looking for food, and this advantage can overcome the disadvantage of competition of finding food[9].

Assume there are n particles in swarm, each of q dimension. The optimization problem can be seen as n particles search a space of q dimension, each particle is referred as $x_i=(x_{i,1},x_{i,2},\cdots,x_{i,q})$, the velocity of each particle is $v_i=(v_{i,1},v_{i,2},\cdots,v_{i,q})$, there are two factors that each particle should concern:

1. The optimal value found by itself in history: $p_i=(p_{i,1},p_{i,2},\cdots,p_{i,q})$, $i=1,2,3,\cdots,n$

2. The optimal value found by the whole swarm: $p_g=(p_{g,1},p_{g,2},\cdots,p_{g,q})$

Each particle update its position and velocity according to the formula below:

$$v_{id}(t) = \omega v_{id}(t-1) + c_1\xi(p_{id}(t-1) - x_{id}(t-1)) + c_2\eta(p_{id} - x_{id}(t-1))$$
$$x_{id}(t) = x_{id}(t-1) + rv_{id}(t) \quad (3)$$

ω is used to keep the original velocity, called inertia weight. c_1 is the weight coefficient of particles found optimal value of their own, usually taken as 2. c_2 is the weight coefficient of particles found optimal value of the whole swarm, usually taken as 2. ξ, η are random numbers evenly distributed in [0,1]. r is the constriction factor, usually taken as 1.

2.3 Verification by simulation function

To test the efficiency of the method above, we adopt a function of multivariate and variables have relationship with each other to do a prediction experiment using the SVM based on PCA, and compare with the method of using SVM directly.

The function is:

$$y = x_1^2 + x_2^{1.5} + x_3 + x_4 + x_5 + x_6 + x_7 + N(0, 1^2)$$
$$x_1 = U[0,2] + N[0, 1^2]$$
$$x_2 = U[0,1.5] + N[0, 1.5^2]$$
$$x_3 = x_1x_2, x_4 = x_2x_3, x_5 = x_2x_4, x_6 = x_3x_5, x_7 = x_2x_5 \quad (4)$$

They are uniform distribution with white noise to simulate the randomness of measurement. Combinations of x_1, x_2 is used to simulate the relationship between variables. We pick 200 data points, in which first 120 points are used to train the model and the rest 80 points are used to predict.

In method 1 we use seven variables directly to construct the training sample to train the model, in method 2 we use the PCA to construct the sample, pick the first three rows of the component matrix whose rate of contribution is 99.72%.

Using the PSO to optimize the parameters in method 1 and method 2, then get ker='*rbf*', C=4000, *e*=0.0105, *p*=9.7andker='*rbf*', C=2000, *e*=0.01, *p*=10 for method 1 MSVM and method 2 PCA-MSVM. The result is as Fig.1:

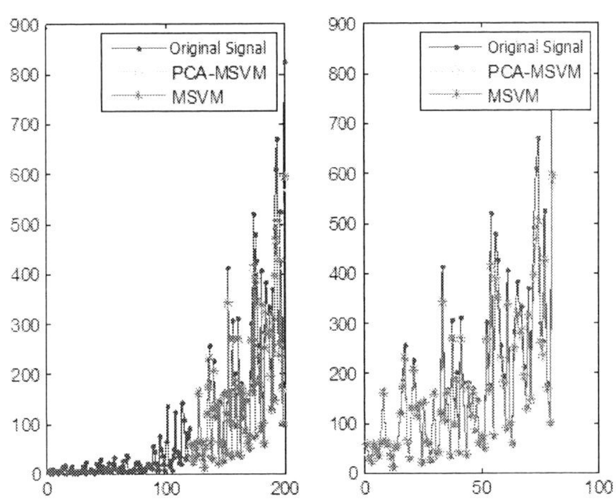

(a)global (b)partial enlarged detail

Fig. 1. simulation result

The average relative error of MSVM is 0.037, and of PCA-MSVM is 0.04, so the PCA-MSVM can ensure the accuracy along with simplifying the sample, which means faster speed of calculation.

3 Experiment Verification

3.1 Whole life fatigue experiment for bearing

The experiment equipment includes: T20-60nF bearing fatigue life testing machine, bearing 30311, LC0401 acceleration sensor, Yihen collector and a computer.

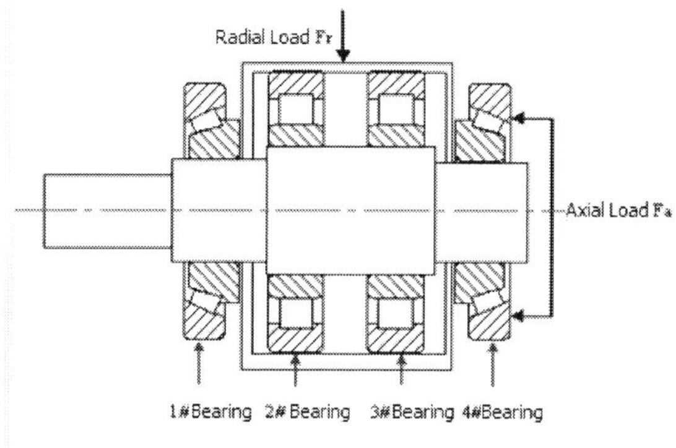

Fig. 2. Test load diagram

The load in experiment is as Fig.2, bearing 1# and 4# are the test bearings, bearing 2# and 3# are the load bearings. During the experiment, we record the acceleration and vibration signal of the four bearings, the load signals, the motor current and rotate speed signals and the vibration signals of the experiment. The size of test and load bearings is shown in Table.1

The experiment system is lubricated by oil circulation, with temperature of bearing cup no more than 90°. We conduct two experiments in total: experiment 3 and experiment 4, we use the data of experiment 4 as training sample, and experiment 3 as test sample. The experiment result is shown in Table.2.

Table 1. Size of bearings

Category	Type	Bore Diameter(mm)	Outside Diameter(mm)	Roll Diameter(mm)	e	Y	Rated dynamic load	Limit Speed(rpm)
Test Bearing	30311	55	120	16.25	0.35	1.7	152	4300
Load Bearing	N312	60	130	19.1	—	—	212	5600

Table 2. Fig. 3. Table.2 Result

Experiment	Type	Radial Load Fr(KN)	Axial Load Fa(KN)	Speed/rpm	Run time/hour	Remove Times	Remark

3	30311	18	15	1500	214	1	4#Bearing Broken
4	30311	0	30	1500	202	2	4#Bearing Broken

3.2 Sample constructed by PCA

In the prediction process of bearings, time domain index is widely used for its simplicity, reliability and direct. Table.3 is some common time domain indexes.

Table 3. Time domain index

$x(i)$ is the signal sequence, $i=1,2,\ldots,N$ is the number of sequence			
$tp_1=\frac{1}{N}\sum_{i=1}^{N}x(i)$	$tp_2=\sqrt{\frac{1}{N}\sum_{i=1}^{N}(x(i))^2}$	$tp_3=\left(\frac{1}{N}\sum_{i=1}^{N}\sqrt{\lvert x(i)\rvert}\right)^2$	$tp_4=\frac{1}{N}\sum_{i=1}^{N}\lvert x(i)\rvert$
$tp_5=\frac{1}{N}\sum_{i=1}^{N}(x(i)-tp_1)^3$	$tp_6=\frac{1}{N}\sum_{i=1}^{N}(x(i)-tp_1)^4$	$tp_7=\frac{1}{N-1}\sum_{i=1}^{N}(x(i)-tp_1)^2$	$tp_8=\max(x(i))$
$tp_9=\min(x(i))$	$tp_{10}=tp_7-tp_8$	$tp_{11}=tp_2/tp_4$	$tp_{12}=tp_8/tp_2$
$tp_{13}=tp_8/tp_4$	$tp_{14}=tp_8/tp_3$	$tp_{15}=tp_5/(tp_7)^{3/2}$	$tp_{16}=tp_6/(tp_7)^2$

The first ten of them are dimensional indexes, the last six are non dimensional indexes. We use the PCA method mentioned above, use the ten dimensional indexes as the input matrix of PCA, here we pick the first l rows that express the 95% information of the original data. The first three rows of component matrix of experiment 3 express the 98.64% information, the experiment 4 is 98.53%.

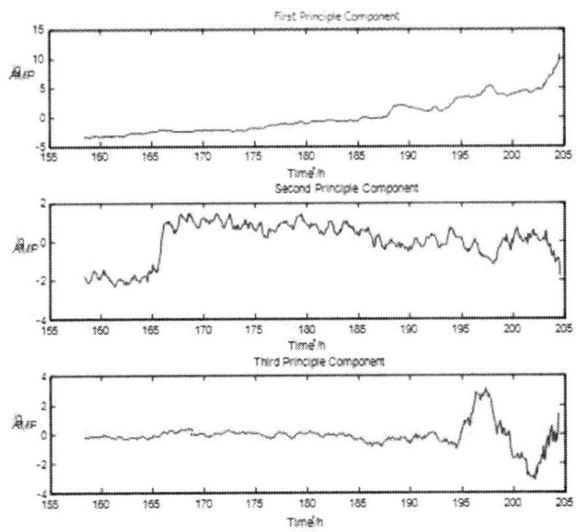

Fig. 3. First three rows of component of exp.3

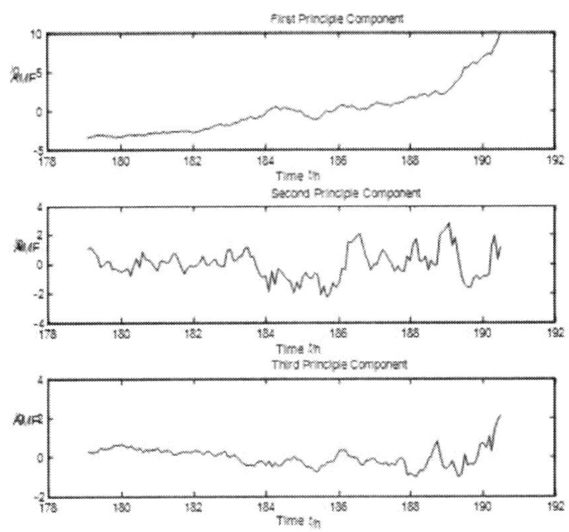

Fig. 4. Fig.4 First three rows of component of exp.4

3.3 SVM bearing residual life prediction based on PSO

First we use the method mentioned above to build training sample by exp.4, test sample by exp.3.

Then we get the parameters of SVM by PSO. The SVM parameters of ordinary sample construction method are: $\ker = 'rbf', C = 47.8, e = 0.013, p = 4.42$, the SVM parameters of sample built by PCA are: $\ker = 'rbf', C = 75, e = 0.002, p = 4$.

Then we train the SVM model, and use it to predict the residual life of bearing in exp.3. The result is shown in Fig.4. Calculating the four kinds of average error of the two prediction methods to estimate the efficiency, result is shown in Table.4.

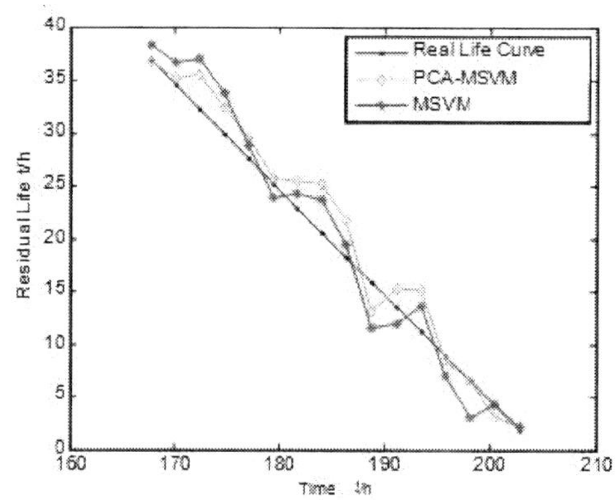

Fig. 5. Prediction result of residual life

Table 4. Prediction error

Prediction Model	Mean Average Error(MAE)	Root Mean Square Error (RMSE)	Normalized Mean Squared Error (NMSE)	Mean Absolute Percentage Error (MAPE)
MSVM	2.1689	2.5613	0.6590	0.1404
PCA-MSVM	1.8605	2.3615	0.5601	0.1186

It can be seen from the diagram that the PCA-MSVM method has a better performance in the prediction when the residual life is short, it is because the PCA method has extracted more information from the original data. As for the prediction error, PCA-MSVM has a better control of average relative error and reduces the redundancy of data, makes the sample simpler and gets a faster speed.

4 Conclusions

The analysis results show that the multivariable support vector machine based on principal component analysis method can greatly facilitate the construction of the sample matrix, reduce the data redundancy on the premise of ensuring the prediction accuracy. Meanwhile, the sample measured and structured by principle component analysis (PCA) can effectively replace the original multi dimension data, the particle swarm optimization method can effectively optimize the performance of SVM. All of above verifies the effectiveness and advantages of the life prediction method of using the multivariable support vector machine based on principal component analysis which is proposed in this paper.

5 Acknowledgments

This research is supported by the Aviation Science Foundation of China (no 20151070005) and the National Natural Science Foundation of China (no 51405028).

Reference:

[1] He Z et al. Mechanical fault diagnosis theory and application [M]. Beijing: Higher Education Press, 2010.6.
[2] Cortes C, Vapnik V. Support-vector networks [J]. Machine Learning, 1995, 20 (3): 273-297.
[3] Qiu H, Lee J, Lin J, et al. Wavelet Filter-Based Weak Signature Detection Method And Its Application On Rolling Element Bearing Prognostics [J]. Journal of Sounds&svibration, 2006, 289 (4-5): 1066–1090.
[4] Shen Z. Remaining Life Predictions of Rolling Bearing Based on Relative Features and Multivariable Support Vector Machine [J]. Journal of Mechanical Engineering, 2013, 49 (2): 183-189.
[5] Lybeck N, Marble S, Morton B. Validating Prognostic Algorithms: A Case Study Using Comprehensive Bearing Fault Data [J]. IEEE Aerospace Conference Proceedings, 2007: 1 - 9.
[6] Chapelle O, Vapnik V, Bousquet O, et al. Choosing multiple parameters for support vector machines [J]. Machine Learning, 2002, 46 (1-3): 2002.
[7] Zhang X. Research on Optimization Method and Application on Mechanical Fault Diagnosis of SVM [D]: Xi'an Jiaotong University, 2011.
[8] Liu Y. Research of Particle Swarm Optimization and Its application [D]:Shangdong Normal University, 2011.
[9] Liu J. The Research of Basic Therory and Improvement on Partical Swarm Optimization [D]: Central South University, 2009.

COMADEM 2016

The 29th International Congress on Condition Monitoring and Diagnostic Engineering Management

20-22 August 2016, Xi'an, China

THE APPLICATION OF VARIATIONAL MODE DECOMPOSITION FOR DIAGNOSING RUB-IMPACT FAULT OF DUAL-ROTOR SYSTEM BASED ON ACCELERATION MEASUREMENTS

Nanfei Wang, Dongxiang Jiang, Te Han, Yizhou Yang and Hongzhi Xu

State Key Laboratory of Control and Simulation of Power System and Generation Equipment, Department of Thermal Engineering, Tsinghua University, Beijing 100084
Email: wnf14@mails.tsinghua.edu.cn
Telephone: +8613601078472

Abstract: In order to improve the energy efficiency and compact structure, the dual-rotor structure including the low pressure rotor and high pressure rotor has been widely used in aero-engine. In the complex operation conditions, the high rotating speed dual-rotor structure is vulnerable to some faults, such as the rotor-stator rubbing, which may cause the strong vibration and even some catastrophic accidents. However, as for aero-engine, it is relatively difficult to obtain the displacement signal, only to measure the acceleration signals. In the paper, aimed at the dual-rotor system test rig, the rub-impact is simulated and the acceleration signals collected from the bearing pedestal are analysed to extract the rubbing faults' characteristics. Multi-component extraction is an available method for vibration signal analysis of rotary machinery, so a novel method of rubbing fault diagnosis based on variational mode decomposition (VMD) is proposed. VMD is a newly developed technique for adaptive signal decomposition, which can non-recursive decompose a multi-component signal into a number of quasi-orthogonal intrinsic mode functions. When the rub-impact fault occurs in the dual-rotor system, the acceleration vibration signals of the dual-rotor system will present amplitude-modulated feature because of the periodic rub-impact between the rotor and the stator. VMD is then applied to detect multiple rubbing-caused signatures for rotor–stator fault diagnosis via practical vibration signal. By using VMD, the rub-impact signal, background signal and noise signal can be separated efficiently and the fault feature of vibration signal of a rotor system with local rub-impact fault can be extracted. Therefore, the VMD method can be applied to the rub-impact fault diagnosis of the rotor systems efficiently.

Keywords: Dual-rotor, variational mode decomposition, rubbing, demodulation, impact

1. Introduction

The clearance between rotor and stator is one of the significant parameters influencing the performance of rotating machinery, such as gas turbine and aero-engine. In order to improve the energy efficiency, it is common practise to minimize the operational rotor-stator clearance. Nevertheless, when the clearance reduces, the probability of the rotor-stator rub-impact increases, which has long been identified as an important contributor to excessive maintenance and in general to engine failure. The rubbing may severely induce complicated vibration of the overall unit and the rotor dynamic instability, and it will reduce the system performance and even lead to the catastrophic accidents.

Also, to improve the structure compact and obtain greater energy efficiency, the dual-rotor system has been widely adopted in aerospace field. Due to complicated dynamics and coupling between dual rotors, the system features complicate nonlinear dynamics, which is crucial to stability of the aero-engine and plane. The dual rotors have co-rotating speeds and are connected by the intershaft bearing—a deep groove ball bearing. The intershaft bearing inner race is mounted on the low-pressure rotor while the outer race is mounted on the high-pressure rotor. The higher-pressure rotor has higher rotational speed. Moreover, as for dual-rotor aero-engine, it is very challenging to obtain vibration displacement signals, and only vibration acceleration signals can be collected.

Complicated nonlinear characteristics is usually associated with a vibration system of a faulty rotor, for instance, chaotic motions often accompanied by the periodic motions. High synchronous vibrations of the rotor may exceed available clearances and lead to rub-impact between rotor and stator. Hence, some signatures, including synchronous vibration, sub-synchronous and super-synchronous vibrations, will be carried in the measured vibration acceleration signals picked up from bearing pedestals under rubbing condition. Many researchers by now have carried out some outlooks on the mechanism and dynamic characteristics of rub in rotors using analytic model or numerical simulations. Muszynska [1] wrote a detailed literature survey on rub-related phenomenon in rotating machinery, such as friction, impacting, stiffening and coupling effects. Isaksson and Frid [2] studied the dynamical behaviour of a rotor rubbing with a non-annular stator and analysed the influences of different stop configurations on the system vibration responses. Zapornel et al. [3] simulated the impacts between the rotor and its housing based on the application of Newton's theory and on the direct calculation of the contact force and investigated the behaviour of rotor systems when the impact occurs. Beatty [4] analysed the rotor response due to the radial rubbing, which is described as a function of the rotor contact circumference based on a combination of analytical and experimental results. Based on finite element (FE) theory with unilateral contact and friction conditions, Behzad et al. [5] developed an algorithm to investigate of the rotor-to-stator rubbing vibration and carried out a case study of the point rubbing between the rotor and an elastic rod by using this algorithm. Popprath and Ecker [6] presented a Jeffcott rotor model with intermittent contact with a stator interacting with the rotor model via non-linear contact forces.

Moreover, the contact and separation process often generated impacts between the rotor and the stator. This is another common defective signature, besides the sub- and super-harmonics caused by the rubbings. Several methods have been proposed to detect all these signatures from the rotor response of signals. Hu et al. [7] proposed a stochastic resonance method for early detecting rub-impact fault of rotor systems. A local rub-impact fault diagnosis method of rotor system based on empirical mode decomposition (EMD) was proposed in [8], where it showed intrinsic mode functions of the vibration signal of a rotor system with local rub-impact fault obviously own the amplitude-modulated characteristics. Peng et al. applied the reassigned scalogram to detect the rubbing-caused impacts in [9]. Lei et al. [10] proposed a method to diagnose rotating machinery with the ensemble empirical mode decomposition (EEMD) to overcome the problem of mode mixing existing in EMD. In view of the dynamics of rotors in retainer bearings is very similar to other problems of rotor-to-stator contact, Markert investigated transient vibrations of elastic rotors in retainer bearings in [11], and he also mentioned the transients from no contact to continuous contact were usually accompanied by more or less intense impacts. A demodulation method based on the local mean decomposition (LMD) was proposed for rub-impact fault diagnosis [12], which can extract the transient fluctuations of the instantaneous frequency of the fundamental harmonic component. As for simulated response displacement signal and practical vibration displacement signal of Jeffcott rotor system, variational mode decomposition (VMD) is applied to detect multiple rubbing-caused signatures for rotor-stator fault diagnosis in [13].

In the paper, VMD method is utilized to decompose the vibration acceleration signals of dual-rotor with rub-impact fault occurred in the outer rotor's disk. After the rub-impact vibration acceleration signals are decomposed by VMD approach, the rub-impact signal, background signal and noise signal can be separated and the rub-impact fault features of the dual-rotor system can be extracted effectively. The analysis results from the experimental signals demonstrate that the rub-impact fault characteristics of the dual-rotor system vibration acceleration signals can be extracted by VMD method efficiently.

2. Variational mode decomposition and its equivalent filter property

2.1 VMD algorithm

The variational mode decomposition (VMD) technique, which is a newly developed methodology for adaptive and quasi-orthogonal signal decomposition [14], is adopted to extract the rubbing fault features of dual-rotor system from the vibration acceleration signal picked up from the bearing pedestal. VMD is a multi-resolution analysis technique employed to non-recursively decompose a non-stationary signal into a number of band-limited intrinsic mode functions (BLIMFs). VMD can decompose a real valued signal f into a discrete number of sub-signals u_k with the specific sparsity properties of its bandwidth in the spectral domain. Each mode is compact around a centre pulsation ω_k and its bandwidth is estimated using H^1 Gaussian smoothness of the shifed signal. The modes and the centre frequencies are extracted by solving the variational optimization problem. The optimization problem is formulated in key three steps. (i) The Hilbert transform is used to make each mode's frequency spectrum one sided or unilateral. (ii) Then, each mode frequency spectrum is shifted to base band by a multiplying the factor $e^{-j\omega_t n}$, where ω_t is the centre frequency of tth mode. (iii) The bandwidth of each mode is computed using the L^2-norm of gradient of the demodulated signal. The original constrained problem to evaluate number of modes and the centre frequencies is given by [14]:

$$\min_{\{u_k\},\{\omega_k\}} \{\sum_k \left\| \partial_t [(\delta(t) + \frac{j}{\pi t}) * u_k(t)] e^{-j\omega_k t} \right\|_2^2 \}$$

$$\text{subject to } \sum_{k=1}^{K} u_k = f \tag{1}$$

where $u_k(t)$ is the signal corresponds to tth mode. f represents the original signal K is the total number of modes. Eq. (1) can be addressed by introducing a quadratic penalty and Lagrangian multipliers. The augmented Lagrangian is expressed as follows:

$$L(\{u_k\},\{\omega_k\},\lambda) = \alpha \sum_{k=1}^{K} \left\| \partial_t [(\delta(t) + \frac{j}{\pi t}) \times u_k(t)] e^{-j\omega_k t} \right\|_2^2 + \left\| f(t) - \sum_{k=1}^{K} u_k(t) \right\|_2^2 + \left\langle \lambda(t), f(t) - \sum_{k=1}^{K} u_k(t) \right\rangle \tag{2}$$

in which α denotes the balancing parameter of the data-fidelity constraint Eq. (2) is then solved with the alternate direction method of multipliers (ADMM) [15]. All the modes gained from solutions in spectral domain are written as

$$\hat{u}_k(\omega) = \frac{\hat{f}(\omega) - \sum_{i \neq k} \hat{u}_i(\omega) + (\hat{\lambda}(\omega)/2)}{1 + 2\alpha(\omega - \omega_k)^2} \tag{3}$$

where the ω_k is computed at the centre of gravity of the corresponding mode's power spectrum. Thus, wiener filtering is embedded in the VMD algorithm that makes it much more robust to sampling and nose.

$$\omega_k = \frac{\int_0^\infty \omega |\hat{u}_k(\omega)|^2 d\omega}{\int_0^\infty |\hat{u}_k(\omega)|^2 d\omega} \tag{4}$$

Complete algorithm of the VMD in detail can be found in [14].

2.2 *Extraction of signal envelops*

A large amount of variation component in signals could be embodied in the envelops of signals, and the carrier wave of envelops is usually taken by the high-frequency components of vibration signal. Hilbert technique is an effective tool to obtain the envelops of signals in fault diagnosis [16, 17]

As for a real signal $x(t)$, $\mathcal{H}\{x(t)\}$ represents the Hilbert transform of $x(t)$, which is defined using the Cauchy principal value (denoted by P) as

$$\hat{x}(t) = \mathcal{H}\{x(t)\} = \frac{P}{\pi} \int_{-\infty}^{\infty} \frac{x(\tau)}{t-\tau} d\tau \tag{5}$$

The analytic signal g(t) of the real signal can be obtained which is defined as:

$$g(t) = x(t) + j\hat{x}(t) \tag{6}$$

where $\hat{x}(t)$ denotes Hilbert transform of the real signal $x(t)$, and the amplitude of analytic signal is expressed as:

$$A(t) = \sqrt{x(t)^2 + \hat{x}(t)^2} \tag{7}$$

This is the envelop of $x(t)$.

3. Application of VMD in rubbing fault features extraction

3.1 Experimental facilities

Fig.1 (a) and Fig.1 (b) show the dual-rotor system test rig and structure diagram, respectively. The inner rotor (1) passing though the outer rotor (2) is connected to a flexible coupling that is driven by a high-speed motor; the outer is driven by a high-speed motor with a belt. Every bearing is supported on one pedestal (13), which is fixed on the foundation by several bolts. The bearing pedestal and membrane coupling (12) can reduce the influence of transverse force due to belt driven. The inner rotor and outer rotor are respectively installed wheel disks to simulate the compressor and turbine load. The both ends of inner rotor are supported by deep groove ball bearing (4) and roller bearing (5), respectively; one end of outer rotor is supported by deep groove ball bearing (6), and the other end is supported on the inner rotor by means of squirrel cage elastic support and roller bearing

(7). Two wheel disks (8, 9) are mounted on the inner rotor, and there are three wheel disks (10, 11, 12) installed on the outer rotor. The elastic support is installed on the wheel disk, the number of which is 11. Vibration acceleration transducers are mounted in 5# and 6# bearing pedestal to pick up acceleration vibration signals.

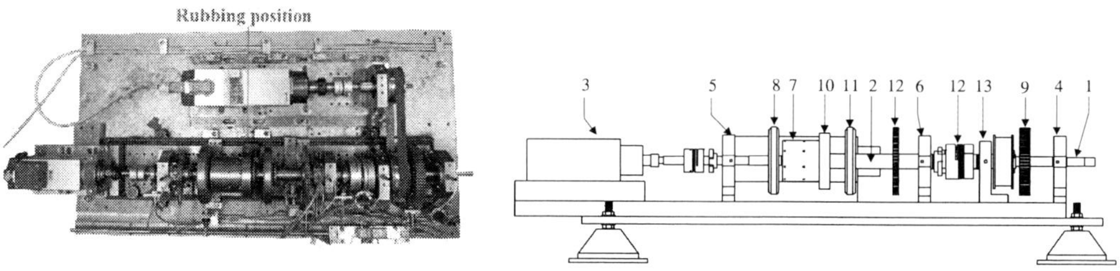

(a) Dual-rotor system test rig (b) Structure diagram of dual-rotor experiment setup

Fig 1 Dual-rotor system test rig and structure diagram

One dimensional finite element model of dual-rotor system (excluding pedestal) is built based on Timoshenko beam element, as presented in Fig.2, which is composed of the inner rotor (node 1 to node 13) with the outer rotor (node 14 to node 20). There are total of four bearings in the model, which are, respectively, located at node 1, node 14, node 9 and node 13, where node 1 and node 13 denote the inter-shaft bearings. The disks, located, respectively, at node 16, node 18, node 3 and node 11, represent the concentrations of high-pressure compressor disks, high-pressure turbine disks, low-pressure compressor disks and low-pressure turbine disks.

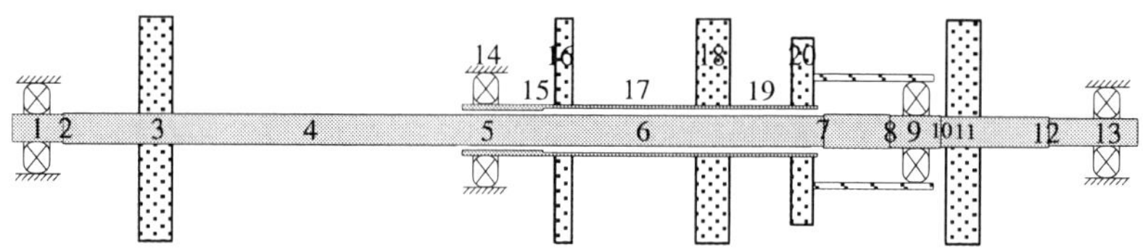

Fig 2 The node map of one dimensional finite element model of dual-rotor system

3.2 Tests using a practical rubbing signal

The vibration acceleration signals are collected from the bearing pedestal of dual-rotor system test rig with a sampling frequency of 5000Hz. The rotation speeds of the outer rotor and inner are set to 3000 rpm (50Hz) and 2400 rpm (40Hz), respectively. The vibration acceleration time waveforms of bearing pedestals (5# and 6# shown in Fig. 1(b)) supported the outer rotor's and inner rotor's bearings are shown in Fig.3 and Fig.4. It is evident from Fig.3 and Fig.4 that the amplitude of outer rotor is greater than that of inner rotor, because the rubbing position is set to the disk of outer rotor, as indicated in Fig.1 (a). Due to the coupling effect of intershaft bearing, rubbing fault characteristics of outer rotor are transferred to inner rotor. Also, in the process of generation and transfer, the fault real-time information may be distorted or mixed with noise and background signals. Misjudgement may be caused by these factors and it is significant to take measures to reduce the influence of interference signals.

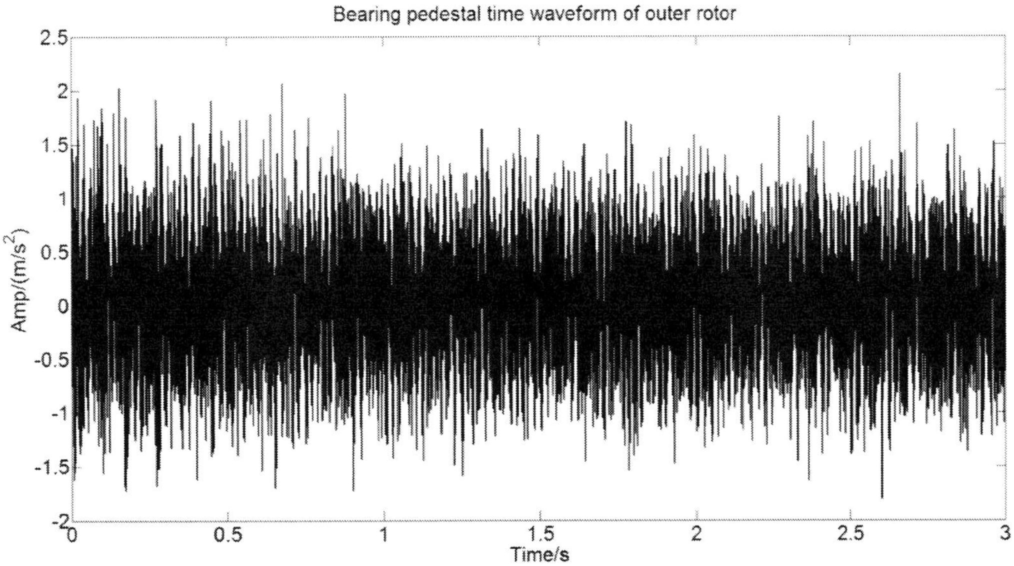

Fig 3 Vibration acceleration signal of bearing pedestal (outer rotor)

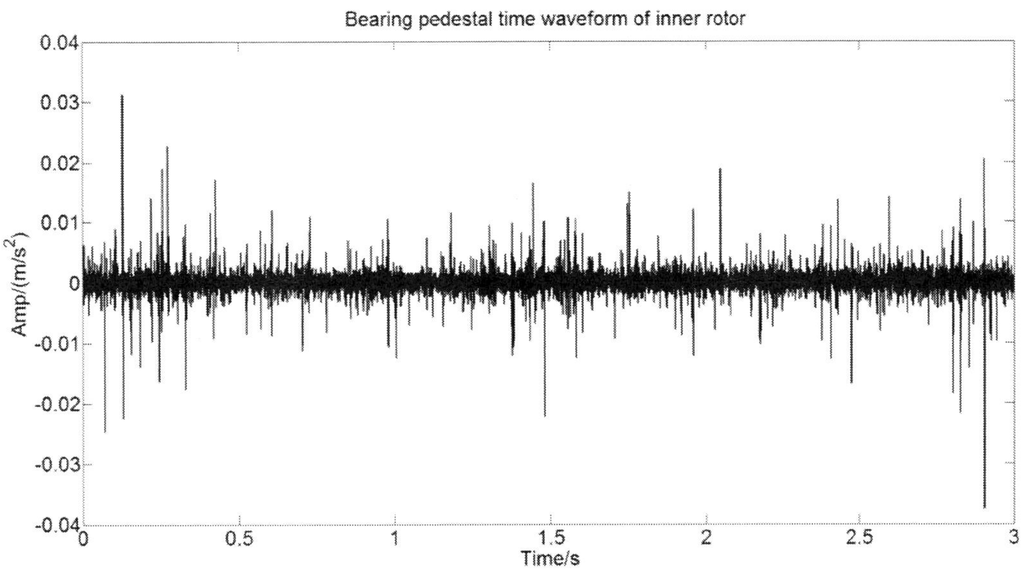

Fig 4 Vibration acceleration signal of bearing pedestal (inner rotor)

VMD uses several parameters to evaluate modes of vibration acceleration signals. These parameters are the time-step of the dual ascent, the number of modes to be extracted (K), the balancing parameter of the data-fidelity constraint (α), the tolerance of convergence criterion (tol), the initial centre frequencies (ω_t) and the number of DC components [14]. Here, the initial values of centre frequencies are randomly assigned. The initial value of other parameters are K=3, alpha=2000 and tol=10^{-7}. The influence of those parameters used in VMD algorithm is investigated in detail in these literatures [13, 14].

VMD is employed to decompose the vibration acceleration signals picked from the bearing pedestal 5# and 6# (see Fig.1 (b)). After observing, it can be seen that the second mode signal has many components with relatively small amplitudes, and contains much higher frequency components. Therefore, the second mode signals are ignored and the rest modes are considered for analyse. It has been discussed in [8], when the rub-impact fault occurs under certain circumstances, the rub-impact signal is submerged in the stronger background signal and accompanied with amplitude-modulated or frequency-modulated characteristics. Hence, it's necessary to perform envelop analysis. The first mode and third mode of the vibration acceleration signals obtained from VMD, corresponding spectral and envelop spectral are illustrated in Fig.5 and Fig. 6, respectively.

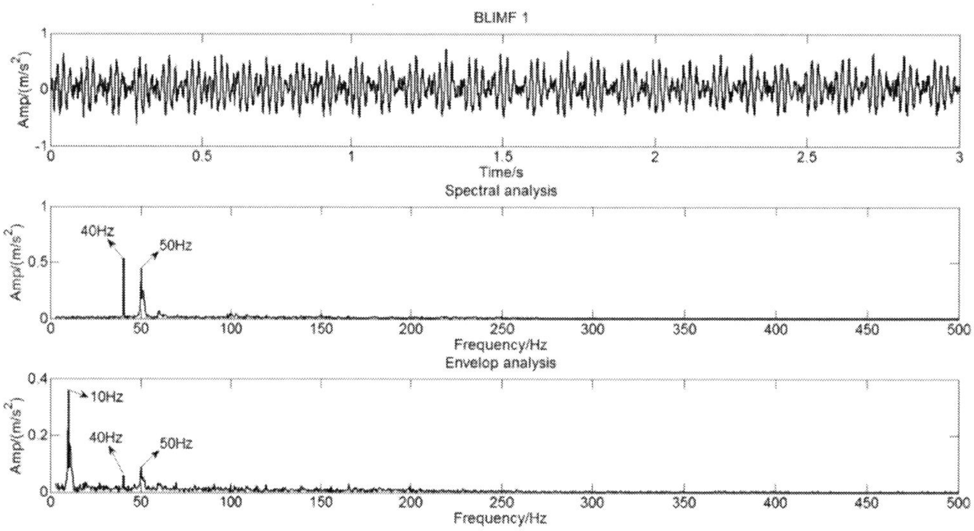

(a) Mode 1 signal, frequency response and envelop analysis

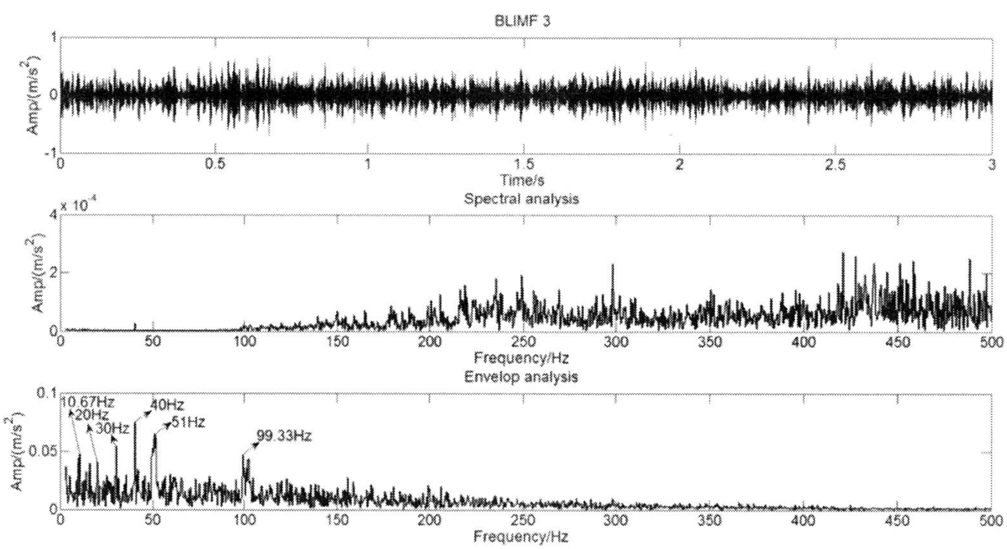

(b) Mode 3 signal, frequency response and envelop analysis

Fig 5 The BLIMFs gained by using VMD (Outer rotor)

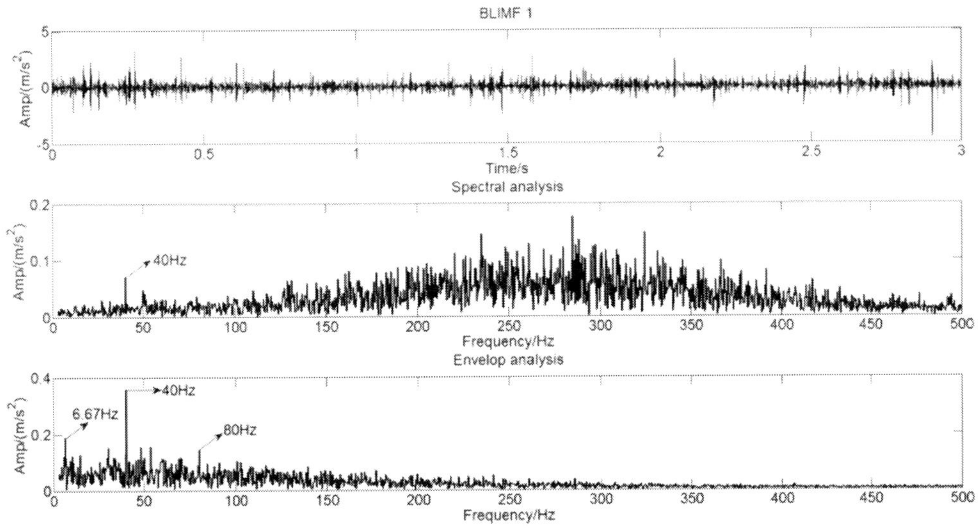

(a) Mode 1 signal, frequency response and envelop analysis

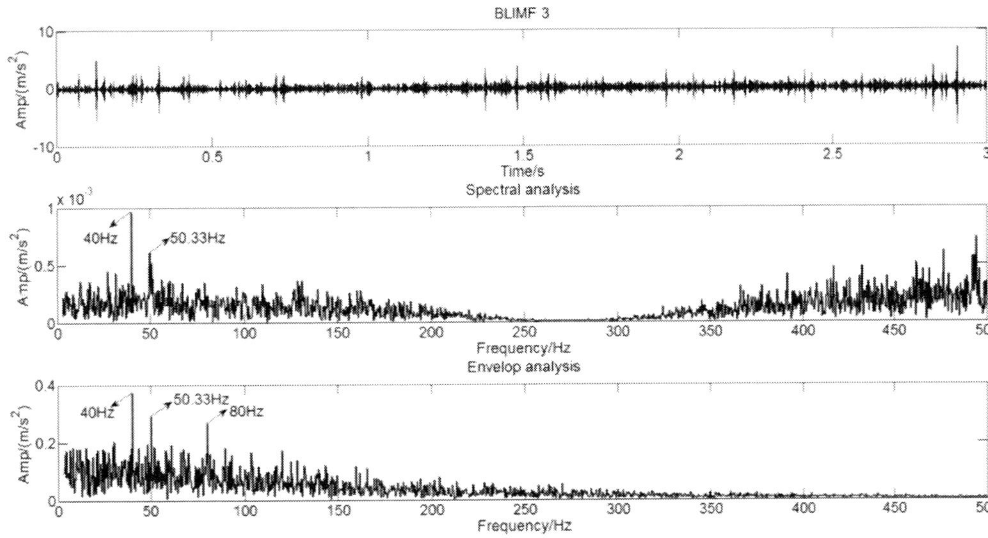

(b) Mode 3 signal, frequency response and envelop analysis

Fig 6 The BLIMFs gained by using VMD (Inner rotor)

It can be seen from Fig. 5(a) that BLIMF 1 contains the periodic components produced by the periodic impact between the rotor and stator. It can be found from Fig.5 and Fig.6 that the spectrums of vibration acceleration under rubbing status not only contain the fundamental frequencies of inner and outer rotor, sometimes difference frequency and superharmonic frequency (such as $2f_L$, f_L and f_H denotes the fundamental frequencies of inner rotor and outer rotor, respectively), but also contain combined frequencies of fundamental frequencies of inner and outer rotor, namely $mf_H \pm nf_L$, $(m,n=1,2,\ldots)$. It can be also observed that the frequency of the detected impacts is the same with that of the $1/6f_L$ harmonic component $(6.67 \approx 1/6\times 40\text{Hz})$, which well illustrates the existing local rubbing malfunction in the machine. Consequently, the above results provide evidence to judge the existence of an early rub-impact in the dual-rotor system. This application, thus further demonstrates the validity of VMD in detecting multiple rubbing-caused signatures, simultaneously.

4. Conclusion

When the rubbing fault occurs in the rotor system, since both the rub-impact signal and the random noise are all the broadband components, whose frequency bands are mixed with the frequency band of the background signal. It is very challenging to determine the frequency range of vibration acceleration signals and extract the rubbing fault features from the vibration signals including noise and background signal by utilizing traditional methods. VMD was a recently proposed novel technique for adaptive signal decomposition. By analysing the vibration acceleration signal collected from the bearing pedestals, it demonstrates the effectiveness of extracting the sub- and super-harmonics and the impacts signatures by using VMD, which shows that VMD method can be applied to the rubbing fault diagnosis of the rotor system.

In addition, it should be noted that using VMD, sometimes the mode mixing problem cannot be avoided due to minor defects of data. Further study will be conducted to investigate this problem as well as limitations of the proposed technique for applications to the healthy monitoring of other mechanical system.

Conflict of interest

There is no conflict of interests in regard to publish of the paper.

Acknowledge

The project is supported by the National Natural Science Foundations of China (No. 11572167). The authors are also grateful to the anonymous reviewers for their valuable comments.

References

[1] Muszynska A. Rotor-To-Stationary Element Rub-Related Vibration Phenomena in Rotating Machinery -- Literature Survey [J]. Shock & Vibration Digest, 1989, 21(3):3-11.
[2] Isaksson J, Frid A. Analysis of dynamical behaviour of a rotor with deflection limiters [J]. Acta Mechanica, 1999, 133(133):1-11.
[3] J. ZAPOMĚL, C.H.J. FOX, E. MALENOVSKÝ. Numerical investigation of a rotor system with disc-housing impact [J]. Journal of Sound & Vibration, 2001, 243(2):215–240.
[4] Beatty R F. Differentiating Rotor Response Due to Radial Rubbing [J]. Journal of Vibration & Acoustics, 1985, 107(2):151-160.
[5] Behzad M, Alvandi M, Mba D, et al. A finite element-based algorithm for rubbing induced vibration prediction in rotors [J]. Journal of Sound & Vibration, 2013, 332(21):5523-5542.
[6] Popprath S, Ecker H. Nonlinear dynamics of a rotor contacting an elastically suspended stator [J]. Journal of Sound & Vibration, 2007, 308(3–5):767-784.
[7] Niaoqing HU, Chen M, Wen X. The application of stochastic resonance theory for early detecting rub-impact fault of rotor system [J]. Journal of Mechanical Engineering, 2001, 17(4):883–895.
[8] Cheng J, Yu D, Tang J, et al. Local rub-impact fault diagnosis of the rotor systems based on EMD [J]. Mechanism & Machine Theory, 2009, 44(4):784-791.
[9] Peng Z K, Chu F L, Tse P W. Detection of the rubbing-caused impacts for rotor–stator fault diagnosis using reassigned scalogram [J]. Mechanical Systems & Signal Processing, 2005, 19(2):391-409.
[10] Lei Y, He Z, Zi Y. Application of the EEMD method to rotor fault diagnosis of rotating machinery [J]. Mechanical Systems & Signal Processing, 2009, 23(4):1327-1338.

[11] Markert R, Wegener G. Transient vibrations of elastic rotors in retainer bearings, in: Conference Proceeding of ISROMAC-7 2, (1998) pp. 764-774.
[12] Wang Y, He Z, Zi Y. A demodulation method based on improved local mean decomposition and its application in rub-impact fault diagnosis [J]. Measurement Science & Technology, 2009, 20(2):28-28.
[13] Wang Y, Markert R, Xiang J, et al. Research on variational mode decomposition and its application in detecting rub-impact fault of the rotor system [J]. Mechanical Systems & Signal Processing, 2015, 60:243-251.
[14] Dragomiretskiy K, Zosso D. Variational Mode Decomposition [J]. Signal Processing IEEE Transactions on, 2013, 62(3):531-544.
[15] Hestenes M R. Multiplier and gradient methods [J]. Journal of Optimization Theory & Applications, 1969, 4(4):303-320.
[16] De Chanvalon, T. Manon. Quantum symmetry groups of Hilbert modules equipped with orthogonal filtrations [J]. Journal of Functional Analysis, 2014, 266(5):3208-3235.
[17] T, Jordão, V. A. Menegatto. Weighted Fourier-Laplace transforms in reproducing kernel Hilbert spaces on the sphere [J]. Journal of Mathematical Analysis & Applications, 2014, 411(2):732-741.

The Fault Diagnosis of Aero-engine Intershaft Bearing Based on EEMD and FastICA

Shenghua Zhang [1] Fengtao Wang [2] Tao Zhang [3] Bei Wang [4] Meng Wang [5]

1School of Mechanical Engineering, Dalian University of Technology, Dalian, China, zhangshenghua2015@163.com

2 School of Mechanical Engineering, Dalian University of Technology, Dalian, China, wangft@dlut.edu.cn

Abstract. For aero-engine intershaft bearing dual-rotor test stand, the vibration signal acquired at the bearing box attenuates seriously and has a low SNR, so it is difficult to extract the fault feature information of the signal directly. This paper presents a method about blind source separation (BSS) based on ensemble empirical mode decomposition(EEMD) and fast independent component(FastICA), the under-determined blind source separation problem is solved by the method of EEMD. First, the signal is decomposed by EEMD. Second, the low-dimensional manifold composition of the IMF components areextracted by the manifold learning method, then the optimal observation signals are constructed, and the number of the vibration source is determined. Finally, the optimal observation signalsare utilized as the input of FastICA analysis. The effectiveness of the proposed method is verified through the practical vibration signal.

Keywords: Intershaft bearing, fault diagnosis, EEMD, FastICA, manifold learning

1 Introduction

As the key components supporting between the low and high pressure rotor of aero-engine, intershaft bearing inner and outer rings rotate simultaneously. Its reliability is critical, which directly affects the performance of aero-engine [1]. Intershaft bearing working in the harsh conditions of high speed, high temperature and high pressure. For the actual aero-engine and aero-engine intershaft bearing test stand, the accelerometer is far away from the intershaft bearing [2−3]. The signal is low SNR resulting from complex signaling pathways. Therefor, it is meaningful to individually study intershaft bearing failure by separating intershaft bearing signal from the complex vibration signal.

EMD is an adaptive signal processing method, which has a good noise reduction for non-stationary, nonlinear signal. However, fault signal in project containing a lot of noise leads the EMD has the "mode mixing" phenomenon [4] resulting in very poor noise reduction effect, and less effect for intershaft bearing early weak fault signal processing. Based on the studying of EMD [5], Wu and Huang Presente Ensemble Empirical Mode Decomposition [6] (EEMD) method. EEMD method solves the problem of cancellation aliasing improving noise reduction performance [7].

Random, intermittent background noise, other interference components and fault signal together form a non-linear, redundant data space in engineering practice. For this type of rendering non-gaussian distribution of the data set in the high-dimensional space, the manifold learning can effectively remove the influence of noise and other interference components to find the essential characteristics of the fault information contained in the data set. It is conducive to the next step about pattern recognition or feature extraction ork [8-9].

FastICA is a fast blind source separation algorithm based on the principle of maximum negative entropy. It can estimate the observed signal into a series of independent components, thereby separating the fault source. It is widely used in fault diagnosis field because of fast convergence and small computation [10-12].

Based on previous studies and the characteristic of early weak fault signal with low SNR of aero-engine ntershaft bearing, first configure virtual multi-channel signal using EEMD, then apply Manifold learning for each component to reduce noise. Finally, separate vibration source signal from the observed signal by FastICA analysis to relise intershat bearing and rotor system fault diagnosis.

2 Theory basis

2.1 EEMD method

The principle of EEMD takes advantage of the uniform distribution statistical characteristics of the gaussian white noise in the frequency range. When the signal is added to a certain amount of white noise, the signal will have continuity over different time scales to reduce mode mixing caused by local intermittent impact component. Specific process of EEMD method is as follows [13]:

(1) Add many white Gaussian noise signal $n_i(t)$ to the original signal $x(t)$:

$$x_i(t) = x(t) + n_i(t) \tag{1}$$

(2)Recorde IMF obtained from EEMD decomposition of $x_i(t)$ as $c_{ij}(t)$, remainder as $r_i(t)$. Among them, $c_{ij}(t)$ for the first additions to white gaussian noise and the first j IMF by EMD decomposition:

$$x_i(t) = \sum_{j=1}^{n} c_{ij}(t) + r_i(t) \tag{2}$$

(3)In order to eliminate the influence of gaussian white noise, calculate the overall average of the IMFs above by utilizing the principle of statistical average uncorrelated random sequences of zero. Obtain IMF as follows:

$$c_j(t) = \frac{1}{N}\sum_{i=1}^{N} c_{ij}(t) \quad (3)$$

2.2 Fast independent component analysis

Fast independent component analys is the improved algorithm of ICA. Restore the optimal source signal estimate from the optimal measurement signal by utilizing the maximum entropy principle to approximate negative entropy, specific algorithm as follows [11].

Firstly, removing mean and whitening process aim at obtain the signal $z(t)$. Select matrix with unit norm, pointing into the iterative method:

$$(1+\alpha)w = E\{zg(w^T z)\} + aw \quad (4)$$

Obtain coefficient by Newton iterative method.When satisfy the constraints $E\{(w^T z)^2\} = \|w\|^2 = 1$ and $E\{zg(w^T z)\} + \beta w = 0$, $E\{G(w^T z)\}$ gets the optimal value, Where G is the integral function. Set $F(w) = E\{zg(w^T z)\} + \beta w$, then $F(w)$ converted into:

$$J\{F(w)\} = E\{zz_T g'(w^T z)\} - \beta I \quad (5)$$

$$E\{zz^T g'(w^T z)\} \approx E\{zz^T\}E\{g'(w^T z)\} = E\{g'(w^T z)\}I \quad (6)$$

Therefore, obtaine approximate Newton iterative algorithm,

$$w \leftarrow E\{zg(w^T z)\} - E\{g(w^T z)\}w \quad (7)$$

$$w = w / \|w\| \quad (8)$$

Repeat (7), until it meets the convergence conditions, thus separating matrix obtained.

3 Process Steps

The vibration signal $s(t)$ of the intershaft bearing is measured. The number of the vibration source and optimal source signal estimation is obtained by the following processing. As figure 1:

(1)A series of IMF components of the signal $s(t)$ are obtained by EEMD method;

(2)The noise of each IMF components is reduced by LTSA manifold algorithm;

(3)Select attractor components which characterize the characteristic of the bearing dynamics, calculate the eigenvalues of covariance, take the number of source signal whose cumulative contribution rate of eigenvalues is more than 85% as n that the number of vibration source signal.

(4)Calculate the correlation coefficient between all attractor components and $s(t)$, take attractor components of the maximum n correlation coefficient to constitute the optimal measurement signal $x(t)$;

(5)Input $x(t)$ to FastICA, the optimal source signal is estimated;

(6)Calculate the hilbert envelope spectrum of the optimal estimate signal.

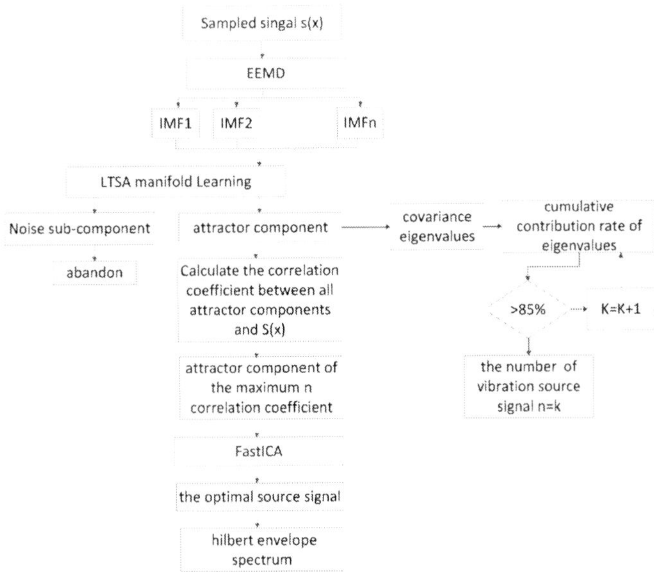

Fig. 1. Diagnostic process

4 Case Analysis

Intershaft bearing is mounted on the dual-rotor test stand, and physical map of test stand is shown in Figure 2.

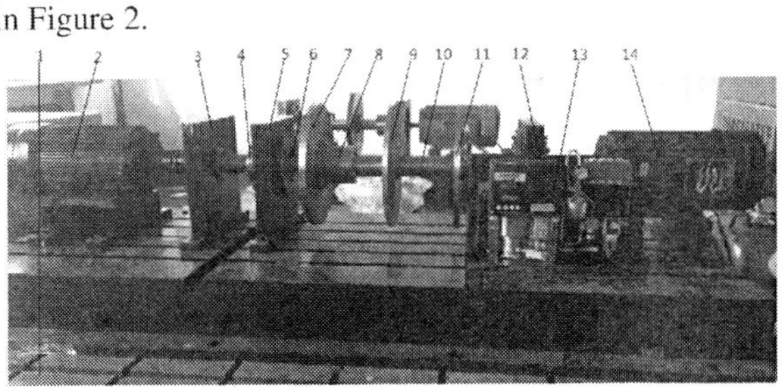

Fig. 2. Aero-engine dual-rotor test stand

Machine a 0.6mm inner ring scratches by electric spark on inner ring of the intershaft bearing NU1013 to analog intershaft bearing inner ring failure.

Put acceleration sensor on 5 main bearing housing to obtain vibration signal, sampled frequency fs=25600Hz, sampled points N=51200, the rotating speed of the low pressure rotor -675rpm, the rotating speed of the high pressure rotor 675rpm, the inner ring fault characteristic frequency 259Hz.

The vibration signal time domain diagram and hilbert envelope spectrum is shown in Figure 3.

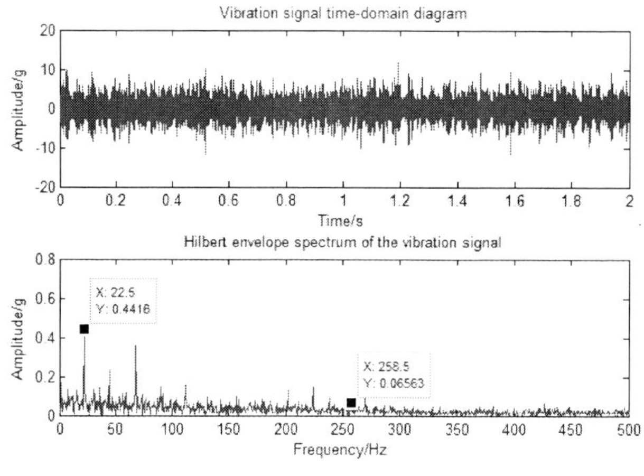

Fig. 3. Time domain signal and Hilbert envelope spectrum

As can be seen from Figure 3, the amplitude of the second harmonic of frequency is high, which can be judged as the fault caused by the misalignment of the rotor. Since the vibration signal contains complex background noise, the amplitude of intershaft bearing fault characteristic frequency that 258.5Hz of $s(t)$ in hilbert spectrum envelope is so small that it is difficult to determine the presence of bearing fault.

The rotor misalignment signal and the intershaft bearing inner failures signal are extracted by the method above. Nine components are got from $s(t)$ through EEMD, and the noise of each IMF is reduced utilizing the LTSA method. Due to space limitations, six IMFs are shown in Figure 4 after LTSA noise reduction.

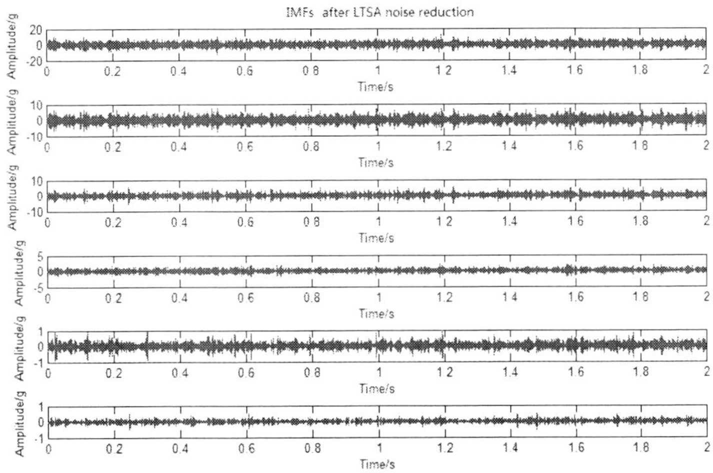

Fig. 4. IMF component time-domain diagram after LTSA noise reduction

Calculate correlation coefficients of all IMFs after LTSA noise reduction with the source signal $s(t)$. The value of each correlation coefficient is shown in Figure 5.

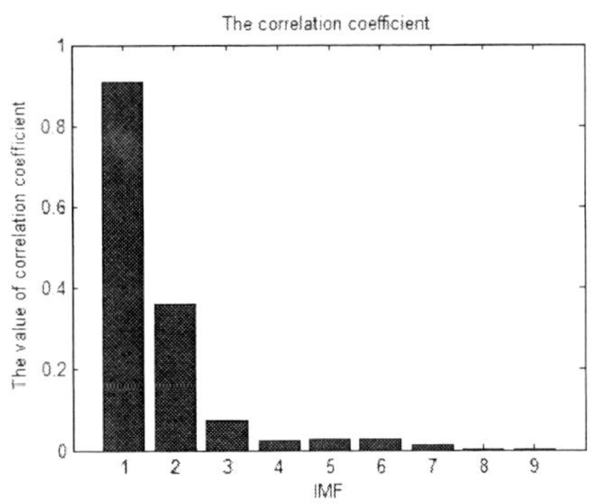

Fig. 5. The correlation coefficient between the IMFs and the source signal

Calculate the eigenvalues cumulative contribution rate of covariance matrix constituted by the IMFs after LTSA noise reduction. The result is shown in Figure 6.

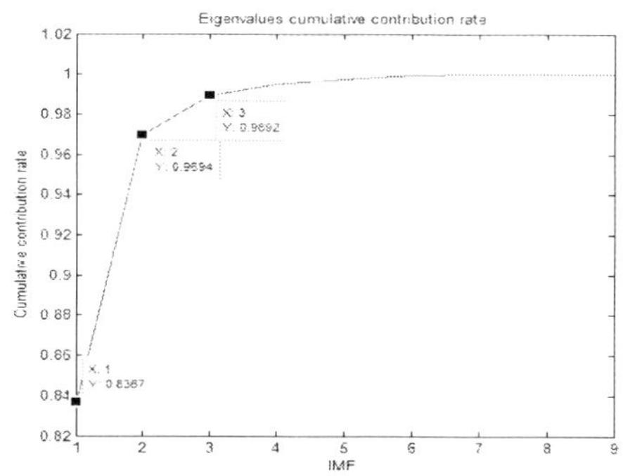

Fig. 6. IMF component eigenvalues cumulative contribution

From Fig 6, the eigenvalues cumulative contribution rate of the first two IMF is-more than 85%, so the number of the vibration source is 2. Select the two previous IMFs after LTSA noise reduction as the optimal observation signal to input into FastICA, the output is the optimal source estimated. Then do hilbert envelope analysis utilizing the optimal source estimated signal. The time-domain diagram and hilbert envelope spectrum of estimated source signal are shown in Figure 7.

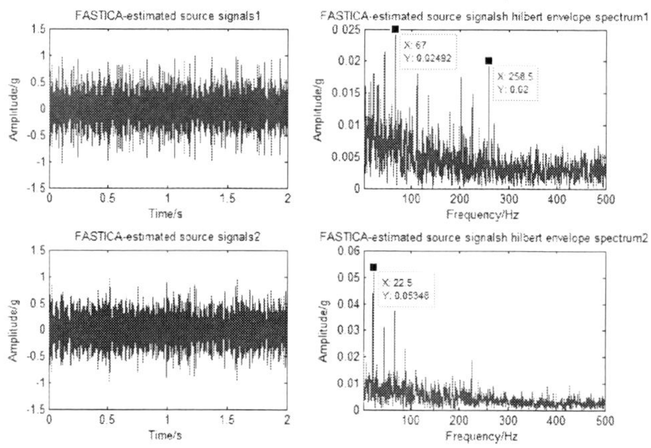

Fig. 7. The time-domain diagram and hilbert envelope spectrum of the estimated source signal

It can be found that $s(t)$ contains two kinds of vibration signal. The estimating source signal 2 is a vibration caused by rotor misalignment and the estimating source signal 1 is intershaft bearing inner ring failures vibration. And in the estimated source signal 1, the SNR of bearing inner ring fault is largely improved, so the bearing inner ring failure can be easily found from hilbert envelope spectrum.

5 The conclusion

This paper provides an effective way for aero-engine intershaft bearing vibration signal fault diagnosis. The rotorandintershaft bearing fault signal are separated from the acquisition vibration signal. The number of vibration source is determined by eigenvaluecumulative contribution rate of the covariance matrix. The SNR of the separated signalis is raised so that it is easier to find the presence of fault.

References

1. Mei HB:Vibration monitoring and diagnosis of rolling bearing-theory—theory· method - system.Beijing: China Machine Press(1996)
2. Liao MF, Ma ZG, Liu YQ, etal: Fault characteristics and diagnosis method of intershaft bearing in aero-engine.J.Journal of Aerospace Power. 28(12):2752-2758(2013)
3. Cao MG, Meng JF: Certain engine body vibration failure analysis.J.Journal of Aeroengine. (04): 15-21(1994)
4. Hu AJ, Sun JJ, Xiang L: Mode Mixing in Empirical Mode Decomposition.J. Journal of Vibration Measurement & Diagnosis. 04:429-434+532-533(2011)
5. Z. Wu and N. E. Huang: A study of the characteristics of white noise using the empirical mode decomposition method. J.Proc. Roy. S oc. London 460A 1597–1611 (2004)
6. Wu, Z.and N. E Huang: Ensemble Empirical Mode Decomposition: A noise-assisted data analysis method.J. Advances in Adaptive Data Analysis. Vol.1, No.1.1-41(2008)
7. Zhou Z, Zhu YS, Zhang YY, etal: Adaptive fault diagnosis of rolling bearings based on EEMD and demodulated resonance.J.Journal of Vibration and Shock. 02:76-80(2013)
8. Wang SP: Research on Fault Feature Extraction Method of Rolling Bearing Based on Manifold.D. Dalian University of Technology(2013)
9. Chen SH: The Research of Classification Method for Rolling Bearing Early Fault Based on Manifold.D. Dalian University of Technology(2014)
10. Wang L: Research on Rolling Element Bearing Fault Diagnosis Methods Based on Manifold Learning.D. Dalian University of Technology(2013)
11. Shi ZW, Zhang CS: Fast nonlinear autocorrelation algorithm for source separation. J. Pattern Recognit. 42(9): 1732-41(2009)
12. Tichavsky P, Koldovsky Z, Oja E.: Performance analysis of the FastICA algorithm and Cramer-Rao bounds for linear independent component analysis(vol 54, pg 1189, 2006). J. IEEE Trans Signal Process 2008; 56(4): 1715-6.(2008)
13. Chen L, Zi YY, He ZJ, C HENG Wei: Research and Application of Ensemble Empirical Mode Decomposition Principle and 1.5 Dimension Spectrum Method. J. JOURNAL OF XI'AN JIAOTONG UNIVERSITY. 05:94-98(2009)

Wavelet-based Control Chart for Early Detection of Gear Failure

Wei Fan[1], Qiang Zhou[1,*] and Zhongkui Zhu[2]

[1]Department of Systems Engineering and Engineering Management, City University of Hong Kong, Hong Kong, China

weifan8-c@my.cityu.edu.hk;q.zhou@cityu.edu.hk

[2]School of Urban Rail Transportation, Soochow University, Suzhou, China

zhuzhongkui@suda.edu.cn

Abstract. Early failure detection is a critical problem in mechanical system. A novel statistical method based on tunable Q-factor wavelet transform (TQWT) is developed in this study. The measured vibration signal is decomposed by TQWT into wavelet coefficients. The energy of each wavelet coefficient is monitored by the control chart. Hilbert envelope spectrum is applied to extract the fault frequency of the reconstructed signal. Both the simulation study and the application study show the effectiveness of the proposed method.

Keywords: Statistical process control, Control chart, Wavelet, Early detection, Gearbox failure

1 Introduction

Early failure detection of gearbox is critical to maintain system safety [1]. However, the fault feature is always surrounded by strong noise in the early-fault-state, which makes the failure detection a key but hard issue for gearbox condition monitoring.

Recently, many different methods have been proposed for fault detection, such as the wavelet analysis [2], spectral kurtosis [3], etc. However, these methods do not perform well in early detection. Statistical control charts also can be used to detect the fault. However, traditional control chart is applied to the time domain signal whose energy is weak, thus is not sensitive to the early failure.

Wavelet transform is a traditional method applied in mechanical fault detection. Recently, a new wavelet transform called tunable Q-factor wavelet transform (TQWT) has been applied in machinery fault diagnosis [4-6]. In this paper, a new statistical method based on TQWT and control chart is developed. Firstly, the TQWT is applied to decompose the vibration signal to wavelet coefficients. Secondly, the energy of each wavelet coefficient is monitored by the control chart. Then the Hilbert envelope spectrum is incorporated to detect the fault signal reconstructed by the out-of-control coefficients.

The remaining parts are organized as follows. Firstly, the theoretical basis of the method is introduced. Then simulations with different signal to noise ratios (SNR) are conducted to verify the proposed method. Then the proposed method is applied to gearbox transient feature extraction. Finally, conclusions are drawn in the last section.

2 Theoretical Basis

2.1 Statistical Process Control

Under the normal working condition, the measured signal only contains background noise, whose amplitude is stable. Once the fault occurs, there will be large amplitudes in the signal. The statistical process control (SPC) method can be used to monitor the amplitude [7]. Control chart is the most widely used technique in SPC.

The one-sided $1-\alpha$upper control limit (UCL) for a chi square distribution statistic A with l degrees of freedom is

$$UCL_A^{\alpha} = \chi_{\alpha,l}^2 \tag{1}$$

2.2 Tunable Q-factor Wavelet Transform

Tunable Q-factor wavelet transform can be tuned according to the oscillatory behavior of the signal being analyzed [8]. Q affects the oscillatory behavior of the wavelet. Redundancy r is the total over-sampling rate of TQWT. Procedure of TQWT is shown in Fig. 1. α is the low-pass scaling parameter, β is the high-pass scaling parameter, J is the maximum number of stages.

After TQWT, the original signal $x(t)$ will be decomposed to the scaling coefficients$a^{\{J\}}$and wavelet coefficients$\boldsymbol{d}$:

$$d = \{d^{\{1\}},\ d^{\{2\}}, \ldots, d^{\{J\}}\} \quad (2)$$

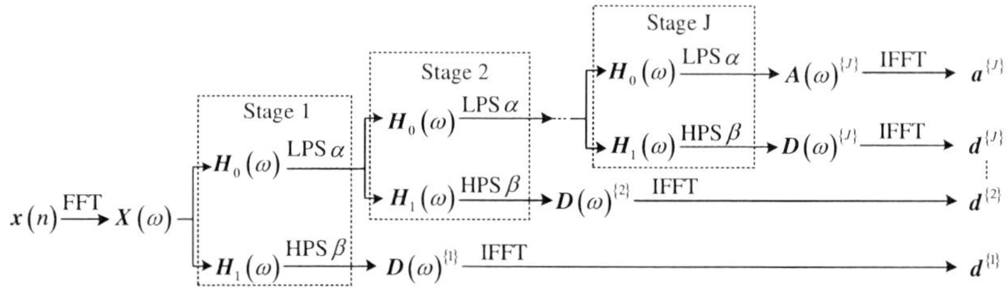

Fig. 1. Procedure of the tunable Q factor wavelet transform.

2.3 Wavelet-based Control Chart For Early Detection

TQWT is applied to decompose the signal before using control chart since the early failure is very weak. The energy of the wavelet coefficient on the j^{th}stage is

$$e^j = \sum_{t=1}^{L_j} d^{\{j\}}(t)^2 \quad (3)$$

whereL_j is the length of $d^{\{j\}}(t)$. The jth stage wavelet coefficient of a Gaussian white noise can be obtained by

$$\boldsymbol{d}^{\{j\}} = \boldsymbol{\Phi}^{\{j\}}\boldsymbol{n} \quad (4)$$

where$\boldsymbol{\Phi}^{\{j\}}$is the jth wavelet decomposition operation, $\boldsymbol{n}$ is the noise. The control chart with UCL is established to monitor the statistic. Hilbert envelope spectrum is applied to analyze the signal reconstructed by the out-of-control coefficients.

3 Simulation Study

The simulation signal $x(t)$is constructed as:

$$x(t) = s(t) + n(t) = A_s \sum_k e^{\frac{-\zeta_0}{\sqrt{1-\zeta_0{}^2}}[2\pi f_0(t-kT_0-\tau_0)]^2} \cos(2\pi f_0(t - kT_0 - \tau_0)) + n(t) \quad (5)$$

where frequency$f_0 = 50\text{Hz}$, damping ratio$\zeta_0 = 0.01$, time index $\tau_0 = 0.1\text{s}$, period $T = 0.2\text{s}$ and $n(t)$ is Gaussian white noise $N(\mathbf{0}, \boldsymbol{\sigma}^2)$. Here we assume $\boldsymbol{\sigma}^2 = \mathbf{I}$.

The normal operation signal is measured to construct the control chart. Firstly, apply TQWT ($Q = 6$, $r = 3$) to decompose each set of signal to 29 stages of wavelet coefficients. Then the UCL with $\alpha = 0.05$ is set up to monitor the fault signal.

To study the effectiveness of the proposed technique, simulated signals with different SNRs are used. For each simulated signal, 10 thousand independent tests are conducted to investigate the detection power. SNR is defined as

$$SNR = 10 \times \lg(P_s/P_n) \tag{6}$$

where P_s and P_n are the energy of the faulty signal and the noise. Detection results are shown in Table 1. Fig. 2 gives the results of the signal whose SNR is -10.116dB.

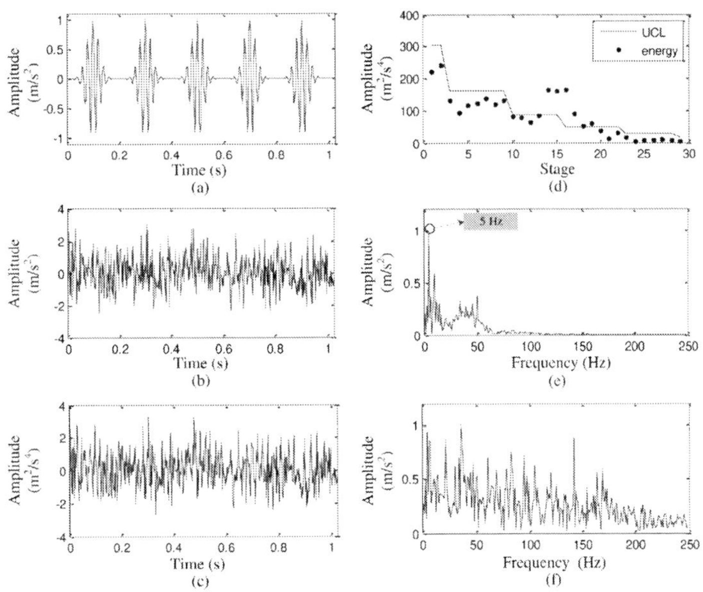

Fig. 2. The results of the proposed method.

Table 1. Monitoring results of signals with different SNRs

A_s (m/s^2)	4	3	2	1.5	1.4	1.3	1.2
SNR (dB)	1.933	-0.559	-4.095	-6.584	-7.2	-7.819	-8.523
Detection power	100%	100%	100%	100%	100%	99.89%	99.79%
A_s (m/s^2)	1.1	1	0.9	0.8	0.7	0.6	0.5
SNR (dB)	-9.273	-10.104	-11.021	-12.045	-13.210	-14.545	-16.124
Detection power	99.04%	96.74%	91.56%	84.07%	72.15%	61.44%	53.64%

Fig. 2 (a)-(c) show the original signal, noise, and the noisy fault signal. Fig. 2 (d) is the control chart, from which we can see that the stages 14-19 are out of control. Fig. 2 (e) shows the envelope spectrum of the reconstructed signal; the fault frequency 5Hz can be identified. Fig. 2 (f) shows the Hilbert envelope spectrum of the noisy signal. Figure 2 show the proposed method outperforms Hilbert envelope spectrum.

4 Application to Early Detection

The proposed method is applied to the experimental data, which were acquired from an automobile transmission gearbox with a tooth-broken fault occurred at the driving gear of the third speed. The normal working condition signals were collected to set up the control chart. The meshing frequency is $f_m = 500\text{Hz}$. The theoretical faulty frequency is 20 Hz, and the sampling frequency is 3000 Hz.

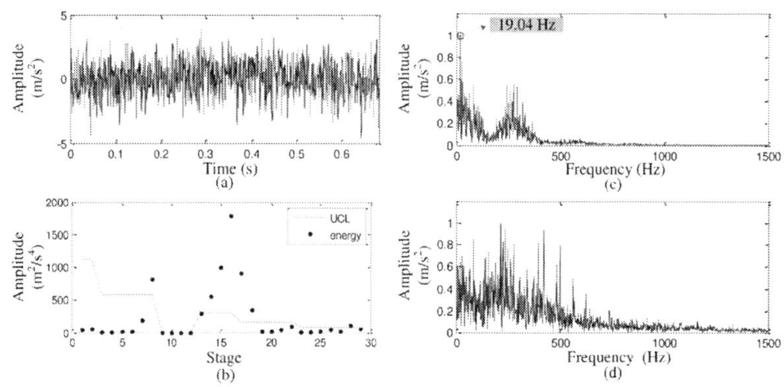

Fig. 3. The results of the proposed method applying to early-fault-detection.

Fig. 3 (a) shows the time domain signal. The meshing component can be decomposed to stages 9 to 12. To avoid the impact of the meshing component, the stages except 9 to 12 should be monitored. Firstly, construct the control chart as shown in Fig. 3 (b), where the stages 14 to 18 are out of control; stages 8 and 28 can be considered as the common cause. Fig. 3 (c) shows the Hilbert envelope spectrum of the reconstructed signal. The fault frequency is identified as 19.04 Hz. Fig. 3 (d) shows the envelope spectrum of the original signal. Fig. 3 shows the proposed method can be applied to the early-fault-detection and outperforms Hilbert envelope spectrum.

5 Conclusion

A novel early-fault-detection method based on TQWT and statistical monitoring method is proposed in this paper. Energies at the decomposed wavelet stages are used as the monitoring statistics to detect the out of control stages. To detect the fault information, the control chart is used in this paper. Hilbert envelope spectrum is also used to detect the main frequency of the reconstructed signal. The effectiveness is verified by both the simulation study and experimental case.

Acknowledgements. The financial support for this work is provided by General Research Fund #11216014 from the Hong Kong Research Grants Council.

References

1. Lei, Y., Lin, J., Zuo, M.J., He, Z.J. Condition monitoring and fault diagnosis of planetary gearboxes: a review. J. Measurement, 48, 292--305 (2014)
2. Yan, R., Gao, R.X., Chen, X. Wavelets for fault diagnosis of rotary machines: a review with applications. J. Signal Processing, 96, 1--15 (2014)
3. Wang, Y., Xiang, J., Markert,R.Liang, M. Spectral kurtosis for fault detection, diagnosis and prognostics of rotating machines: A review with applications. J. Mechanical Systems and Signal Processing, 66, 679--698 (2016)
4. Cai, G., Chen, X., He, Z. Sparsity-enabled signal decomposition using tunable Q-factor wavelet transform for fault feature extraction of gearbox. J. Mechanical Systems and Signal Processing, 41(1), 34--53 (2013)
5. He, W.P., Zi, Y.Y., Chen, B.Q., Wang, S.B., He, Z.J. Tunable Q-factor wavelet transform denoising with neighboring coefficients and its application to rotating machinery fault diagnosis. J. Science China Technological Sciences, 56(8), 1956--1965 (2013)
6. Ding, B., Tong, C., Xin, W., Wang, S., Chen, X.:Sparsity-enabled denoising method based on tunable Q-factor wavelet transform for bearing fault diagnosis. In:Proceedings of the 2nd International Conference of Structural Health Monitoring and Integrity Management, pp. 123--126. CRC Press, (2015)
7. Xie, M., Goh, T.N., Ranjan, P. Some effective control chart procedures for reliability monitoring. J. Reliability Engineering & System Safety, 77(2), 143--150 (2002)
8. Selesnick, I.W. Wavelet transform with tunable Q-factor. J.IEEE Transactions on Signal Processing, 59(8): 3560--3575 (2011)

RESEARCH ON A FAULT DIAGNOSIS METHOD BASED ON DT-CWT AND SVD FOR WIND-TURBINE GEARBOX

Wang Xingda[1],Gu Yuhai[2],He Linfeng[3],Xu Xiaoli[4]

[1]Beijing Information Science and Technology University, The ministry of education key laboratory of modern measurement and control technology, No.12 Xiaoying east road, Qinghe, Haidian district, Beijing, 100192, China.
wxdbistu@foxmail.com

[2]Beijing Information Science and Technology University, The ministry of education key laboratory of modern measurement and control technology, No.12 Xiaoying east road, Qinghe, Haidian district, Beijing, 100192, China.
guyuhai@bistu.edu.cn

[3] Beijing Information Science and Technology University, The ministry of education key laboratory of modern measurement and control technology, No.12 Xiaoying east road, Qinghe, Haidian district, Beijing, 100192, China.
left921121@163.com

[4]Beijing Information Science and Technology University, The ministry of education key laboratory of modern measurement and control technology, No.12 Xiaoying east road, Qinghe, Haidian district, Beijing, 100192, China.
xuxiaoli@bistu.edu.cn

Abstract. Aiming at the difficulty to extract the fault feature frequencies from vibration signal of wind turbine transmission system, a new fault feature extraction and denoising method based on the dual tree complex wavelet transform (DT-CWT) and the singular value decomposition (SVD) is proposed. Because of the non-stationary and modulation characteristics of fault signal, the original vibration signal is decomposed into several different frequency band components by dual-tree complex wavelet transform. After that, for the layered signal containing fault characteristic frequency, Hankel matrix is constructed and decomposed by using the SVD to eliminate noise and reconstruct signal. Finally, the fault characteristic frequency can be identified accurately by the Hibert envelope spectrum. With the simulation and experiment, it shows that this method is effective for wind turbine transmission fault feature extraction.

Keywords: Dual-tree complex wavelet transform, vibration signal, fault characteristic frequencies, Hankel matrix, singular value decomposition, fault diagnosis

1 Introduction

Gear box is an important drive part of the wind turbine generator system and its working status will directly affect the working condition and the production efficiency of the wind generator system. Because of its complex working environment, it is very

Foundation-supported program: National Natural Science Foundation-supported Program (51275052); Key Program Funded by Beijing Municipal Natural Science Foundation (3131002).

prone to failure. Gear is the key part of the gear box, and it is also a frequent failure part. Since the wind turbine gear failure mode is quite complicated, the fault signal is nonlinear and non-instability and the early fault signal is covered by the strong background noise, which causes great difficulties to the gear fault diagnosis[1]. How to extract accurately the fault characteristic frequency from the vibration signal with a large amount of noise is the key to the gear fault diagnosis. Therefore, it has important theoretical significance and engineering application value to research the fault feature information extraction method of gear box.

With the continuous research to gear fault diagnosis, several signal processing methods have emerged. Such as: Zhao Xue Zhi and someone else proposed the singular values differential spectrum theory, which is applied to the practical engineering and successfully locate the fault site[2-3]; Li Qiang proposed the diagnosis method of the combination spectral kurtosis algorithms and SVD, which can find the fault frequency accurately[4].

In this paper, a fault feature extraction method based on the DT-CWT and the SVD is proposed. Experiments show that this method can effectively extract the fault signal from the weak signal.

2 Dual tree Complex Wavelet Transform

DT-CWT uses two parallel wavelet transform to realize signal decomposition and reconstruction, which are called real tree and imaginary tree. The decomposition and reconstruction process of DT-CWT is shown as Fig.1. As shown in the figure, the h0 and h1 respectively represent the low pass filter and high pass filter of the real tree, meanwhile the h0′ and h1′ are the corresponding filter in the process of real tree reconstruction. The g represents the filter of the imaginary tree and others are in the same way with real tree. This wavelet decompose algorithm can make the dual-tree complex wavelet transform have the approximate shift invariance[5-6].

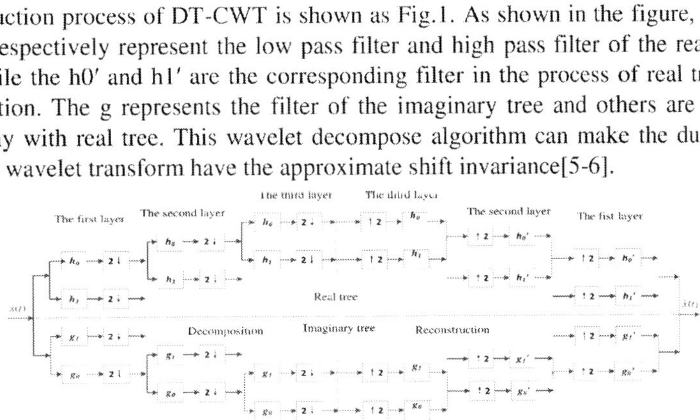

Fig. 1. Decomposition and reconstruction process using dual-tree complex wavelet transform

3 Singular Value Decomposition

Define the Hankel matrix of time sequence X = (x (1), x (1), ..., x (N)) :

$$H = \begin{bmatrix} x(1) & x(2) & \cdots & x(n) \\ x(2) & x(3) & \cdots & x(n-1) \\ \cdots & \cdots & \cdots & \cdots \\ x(N-n+1) & x(N-n+2) & \cdots & x(N) \end{bmatrix} \quad (1)$$

The restrictions in the formula are as follows: 1<n<N, m=N-n+1,H Rm n. The matrix is called reconstruction attractor track matrix, and the SVD of the signal is defined as SVD of the reconstruction attractor track matrix.

The SVD theory of matrix is as follows:

For a real matrix $H \in R^{m \times n}$ whether matrix row and column are related or not, there must be a pair of orthogonal matrix U and V (U and V are given by (2)and(3)) making:

$$U = \left(u_1, u_2, \cdots, u_m\right) \in R^{m \times m} \quad (2)$$

$$V = \left(v_1, v_2, \cdots, v_n\right) \in R^{n \times n} \quad (3)$$

$$H = USV^T \quad (4)$$

S is given by (5).

$$S = \begin{cases} (diag(\sigma_1, \sigma_2, \cdots, \sigma_q), 0) & m \le n \\ (diag(\sigma_1, \sigma_2, \cdots, \sigma_q), 0)^T & m > n \end{cases} \quad (5)$$

Where 0 represents a zero matrix, and q can be determined by q=min(m,n). The other informations in (5) are: $\sigma_1 \ge \sigma_2 \ge \cdots \ge \sigma_q$, $S \in R^{m \times n}$. The $\sigma_i (i=1,2,\cdots q)$ is called singular values of the matrix H.

The Hankel matrix constructed by the signal containing noise is full-rank, and the q-k singular values are obviously less than the first k, namely singular value suddenly changes in point k, which represent the main component of the signal. Therefore, the essence of the singular value method to denoise is to select the main components of the signal and reconstruct, so how to select the value of the k is the key of SVD. In this paper, the value of the k is determined by using the singular value difference spectrum theory proposed by literature [3].

Set a sequence S obtained by SVD of the signal as:$S=(\sigma_1, \sigma_2 \ge \cdots \ge \sigma_q)$, $\sigma_1 \ge \sigma_2 \ge \cdots \ge \sigma_q$. And a formula can be constructed as follows:

$$b_i = \sigma_i - \sigma_{i+1}, \quad i = 1, 2, \cdots, q-1 \quad (6)$$

The sequence B=(b1,b2,···bq-1) constituted by all of bi is called singular value spectrum, which describes the variation of adjacent singular values. When the difference between two adjacent singular values is large, there must be a peak b_k in the difference spectrum according to the definition of spectral difference, which means singular value sequence at position k occurred the greatest mutation. The maximum point mutation obviously represents a boundary between the ideal signal and noise, and the differential spectra curve will show a significantly larger value. The values prior to the largest one represent useful component of the signal and the others correspond to the noise signal.

4 Fault Diagnosis Method Based on DT-CWT and SVD

In order to successfully extract the weak fault signal of gear box from vibration signal with strong noise, firstly, the original vibration signal is decomposed into multilayer signals through DT-WCT, and different frequency components of the signal will be obtained. Secondly, the Hankel matrix is constructed by the component which contains obvious fault information, then the singular value spectrum can be obtained after SVD and the singular number will be determined by the singular value difference spectrum. After that, the signal after denoising can be reconstructed by SVD. Finally, the fault characteristic frequency can be identified accurately by the Hilbert envelope spectrum[8]. The flow chart of the method is shown in Fig. 2.

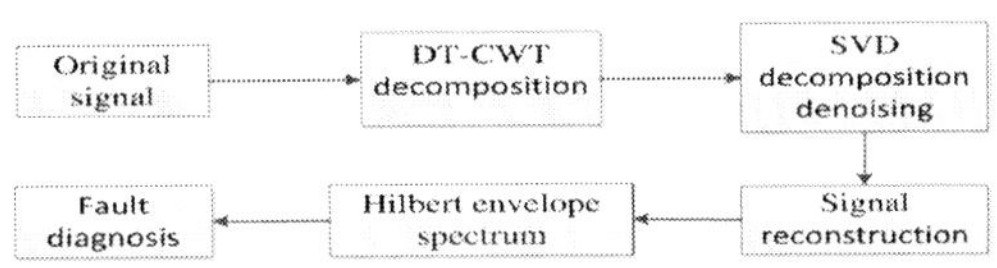

Fig. 2. Feature extraction method based on DT-CWT and SVD

5 Experimental Data Analysis

The experimental system is composed of a planetary gear box test bed, a piezoelectric acceleration sensor, a data acquisition instrument and a computer. The gear of simulation pitting fault is installed in the gearbox, and the experimental data is introduced into the computer through the signal acquisition instrument, and then the data is analyzed. The fault frequency of the gear is calculated as 13.625Hz.

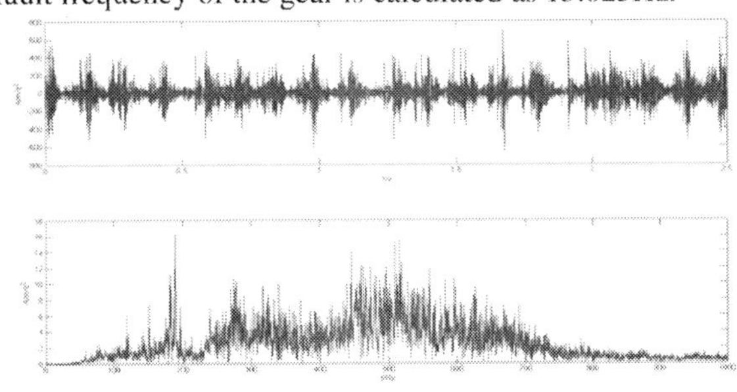

Fig. 3. Waveform and spectrum of gear fault

Fig.4 is a time-domain waveform and frequency spectrum of the vibration signal of the gear with pitting fault, and it can be seen strong signal interference noise from the figure waveform and spectrum. The original signal can be decomposed into four layer signals through DT-CWT, then the layered signals can be obtained by single branch reconstruction, and they are respectively a4, d4, d3, d2, d1, as shown in Fig.5.

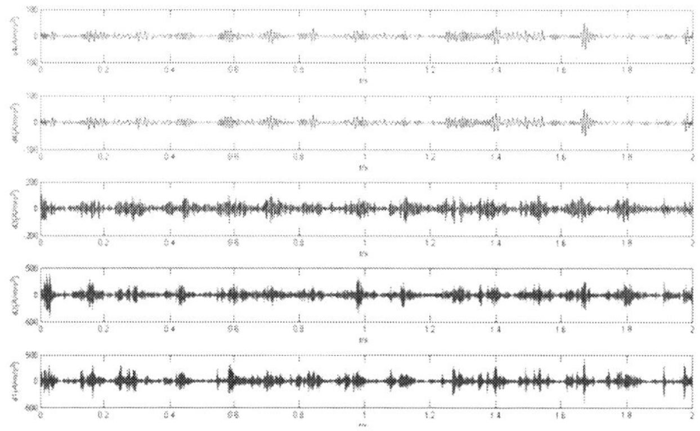

Fig. 4. Waveform of DT-CWT decomposition

Because the d3 has obviously periodic impact signal components, the d3 is taken as the research object to construct Hankel matrix, which will be decomposed by SVD, then the singular value difference spectrum is calculated. Since the mutation of the singular value occurs in the first few points, in order to clearly observe the mutation of the singular value, the first 100 singular values and the difference spectrum display in the same coordinate system, as shown in Fig.5.

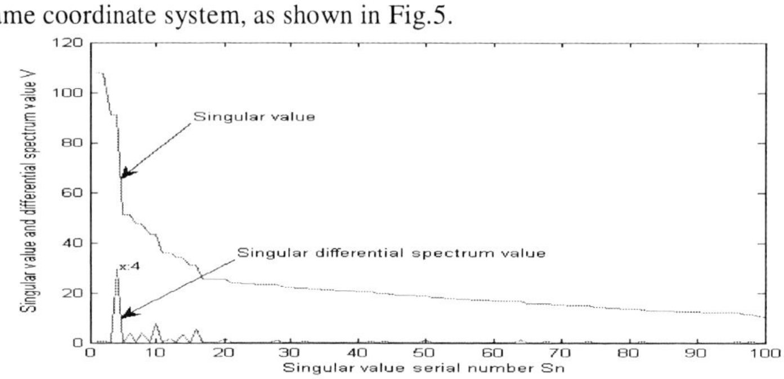

Fig. 5. The former100 points of singular value and singular value difference spectrum

There is a peak value in the fourth singular value, so the front four singular values are preserved to reconstruct the signal, and then the envelope spectrum of the Hilbert envelope is obtained, as shown in Fig.6. It can be seen clearly from Fig.6 the 13.5Hz frequency and its frequency, and this frequency is very close to gear fault characteristic frequency, so it can be judged gear fault.

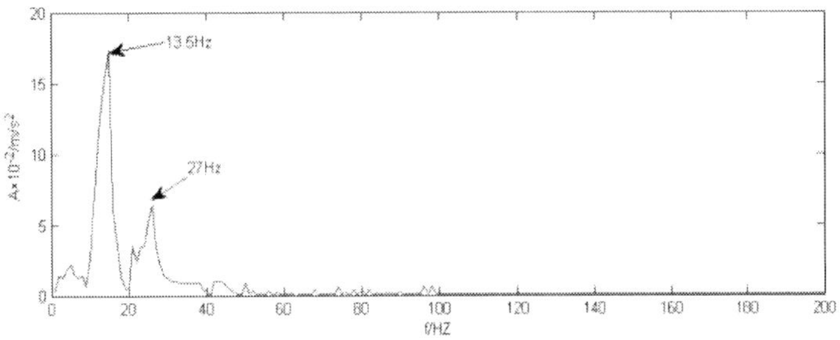

Fig. 6. Hilbert envelope spectrum after SVD reconstruction

6 Conclusion

This paper studied the application of DT-CWT and SVD in fault diagnosis of wind turbine gearbox. The component of different frequency bands can be obtained from gear vibration signal through DT-CWT, which has the advantages of approximate shift invariance, anti-aliasing and effective noise reduction, and each sub signal contains the information of the original signal in different frequency bands. The useful components can be further determined by SVD to realize the separation of the serviceable components and the noise in the signal. Experimental and simulation results show that the combination of DT-CWT and SVD can effectively determine the characteristic frequency of gear fault.

References

1. ZhangChao, Cheng Jianjun, Guoxun. A gear fault diagnosis method based on EMD energy entrgy and SVM. J. Journal of Vibration and Shock,2010,29(10):216-220+261.
2. Zhao Xuezhi, Ye Bangyan, Chen Tongjian. Extraction method of faint fault feature based on wavelet-SVD difference spectrum.J. Journal of Mechanical Engineering,2012,48(7):37-47
3. Zhao Xuezhi, Ye Bangyan, Chen Tongjian. Difference spectrum theory of singular value and its application to the fault diagnosis of headstock of lathe. J. Journal of Mechanical Engineering,2010,46(1):100-108.
4. Li Qiang. Feature extraction method of gearbox based on spectral kurtosis and SVD. J. Vibration and Shock,2014,V.33:116-119.
5. Seleniek I W. Hilbert transform pairs of wavelet bases. J. IEEE Signal Processing Letters,2001,8(6):170-173.
6. Selenick I W, Baraniuk R G, Kingsbury N G. The dual-tree complex wavelet transform. J. IEEE Digital Signal Processing Magazine,2005,22(6):123-151.
7. Xu Yonggang, Meng Zhichao, Lu ming. Gear fault diagnosis based on dual-tree complex wavelet transform and singular value difference spectrum. J. Journal of Vibration and Shock,2014,V.33;No.201301:11-16+23
8. Shao Keyong, Cai Miaomiao, Lixin Rolling bearing fault diagnosis based on wavelet analysis and difference spectrum theory of singular value. J. Manufacturing Automation, 2013,v.3508:79-82

Ball Screw System Performance Evaluation based on Denoised AutoEncoder

Liang Guo, Hongli Gao, Haifeng Huang
School of Mechanical Engineering
Southwest Jiaotong University
ChengDu, SiChuan, China
guoliang2248@my.swjtu.edu.cn

Weixiong Liu
Air-breathing Hypersonic Technology Research Center
China Aerodynamics Research and Development Center
MianYang, SiChuan, China

Abstract—**Ball screw is widely used in many precision machinery. And its performance is valuable to assure safe operation. So it is important to study on condition-based maintenance technique of ball screw. In this paper, we introduce a new ball screw performance evaluation based on deep learning model. First, the ball screw system acceleration performance degradation experiment is conducted to acquisition the experiment data. Then, the time, frequency and time-frequency domain features are extracted to represent the characteristic of vibration signal. At last, the denosied autoencoder deep neural network is applied to learn the deeper features and recognition the ball screw performance. Some contrast experiments are performed, the results demonstrates the proposed method is effective in ball screw system performance evaluation.**

Keywords- Ball screw system, performance evaluation, denoised autoencoder, Deep neural network Introduction

I. INTRODUCTION

The ball screw system is one of the most critical components in advanced manufacturing and is expected to perform with high accuracy. However, any potential failures or degradation of mechanical parts in the system would affect its efficiency and position precision[1]. It is obvious that the condition based maintenance of ball screw plays a crucial role to keep high precision operation. Meanwhile, the vibration signal is an important information carrier to machinery condition. So the performance evaluation based on vibration signal is an important method.

Recently, there is few research result about the ball screw system performance evaluation. But the machinery equipment performance evaluation is an always a hot research topic during the past decades. Usually, there are two kinds of approaches: physical based method and data driven based method[2].

The physical based method needs to build the research object's rigorous mathematics model to describe the physical property and failure mode. The error value of measurement and modeling is the feature indicator to the performance. Sukhwan Choi e.t.[3] build the gear failure model based on fracture mechanics. GPAT is applied to build the gear failure model to predict the residual service life of the crack gear. Ray A e.t.[4] proposed the gear dynamic model of endurance crack changes over time. And the extended Kalman filter is applied to calculate the faulted condition and remaining useful life on line. Orsagh R F e.t.[5] use Yu-Harris bearing life equation to predict the beginning time of crack. At the same time, The Kotzalas-Harris series model is applied to estimate the remaining useful life. Physical based approaches are simple calculation. But in practical application, it is difficult to build the detail and accurate physical of research object.

Data-driven performance evaluation method collects the condition data with little priori knowledge to evaluate the equipment performance. The function relationship between history data and performance condition is built to predict the monitoring object. There are three mainly steps to estimate those models: data acquisition, feature extraction and selection and pattern recognition. The vibration signal is the most commonly sued sensor data for its strong representation characteristic. Recently, the time, frequency and time-frequency domain feature extraction methods are usually combined to construct the feature vector. Too many of features will be redundancy. So the feature vector is generally reduced dimension before inputting into the recognition model. Principal Component Analysis (PCA) is the common way to feature dimension reduce but it is linear processing algorithm. Pattern recognition is the important step for performance evaluation. The statistical learning based and neural network based methods are usually used. T.Benkedjouh e.t.[6] extract the time, frequency and time-frequency domain features of bearing vibration signal. Then those features are reduced dimension by ISOMAP nonlinearly. At last, the selection features are inputted the Support Vector Machine (SVM) to calculate the remaining useful life. Wahyu Caesarendra e.t.[7] propose the performance degradation assessment model based on relevance vector machine and logistic regression. The logistic regression computes the bearing all life cycle state value set. Then the state value set and the bearing sample set data are used to train the relevance vector machine for bearing performance evaluation. Zhigang Tian[2] collect the measurement values at the current moment and a moment before. The Weibull failure function is applied to fit the measurement value as a history data. Finally, those data is used to form the artificial neural network training sample set. And result demonstrate that the proposed method is robust to the noise signal.

Deep neural network is a breakthrough in machine learning research area[8]. The technique has achieved the state-of-the-art results in the fields of image recognition, speech recognition, natural language processors and so on. And it is constructed with input layer, multi-hidden layers and output layer that attempt to learn high-level abstraction of input data. The aim of those models is to learn the intrinsic representation of unlabeled large scale original data. Usually, the training processing is composed of unsupervised pre-training and fine-tuning supervised training. Recently, those algorithms are also used to process the mechanical vibration signal and applied to bearing fault diagnosis.

In this paper, we proposed a new ball screw system performance evaluation method based on deep neural network. Ball screw has a long design life, so it is difficult to acquire the whole ball screw life data. This paper built a ball screw acceleration performance degradation testing bench to get enough raw data. The raw time domain signal is chaotic and redundancy. So the time, frequency and time-frequency domain features are extracted to represent the characteristic of vibration signal as the input feature set. Then, the compact and identifiability high-level features are learned through multi-layers unsupervised learning. At last, the ball screw performance condition is calculated with those learned features and Softmax regression calssifier.

The rest of the paper is organized as follows: section two will descript the experiment platform and experiment setup. The denoised autoencoder and the proposed method are shown in section three. In section four, we will give the experiment results. At last, section five describes the result discussion and future work.

II. BALL SCREW EXPERIEMNT

The experiment used an experiment platform which is designed and assembled in-house (As shown in Fig. 1). The mechanical system is mainly consisted of three parts: a transportation part, a load part and a measurement part.

The transportation part includes the servo Motor with a ball screw system. The driven motor is Beckhoff AM3063-OH30-0000 with the maximum power 2KW and speed 1,500 rpm. And the rigid shaft couplings is used to connect the motor shaft and screw shaft. The bearing support the screw rotating shaft. And the drive end bearing is fixed while the other is active. A motion platform is fixed with the feed screw nut that is supported by guide rail and slide block.

The load part can simulate the load movement of ball screw system which includes the magnetic powder brake and the force transmission system. The magnetic powder brake delivers the axial load. Then the force is passed to the motion platform through the pinion and rack. And the load force can be adjusted by the drive current value of magnetic powder brake.

The ball screw's operation performance is determined by instantaneous measures of the axis force applied on the ball screw, the rotation speed of the shaft. Each of these two measures is acquired with the torque sensor and motor encoder. The characteristic of ball screw performance degradation is based on the vibration signal of ball screw system. So three vibration sensors are positioned at three different location (two bearing end and screw nut). And to get richer vibration information of screw nut, the vibration sensor is three-axis. All of those sensors are acquired at a frequency equal to 5,000Hz.

The detail experiment condition is shown in table 1. The experiment cover two iteration steps: accelerated wear and operation condition simulation. During the accelerated wear stage, the ball screw system is run with 100N load and 1,000 rpm speed. Every 30 minute later, the system is switched into the operation condition simulation stage. At this stage, the different working condition is simulated and the relevant sensor data is acquired.

Table 1. Experiment condition

Accelerated wear		Operation condition	
Load(N)	100	Load(N)	[null,50,100]
Speed(rpm)	1000	Speed(rpm)	[100,300,800]
Interval(min)	30	Distance(mm)	200

The experiment lasted 37 days (From March 24th,2015 to April 29th, 2015). And finally, the maximum amplitude of vibration signal reach up to 3.8g, then the experiment is stopped. Fig 3 is the vibration signal of ball screw system at the initial operating stage. Fig4 is corresponding data at the late operating stage. And (a) is the fixed end bearing vibration, (b) is the active end bearing vibration, (c) is the z directions of screw nut, (d) is the y directions and (e) is the x directions. As shown in the Fig2 and Fig3, the amplitude and shape of vibration signal is significant changes.

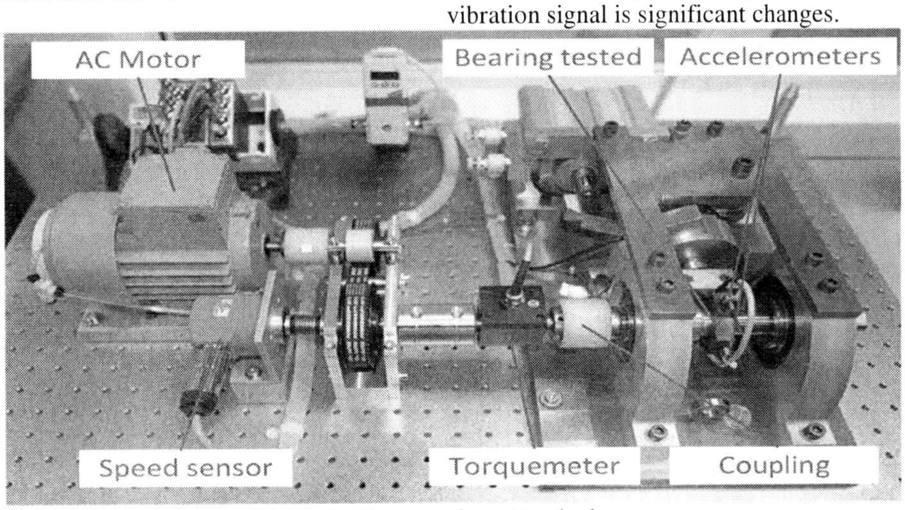

Figure 1. The experiment testbed

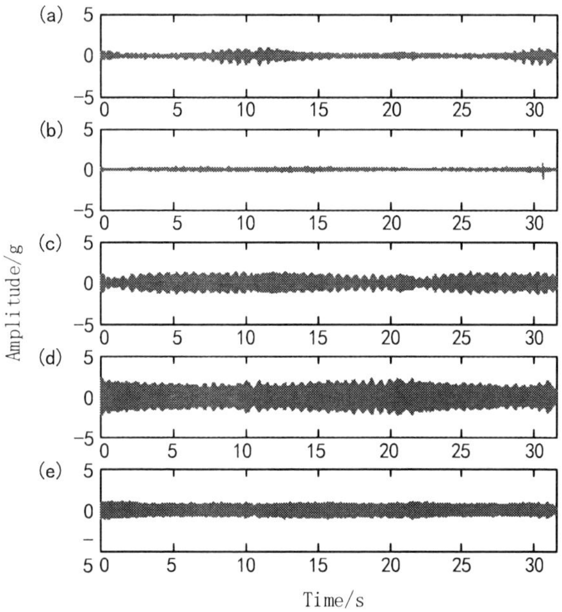

Load: null Speed: 100rpm Time: March 24th,2015

Figure 2 The vibration signal at the initial operating stage

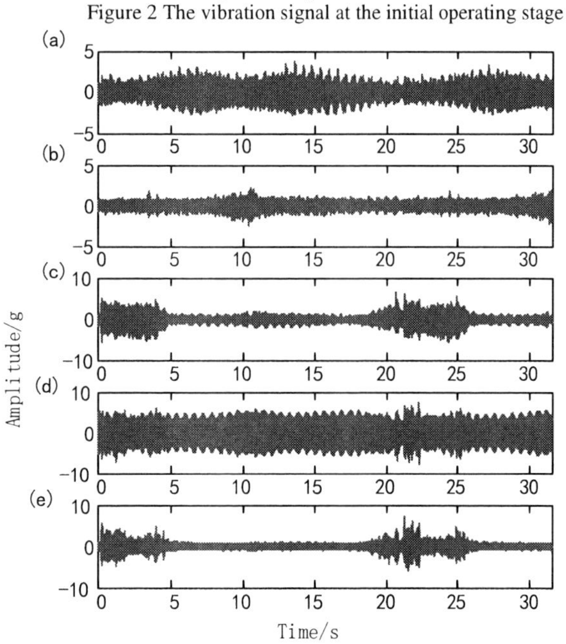

Load: null Speed: 100rpm Time: April 29th, 2015

Figure 3 The vibration signal at the late operating stage

III. THE PROPOSED METHOD

The core issue of ball screw system performance evaluation is accuracy condition recognition. The deep neural network based on autoencoder structure is applied. And the denoised autoencoder model is applied to enhance the system robust.

Autoencoder

The deep neural network is a typical representation learning algorithm. These algorithms try to construct the multi-hidden layers model with the vast unlabeled data and a small number of labeled data. And every layer data transform is constructed of some simple but nonlinear models. The combinations of those models can fit any complex nonlinear transform function.

As shown in Fig4., the trainging process of deep neural network is composed of two stages: layered pre-training and fine-turning. At the stage of layered pre-training, the hidden layer output set of current submodel is the input set of the next submodel. Meanwhile, the input set of the first submodel is the raw signal. The parameters of layered pre-training are merged into the multi-layers completion model as the initial value. Then, the whole parameters are optimized with the labeled training samples set. The processing is called fine-turning.

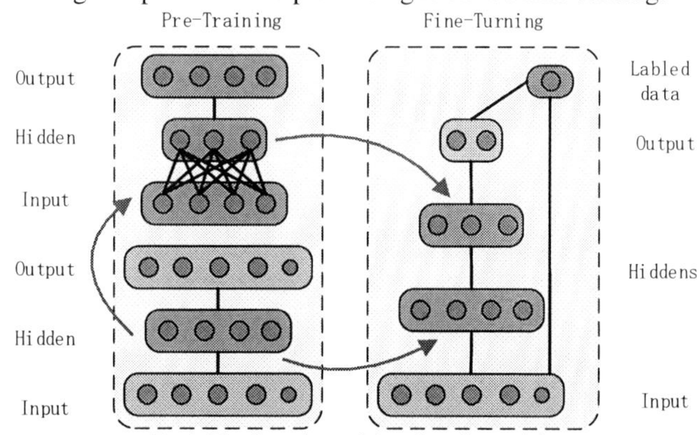

Fig 4. The flowchart of deep learning training

Deep Autoencoder is a typical deep neural network model. And the basic unit, autoencoder, is shown in Fig.5 which is made up of encoder and decoder. From the visual point, tt is similar to BP neural network architecture. Actually, the cost function is different. The output of autoencoder is compared with its input value. So cost function of autoencoder is,

$$J(W,b;x_u)=\frac{1}{2}\left\|x_u-\hat{x}_u\right\|^2 \tag{1}$$

Where W is the weighting parameter; b is the deviation threshold; x_u is the input value; and $\hat{x}_u$ is the output value.

In the encoder stage, the input value is the hidden layer's output value of the previous submodel.

$$a=f(z^1) \tag{2}$$

$$z^1=W^1*x_u+b^1 \tag{3}$$

Where x_u is the input value; a is the code data vector; W^1 is the weighted parameters between input layer and hidden layer; b^1 is the deviation threshold; $f(\bullet)$ is the activation function.

In the decoder stage, the neural reconstructs the output value which is the closest to the input value.

$$\hat{x}_u=f(z^2) \tag{4}$$

$$z^2=W^2*a+b^2 \tag{5}$$

Where $\hat{x}_u$ is the input value; a is the code data vector; W^2 is the weighted parameters between hidden layer and output layer; b^2 is the deviation threshold.

From Eq(1), the training processing of autoencoder is a unsupervised learning without labeled samples. Meanwhile, the encoder part and the deconder part are mutually inverse.

Usually, a sparse penalty term is used to the cost function. This term can well balance the reconstruction error and the number of activation function. So Eq(1) is changed to,

$$J(W,b;x_u)=\frac{1}{2}\left\|x_u-\hat{x}_u\right\|^2+\lambda\sum KL(\rho\,\|\,\hat{\rho}_j) \tag{6}$$

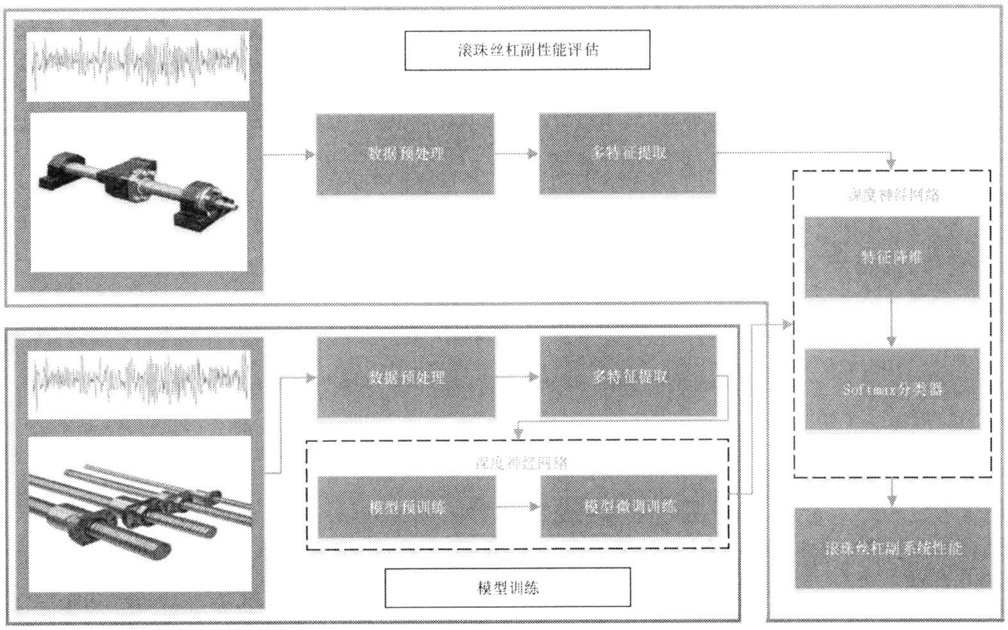

Fig 5. The autoencoder struction

Where λ is the penalty parameter; $KL(\rho\|\hat{\rho}_j)$ is the Kullback-Liebler divergence between ρ and $\hat{\rho}_j$; ρ is the fix sparse coefficient; $\hat{\rho}_j$ is the activation of the j-th neural.

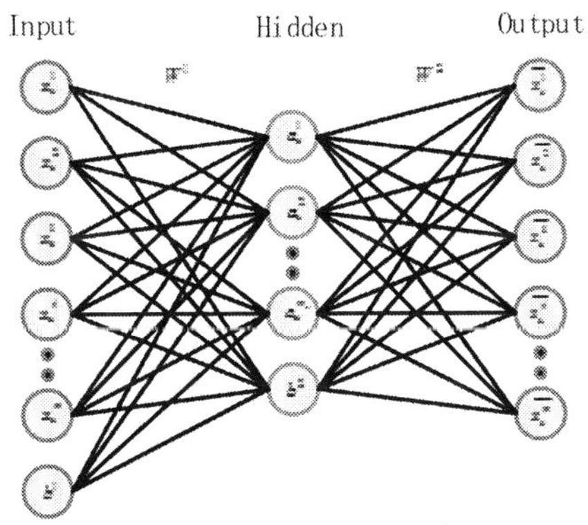

Fig 6. The autoencoder struction

Denosied autoencoder

Denoised autoencoder is an extension of autoencoder in order to improve the model robustness. Compare to the classical autoencoder, the input value of denosied autoencoder is random replaced the noise value as shown in Fig.6. And the noise value meet some certain probability distribution model, $\hat{x} \sim q_D(\hat{x}\,|\,x)$. So the output value of the output layer is based on the noised input value. But when the cost function is calculated through Eq.(6), the input vale is x_u rather than the noised value $\hat{x}$.

At the meantime, the processing of nosied input value is made in the stage of unsupervised pre-training rather than the fine-tuning and model calculation.

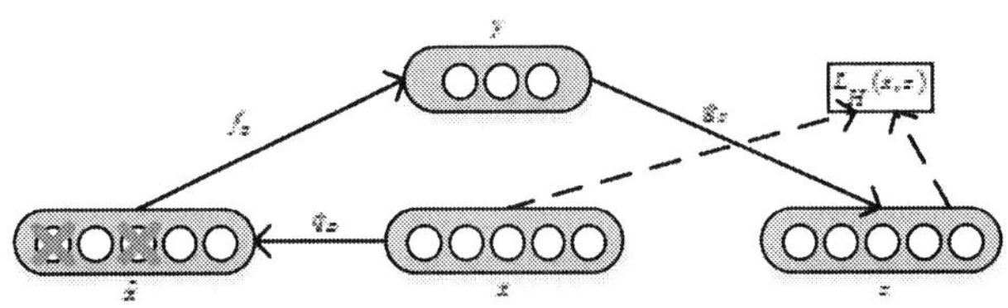

Fig 7. The autoencoder struction

The proposed method

Based on the aforementioned, this paper proposed a new ball screw performance evaluation method based on multi feaures fusion and deep learning. Fig.5 is the flowchart of the proposed method. And for estimate the performance condition of ball screw, a multi-class classifier, softmax regression, is added into the deep denoised autoencoder model.

From the figure, the method is consist of two stage: model training and performance evaluation.

At the stage of model training, there are five steps:

(1) Data acquisition: multi performance condition of ball screw is needed.;
(2) Pre-processing: the collected data is preprocessing with the dispose of trend term;
(3) Feature extraction: the time, frequency and time-frequency features are extracted;
(4) Pre-training: the layered submodel is pre-trained with unlabeled data;
(5) Fine-tuning: the whole deep denosied autoencoder mdoel parameters are optimized with labeled data.

At the stage of performance evaluation, there are four steps:

(1) Data acquisition: the vibration signal of current operation performance of ball screw is collected;
(2) Pre-processing: the processing is same as the stage of model training;
(3) Feature extraction: the time, frequency and time-frequency features are extracted;

(4) Performance evaluation: the ball screw system's performance evaluation is calculated through the trained deep denoised autoencoder.

The acquired vibration signal comes from three different locations (two axial bearings and screw nut).

IV. EXPERIMENT RESULTS

The operation condition of ball screw system is divided into three different performance categories: normal, degeneration and failure. Depends on the characteristic of ball screw system vibration signal. There are 80 group data sets at the normal performance condition. 48 group data sets locate in the degeneration section. And the data sets of failure phase are 32. And during every performance conditions, there are nine different operation conditions depends on the varying load and speed.

Feature Extraction

The raw signal is vast but less information, it is difficult to establish the relationship between vibration signal and ball screw system performance condition. Feature extraction is an effective means to that problem. In this paper, the time, frequency and time-frequency domain features are extracted.

Time domain features are almost based on the statistical analysis. And different features have different indicator for various fault types. Such as, Peak-to-Peak value is sensitive to impact signal types. The RMS (Root Mean Square) is a distinct indicator to the slowly changing fault type, like wear. In addition, the discrimination number is less sensitive to different equipment than dimensional features. So there are nine different features extracted which are Peak-to-Peak, Mean, RMS, Skewness, Kurtosis, Crest Factor, Crest Factor, Shape Factor, Impulse Factor and Margin Factor.

Frequency spectral kurtosis is a new frequency feature value. The Fourier transform of discrete-time signal $x(n)$ is $X(m)$. Then the frequency spectral kurtosis is defined as,

$$K_x(m) = \frac{E\{|X(m)|^4\}}{E\{|X(m)|^2\}^2} - 2 \qquad (7)$$

Where m is the frequency value, $K_x(m)$ is the frequency spectral kutosis.

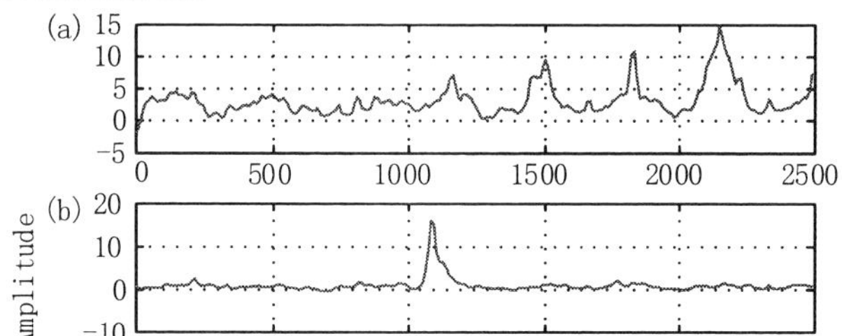

Fig 8 Frequency spectral kurtosis of different performance codnition

Fig. 8 plots the frequency spectral kurtosis from the different performance condition vibration signal. From the figure, there is a big difference between those waveform that is from three different operation performance condition. Then the correlation coefficient is calculated to represent the frequency feature.

Time-frequency domain feature is an important and effective indicator because of its ability of time-frequency characteristic representation. Wavelet transform (WT) and empirical model decomposition (EMD) are two kinds of typical time-frequency analysis technique. This paper calculates the energy value of transform coefficients as the time-frequency features. Fig.9~11 are the wavelet packet transform plane, those figures show that the spectral distribution is varying with the time and performance condition. And the average energy value of transform coefficients within the time interval 0.5s and the frequency interval 500Hz are computed as the time-frequency features.

So there are 70 features in each group of vibration signal. And the dimension of input features set is 350 (70*5 group raw vibration signal).

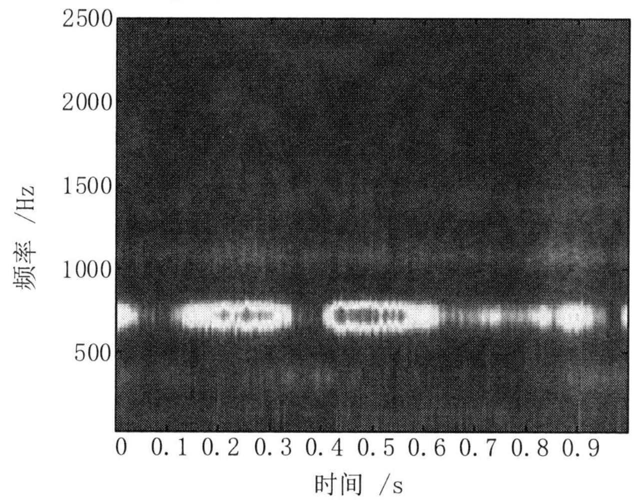

Load: null Speed: 100rpm Time: March 24th,2015

Figure 9 The vibration signal at the initial operating stage

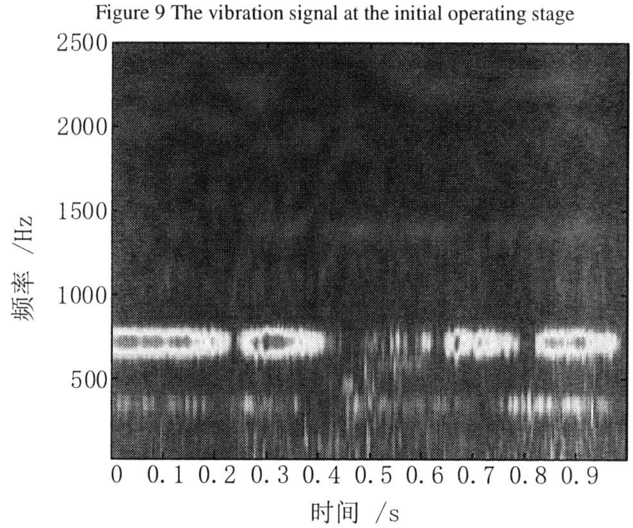

Load: null Speed: 100rpm Time: March 24th,2015

Figure 10 The vibration signal at the initial operating stage

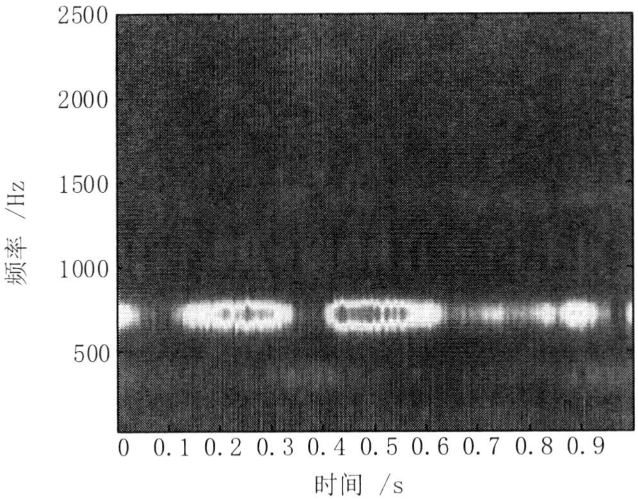

Load: null Speed: 100rpm Time: March 24th,2015

Figure 11 The vibration signal at the initial operating stage

Evaluaiton result

Depending on the dimension of input feature set, the number of neural in every layer is 350-10-50-30-3. It means that the deep denoised autoencoder has one input layer, three hidden layer and one output layer. The training samples set is 19200, and the testing set is 12800. At the training stage, the regularization coefficient is 0.01, the rate of random noise of input value is 0.5, the number of iteration is 500. And the parameters of recognition stage is same as it.

Tab.2 is the ball screw performance evaluation results of the proposed method and the comparing algorithms. From the table, the best recognition rate is 93.9% which comes from the proposed method. And the deep learning from comparing algorithms is the classical deep autoencoder. The detail parameters are same as the proposed method. The method based on PCA reduces the dimension of input set from 350 to 30, then the performance condition is recognized by a three layer BP neural network. Meanwhile, the 350 features are inputted into the SVM model as another comparing model.

Table 3. Ball screw system performance recognition results

	Proposed algorithm				Comparing algorithms		
	Normal	Degeneration	Failure	Accuaracy	Deep learning	PCA	SVM
Normal	0.950	0.036	0.014	0.950	0.944	0.928	0.936
Degeneration	0.023	0.927	0.049	0.927	0.911	0.898	0.911
Failure	0.012	0.059	0.929	0.929	0.910	0.889	0.894
Accuaracy	0.950	0.927	0.929	**0.939**	0.927	0.911	0.920

V. CALCULATION

A new ball screw system performance evaluation method based on deep denoised autoencoder is proposed in this paper. A special experiment testbed is designed and manufactured to obtain the massive samples involving different performance condition under various operation condition. Through the experiment results from the proposed method and some comparing algorithms. The proposed method has the best recognition rate of ball screw system performance. In the meantime, the proposed deep denosied autoencoder is superior to the deep autoencoder.

In the proposed method, the raw input signal is the multi-features set. And the proposed deep learning algorithm relearns the features set. To take full advantage of feature learning characteristic of deep learning, the author would investigate another method that the raw vibration signal is inputted into the deep learning model directly.

ACKNOWLEDGMENT

This work was supported by the National science funding(NSF51275426) and Sichuan foreign cooperation fund (514230251).

REFERENCES

[1] Lee, Won Gi, Jin Woo Lee, Min Sung Hong, Sung-Ho Nam, YongHo Jeon, and Moon G. Lee. "Failure Diagnosis System for a Ball-Screw by Using Vibration Signals." Shock and Vibration 2015 (2015).

[2] Tian, Zhigang. "An artificial neural network method for remaining useful life prediction of equipment subject to condition monitoring." Journal of Intelligent Manufacturing 23, no. 2 (2012), pp. 227-237.

[3] Choi, Sukhwan, and C. James Li. "Model based spur gear failure prediction using gear diagnosis." In ASME 2005 International Mechanical Engineering Congress and Exposition, pp. 819-823. American Society of Mechanical Engineers, 2005.

[4] Ray, Asok. "Stochastic modeling of fatigue crack damage for risk analysis and remaining life prediction." Journal of dynamic systems, measurement, and control 121, no. 3 (1999): 386-393.

[5] Orsagh, Rolf F., Jeremy Sheldon, and Christopher J. Klenke. "Prognostics/diagnostics for gas turbine engine bearings." In ASME Turbo Expo 2003, pp. 159-167.

[6] Si, Xiao-Sheng, Wenbin Wang, Chang-Hua Hu, Dong-Hua Zhou, and Michael G. Pecht. "Remaining useful life estimation based on a nonlinear diffusion degradation process." Reliability, IEEE Transactions on 61, no. 1 (2012): 50-67.

[7] Caesarendra, Wahyu, Achmad Widodo, and Bo-Suk Yang. "Application of relevance vector machine and logistic regression for machine degradation assessment." Mechanical Systems and Signal Processing 24, no. 4 (2010): 1161-1171.

[8] Hinton, Geoffrey E., and Ruslan R. Salakhutdinov. "Reducing the dimensionality of data with neural networks." Science 313, no. 5786 (2006): 504-507.

Dynamical Linear Models for Condition Monitoring with Multivariate Sensor Data

Erik Vanem[1 2], Ingrid Kristine Glad[2], and Geir Olve Storvik[2]

[1] DNV-GL Strategic Research & Innovation, Høvik, Norway,
Erik.Vanem@dnvgl.com,
[2] University of Oslo, Department of Mathematics, Oslo, Norway
{erikvan, glad, geirs}@math.uio.no

Abstract. This paper presents an application of dynamical linear models for anomaly detection and condition monitoring of ship machinery systems based on multivariate sensor data. Various model alternatives are specified and fitted to a set of training data before they are applied to a test set. Sequential monitoring based on statistical tests are applied to detect model breakdown as an indication of deviation from normal conditions. The framework is very flexible and allows for a range of different candidate models to be specified. In this paper, some of the estimated models perform rather poorly, but the best ones did quite well in flagging anomalies in the data streams. Hence, it is demonstrated that dynamical linear models may be utilized for anomaly detection and condition monitoring with multivariate sensor data. However, identification of the best model structure is challenging and requires representative training data and careful consideration in the model specification.

Keywords: Condition monitoring, anomaly detection, dynamical linear models, ship machinery systems, Bayesian time-series, sensor data

1 Introduction

The safety and reliability of ships and ship systems is a main concern within the maritime industries and major shipping accidents can have catastrophic consequences in terms of lives lost, environmental pollution and economic losses. There are many regulatory instruments in place to ensure that the risk level of maritime transport is within an acceptable level.

Notwithstanding a strong and continued focus on safety and reliability of shipping, maritime accidents continue to occur and recent studies have indicated that the frequency of occurrence of major maritime accidents may even have increased in the last decade [4]. Even though many of the accidents are ascribed to human errors [13], the reliability of the ship and its various technical subsystems greatly influence the overall safety and risk of shipping. Hence, it is of paramount importance for the maritime industries to implement good maintenance strategies and to continuously monitor the condition of the ship and different ship systems and components.

Ship machinery systems provide safety critical functions to the ship related to propulsion and manoeuvrability as well as electrical power production for other ship systems. Thus, any failure in the ship machinery system may potentially be very critical to the safety of the ship. Loss of propulsion and manoeuvrability may lead to drift grounding and collision accidents with potential loss of stability and sinking of the vessel as ultimate consequences [17]. Loss of electric power production may lead to blackout and loss of other safety-critical functions from other ship systems [10] and failures in ship machinery system may cause engine room fires which is one of the most frequent category of fire incidents for merchant ships [18]. The condition and performance of the machinery system is also important for the overall fuel efficiency of the vessel and may have a large impact on the economics and environmental footprint of ship operations. Hence, the condition of the ship machinery system is critical to the safety and economics of the vessel and needs to be carefully monitored.

The shipping industry is currently experiencing a shift towards digital solutions for monitoring of equipment and systems, and more and more sensors are being installed on components and systems onboard ships. These provide real-time information about the performance and condition of the monitored systems and may be utilized for fault detection, diagnosis and prognosis. A critical first step in such a setup is to automatically detect deviations from normal operational conditions, which may indicate failure of the system or even an impending failure that has not yet occured. It is assumed that statistical methods can be exploited in order to detect anomalies in the, often high-dimensional, sensor data streams as early as possible. This paper presents an application of one such statistical approach for anomaly detection, based on Dynamical Linear Models (DLMs) for describing and predicting the sensor time series. The subsequent sections of this paper give a brief outline of the dynamical linear model framework and suggest how it can be utilized in an anomaly detection scheme. Then, the results of applying the methodology to a particular set of sensor data collected from an ocean going ship is presented.

2 The Dynamical Linear Model

A thorough introduction to the theory and application of dynamical linear models is given in [20], and only a brief introduction will be given below. Letting $\boldsymbol{Y}_t$ denote the r-dimensional vector of observed signals at time t, and $\boldsymbol{\theta_t}$ be the corresponding m-dimensional state-vector of an unobserved, latent process, the DLM model equations are on the following form:

$$\begin{aligned} \boldsymbol{Y}_t &= \boldsymbol{F}_t^{'}\boldsymbol{\theta}_t + \boldsymbol{\nu}_t, \qquad & \boldsymbol{\nu}_t &\sim N(\boldsymbol{0}, \boldsymbol{V_t}) \\ \boldsymbol{\theta}_t &= \boldsymbol{G}_t\boldsymbol{\theta}_{t-1} + \boldsymbol{\omega}_t, \qquad & \boldsymbol{\omega}_t &\sim N(\boldsymbol{0}, \boldsymbol{W_t}). \end{aligned} \tag{1}$$

The first line of eq. (1) are often referred to as the observation equation and the second line is called the evolution, state or system equation. Such a model is fully specified by the quadruple $\{\boldsymbol{F}_t, \boldsymbol{G}_t, \boldsymbol{V}_t, \boldsymbol{W}_t\}$, as follows

- $\boldsymbol{F}_t$ is an $(m \times r)$ matrix, sometimes referred to as the regression matrix
- $\boldsymbol{G}_t$ is an $(m \times m)$ matrix, sometimes referred to as the state evolution matrix
- $\boldsymbol{V}_t$ is an $(r \times r)$ variance matrix for the observational variances
- $\boldsymbol{W}_t$ is an $(m \times m)$ variance matrix for the evolution variances

The error sequences $\boldsymbol{\nu_t}$ and $\boldsymbol{\omega_t}$ are assumed to be independent and mutually independent, and will be further assumed to be multivariate normally distributed with zero mean. The model further assumes a conditional independence structure so that, conditioned on the state-vector $\boldsymbol{\theta_t}$, the observations at time t, $\boldsymbol{Y_t}$ are independent of the history of observations $D_{t-1} = \{\boldsymbol{y}_{t-1}, \boldsymbol{y}_{t-2}, \ldots, \boldsymbol{y}_1, D_0\} = \{\boldsymbol{y}_{t-1}, D_{t-2}\}$. Moreover, given the whole history until time t, all model information of the future is contained in $\boldsymbol{\theta}_t|D_t$.

In order to fully specify such a model, one also needs an initial prior for the state vector at time $t = 0$, i.e. assuming a multivariate normal distribution with a mean vector $\boldsymbol{m}_0$ and an initial variance matrix $\boldsymbol{C}_0$:

$$(\boldsymbol{\theta}_0|D_0) \sim N(\boldsymbol{m}_0, \boldsymbol{C}_0). \tag{2}$$

2.1 Updating and forecasting with a dynamical linear model

Having established such a dynamical linear model, one may obtain, at each time t, posterior distributions for the state vectors and forecast distributions for future observations within a Bayesian setting, i.e. by applying Bayes' theorem, as follows.

i) Posterior at time $t-1$:
For some mean vector $\boldsymbol{m}_{t-1}$ and variance matrix $\boldsymbol{C}_{t-1}$,

$$(\boldsymbol{\theta_{t-1}}|\boldsymbol{D_{t-1}}) \sim N\left(\boldsymbol{m}_{t-1}, \boldsymbol{C}_{t-1}\right). \tag{3}$$

ii) Prior at time t:

$$(\boldsymbol{\theta_t}|D_{t-1}) \sim N(\boldsymbol{a}_t, \boldsymbol{R}_t), \tag{4}$$

where $\boldsymbol{a}_t = \boldsymbol{G}_t\boldsymbol{m}_{t-1}$ and $\boldsymbol{R}_t = \boldsymbol{G}_t\boldsymbol{C}_{t-1}\boldsymbol{G}_t^{'} + \boldsymbol{W}_t$.

iii) One-step ahead forecast for time t:

$$(\boldsymbol{Y}_t|D_{t-1}) \sim N(\boldsymbol{f}_t, \boldsymbol{Q}_t), \tag{5}$$

where $\boldsymbol{f}_t = \boldsymbol{F}_t^{'}\boldsymbol{a}_t$ and $\boldsymbol{Q}_t = \boldsymbol{F}_t^{'}\boldsymbol{R}_t\boldsymbol{F}_t + \boldsymbol{V}_t$.

iv) Posterior at time t:

$$(\boldsymbol{\theta_t}|\boldsymbol{D_t}) \sim N\left(\boldsymbol{m}_t, \boldsymbol{C}_t\right) \tag{6}$$

with $\boldsymbol{m}_t = \boldsymbol{a}_t + \boldsymbol{A}_t\boldsymbol{e}_t$ and $\boldsymbol{C}_t = \boldsymbol{R}_t - \boldsymbol{A}_t\boldsymbol{Q}_t\boldsymbol{A}_t^{'}$, and where $\boldsymbol{A}_t = \boldsymbol{R}_t\boldsymbol{F}_t\boldsymbol{Q}_t^{-1}$ and $\boldsymbol{e}_t = \boldsymbol{Y}_t - \boldsymbol{f}_t$

The vector $\boldsymbol{e}_t$ is the one-step ahead forecast errors and the matrix $\boldsymbol{A}_t$ is a matrix of adaptive coefficients. Looking further ahead from time t, for $k \geq 0$, the k-step ahead forecast distributions for $\boldsymbol{\theta}_{t+k}$ and $\boldsymbol{Y}_{t+k}$ given current information D_t are given by, for the state and forecast distributions, respectively:

$$(\boldsymbol{\theta}_{t+k}|D_t) \sim N\left(\boldsymbol{a}_t(k), \boldsymbol{R}_t(k)\right) \tag{7}$$

$$(\boldsymbol{Y}_{t+k}|D_t) \sim N\left(\boldsymbol{f}_t(k), \boldsymbol{Q}_t(k)\right) \tag{8}$$

with the following moments defined recursively, from starting values $\boldsymbol{a}_t(0) = \boldsymbol{m}_t$ and $\boldsymbol{R}_t(0) = \boldsymbol{C}_t$:

$$\boldsymbol{f}_t(k) = \boldsymbol{F}_t^{'}\boldsymbol{a}_t(k) \quad \text{where} \quad \boldsymbol{a}_t(k) = \boldsymbol{G}_{t+k}\boldsymbol{a}_t(k-1)$$

and

$$\boldsymbol{Q}_t(k) = \boldsymbol{F}_t^{'}\boldsymbol{R}_t(k)\boldsymbol{F}_t + \boldsymbol{V}_{t+k} \quad \text{where} \quad \boldsymbol{R}_t(k) = \boldsymbol{G}_{t+k}\boldsymbol{R}_t(k-1)\boldsymbol{G}_{t+k}^{'} + \boldsymbol{W}_{t+k}.$$

There are many versions of the dynamical linear model, and in general the model parameters are allowed to vary in time. In these preliminary investigations, however, a constant DLM is assumed, i.e. a model where the model parameters are constant in time $\{\boldsymbol{F}_t, \boldsymbol{G}_t, \boldsymbol{V}_t, \boldsymbol{W}_t\} = \{\boldsymbol{F}, \boldsymbol{G}, \boldsymbol{V}, \boldsymbol{W}\}$. Moreover, the parameters are estimated from a training data set and consequently assumed as known.

2.2 Monitoring of dynamical linear models

One important feature of any statistical model used for condition monitoring is the ability to monitor the model and to detect, as early as possible, any departure from the model. With a DLM this can be done by Bayes' factors. For example, assuming a null model M_0 and an alternative model M_1 for the signal time series. Both models provide a predictive distribution for $\boldsymbol{Y}_t$ conditioned on the available information D_{t-1}, i.e. for $i = 0, 1$

$$p_i(\boldsymbol{Y}_t|D_{t-1}) = p(\boldsymbol{Y}_t|D_{t-1}, M_i). \tag{9}$$

The Bayes' factor for M_0 versus M_1 based on the observed value of $\boldsymbol{Y}_t$ can then be defined as

$$H_t = \frac{p_0(\boldsymbol{Y}_t|D_{t-1})}{p_1(\boldsymbol{Y}_t|D_{t-1})}. \tag{10}$$

In this study, the sequential probability ratio test (SPRT, [19]) is applied, which is based on the SPRT index and is similar to the Bayes' factor. This index is defined as the natural logarithm of the likelihood ratio for M_1 versus M_0 (the reciprocal of eq. 10). The SPRT is additive and is updated as subsequent observations arrive. For a sequence of errors, $e_1, \ldots, e_n$, the SPRT index can be calculated as

$$l_n = \ln L_n = \ln \prod_{t=1}^{n} \frac{p_1(e_t|D_{t-1})}{p_0(e_t|D_{t-1})}. \tag{11}$$

The sequential probability ratio test [19] builds evidence for or against the null model and as soon as enough evidence is obtained to make a decision (accept or reject the alternative model), the SPRT index is reset to zero. Two threshold values, A and B need to be specified. If the SPRT index reaches the upper threshold an alarm is flagged for possible model breakdown. If the lower threshold is reached, the null model is accepted as adequate. In either case, the SPRT index is reset to zero and the monitoring continues until any of the thresholds are reached again. The limits can be related to the acceptable probabilities of false alarms (α) and missed alarms (β) as follows

$$A = \ln\left(\frac{\beta}{1-\alpha}\right) \quad \text{and} \quad B = \ln\left(\frac{1-\beta}{\alpha}\right). \tag{12}$$

2.3 An R package for dynamical linear models

Some of the basic framework of the dynamical linear models with Gaussian noise terms has already been implemented in the *R* package *dlm* [12, 11]. This package contains functions for the specification of standard dynamical linear models as well as routines for estimating the parameters in such models by maximum likelihood. Having specified a model, the packages also allows the filtered, smoothed and forecast distributions to be calculated as well as the evaluation of the log-likelihood of a set of data for a specified state space model. Moreover, the raw or standardized residuals from the one-step ahead forecast errors can be provided. This *R* package has been utilized in this study.

3 Data

For the purpose of this exercise, a set of five sensor signals collected from an ocean going vessel over a period of a few months is used. The sampling interval is every 30 seconds. The signals measure the loads on a ship machinery system with four gensets (first four signals) as well as the vessel's speed over ground (fifth signal). Each of the five time series contain a total of 23 337 observations and the data is divided in a training dataset for fitting the dynamical linear models and a test set for applying the model to. 66% of the data are used for training, i.e. 15 503 observations and the remainder of the time series are reserved for testing the fitted models. The raw time series are shown in Figure 1. The vertical red line represents the splitting of the data in a training set and a test set, with the training data to the left of this line.

In order to have a better understanding of the data and the relationship between the different sensor signals, it is interesting to look at the cross-correlation between the time series. This is shown in the correlation plots in figure 2, where separate plots are shown for the training data and the test data. One feature that becomes immediately clear is that the correlation structure is very different in the two subsets of data. This is of course a cause for concern, and if the models are to be estimated based on the training data to exploit the correlation between signals, it cannot be expected to describe the test data well if the structure in the two datasets is very different.

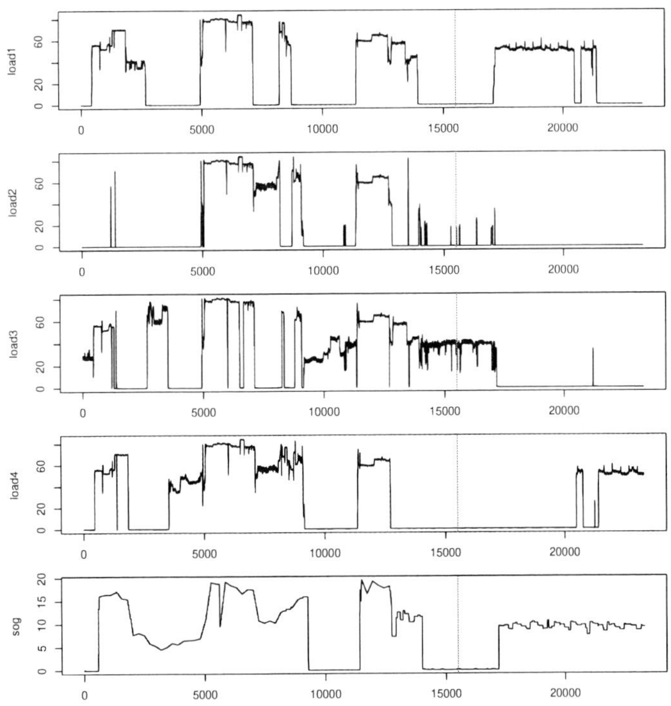

Fig. 1. The five raw sensor signals

4 Dynamical Linear Models for the sensor signal data

Some basic dynamical linear models have been fitted to the training data by maximum likelihood. In the following, models resulting from various model specifications will be explored. Obviously, the framework of dynamical linear models is very flexible and allows for a large number of model candidates, but only a limited set of candidate models are tried out in this study. As outlined above, all the models are fully specified by the quadruple $\{\boldsymbol{F}_t, \boldsymbol{G}_t, \boldsymbol{V}_t, \boldsymbol{W}_t\} = \{\boldsymbol{F}, \boldsymbol{G}, \boldsymbol{V}, \boldsymbol{W}\}$, where the model parameters are assumed to be time-invariant.

4.1 Model specifications and estimation

Model alternative 1. The first model alternative assumes a latent process with four hidden states. The five observed signals are then related to these states by the observation equation. $\boldsymbol{F}_1$ is now a (4×5)-dimensional sparse matrix that relates the signals from each genset to the corresponding genset-state and the speed over ground signal to the sum of all the genset states. A particular form of the variance matrix is also assumed, and the errors are assumed to be equal for

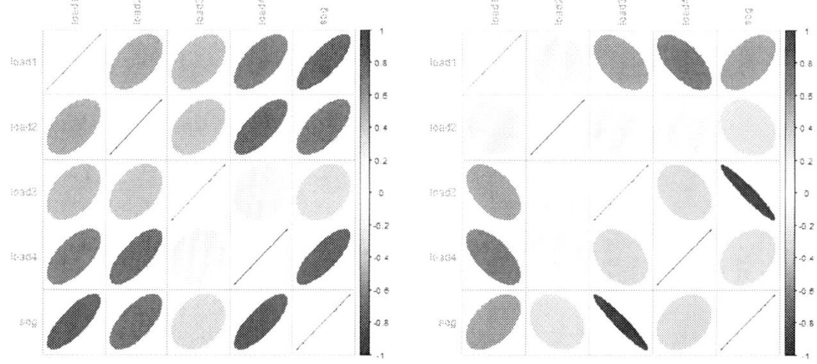

Fig. 2. Correlation matrix for the sensor signals; training data (left) and test data (right)

the four loads but different for the speed. Moreover, the covariance between any two loads is assumed identical, and the covariance between the various loads and the speed is assumed the same. Hence the matrices of the observation equation for model alternative 1 take the following form

$$\boldsymbol{F}_1' = \begin{bmatrix} \alpha & 0 & 0 & 0 \\ 0 & \beta & 0 & 0 \\ 0 & 0 & \gamma & 0 \\ 0 & 0 & 0 & \delta \\ \zeta & \eta & \iota & \kappa \end{bmatrix}, \quad \boldsymbol{V}_1 = \begin{bmatrix} \sigma_l^2 & \rho_l & \rho_l & \rho_l & \rho_s \\ \rho_l & \sigma_l^2 & \rho_l & \rho_l & \rho_s \\ \rho_l & \rho_l & \sigma_l^2 & \rho_l & \rho_s \\ \rho_l & \rho_l & \rho_l & \sigma_l^2 & \rho_s \\ \rho_s & \rho_s & \rho_s & \rho_s & \sigma_s^2 \end{bmatrix}. \tag{13}$$

The evolution of the genset states are described by the system equation. A rather simple model for the evolution of the states is assumed where the expected value for each of the states are a linear combination of the previous states. This reflects the interchangeability of the gensets to maintain the propulsion of the vessel. For the evolution noise terms, it is assumed that the variances for all the genset states are equal and that the covariances between all two genset states are equal. Hence, the evolution-matrices are on the following form:

$$\boldsymbol{G}_1 = \begin{bmatrix} a & b & b & b \\ b & a & b & b \\ b & b & a & b \\ b & b & b & a \end{bmatrix}, \quad \boldsymbol{W}_1 = \begin{bmatrix} \sigma_\omega^2 & \rho_\omega & \rho_\omega & \rho_\omega \\ \rho_\omega & \sigma_\omega^2 & \rho_\omega & \rho_\omega \\ \rho_\omega & \rho_\omega & \sigma_\omega^2 & \rho_\omega \\ \rho_\omega & \rho_\omega & \rho_\omega & \sigma_\omega^2 \end{bmatrix}. \tag{14}$$

The model is fully specified by assuming a prior distribution for the state vector. In this study, these are assumed to be normal distributions with zero means and diagonal variance matrix with variances of 100 for all model alternatives. Hence, this model alternative contains 16 parameters to be estimated from the training data, as presented in table A1 in the appendix.

A special case of model alternative 1, model 1b, with identical relation between the various gensets and the underlying hidden states, is also investigated. Hence, the parameters $\alpha, \beta, \gamma, \delta$ and $\kappa, \iota, \eta, \zeta$ takes the same values, say α and κ. The remainder of the model specification is the same and the estimated model parameters are presented in table A2 in the appendix.

Model alternative 2. Another model alternative assumes that the five recorded signals are all independent and relates the signals to five unobserved states without any interactions. In such a model, the five signals are actually modelled independently and $\boldsymbol{F}_2$ and $\boldsymbol{G}_2$ will both be unit diagonal. Moreover, the variance matrices $\boldsymbol{V}_2$ and $\boldsymbol{W}_2$ will be diagonal with only non-zero variances along the diagonal. The different signals and underlying states are allowed to have different variances and therefore such a models is specified by ten parameters, one parameter for each diagonal element in the two variance matrices. The estimated model parameters are presented in table A3 in the appendix.

A further simplification from model alternative 2 is to assume that all the load signals have equal variances and that only the speed signals and underlying hidden states have different variance. Such a simple model, model alternative 2b, is completely specified by four parameters, the observational and system variances of the loads and speed measurements, respectively. The estimated model parameters are summarized in table A4 in the appendix.

Model alternative 3 The third model alternative again assumes five hidden states; one for each recorded signals. However, the observations are allowed to be dependent, and non-zero covariances are estimated in both the observational and the system covariance matrices. Again, the $\boldsymbol{F}_3$-matrix is a five-dimensional unit diagonal, and the $\boldsymbol{V}_3$-matrix are described by four parameters: the variance for the load signals and the speed signal as well as the covariance between any two loads and between the speed and any of the loads. However, the gensets are assumed equal. The same structure is assumed for the $\boldsymbol{W}_3$-matrix which is specified by another four parameters. Finally, the system equation is described by the $\boldsymbol{G}_3$-matrix which relates the states to its values at the previous time-step and also relates each loads-state to the speed-state. The estimated parameters are presented in table A5 in the appendix, with

$$\boldsymbol{G}_3 = \begin{bmatrix} a & 0 & 0 & 0 & c \\ 0 & a & 0 & 0 & c \\ 0 & 0 & a & 0 & c \\ 0 & 0 & 0 & a & c \\ c & c & c & c & b \end{bmatrix}. \tag{15}$$

Model alternative 4 A final model alternative assumes only one hidden state, which could be construed to related to the operational model of the vessel, and relates all the measured signals to this underlying state. In this case, $\boldsymbol{F}_4$ reduces to a 5-dimensional vector, $\boldsymbol{F}_4^{'} = (f_1, f_2, f_3, f_4, f_5)$ and $\boldsymbol{G}_4$ and $\boldsymbol{W}_4$ reduce to scalars, say g and σ_ω^2. The $\boldsymbol{V}_4$-matrix is modelled with seven free parameters, i.e. a different observational variance for each of the signals (different values along the diagonal, $\sigma_1^2 \cdots \sigma_5^2$) and with a similar observational covariance between the different loads signals (ρ_l). The last parameter describes the covariance between any of the loads signals and the speed signal (ρ_s). The estimated model parameters are presented in table A6 i the appendix.

A generalization of model alternative 4, where all the elements in the $\boldsymbol{V}$-matrix are allowed to vary freely is investigated as model 4b. This yields a covariance matrix with 15 free parameters. Together with the five f_i's and the g and σ_ω^2 this gives a total of 22 parameters to estimate for this model. These are presented in table A7 the appendix.

4.2 AIC and RMSE for Model selection

The Akaike Information Criterion (AIC) [1] is on the form of eq. 16, where L is the likelihood of the model and k is the number of free parameters in the model. The preferred model according to this criterion is the one with the lowest AIC score. The log-likelihood and AIC for all the candidate models, for the training data, are summarized in table 1. For the test data, the log-likelihood as well as the root mean square errors (RMSE) based on the one-step ahead predictions from each of the candidate models are calculated and shown in the table.

$$AIC = 2k - 2\ln(L) \tag{16}$$

According to the calculated AIC and RMSE the models with one hidden state for each signal are superior. The three model alternatives with this overall structure are all significantly better than any of the other alternatives, for both the training data and the test data. Overall, model alternative 2 has the lowest AIC and RMSE and also highest log-likelihood for both datasets and would be selected as the best one according to the AIC and RMSE criteria. This model assumes independence between the various sensor signals and in fact models the five time series independently. Clearly, the other model alternatives, which tried to take the relationship between the signals into account, failed to adequately model the cross-correlation structure in the data. One explanation for why this failed could be the heterogeneous dependence structure in the data, as illustrated by the correlation plots in figure 2.

	Model 1	Model 1b	Model 2	Model 2b	Model 3	Model 4	Model 4b
k	16	10	10	4	10	14	22
$\ln(L)_{train}$	−158269.7	−87242.48	−36679.75	−41478.78	−42676.28	−206043	−211682.2
AIC_{train}	316571.4	174505	73379.49	82965.57	85372.57	412114	423408.4
$\ln(L)_{test}$	−95641.11	−34997.89	−5249.045	−6381.872	−8204.369	−106050.3	−106462.9
RMSE_{test}	11.56	2.51	0.728	0.728	0.737	20.84	22.82

Table 1. AIC and RMSE for model selection

4.3 Model filtering and forecasts

Having established a dynamical linear model, it can be used to derive filtering and forecast distributions. The conditional distribution of the unobservable state vector $\boldsymbol{\theta}_t$ given all observations up until time t, $\boldsymbol{y}_1, \boldsymbol{y}_2, \ldots, \boldsymbol{y}_t$, is referred to as the filtering distribution. The forecast distribution refers to predictions at the observation level of future measurements.

Models with a four-dimensional state vector. The first candidate models assumed four hidden states. In Figure 3 (left), the measured signals are shown together with the one-step ahead forecasts of the corresponding signals from model 1. Also the sum of all four loads are shown together with the sum of the forecasts for the loads. The results indicate that the model does a reasonable job in predicting the speed signal but that it has severe problems with the loads. Comparing the filtered states to the corresponding load signals (not shown), it is observed that there is a one-to one correspondence between these. This is by design and follows from the zero-elements of the $\boldsymbol{F}_1$-matrix.

Model alternative 1b is a special case of model 1 with some additional constraints on some of the elements of the $\boldsymbol{F}_2$-matrix. It is noted that this model gives reasonable predictions for the various loads signals, but fails to predict the speed very well (plots not shown).

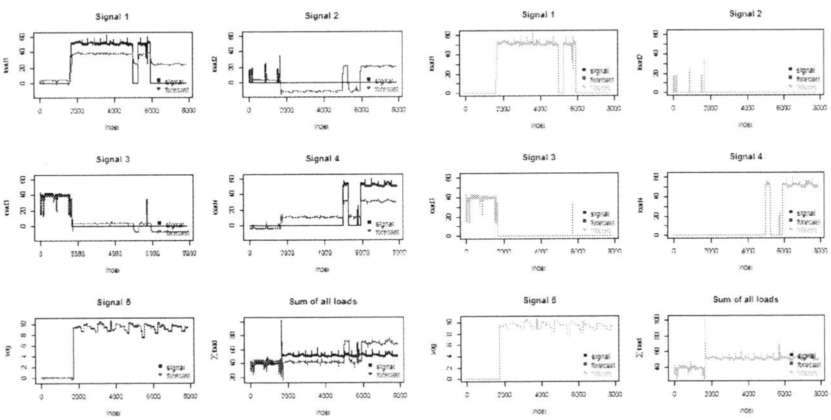

Fig. 3. Predictions from model alternative 1 (left) and 2 (right)

Models with full-dimensional state vectors With model alternatives 2 - 3 the system is modelled with a five-dimensional hidden state-vector and model 2 and 2b assumes that the signals can be modelled independently without any interaction or correlation between the signals. The signals are plotted together with the corresponding filtered and forecast distributions from model 2 in figure 3 (right). The confidence bands of the filtered signals are not shown but are very narrow, due to the very small observational variances. It is noted that the predicted and filtered values from models 2 and 2b are similar (identical) and that the only difference between these models is the spread of the filtering and forecast distributions. In model 2b these are assumed to be equal for all load-signals, but model 3 allows them to be different for different load-signals.

Also model alternative 4 consist of five hidden states, but allows for some interaction between the states. The filtered and predicted values from this model are not shown but the plots are very similar to the plots for models 2 and 2b.

Models with one-dimensional state vectors. The remaining model alternatives assume only one hidden state and tries to relate all the sensor signals to this unobservable state of the vessel. However, such models fail to capture the behaviour of all the sensor signals and in fact does a reasonable job only for the speed signal. Model alternative 4b is a slight improvement compared to model 4, but also this fails to capture the characteristics of the load signals.

k-step ahead forecasts. With a fitted model, it is quite straightforward to forecast the time series further into the future and to obtain the k-step ahead forecast distributions for future observations. The k-step ahead forecasts for the various sensor signals for $k = 100$ for model alternative 2 is shown in figure 4 (left). The 200 last observations in the test data are plotted together with the mean forecast as well as the 95% confidence interval for 100 future time points. As expected, the confidence interval grows wider as the uncertainty about future observations increases further into the future. For models that include interactions between the signals, the point forecasts of individual signals may be different from a constant. This can be seen in figure 4 (right) where the 100-step ahead forecasts for model alternative 3 are shown. The forecasts are quite different from the independent model, in particular for the speed signal. This illustrates the potential effect on the forecasts of modelling the signals dependently. However, for this particular dataset, with very different correlation structure in the training data compared to the test data, these interactions make the models worse and the models with independent states and signals perform best.

4.4 Predictive performance and monitoring

The display of the filtering and predicted values for each of the model alternatives suggest that the attempts on model reduction, either to four or only one hidden states leads to poor models. Hence, it might be necessary to use a model with five hidden states for the data at hand. Nevertheless, the predictive performance of the different model alternatives can be monitored by way of Bayes' factors or sequential probability ratio tests.

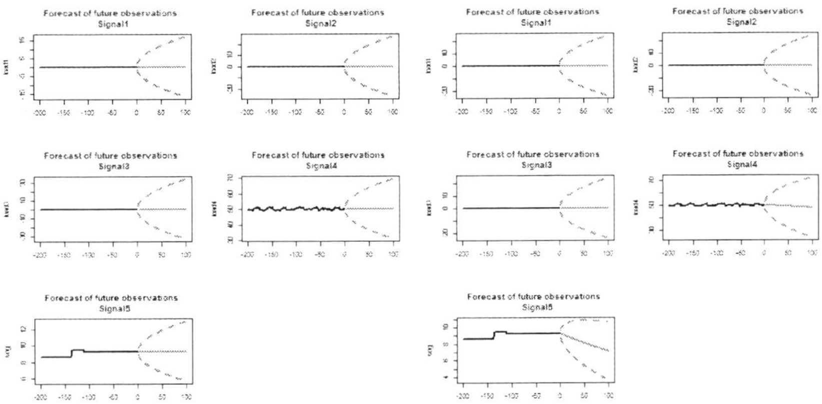

Fig. 4. Forecasts of future observations from models 2 (left) and 3 (right)

Forecast errors. According to the model assumptions, the one-step forecast errors, $\boldsymbol{e}_t = \boldsymbol{Y}_t - \boldsymbol{f}_t$. should be normally distributed with zero mean and variance matrix $\boldsymbol{Q}_t = \boldsymbol{F}_t^{'}\boldsymbol{R}_t\boldsymbol{F}_t + \boldsymbol{V}_t$ (see eqs. 3-6). Accordingly, the standardized one-step forecast errors, $\tilde{\boldsymbol{e}}_t = \boldsymbol{e}_t/\boldsymbol{Q}_t^{1/2}$, should be uncorrelated and distributed according to the standard normal distribution.

In order to track the predictive performance of the various models, it is possible to observe the one-step forecast errors and compare those to the underlying assumption. Traceplots of the one-step forecast errors from model alternative 1, for the test data, are shown in figure 5. The empirical densities of the forecast errors for each signal is also shown and compared to the standard normal density. The density plots clearly demonstrates that the standardized errors are not standard normally distributed and it is also clear that the errors are not uncorrelated. Hence, this model is not capturing the structure in the data very well. Similar plots for model 1b are not shown, but one finds that model 1 does a reasonable job in predicting the vessel speed whereas model 1b performs well for the various load signals. However, neither of the models are able to capture the joint behaviour of all signals.

Model alternatives 2 and 3 contain five hidden states and was found to give better descriptions of the data. The one-step forecast errors from model 2 are shown in figure 6 and confirm that the one-step errors resemble zero-mean noise. However, their variances are much smaller than what the model gives, as can be seen from the comparison with the standard normal distribution (normal distribution with $\sigma = \sqrt{5}$ for the sum of the errors from all signals). Nevertheless, apart from a few large forecast errors for the parts of the data where there are large jumps, the forecast errors are small and centred around zero. The large variances that are estimated for the models can be ascribed to these jumps, indicating that the models actually perform quite well for the parts of the

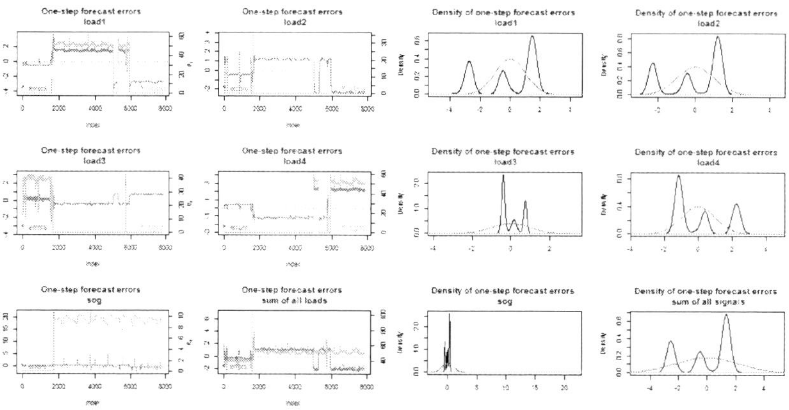

Fig. 5. Traceplots and densities of the one-step forecast errors from model 1 on the test data

data without any jumps. Similar plots for models 4 and 4b (not shown) clearly demonstrate that these model alternatives, assuming only one hidden state, do not represent the data well.

Sequential model monitoring. A dynamical linear model may be monitored sequentially to discover any departure from the model in order to detect for anomalies or faults. This could be done by measuring the actual sensor values, but could also be done by monitoring the forecast errors. Letting the null hypothesis be that the established DLM adequately describes the process, the forecast errors may be subject to sequential testing to discover if and when the model breaks down.

Under the null model, M_0, the standardized forecast errors, $\tilde{\boldsymbol{e}}_t$ are uncorrelated standard normally distributed variables and various alternative models can easily be specified, e.g. as a change in level (mean), in spread (variance) or in covariance between the errors. In such situations the predictive densities for the errors become $p_0(\tilde{\boldsymbol{e}}_t|D_{t-1}, M_0) \sim N(\boldsymbol{0}, \boldsymbol{I})$ and $p_1(\tilde{\boldsymbol{e}}_t|D_{t-1}, M_1) \sim N(\boldsymbol{\mu}, \boldsymbol{\Sigma})$ for some mean vector $\boldsymbol{\mu}$ and variance matrix $\boldsymbol{\Sigma}$. Having established one or more alternative models, sequential probability ratio tests can be performed by monitoring the desired SPRT indices.

In the following, alternative hypotheses for a positive and negative mean level change, respectively, will be formulated and the SPRT index for the forecast error for each sensor signal will be monitored. The alternative levels $\pm\mu_i$ must be specified and in this study this is assumed to be the same value for each signal, with $\mu = 1$. Moreover, under both the null and the alternative hypotheses the errors are assumed uncorrelated across the signals. Hence, the SPRT indices for the two tests take the following form (assuming now unit variance for the

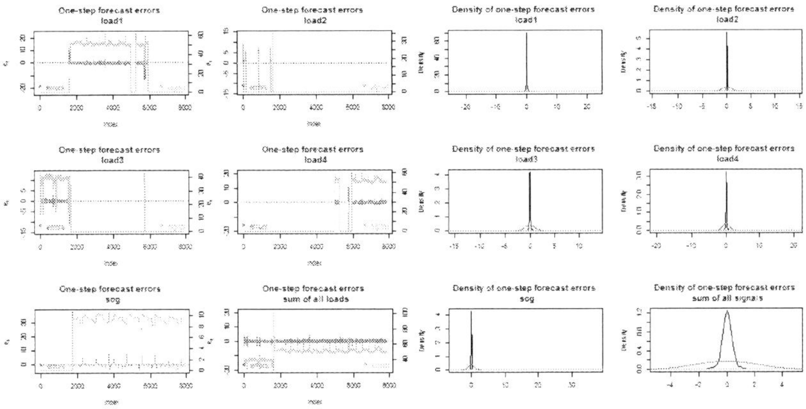

Fig. 6. Traceplots and densities of the one-step forecast errors from model 2 on the test data

standardized forecast errors)

$$\begin{aligned} l_n^+ &= \mu \sum_{i=1}^{n} \left(\tilde{e}_i - \frac{\mu}{2}\right) \\ l_n^- &= \mu \sum_{i=1}^{n} \left(-\tilde{e}_i - \frac{\mu}{2}\right) \end{aligned} \quad . \tag{17}$$

The model alternatives with the same number of hidden states were found to perform similarly, and only the results for forecast errors from models 1 and 2 are shown in figure 7, for both the positive and negative mean level tests. The first five plots show the SPRT for the individual signals and the last plot show the SPRT for the joint signals under the assumption that the errors in each signal is uncorrelated (under both hypotheses). Whenever the SPRT index reaches the upper threshold, an alarm is flagged to warn about possible model breakdown, indicated by a red circle in the figures.

For model alternative 1, it is observed that the model breaks down continuously for most of the load signals. This is expected, since the model was not able to describe the data well. However, only a few alarms are given for the speed signal, most notably at the point of sudden jumps in the time series. For model alternative 2, much fewer alarms are raised, indicating that the forecasts are acceptable for all signals. Again, a few alarms are flagged at places where the time series makes sudden jumps. The total number of alarms for each model is summarized in table 2. These results confirm the conclusion from the AIC based model selection that models 2, 2b and 3 are the best ones; They include a reasonable number of alarms and are able to correctly detect when there are large sudden jumps in the data or other surprising observations, but are able to forecast the signals well for reasonable sensor values.

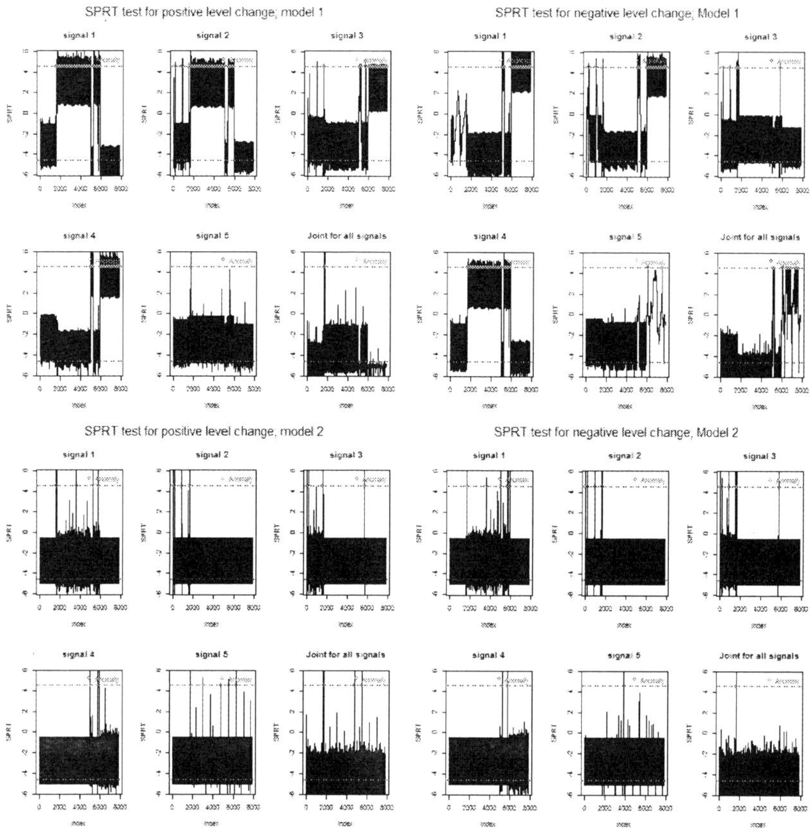

Fig. 7. SPRT for model monitoring; tests for positive level change (left) and negative level change (right) for model 1 (top two rows) and model 2 (bottom two rows)

5 Discussion

The approach presented in this paper utilizes dynamical linear models for monitoring of multivariate sensor signals. Based on the forecast distributions of future observations and comparing these with the sensor measurements as they arrive, on-line model monitoring can be performed by sequential tests. However, a crucial prerequisite for this approach to work well is the ability to establish a model that describes the sensor data well. This study has revealed that this can be challenging but achievable.

The framework of DLM offers the possibility of dimension reduction in that a set of high-dimensional sensor data can be described by a lower-dimensional state-vector. However, among the candidate models explored in this study, the ones without any dimension reduction were, by far, superior to the ones with

	Mod 1	Mod 2	Mod 3	Mod 3b	Mod 4	Mod 5	Mod 6
N	7834	7834	7834	7834	7834	7834	7834
Positive mean change	2182	1138	39	39	37	1842	6774
Negative mean change	2083	85	33	34	35	2399	2536
Total alarms	4265	1223	72	73	72	4241	3013

Table 2. Number of anomalies detected with each model alternative (total number of anomalies for all five sensor signals)

lower-dimensional state-vectors. In fact, modelling the multivariate sensor data as independent signals without any interaction yields the best results. Obviously, to be able to establish models that can exploit the correlation structure between signals there is a need for a homogeneous structure in the data throughout, and this was not the case for the data used in this study. As can be clearly seen in figure 2, the correlation is very different in the training data from the test data. This can explain why none of the models that included interactions between the signals were adequate. However, the DLM that modelled the time series independently were found to perform quite well.

In this paper DLM is applied to a dataset containing 5 sensor signals. However, sensor monitoring of typical ship systems may consist of hundreds of sensors and it remains to be seen how well the proposed approach scales up to higher dimensions. Dimension reduction is provided as an integral part of the framework of DLM and in this paper, dimension reduction techniques were applied by reducing the dimension of the state-space compared to the sensor signal space. However, the models which incorporated dimension reduction did not perform well on this particular dataset. For this exercise DLMs without dimension reduction performed quite well, but it is questionable whether DLMs with no dimension reduction will scale well and be computationally effective in much higher dimensions. This study did not explore how the proposed approach performs in higher dimensions and this is identified as an interesting and important topic for further research.

One feature of the sensor data that is not captured well by any of the models is the large jumps in the data appearing at irregular intervals. Between such jumps, the best model alternatives describe the data reasonably well, but the jumps themselves are not captured. There are different ways to extend the models to account for such jumps. It is possible to filter the data streams to different operational modes and to fit separate models for each state, for example, a transit mode, an approach to port mode, a berthed mode, dynamical positioning mode etc. This would require the various modes of the ship to be known, or it could be detected from the sensor data. Such an approach was adopted by [15] where various filters were applied to sensor data streams in order to isolate different steady-state conditions from continuous measurements for further analysis.

Another, perhaps more elegant way of modelling the large jumps could be to extend the hierarchical models with an additional layer to account for the jumps. Then, the observations could be modelled with a hidden layer of latent variables $\boldsymbol{\theta}_t$ which is again modelled with an additional hidden layer, say γ_t, which could depend on whether or not there are any jumps at time t. For example, the dichotomous latent variable $\boldsymbol{\gamma}_t$ could be introduced, letting $\gamma_t = \{0, 1\}$ depending on whether there are any jumps at time t or not. One could then let the evolution of the second layer of latent variables, $\boldsymbol{\theta}_t$, depend on $\boldsymbol{\gamma}_t$ in a manner that is incorporated in the evolution matrix, i.e. $\boldsymbol{\theta}_t = \boldsymbol{G}_{\gamma_t, \gamma_{t-1}} \boldsymbol{\theta}_{t-1}$. Alternatively, the system variance could be inflated at times where there is a jump in the observations, using a multiplier bigger than one, i.e. $\boldsymbol{W}_t(\gamma_t) = \boldsymbol{\alpha}(\gamma_t) \boldsymbol{W}_t$.

Obviously, completely different approaches to anomaly detection could have been used for the time series of sensor signals and dynamical linear models is merely one approach. Other promising approaches could be autoassociative kernel regression (AAKR) [5, 2], principal component analysis (PCA) [7, 6] and Bayesian networks [9]. A recent application of AAKR on a similar dataset from the same ship demonstrates that AAKR is a promising option for anomaly detection based on sensor signals. [3]. Compared to the AAKR approach, the DLM method has the advantage of taking temporal structures into account. On the other hand, the AAKR is a more purely data-driven approach and requires less model specification and tailoring to the specific dataset and problem at hand. This could be an advantage if more generic anomaly detection methods are needed for application to various systems and sensor signals without much modifications. On the other hand, if detailed domain knowledge are available regarding the particular system being monitored or the structure of the sensor signals, the DLM provides means to model such knowledge explicitly.

One major challenge in modelling the sensor data investigated in this study is the sudden jumps in the signals. These may partly be an effect of changes in the operational mode of this ship and display the response of the system to manoeuvres on the bridge. Clearly, such sudden shifts are difficult to model and even harder to predict. Other extraneous factors that may influence the sensor signals, and that is perhaps particular to ship systems, are the impacts from the ship environment. Such impacts may introduce larger variability in the recorded sensor signals and need to be captured by the statistical models used for anomaly detection. Clearly, one wants to distinguish between anomalous behaviour of the ship system and anomalous environmental conditions, but the environmental conditions could be used as covariates in a regression-type component of the model. Previously, it has been shown that up to 50% of the variation in the efficiency of a ship machinery system can be explained by such covariates, suggesting that these effects should be taken into account [16]. If such operational conditions are found to be very influential on the monitoring signals and, consequently, on the sensor based anomaly detection, methods for conditional or contextual anomaly detection might be required [14, 8].

Another implication of the varying operational conditions for sensor based monitoring of ships is that a short record of training data may not be sufficient if it does not cover all (or at least the most frequent) possible operational conditions. In this study, a rather limited record containing three months of data has been available and this is presumably too short. Indeed, from figure 1 it is evident that the characteristics of the test data is different from the training data. Thus, the limited duration of training data that has been available for this study could possibly explain the very different correlation structure in the training data compared to the test data, as illustrated in figure 2. Possibly, with more training data that are more complete with regards to the operational modes that are covered, and a more homogeneous correlation structure across subsets of data, the model alternatives that incorporated dimension reduction would perform better.

6 Conclusions

The study presented in this paper has applied the Bayesian framework of Dynamical Linear Models to a stream of multivariate sensor signals collected from an ocean going ship for the purpose of detecting anomalies in the signals. Such anomalies can be construed as indicators of faults, either in the system itself or in the sensors. Different model alternatives within the DLM framework have been specified with different interactions between the various sensor signals and underlying unobserved processes. Sensor signal predictions from the models are then compared to the actually observed signals and sequential probability ratio tests are employed in order to classify anomalous sensor readings. As it turned out, only one group of models were successful in providing reasonable predictions of the sensor signals, i.e. the ones employing full-dimensional state vectors. Moreover, the models that performed best within this group are the ones that modelled the different sensor signals independently of each other. Even though these models were found to perform reasonably well, they do not exploit any correlation structure between the signals to classify faults. Ideally, one would prefer to utilize such correlations in order to improve the models and further research will be carried out with other data streams in order to develop such models.

Notwithstanding, the condition monitoring framework presented in this paper performs well for the DLMs with full-dimensional state-space. For these models, the sequential model monitoring based on the SPRT concludes that the models mostly performs well, but raises a few alarms when there are sudden jumps or unexpected sensor readings in the data stream that indicate breakdown of the model. After such jumps, the models are able to swiftly recover to the new level of the sensor values to continue to monitor the system. This is encouraging and indicates that the framework presented in this paper is applicable for condition monitoring based on sensor data as long as a suitable dynamical linear model can be specified.

Acknowledgement

The study presented in this paper is partly carried out within the centre for research-based innovation, BigInsight.

References

1. Akaike, H.: A new look at the statistical model identification. IEEE Transactions on Automatic Control 19, 716–723 (1974)
2. Baraldi, P., Di Maio, F., Turati, P., Zio, E.: Robust signal reconstruction for condition monitoring of industrial components via a modified auto associative kernel regression method. Mechanical Systems and Signal Processing 60-61, 29–44 (2015)
3. Brandsæter, A., Manno, G., Vanem, E., Glad, I.K.: An application of sensor based anopmaly detection in the maritime industry. In: Proc. IEEE PHM2016. IEEE Reliability Society (June 2016)
4. Eleftheria, E., Papanikolaou, A., Voulgarellis, M.: Statistical analysis of ship accidents and review of safety level. Safety Science 85, 282–292 (2016)
5. Garvey, J., Garvey, D., Seibert, R., Hines, J.W.: Validation of on-line monitoring techniques to nuclear plant data. Nuclear Engineering and Technology 39, 149–158 (2007)
6. Goulding, P., Lennox, B., Sandoz, D., Smith, K., Marjanovic, O.: Fault detection in continuous processes using multivariate statistical methods. International Journal of Systems Science 31, 1459–1471 (2000)
7. Hawkins, D.M.: The detection of errors in multivariate data using principal components. Journal of the American Statistical Association 69, 340–344 (1974)
8. Hayes, M.A., Capretz, M.A.: Contextual anomaly detection framework for big sensor data. Journal of Big Data 2, 1–22 (2015)
9. Hill, D.J., Minsker, B.S., Amir, E.: Real-time Bayesian anomaly detection in streaming environmental data. Water Resources Research 45, W00D28(1–16) (2009)
10. Mindykowski, J., Tarasiuk, T.: Problems of power quality in the wake of ship technology development. Ocean Engineering 107, 108–117 (2015)
11. Petris, G.: An R package for dynamical linear models. Journal of Statistical Software 36, 1–16 (2010)
12. Petris, G., Petrone, S., Campagnoli, P.: Dynamical Linear Models with R. Springer (2009)
13. Psarros, G.A.: Comparing the navigator's response time in collision and grounding accidents. In: Proc. 34th International Conference on Ocean, Offshore and Arctic Engineering (OMAE 2015). American Society of Mechanical Engineers (ASME) (May-June 2015)
14. Song, X., Wu, M., Jermaine, C., Ranka, S.: Conditional anomaly detection. IEEE Transactions on Knowledge and Data Engineering 19, 631–645 (2007)
15. Trodden, D., Murphy, A., Pazouki, K., Sargeant, J.: Fuel usage data analysis for efficient shipping operations. Ocean Engineering 110, 75–84 (2015)
16. Vanem, E., Brandsæter, A., Gramstad, O.: Regression models for the effect of environmental conditions on the efficiency of ship machinery systems. In: Proc. ESREL 2016. European Safety and Reliability Association(ESRA) (September 2016)
17. Vanem, E., Rusås, S., Skjong, R., Olufsen, O.: Collision damage stability of passenger ships: Holistic and risk-based approach. International Shipbuilding Progress 54, 323–337 (2007)
18. Vanem, E., Skjong, R.: Fire and evacuation risk assessment for passenger ships. In: Proc. 10th International fire science and engineering conference (Interflam) 2004. vol. 1, pp. 365–374 (July 2004)
19. Wald, A.: Sequential tests of statistical hypotheses. The Annals of Mathematical Statistics 16, 117–186 (1945)
20. West, M., Harrison, J.: Bayesian Forecasting and Dynamic Models. Springer-Verlag, second edn. (1997)

Appendix: Estimated model parameters

α	β	γ	δ	κ	ι	η	ζ	σ_l^2	σ_s^2	ρ_l	ρ_s	a	b	σ_ω^2	ρ_ω
0.78	0.99	2.11	1.31	−0.54	−0.30	0.42	0.39	77.3	0.041	1.91	0.92	1.00	2.64×10^{-6}	0.69	0.54

Table A1. Parameter estimates for model alternative 1

α	κ	σ_l^2	σ_s^2	ρ_l	ρ_s	a	b	σ_ω^2	ρ_ω
0.600	0.0407	0.0175	17.09	0.0175	−0.445	0.998	0.000499	6.72	−0.819

Table A2. Parameter estimates for model alternative 1b

$\sigma_{\nu_1}^2$	$\sigma_{\nu_2}^2$	$\sigma_{\nu_3}^2$	$\sigma_{\nu_4}^2$	$\sigma_{\nu_5}^2$	$\sigma_{\omega_1}^2$	$\sigma_{\omega_2}^2$	$\sigma_{\omega_3}^2$	$\sigma_{\omega_4}^2$	$\sigma_{\omega_5}^2$
1.14×10^{-8}	2.09×10^{-9}	3.22×10^{-11}	1.55×10^{-9}	4.68×10^{-10}	1.03	2.69	4.24	1.34	0.049

Table A3. Parameter estimates for model alternative 2

$\sigma_{\nu_l}^2$	$\sigma_{\nu_s}^2$	$\sigma_{\omega_l}^2$	$\sigma_{\omega_s}^2$
4.77×10^{-8}	4.16×10^{-14}	2.325	0.0485

Table A4. Parameter estimates for model alternative 2b

$\sigma_{\nu_l}^2$	$\sigma_{\nu_s}^2$	ρ_{ν_l}	ρ_{ν_s}	a	b	c	$\sigma_{\omega_l}^2$	$\sigma_{\omega_s}^2$	ρ_{ω_s}
0.0931	0.00218	−0.031	−0.000169	0.999	0.995	0.000333	2.242	0.0697	−0.0287

Table A5. Parameter estimates for model alternative 3

f_1	f_2	f_3	f_4	f_5	g	σ_ω^2	σ_1^2	σ_2^2	σ_3^2	σ_4^2	σ_5^2	ρ_l	ρ_s
1.13	0.929	1.01	1.23	0.320	0.9999	0.220	320	473	855	275	4.22	0.67	−20.5

Table A6. Parameter estimates for model alternative 4

f_1	f_2	f_3	f_4	f_5	g	σ_ω^2	V_{11}	V_{22}	V_{33}	V_{44}
0.976	0.773	0.856	1.055	0.287	0.9999	0.596	424	598	967	413

V_{55}	V_{12}	V_{13}	V_{14}	V_{15}	V_{23}	V_{24}	V_{25}	V_{34}	V_{35}	V_{45}
0.000271	−0.153	0.143	−0.0524	0.283	0.0519	0.247	0.0952	−0.233	0.231	−0.0269

Table A7. Parameter estimates for model alternative 4b

A MULTI-SENSOR DEGRADATION MONITORING METHORD FOR BALL SCREW BASED ON DEEP LEARNING

Li Zhang and Hongli Gao

Department of Mechanical Engineering,
Southwest Jiaotong University,
Chengdu 610031, China

Li_Zhang3836@163.com
hongli_gao@home.swjtu.edu.cn

Wei Yue

Air Force Chengdu Bureau
Chengdu, China

41710041@qq.com

Abstract: The performance degradation of ball screw will directly affect the machining accuracy. Therefore, monitoring the degradation degree of ball screw is important in improving the machining quality and stability. Single signal collected by one sensor cannot well represent the global degradation characteristic of ball screw, and the research in monitoring the degradation degree of ball screw by multiple sensors is lack. Aiming at this problem, a method based on multi-sensor information fusion and deep learning is proposed in this paper. Vibration signals are acquired by a plurality of acceleration sensors in multiple positions and directions of ball screw. Positions include ball nut and two bearing seats, and directions include the axial and radial directions of screw. The vibration signals collected in 7 different degradation periods. Multiple signals can more comprehensively reflect the performance degradation of ball screw, and the information is more reliable, accurate and redundant compared with single signal. In the proposed method, deep neural networks with deep architectures are established by utilizing deep learning theory and fused vibration frequency spectra. The input frequency spectra of deep neural networks are the fusion of frequency spectra collected by all acceleration sensors. 60% samples collected by each sensor were randomly selected as the training sample, and the others as the testing sample. The layer of deep neural network is set to 5. Deep neural networks are served as fusion technology to fuse all the vibration frequency spectra and to adaptively extract available degradation characteristics and automatically identify the degradation degree of ball screw. Finally, a test is designed to collect the degradation data of ball screw in different degradation period. The proposed method is applied to classify the degradation degree of ball screw in the test to demonstrate its effectiveness.

Keywords: ball screw; deep learning; multi-sensor information fusion; degradation monitoring

1 Introduction

As an important component of the drive system in machine, ball screw convert the rotary motion to translational motion. Its performance degrade continuously because the existence of friction and wear, which will directly affect the machining accuracy and processing stability. Condition monitoring could provide decision in maintenance and repair by recording, classifying and evaluating the running state of equipment, is of great engineering significance. Effective condition monitoring of ball screw could increase the stability and reliability of machine, and avoid economic losses caused by sudden failure in machining process.

However, traditional monitoring method requires priori knowledge based feature extraction and expression, and also rely on signal processing techniques and diagnostic experiences. And as a kind of complex mechanical equipment, degradation signals collected by multi-sensor is more comprehensive and accurate than single-sensor signal for ball screw. Therefore, deep learning theory and multi-sensor data fusion technology can be used in degradation condition monitoring to adaptively estimate the degradation degree of ball screw, and corresponding prevention measures could be made to effectively prevent the machine downtime.

Hinton[1] proposed the concept of deep learning for the first time in 2006 through the study on handwritten numeral recognition. During the past 10 years, deep learning had a lot of research in many fields like computer vision[2-3], speech recognition[4] and information retrieval[5-6]. There are also many studies have applied deep learning in fault diagnosis and condition monitoring[7-8]. Verma et al.[9] proposed the intelligent condition based monitoring of rotating machine using sparse auto-encoder. Feng Jia et al.[10] proposed a novel intelligent diagnosis method that DNNs are utilized to implement both fault feature extraction and intelligent diagnosis and achieved good results. The previous studies show that deep learning usually is more effective compared with traditional machine learning. By establishing deep model, deep learning get rid of the dependency on signal processing techniques and diagnostic experiences,

Paper ID: COMADEM2016_114

and can extract features from the frequency domain adaptively to realize intelligent diagnosis. Deep learning become a new development direction of fault diagnosis and condition monitoring.

Since the concept of information fusion has been formally put forward in 1973, the multi-sensor information fusion technology was widely used in military, robot, image processing, speech recognition and some other fields, and also used in fault diagnosis and condition monitoring. Yaguo Lei et al.[11] utilize ANFIS to fuse all features from different sensors, and achieve much higher accuracies in detecting planetary gearbox faults compared with individual sensors. M.S.Safizadeh et al.[12] present a new method for bearing fault diagnosis using the fusion of vibration signal and load signal respectively obtained by accelerometer and load cell, and experiment is designed to show its effectiveness. Multi-sensor data fusion technology can express the equipment state more comprehensively and accurately, which is an important theoretical basis and development direction of artificial intelligence.

This paper proposed a degradation monitoring method for ball screw based on deep learning and multi-sensor data fusion. The rest of the paper is organized as follow: Section 2 explains the theory and basic steps of DNNs, and also has a brief introduction to the multi-sensor data fusion technology. Section 3 introduces the implementation steps of the proposed multi-sensor degradation monitoring method for ball screw in detail. In section 4, the effectiveness of the proposed method is verified using the degradation vibration signals collected by multiple accelerometers under various operating conditions. The frequency spectra dataset includes thousands of signal samples in different degradation degrees, locations and operating conditions. And the proposed method is compared with the unfused frequency spectra monitoring method. Section 5 summarizes the research and draw conclusions.

2 Background

2.1 A brief introduction to DNNs based on SAE

Deep neural network (DNN) have multiple hidden layers and each hidden layer computes a non-liner transformation of the previous layer[13], which makes deep network can adaptively extract features, thus DNN have greater representational power compared with shallow network. Greedy layer-wise unsupervised pre-training using unsupervised learning algorithm such as stacked auto-encoder is the key ingredient of deep learning. Then, load the higher level representation into softmax classifier to classify the various degradation condition of ball screw. SAE and softmax classifier formed the deep architecture and back propagation (BP) algorithm is used to fine-tune all the parameters by supervised training.

2.1.1 Stacked Auto-encoder

1) Auto-encoder.

Auto-encoder is a kind of 3 layers neural network which used in deep architecture as unsupervised learning algorithm, and the output target of this neural network is the input data. The auto-encoder is composed of encoder network and decoder network as shown in Fig. 1. The encoder network transform the input data of high dimension space into codes in a low-dimensional space and the decoder network reconstruct the input data from the corresponding codes. The hidden layer of auto-encoder, i.e., codes in low-dimensional space, is treated as the input data of the next auto-encoder.

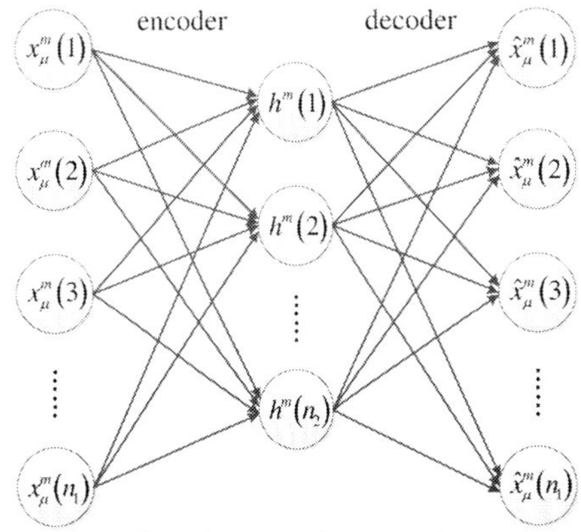

Fig.1. Structure of auto-encoder

A degradation condition dataset $\{x_\mu^m\}_{m=1}^M$ without label is given as the input dataset. Each measured signal x_μ^m is transformed into code (i.e., the hidden layer) by encoder function f_θ, and the process of encoding is defined as:

$$h^m = f_\theta\left(x_\mu^m\right) = s_f\left(Wx_\mu^m + b\right) \quad (1)$$

Where s_f is the encoder activation function, W and b are the weight matrix and bias vector of encoder network, respectively.

The decoder network is defined as a reconstruction function $g_{\theta'}$ to map the h^m from hidden layer to high-dimensional space $\{\hat{x}_\mu^m\}_{m=1}^M$ which is the reconstruction of $\{x_\mu^m\}_{m=1}^M$. The process of decoding is defined as:

$$\hat{x}_\mu^m = g_{\theta'}\left(h^m\right) = s_g\left(W^T h^m + d\right) \quad (2)$$

Where s_g is the decoder activation function, W^T and d are the weight matrix and bias vector of decoder network, respectively.

The aim of auto-encoder training is to find the parameter sets $\{W,b\}$ and $\{W^T,d\}$ to minimize the reconstruction error:

$$\begin{aligned} L\left(x_\mu,\hat{x}_\mu\right) &= \frac{1}{M}\left\|x_\mu - \hat{x}_\mu\right\|^2 \\ &= \frac{1}{M}\left\|x_\mu - g_{\theta'}\left(f_\theta\left(x_\mu\right)\right)\right\| \end{aligned} \quad (3)$$

The reconstruction error relate to all the training samples and has nothing to do with the input sequence.

2) Stacked Auto-encoder.

N auto-encoders stacked to pre-train a higher performance DNN with N hidden layers, and the structure of spaced auto-encoder is shown in Fig. 2. The original signal x_μ is the input layer of AE1 to produce the hidden layer h_1 which is a higher level representation of x_μ, and the hidden layer h_1 of AE1 is the input layer of AE2 to produce the hidden layer h_2 which is a higher level representation of h_1, and this process is continued until the hidden layer h_N of AEN is obtained. The representation of each level is more abstract than previous level. All the hidden layers $\{h_n\}_{n=1}^{N}$ of stacked auto-encoders form the hidden structure of DNN. The data for the entire stacked auto-encoder process are unlabeled and the stacked auto-encoder is unsupervised training.

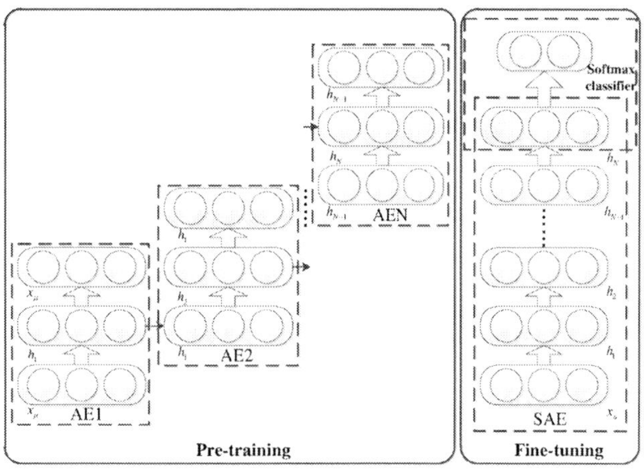

Fig.2. The training process of SAE based DNN

2.1.2 Classification using Softmax Regression

After getting the SAE with feature extraction and dimension reduction, the softmax classifier with classification capability can be added to the top layer of the stacked autoencoder neural network to construct a complete feature extraction and classification model.

Softmax classifier is a generalized logistic regression where class labels can take multiple values[14]. For the training set $\{(x^1, y^1),\ldots(x^m, y^m)\}$, where m is the number of training samples, and $y^m \in \{1, 2, \ldots, k\}$ are the number of classes. For a given test input x', the hypothesis estimates the probability $p(y = j|x')$, $j = 1,\ldots,k$. Thus the hypothesis output a k dimensional vector with k estimated probabilities.

The softmax regression model's sparse cost function with coefficient parameters sparse penalty term is:

$$J_{sp}(\theta;x,y) = -\frac{1}{m}[\sum_{i=1}^{m}\sum_{j=1}^{k} I\{y^i = j\} \log\frac{e^{\theta_j^T x^i}}{\sum_{l=1}^{k} e^{\theta_l^T x^i}}] + \frac{\lambda}{2}\sum_{i=1}^{m}\sum_{j=1}^{k}\theta_{ij}^2 \qquad (4)$$

Where θ is the model parameter, m and k are the number of samples and classes, λ is the weight penalty coefficient and $I\{\bullet\}$ is the indicator function that $I\{true\ statement\} = 1$ and $I\{false\ statement\} = 0$.

In the process of training model, partial derivatives for cost function are needed.

$$\nabla_{\theta_j} J_{sp}(\theta) = -\frac{1}{m}\sum_{i=1}^{m}[x^i(I\{y^i = j\} - p(y^i = j \mid x^i;\theta))] + \lambda\theta_j \qquad (5)$$

2.1.3 Fine tuning

After adding the softmax classifier to the stacked autoencoder neural network, the next step is the supervised learning process, which will be accomplished by the back propagation (BP) training algorithm. This supervised training process adjusts parameters of all layers to reduce the training error and improve the classification accuracy, and this process is called fine tuning. This supervised process use labeled data to train the DNN model. Fine-tuned SAE and softmax classifier are respectively used as feature extraction module and feature classification module of the DNN as shown in Fig. 2.

Back propagation (BP) algorithm update weight matrix by gradient descent method until the cost function reaches the set stop threshold, which means the fine tuning of the DNN is finished.

2.2 Multi-sensor information fusion technology

Multi-sensor information fusion is a kind of information processing process by using the computer technology to analyze and synthesize the information of multiple sensors automatically in a certain principle to complete the decision and estimation.

Monitoring the condition of equipment using single sensor will reduce the accuracy or even appear error because the uncertainty of single-sensor data and the complexity of equipment. The development of multi-sensor information fusion technology provided a new way in solving the uncertainty of condition monitoring for complex machine such as ball screw. Multi-sensor information fusion can provide more degradation state information from different sides to improve the monitoring accuracy and reliability, and is a new development direction of artificial intelligence.

Information fusion can be divided into data level fusion, feature level fusion and decision level fusion. Each fusion method has its advantage and disadvantage. The fusion method can be chosen according to the demand and sensor types.

3 Degradation condition classification method based on deep learning and multi-sensor information fusion

An intelligent degradation condition classification method based on deep learning and multi-sensor information fusion of ball screw is proposed in this study, and can adaptively extract degradation characteristics from raw signals and automatically distinguish the degradation degree. In this method, fused spectra are used as the input data of DNN constructed by unsupervised learning and supervised learning.

As shown in Fig. 3, the proposed method can be divided into four steps: sample acquisition and data preprocessing, unsupervised pre-training, supervised fine-tuning and degradation condition monitoring.

Firstly, after getting the time domain signals $\left\{x_{(1)}^{i},\ldots,x_{(n)}^{i}\right\}_{i=1}^{M}$ from n different sensors, calculate frequency spectra $\left\{f_{(1)}^{i},\ldots,f_{(n)}^{i}\right\}_{i=1}^{M}$ corresponding to these time domain signals. Next is the feature-level data fusion, which do parallel superposition of the n spectra to increase the dimension of the feature space indicated by $\left\{F^{i}\right\}_{i=1}^{M}$, and $F^{i}=f_{(1)}^{i}\cup\ldots\cup f_{(n)}^{i}$. The fused spectra are directly as raw signals of DNN.

Secondly, establish a DNN with multiple hidden layers. The number of input nodes is the dimension of the fused spectra, and equal to the sum of spectral dimension of the n single sensor signals. The relation can be expressed as $dim(F^{i})=\sum_{j=1\ldots n} dim(f_{(j)}^{i})$, dim represents spectral dimension. Determine the number of hidden layers N and the number of nodes $\left\{Z_{i}\right\}_{i=1}^{N}$ in each hidden layer of the DNN. Then, the unlabeled training set $\left\{F^{i}\right\}_{i=1}^{M}$ are used to pre-train the DNN layer by layer, and the number of hidden layer inside the DNN equal to the number of hidden layer in auto-encoder. The unlabeled training set $\left\{F^{i}\right\}_{i=1}^{M}$ is utilized as input data and output target to train the first auto-encoder with Z_1 hidden layer nodes. The hidden layer of the first auto-encoder is as the first hidden layer h_1 of DNN and the trained parameters $\left\{W_1,b_1\right\}$ of the first auto-encoder are used to initialize the parameters of h_1. Then, utiliz h_1 as input data and output target to train the second auto-encoder, the second hidden layer h_2 of DNN are obtained and the parameter of h_2 are initialized. Continue the training steps until the Nth auto-encoder is trained and the fused frequency spectra are coded into h_N with Z_N dimension. All the hidden layers $\left\{h_n\right\}_{n=1}^{N}$ of DNN are pre-trained.

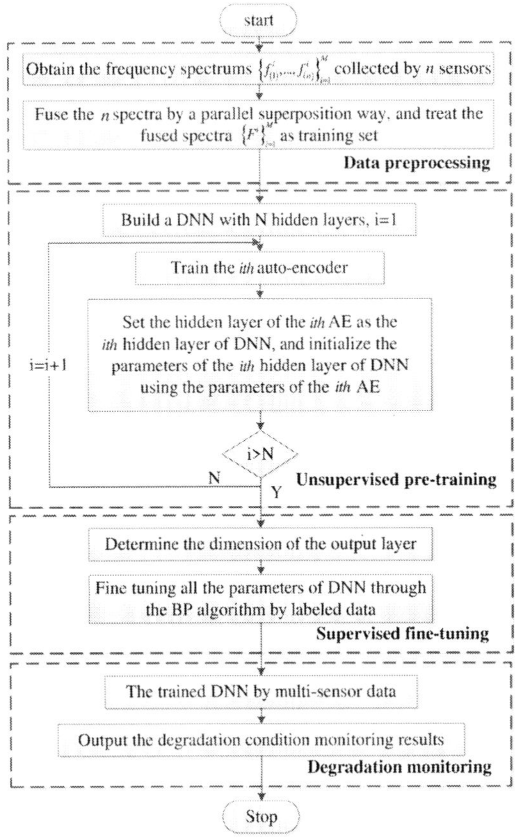

Fig.3. Flowchart of the proposed method

The next step is to determine the dimension of the output layer corresponding to the degradation condition grade. Fine tuning all the parameters of DNN by supervised back propagation (BP) training algorithm utilizing the labeled data. The target is to minimizing the error between output calculated from the fused frequency spectra $\left\{F^{i}\right\}_{i=1}^{M}$ and the degradation condition grade label.

Finally, the trained DNN can be used to monitor the degradation grade of ball screw by collecting multi-sensor vibration signal.

The proposed method transforms the frequency spectra into a higher level representation by encoding, which is more comprehensive, accurate and redundant by fusing the frequency spectrum of multiple sensor signals. Traditional BP neural network usually get a local optimal solution rather than global optimal solution. The deep learning theory add the unsupervised pre training process to calculate the hidden layers and initialize the parameters of DNN to avoid local optimum. This method enables the trained DNN adaptively mine essential characteristics from fused frequency spectra and establish the non-liner mapping between the fused frequency spectra and degradation grade to realize intelligent degradation monitoring.

4 Degradation monitoring for ball screw using the proposed method

4.1 Application study description

Ball screw is an important precision transmission parts of machine tool transmission system, its performance degradation will lead to lower machining precision, and even downtime resulting in big economic losses. Therefore, monitoring the degradation degree of ball screw is important in improving machining stability and machining efficiency.

Fig.4. Acceleration performance degradation test bench of ball screw

In order to study the degradation characteristics of ball screw, FFZD4010R-3 type of ball screw is chosen to install on the acceleration performance degradation test bench as shown in Fig.4. The test bench is composed of driving motor, ball screw, gear and rack device, slide guide and magnetic powder brake. The test bench is used to simulate the performance degradation during processing of ball screw. Meanwhile, multiple sensors are used to collect degradation signals under different degradation conditions, and the degradation grade can be judged through analyzing the degradation data. Acceleration sensors are selected to collect the vibration signals which is always of good response to the device status. The objective of this application test is to classify different degradation condition grade of ball screw based on the multi-sensor vibration signals.

4.2 Data acquisition and data preprocessing

Five groups of vibration signals are acquired by five acceleration sensors installed on multiple positions and directions of ball screw. Positions include ball nut and the two bearing seats, and directions include the axial and radial directions of screw. The test bench simulates the whole performance degradation process of ball screw from new to failure.

In order to distinguish and calculate, divide the whole performance degradation process of ball screw into 7 grades, and any degradation condition included in the 7 degradation grades. Vibration signals of each degradation grade are collected in 9 working conditions, which are the all cross combinations of three speed (100r/min, 300r/min, 800r/min) and three axial load (0kN, 1kN, 2kN). Collect 4950 samples in each degradation grade, namely 600 samples in the first 6 working conditions and 450 samples in the last 3 working condition. Five sensors collect signals at the same time, and each sensor collect 6930 samples for total 7 degradation grades. In conclusion, the degradation data of ball screw contains 34650 samples, which are collected in 9 working conditions, 7 degradation grades and 5 sensors installed in different positions and directions.

The degradation data should be pre-processed after collecting. The spectrum of each data sample is calculated respectively, and the spectrum is a vector of length 512. Hence there are 5 spectrum matrixes of 6930 lines and 512 columns. These 5 sensors spectrum matrixes are fused according to parallel superposition approach, getting a spectrum matrix of 6930 lines and 2560 columns. Thus the number and dimension of input data are 6930 and 2560 respectively. The fused spectrum matrix is the input to DNN. Degradation grades of ball screw are represented by 7 seven-dimension vector, and the nth value of the vector is 1 whereas the other values are 0 when the ball screw is in the nth degradation grade.

4.3 DNN based degradation grade monitoring

The trained DNN classifier model has five layers namely one input layer, one output layer and three hidden layers. The neuron number of the input layer is determined by the dimension of the input spectrum dataset, the neuron number of the output layer equals to the number of defined degradation grade, and the neuron number of the first hidden layer, the second hidden layer and the third hidden layer are 1024, 250 and 40 respectively. The maximum number of training epochs used in the SAE training process and back propagation learning process are respectively set to 200 and 700, the learning rate and momentum are both 0.07 and the batch size is 63. The active functions of the DNN are hyperbolic tangent function. In the trial, 60% of the data samples are randomly selected for training, and the others are used for testing. The weights of DNN are initialized randomly and the biases are initialized to zero. Trials repeated twenty times. The degradation classification results using the proposed method are shown in Fig.5. The training accuracies of all the trials are over 99% and the testing accuracies are all over 93%, which indicate the proposed method is effective and is able to distinguish the seven degradation condition grades of ball screw in a high accuracy. This is mainly because the DNN has the capability to learn high complexity non-liner relationship between the input data set and 7 degradation states.

For comparison, we also classify the degradation grade of ball screw utilizing the unfused spectra. Original signals and parameters used in training are same except the dimension of input dataset. The dimension of the proposed method is four times larger than the unfused spectra method. Trials also repeated twenty times. The degradation classification results using the unfused spectra method are shown in Fig.6. Most of the training accuracies are around 87% and testing accuracies are around 76%, and the classification accuracies is smaller than the proposed method obviously. The average accuracies and the standard deviation of the proposed method are 97.07% and 0.47%, whereas the average accuracies and the standard deviation of the unfused spectra method are 81.49% and 0.57% respectively. The proposed method has bigger average accuracies and smaller standard deviation, which means is more effective and stable than the unfused spectra method and has more superiority.

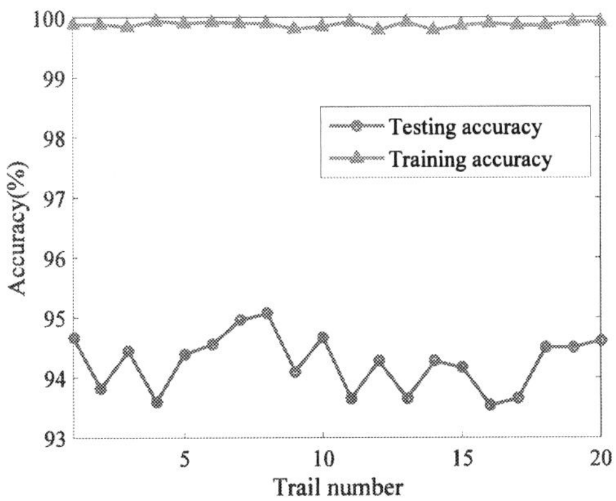

Fig.5. Classification results using the proposed method

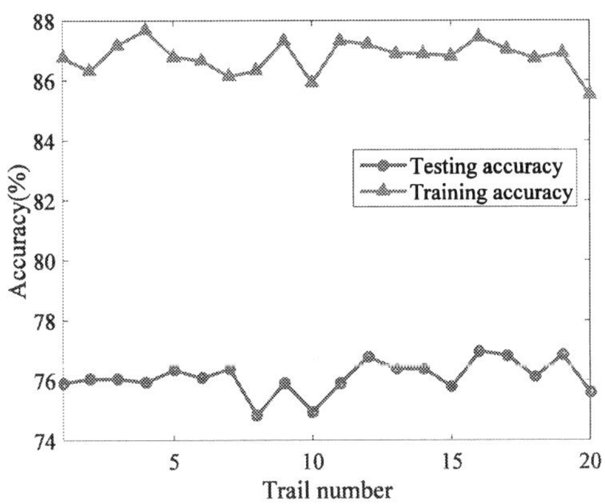

Fig.6. Classification results using the unfused spectra method

5 Conclusions

This paper presented a novel degradation monitoring method for ball screw using multi-sensor data fusion technology and deep learning theory. This method was divided into four steps: (1) calculating the fused frequency spectra by parallel superposition, (2) pre-training the DNN layer by layer by unsupervised algorithm SAE, (3) fine-tuning parameters of DNN by supervised BP algorithm, (4) monitoring the degradation grade of ball screw. The effectiveness of the proposed method was verified using the fused spectra datasets which contain massive samples of ball screw. The results shows that the proposed method is able to mine degradation characteristics from the multi-sensor frequency spectra adaptively and classify the degradation grade effectively. The proposed method depend less on signal processing technology and diagnosis experience compared with traditional method. Monitoring results of the proposed method were also compared with the unfused spectra method and the proposed method is proved to have higher accuracy and stability. The study results indicate that the fused data is more comprehensive and accurate in reflecting the degradation condition of ball screw.

In future study, it will be interesting to study on choosing parameters of DNN such as number of layers, number of nodes, learning rate, etc. to improve both the efficiency and accuracy in degradation monitoring.

References

[1] Hinton G E, Salakhutdinov R R. Reducing the dimensionality of data with neural networks[J]. Science, 2006, 313(5786): 504-507.

[2] Ekanadham C, Reader S, Lee H. Sparse deep belief net models for visual area V2[J]. Advances in Neural Information Processing Systems, 2008, 20.

[3] Ranzato M, Susskind J, Mnih V, et al. On deep generative models with applications to recognition[C]//Computer Vision and Pattern Recognition (CVPR), 2011 IEEE Conference on. IEEE, 2011: 2857-2864.

[4] Lee H, Pham P, Largman Y, et al. Unsupervised feature learning for audio classification using convolutional deep belief networks[C]//Advances in neural information processing systems. 2009: 1096-1104.

[5] Le Cun Y, Chopra S, Marc'Aurelio Ranzato, et al. Energy-Based Models in Document Recognition and Computer Vision[C]//ICDAR. 2007, 7: 337-341.

[6] Glorot X, Bordes A, Bengio Y. Domain adaptation for large-scale sentiment classification: A deep learning approach[C]//Proceedings of the 28th International Conference on Machine Learning (ICML-11). 2011: 513-520.

[7] Tamilselvan P, Wang P. Failure diagnosis using deep belief learning based health state classification [J]. Reliability Engineering & System Safety, 2013, 115: 124-135.

[8] Tran V T, Al Thobiani F, Ball A. An approach to fault diagnosis of reciprocating compressor valves using Teager–Kaiser energy operator and deep belief networks [J]. Expert Systems with Applications, 2014, 41(9): 4113-4122.

[9] Nishchal K Verma, Vishal Kumar Gupta, Mayank Sharma, et al. Intelligent condition based monitoring of rotating machines using sparse auto-encoders[C]//Prognostics and Health Management (PHM), 2013 IEEE Conference on. IEEE, 2013: 1-7.

[10] Feng Jia, Yaguo Lei, Jing Lin, et al. Deep neural networks: A promising tool for fault characteristic mining and intelligent diagnosis of rotating machinery with massive data[J]. Machine System and Signal Processing, 2016, 72-73:303-315.

[11] Lei Y, Lin J, He Z, et al. A method based on multi-sensor data fusion for fault detection of planetary gearboxes[J]. Sensors, 2012, 12(2): 2005-2017.

[12] Safizadeh M S, Latifi S K. Using multi-sensor data fusion for vibration fault diagnosis of rolling element bearings by accelerometer and load cell[J]. Information Fusion, 2014, 18: 1-8.

[13] Schmidhuber J. Deep learning in neural networks: An overview[J]. Neural Networks, 2015, 61: 85-117.

[14] Duan K, Keerthi S S, Chu W, et al. Multi-category classification by soft-max combination of binary classifiers[M]//Multiple classifier systems. Springer Berlin Heidelberg, 2003: 125-134.

ONLINE MONITORING AND FAULT DIAGNOSIS SYSTEM OF SMART BEARING

Qiang Zeng[a], Fang Guo[a], Fei Liu[a] and Yimin Shao[a]*

[a]The State Key Laboratory of Mechanical Transmissions, Chongqing University, Chongqing, 400044, P.R. China
*Corresponding author. Tel.: +86-023-6511-2520; fax: +86-023-6518-6195; e-mail: ymshao@cqu.edu.cn

Abstract. Bearing is one of the most widely used mechanical components. Its condition affects the performance and accuracy of the mechanic, and its fault can lead to casualties and great economic losses. In traditional condition monitoring and fault diagnosis system, sensors are placed on bearing housing which may be away from the fault sources. Thus, it is difficult to obtain information of incipient fault. To solve this problem, a smart bearing online monitoring and fault diagnosis system is designed and developed based on "smart bearing local strong signal concept". The system includes a set of integrated build-in sensors hardware component and a wireless monitoring software system based on LabVIEW platform. Experiment results show that the system is more accurate and reliable than traditional monitoring methods, which provides a new way for bearing online monitoring and incipient fault diagnosis.

Keywords: smart bearing, fault diagnosis, wireless monitoring, local strong signal

1 Introduction

Benefiting from simple structure, low cost, high running speed, large operating temperature range and easiness to maintenance, bearing is widely used in rotating machinery. Due to the discreteness of bearing's life and bad severe operation condition, bearing is one of the components which are susceptible to damage. Bearing's condition affects the performance and accuracy of the mechanic. And its fault can lead to casualties and great economic losses. Statistics indicate that 30%-40% failures of rotating machinery are caused by the bearing [1], and only 10%-20% bearings meet the designed life[2]. To avoid the loss caused by bearing fault, it is urgent and necessary to monitor and diagnose the fault.

Bladon, Keith[3] etc. built a wayside monitoring alliance, which can monitor the high-speed Train Bearing, FAG and SKF[4] used out hung structure of micro-sensor to monitor bearings. Overtemperature may lead faults like cut-axis, bearing glue and wear. To solve this problem Rodrick K. Draney[5] etc. designed a build-in sensor to measure the bearing's temperature, Yimin Shao[6] designed a smart bearing with temperature sensors. Abhijit P. Suryavanshi[7], Robert X. Gao[7,8] and H Gao[9] used integrated sensor which is embedded into the grooving outer ring of bearing. In these bearing monitoring approaches, most sensors are away from the fault sources or only acquire few parameters. During actual operation, high-speed trains always work under speed variation condition, during which the RMS and kurtosis values change dramatically. It may provide misleading messages to the diagnosis, and affect the fault diagnosis accuracy. In this paper, to solve these problems, a monitoring and fault diagnosis system of smart bearing is designed. The smart bearing not only has the function of the normal bearing, but also has the function of sensor, which monitors the inner signals of bearing, such as vibration, rotating speed and temperature.

2 Smart bearing and its parameters

In current bearing monitoring systems, problems like measuring points are away from the fault sources, measurement parameters are scarce, and speed variation condition still affects the accuracy of incipient fault diagnosis. To have more comprehensive and accurate signals of the inner bearing, a sensor-integrated smart bearing is designed, which can collect the signal of speed variation condition and lay the foundation of incipient fault diagnosis. Smart bearing is designed under the limitation of bearing structure. Based on "smart bearing local strong signal concept", the smart bearing sensors are integrated in the bearing, which can collect the vibration, rotating speed and temperature signal near the fault sources and can be used to diagnose the incipient fault under speed variation condition and prevent the cut-axis accident caused by overtemperature.

The monitoring and diagnosis system of smart bearing includes a hardware component, which is composed of integrated measurement modules and signal conditioning board, and a wireless monitoring software system based on LabVIEW platform. The microcomputer can process the signal of the smart bearing and monitor the condition of the bearing.

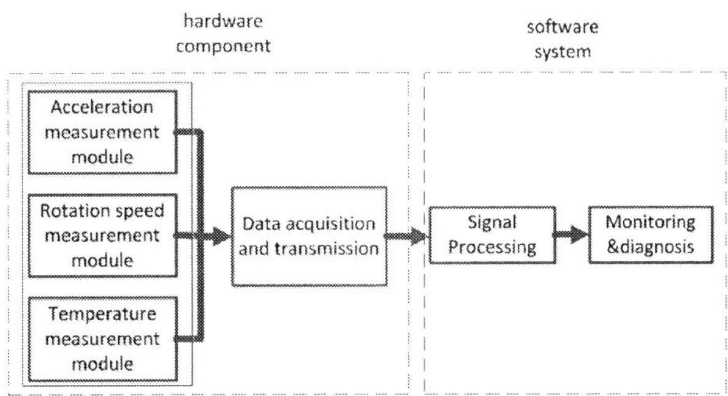

Fig. 1. The diagram of monitoring and fault diagnosis system of smart bearing

Fig. 2. Photo of smart bearing

Acceleration measurement module uses MEMS accelerometer to measure the triaxial acceleration of the bearing. It measures acceleration with a range ±16 g and bandwidth 0.5~1600Hz. The speed measurement module is composed of the Hall sensor and the magnet ring. It measures rotation speed with a range 0~2000rpm and measuring accuracy is 1%. The temperature measurement module uses Pt-thin-film resistor temperature sensor. It measures temperature with a range -50~300 °C and measuring accuracy of 1%. Three measurement modules are integrated in bearing.

The acquisition and wireless transmission module has a built-in anti-aliasing filter within each channel. It can sample the data with 24 bit precision and 512 00Hz maximum sampling frequency, and can be used for wireless transmission.

3 Diagnostic system using smart bearing

Kurtosis is a dimensionless parameter, which is independent of size and load. It is particularly sensitive to impact. In the incipient fault of rolling bearings, kurtosis value increased significantly. It can provide an early prediction for bearing fault.

$$K = \frac{1}{N}\sum_{j=1}^{N}[\frac{x_j - \mu}{\sigma}]^4 \tag{1}$$

Where x_j is vibration acceleration signal, N is the length of x_j, μ is the mean value of x_j, σ is the standard deviation of x_j.

Under speed variation condition, kurtosis shows great dependence of speed, which makes the incipient fault diagnosis difficult. To make it more suitable for diagnosis under speed variation condition, the non-stationary signal in time domain is resampled into stationary signal in angular domain with rotation speed impulses.

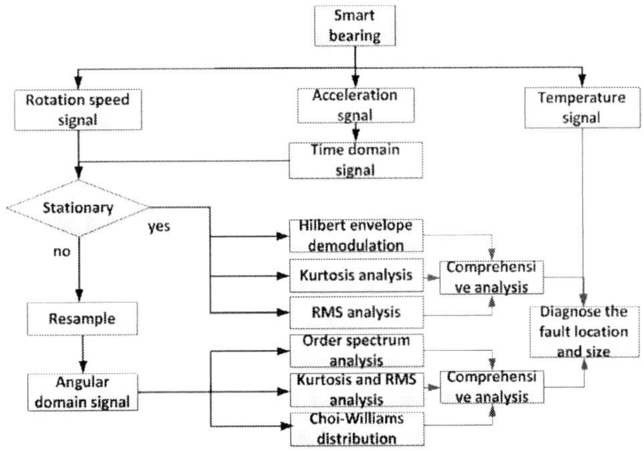

Fig. 3. The fault diagnosis process using smart bearing

Fig. 3 illustrates the diagnosis process of smart bearing. First, if the time domain signal is stationary, the Hilbert envelope demodulation, kurtosis and RMS in time domain will be analyzed comprehensively with the threshold to diagnose the fault location and fault size. If the signal is non-stationary, the signal will be resampled into an angular domain stationary signal. Then the order spectrum, kurtosis and RMS in angular domain will be analyzed comprehensively with the threshold to diagnose the fault location and fault size. The temperature signal is used to monitor the bearing's temperature. The thresholds of different fault size are obtained from experiments.

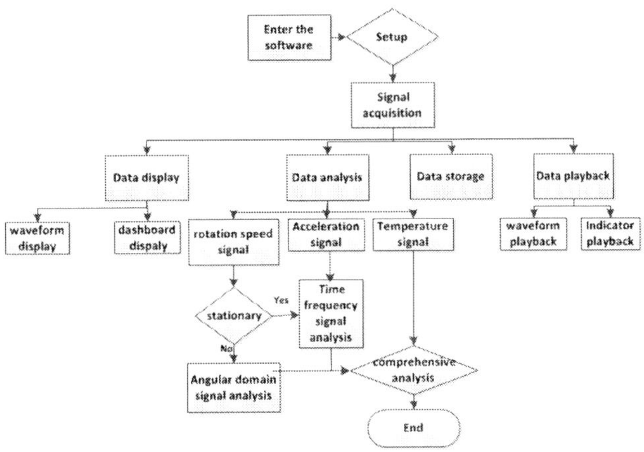

Fig. 4. Function modules of monitoring software

The monitoring software can monitor parameters like kurtosis and RMS in time and angular domain, temperature and rotation speed etc. The monitoring software is composed of data acquisition module, data display module, data analysis module data storage module and data replay module.

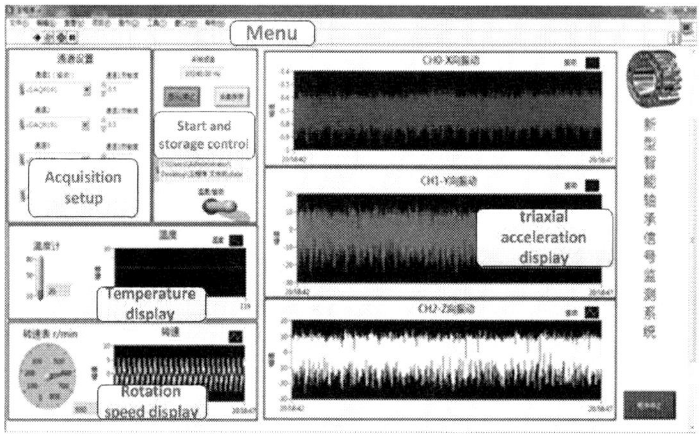

Fig. 5. The main interface of the monitoring system

4 Experimental results

For validation, data is collected from normal bearing and fault bearing with small size fault located in inner ring, and then compared with each other. The testing system is illustrated as **Fig. 6**. To verify the superiority of sensor-integrated smart bearing based on "smart bearing local strong signal concept", two more accelerometers are placed at different location. A1 is the accelerometer integrated in smart bearing, A2 is the accelerometer placed at the outer ring and A3 is the accelerometer placed at the bearing housing.

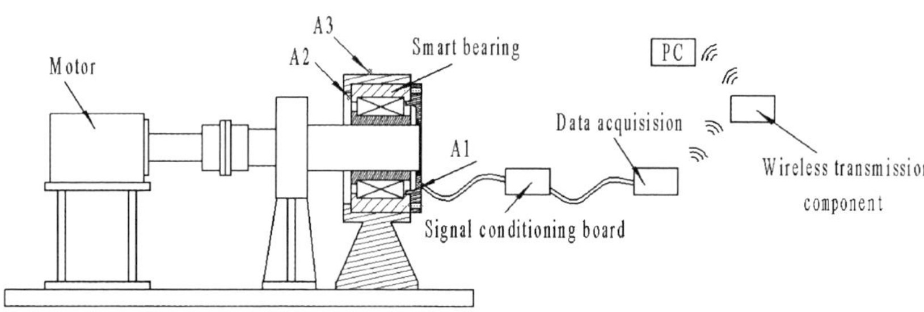

Fig. 6. Diagram of smart bearing testing system

Fig. 7. The inner ring with small size fault

Fig. 8 shows the data collected from speed-up and speed-down process. The maximum rotation speed of this process is 540rpm.

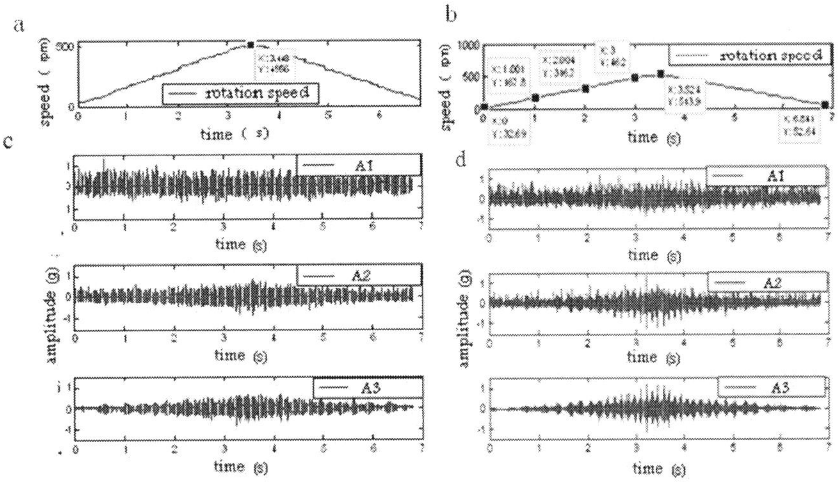

Fig. 8. The rotation speed signal and acceleration signal of normal and fault bearing under speed variation condition

Fig. 8(a) and **Fig. 8**(b) separately show the rotating speed of normal bearing and fault bearing. Both bearing reach the maximum speed around 3.5s. **Fig. 8**(c) and **Fig. 8**(d) separately show the acceleration signals of normal bearing and fault bearing. In **Fig. 8**(c) and **Fig. 8**(d), during 0-2s and 6-7s the amplitude of A1 is larger and A3 is smaller. A1 is closed to fault source and away from the surrounding interference, which benefits the incipient fault diagnosis.

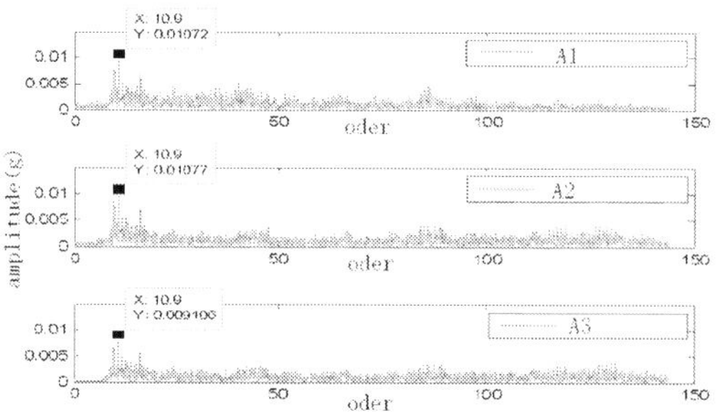

Fig. 9. The order-spectrum of fault bearing under speed variation condition

The fault is located in inner ring and the fault characteristic order calculated as 10.8. In **Fig. 9**, all sensors reach the maximum amplitude at 10.9, which matches the calculation result well. Among them the amplitude of A3 is smaller than others.

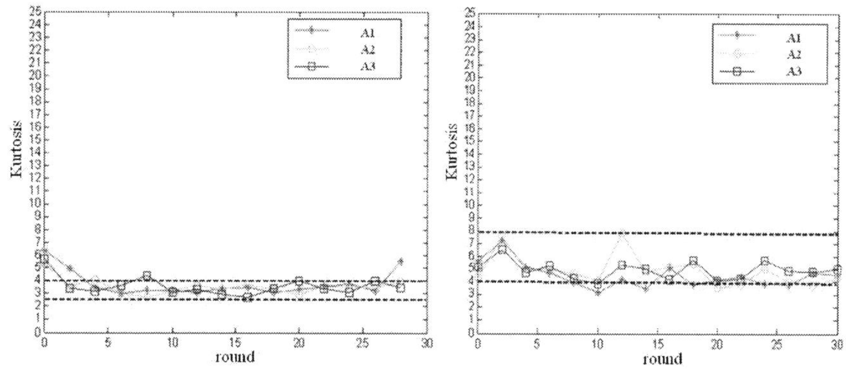

Fig. 10. Angular domain kurtosis of normal and fault bearing under variation condition

Fig. 10 (a) is the angular domain kurtosis of normal bearing, which mainly distributed between 2.5 and 4. **Fig. 10** (b) is the angular domain kurtosis of fault bearing, which mainly distributed between 4 and 8. The fault bearing and normal bearing can be easily separated by angular domain kurtosis under speed variation condition.

5 Conclusion

This paper introduced an online monitoring and diagnosis system of smart bearing. To solve the problems like few monitoring parameters, difficulties of sensor integration, and the sensors placed away from the fault source, a system with hardware component and wireless monitoring software based on LabVIEW platform is designed. It collects time domain signal and angular domain signal, and monitors the parameters like RMS, kurtosis, rotation speed and temperature. The experimental results indicate that the monitoring system is more accurate and reliable than the traditional monitoring methods, especially under the speed variation condition. It provides a new way for bearing online monitoring and incipient fault diagnosis. This system can be used in rotating machinery with speed variation condition such as high-speed railway trains, urban railway trains, and mine trucks.

Reference

1. J. Morel (2002). Vibratory monitoring and predictive maintenance. Measurement and Control. Vol. RD, Art. R6100.
2. Berry, J. E (1991). how to track rolling element bearing health with vibration signature analysis. Sound and Vibration.24-35.
3. Bladon, Keith, Rennison, David (2004). Predictive condition monitoring of railway rolling stock. Conference on Railway Engineering, 22.1-22.12.
4. Du yinghui (2002). Bearing with integrated sensor. bearing. (in Chinese)
5. Rodrick K. Draney (2008). High temperature sensor for bearing health monitoring. IEEE,.
6. Yimin Shao, Liang Ge, Jieping Fang (2008). Fault diagnosis system based on smart bearing. International Conference on Control, Automation and Systems. 1084-1089.
7. Abhijit P. Suryavanshi and Robert X. Gao(2001). Vibration Analysis of a Bearing Integrated Sensor Module. Proceedings of SPIE. Vol. 4563, 75-83.
8. Holm-Hansen B T, Gao R X(1997). Integrated micro-sensor module for a smart bearing with on-Line fault detection capabilities. IEEE Instrumentation & Measurement Technology Conference. Vol.2, 1160-1163.
9. H Gao, Q Lv, LG Qu(2005). Influence of Groove Slotting for Micro-Sensor Embedding on Smart Bearing. Key Engineering Materials. 297-300:562-566.

ROLLING BEARING FAULT DIAGNOSIS BASED ON LOCAL MEAN DECOMPOSITION AND MULTI-SCALE SPECTRUM ENTROPY ANALYSIS

Meng Zong Zhao Dongfang Li Jing Zhang Yuanyuan and Fan Fengjie*

School of Electrical Engineering, Yanshan University, Qinhuangdao, Hebei, China
E-mail:ffjmz@126.com

Abstract. Combining morphological spectrum entropy with multi-scale entropy, a new fault feature parameter – multi-scale spectrum entropy is proposed in this paper. Then, a method for rolling bearing fault diagnosis based on local mean decomposition(LMD) and multi-scale spectrum entropy analysis is studied, combining the multi-scale spectrum entropy with local mean decomposition. First, wavelet half soft threshold denoising is conducted on the rolling bearing vibration signal. Then the noise reduced signal is decomposed adaptively by LMD, and a new signal is recombined according to the correlation degree of each PF components and the noise reduced vibration signal. Finally, the rolling bearing fault feature extraction and diagnosis is realized by getting the multi-scale spectrum entropy of the recombined signal. Applying this method to rolling bearing fault diagnosis, the analysis results show that the method proposed in this paper can extract the fault features of rolling bearing fault vibration signal effectively, and realize the fault type diagnosis of rolling bearing accurately.

Keywords: Rolling bearing, fault diagnosis, multi-scale spectrum entropy, local mean decomposition, morphological spectrum entropy

1 Introduction

Rolling bearing is a kind of mechanical part that widely used in modern mechanical equipment, and it is also the most fault-prone mechanical part, so rolling bearing condition monitoring and fault diagnosis is of great significance[1].

The rolling bearing vibration signal contains abundant fault feature information, extracting feature information from vibration signal is the key to rolling bearing fault diagnosis[2]. As an adaptive method for non-stationary and nonlinear signal analysis[3], local mean decomposition is widely used in machinery fault diagnosis field[4,5,6]. Morphological spectrum can intuitively reflect the characteristics composition of signal under different scales, and morphological spectrum entropy can quantitatively describe the morphological characteristics complexity of the signal[7], multi-scale entropy can measure the complexity of time series under different time scales[8].

Based on the idea of multi-scale entropy, combining morphological spectrum entropy with multi-scale entropy, a new fault feature parameter—multi-scale spectrum entropy is proposed in this paper. Multi-scale spectrum entropy can be used to measure the morphological characteristics complexity of time series under different time scales. Combining LMD with multi-scale spectrum entropy, a new method for rolling bearing fault diagnosis based on LMD and multi-scale spectrum entropy analysis is proposed in this paper.

2 Multi-scale spectrum entropy

Based on the idea of multi-scale entropy, introducing time scale factor into morphological spectrum entropy, this paper puts forward the concept of multi scale spectrum entropy, it can measure the morphological characteristics complexity under different time scales. The calculation method is summarized as below:

1. $[x_1, x_2, \ldots, x_N]$ is a continuous time series, conducting coarse graining transform on the original time series, and a new time series is got:

$$f(\tau) = y_j(\tau) = \frac{1}{\tau} \sum_{i=(j-1)\tau+1}^{j\tau} x_i \quad (1)$$

In the formula, $1 \le j \le N/\tau$, $\tau = 1,2\ldots$ is the time scale factor. the original time series is divided into a coarse grain sequence that contains τ sections, and the length of every coarse grain series is N/τ .

2. Calculating the morphological spectrum entropy of each coarse grain sequence, and τ morphological spectrum entropies of the coarse grain sequences are got, this method is called as multi-scale spectrum entropy.

Combining LMD with multi-scale spectrum entropy, this paper puts forward a method for rolling bearing fault diagnosis. It's specific steps are as below:

1. Dealing with the noise of rolling bearing vibration signal, using the wavelet half soft threshold denoising method.
2. Conducting LMD operation on the noise reducted vibration signal, and several PF components are got.
3. Calculating the correlation coefficients of each PF components and the noise reducted signal. Choosing the PF components whose correlation coefficient is larger than a certain value, and discarding the PF components whose correlation coefficient is less than the certain value to reconstruct the signal.
4. Then calculating multi-scale spectrum entropy of the reconstructed signal, and the working status and fault types of rolling bearing can be judged.

3 Experimental research

Experimental data is got from the rotating machinery fault simulation test rig of Case Western Reserve University. Single point failures were build on the bearings, using electrical discharge machining technology. Fault diameter is 0.1778mm, sampling frequency is 12000Hz. Bearing states including normal state, rolling element fault, inner ring fault and outer ring fault, the time domain waveform of different states are shown in Fig. 1. To deal with the noise of rolling bearing vibration signal, wavelet db5 is selected as the wavelet function, and wavelet half soft threshold is selected as the threshold, then conducts 3 layers decomposition on the vibration signal. The time domain waveform diagrams after denoising are shown in Fig. 2. The noise reducted vibration signal is decomposed into several PF components by LMD. Then, in order to select the components that contain main fault feature information. It calculates the correlation coefficients of each PF components and the noise reducted signal, and puts the value of each correlation coefficient into comparision. This paper sets the threshold value c=0.100, it provisions that the PF components whose correlation coefficient is smaller than 0.100 shouled be given up, and the PF components whose correlation

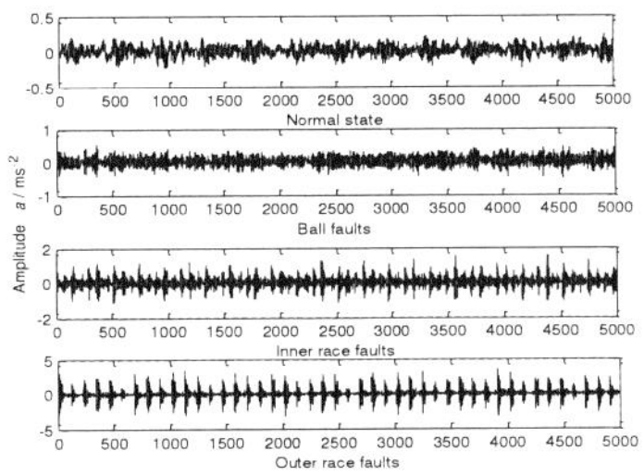

Fig. 1. Waveforms before denoising

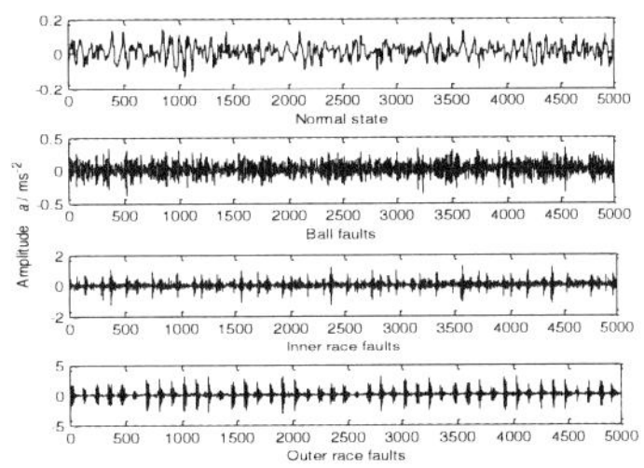

Fig. 2. Waveforms after denoising

coefficient is not smaller than 0.100 will be reserved. Thus, the reconstructed signal is a sum of PF components that has been reserved. Calculating the correlation coefficients of the four different state signals, and reconstructing the signals. The reconstructed signals are shown in Fig. 3. It can be seen from Fig. 3 that there is no obvious difference between the waveforms of the four reconstructed signals. Then, calculating morphological spectrum and morphological spectrum entropy of the four reconstructed signals. The morphological spectrums are shown in Fig. 4. In the calculation processs of morphological spectrum entropy, it selects 3 groups data randomly from each state as samples and gets the average value of morphological spectrum entropy. The results are shown in table 1.

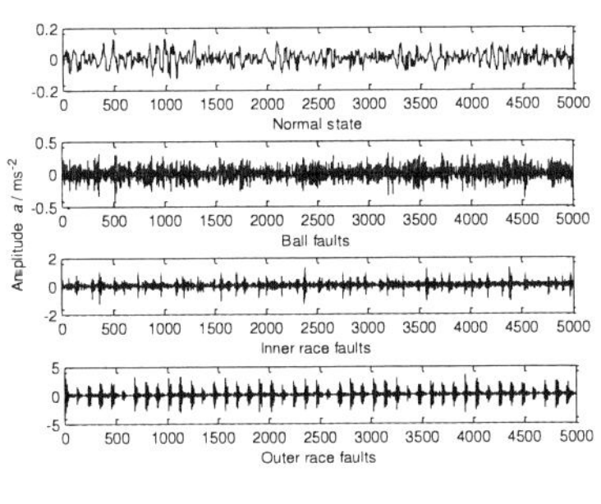

Fig. 3. The reconstructed signals

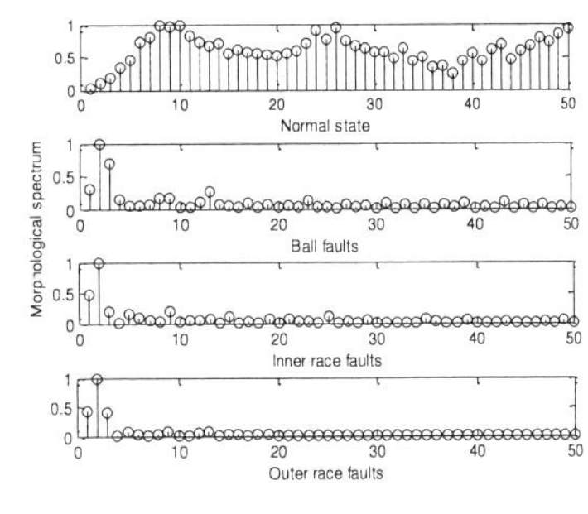

Fig. 4. Normalized morphological spectrum

It can be seen from the morphological spectrum and the morphological spectrum entropy, that if the status of rolling bearings are different, the morphological spectrum and morphological spectrum entropies are different also, and inversely, the morphological spectrum and the morphological spectrum entropies are similar to each other. Because the morphological spectrum entropies are got under a single time scales, although it is able to distinguish the rolling bearing working status of different fault types, the effect is not obvious. Next, calculating the multi-scale entropy of the four reconstructed signals, establishing the function relationship of entropy values and

Table 1. The morphological spectrum entropy of rolling bearing signal

morphological spectrum entropy	Sample 1	Sample 2	Sample 3	Average
Normal state	3.8341	3.8305	3.8035	3.8227
Ball faults	3.0487	3.0219	3.0043	3.0250
Inner race faults	3.1277	3.1606	3.2019	3.1634
Outer race faults	2.2368	2.2754	2.2097	2.2406

time scale factors, their curves are shown in Fig. 5. It can be seen that the multi-scale entropies of 3 status tend to be a similar value with the increase of time scale factor. However, curves of multi-scale entropy have certain overlap between the different working status, the effect is not perfect. With the increase of time scale factor, there is no certain regularity presented in the change of the multi-scale entropy value, and it is not conducive for further analysis. Next, the multi-scale spectrum entropy are calculated. Establishing the function relationship between multi-scale spectrum entropy and time scale factor, the curves are shown in Fig. 6. It can be seen that multi-scale spectrum entropy can obviously distinguish the different status of rolling bearings. It can not only reflect the morphlogical characteristics complexity of time series, but also contain the information under different time scales. With the increase of time scale factor, the multi-scale spectrum entropies of each state signal tend to be a fixed value that different with each other, it is conducive for further analysis. Therefore, the multi-scale spectrum entropy has its superiority in rolling bearing fault feature extraction.

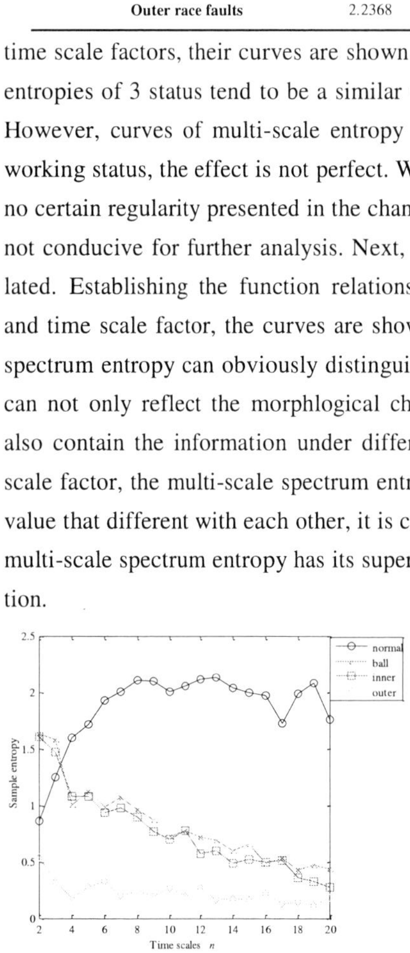

Fig. 5. Multi-scale entropy

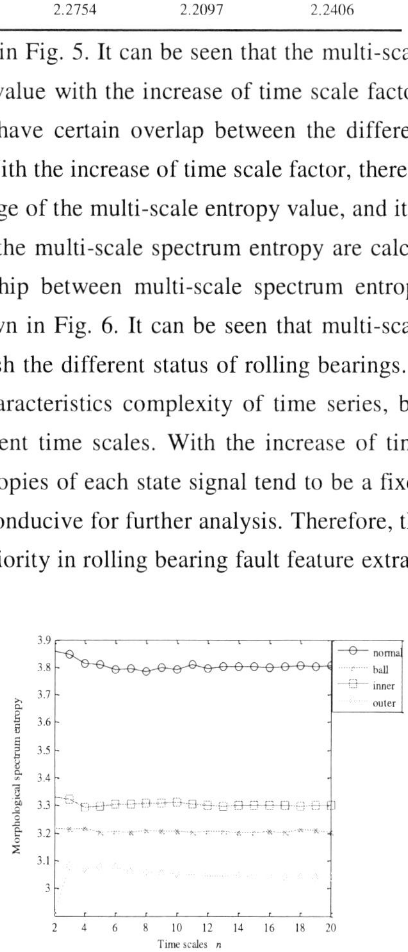

Fig. 6. Multi-scale spectrum entropy

4 Conclusions

The concept of multi-scale spectrum entropy is proposed in this paper, combining morphological spectrum entropy with multi-scale entropy. Then, a method for rolling bearing fault diagnosis based on LMD and multi-scale spectrum entropy analysis is studied. The experimental results show that the proposed method can extract the fault features of rolling bearing fault vibration signal effectively, and realize the fault type diagnosis of rolling bearing accurately by multi-scale spectrum entropy analysis.

Acknowledgements. This work is supported by the National Natural Science Foundation of China (No.51575472), The Natural Science Foundation of Hebei Province of China (No.E2015203356), the scientific research program of Hebei Education Department(No.ZD2015049) and the scientific research program for Talents Returning from Overseas of Hebei Province(No.C2015005020).

References

1. Jin Zhang, Zhiping Feng, Fulei Chu.: Extraction of rolling bearing fault feature based on Time-wavelet energy spectrum. J. Journal of Mechanical engineering. 47 (17), 44-49 (2011)
2. Lotfi Saidi, Jaouher Ben Ali, Farhat Fnaiech.: Bi-spectrum based-EMD applied to the non-stationary vibration signals for bearing fault diagnosis. J. ISA Transcations. 53, 1650-1660 (2014)
3. Smith J S.: The Local mean decomposition and its application to EEG perception data. J. Journal of the Royal Society Interface. 2 (5), 443-454 (2005)
4. W.Y.Liu, W.H.Zhang, J.G.Han.: A new wind turbine fault diagnosis method based on the local mean decomposition. J. Renewable Energy. 48, 411-415 (2012)
5. Junsheng Cheng, Yi Yang, Yu Yang.: A rotating machinery fault diagnosis method based on local mean decomposition. J. Digital Signal Processing. 22, 356-366 (2012)
6. Zong Meng, Yachao Wang, Xiaoyan Wang.: Rotating machinery fault diagnosis method based on the local mean decomposition and extreme points extension. J. Acta Metrologica Sinica. 35 (5), 469-474 (2014)
7. Wanlu Jiang, Zhi Zheng, Yong Zhu.: Demodulation for hydraulic pump fault signals based on local mean decomposition and improved adaptive multi-scale morphology analysis. J. Mechanical systems and signal processing. 58-59, 179-205 (2015)
8. Alok Verma, Somnath Sarangi, M.H.Kolekar.: Stator winding fault prediction of induc-tionmotors using multiscale entropy and grey fuzzy optimization methods. J. Computers and Electrical Engineering. 40, 2246-2258 (2014)

ROTATING MACHINERY FAULT DIAGNOSIS BASED ON VARIATIONAL MODE DECOMPOSITION AND BLIND SOURCE SEPARATION

Fan Fengjie Zhao Dongfang Xiong Jingming Wei Chao and Meng Zong*

1 Key Laboratory of Measurement Technology and Instrumentation of Hebei Province, Qinhuangdao,Hebei,China

E-mail:mzysu@ysu.edu.cn

Abstract. A blind source separation method based on Variational Mode Decomposition(VMD) is studied in this paper. First, Variational Mode Decomposition method is used to decompose the observed signal; then, a new observed signal is composed by the variational mode components and the original observed signal, thus, it turns the underdetermined blind source separation issue into positive definite issue; finally, blind source separation is conducted on the new observed signal. Compared with the underdetermined blind source separation method based on Empirical Mode Decomposition(EMD), the experimental results verify the validity and feasibility of this method, and it puts forward a new method for underdetermined blind source separation.

Keywords: rotating machinery, fault diagnosis, variational mode decomposition, blind source separation

1 Introduction

Rotating machinery is a kind of mechanical equipment that widely used in industries[1,2]. The malignant accidents caused by rotating machinery fault caused great loss to the country. Therefore, effective detection and diagnosis for rotating machinery is of great significance. In fault diagnosis field, blind source separation has its own unique advantages on fault source separate and diagnosis[3]. However, in practical engineering, there is a large number of underdetermined blind source signal separation issue. The traditional blind source separation method must be implemented in the case of positive definite or super-determined. Once the number of observed signal is less than the number of source signal, the traditional blind source separation method will be powerless [4,5].

In 2014, Dragomiretskiy K et al proposed a new adaptive signal processing method -variational mode decomposition[6], this method determines the frequency center of each component and its bandwidth by iterativing search the optimal solution of variational mode in the acquisition process of decomposition components. Compared with EMD, VMD have more advantages on weak fault feature extraction[7].This paper puts forward a kind of underdetermined blind source separation method based on VMD, to solve underdetermined blind source separation issue. The experimental results show that this method is effective for the separation of underdetermined blind source signals,and provides an effective method for underdetermined blind source separation.

2 VMD-BSS Fault Source Separation Method

2.1 Variational mode decomposition Method

Compared with EMD algorithm, the difference of VMD algorithm is that the intrinsic mode function (IMF) is redefined as a frequency-modulated and amplitude-modulated signal, which is defined as variational mode component.

The VMD algorithm obtains the variational mode component by transferring the signal decomposition process into variational mode framework, and looking for the optimal solution that constrained the variational modal model to realize signal decomposition adaptivly, it gets rid of the loop screening stripping method that used in EMD algorithm. The frequency center and bandwidth of every variation mode component updated constantly in the terative process, and then it realized the frequency band adaptive subdivision based on the frequency domain characteristics of the actual signal. Finally, a number of narrow band variational mode components are obtained. The specific implementation process of variational mode decomposition can be got from reference[6].

2.2 VMD-BSS Method

The basic idea of the VMD-BSS method is decomposing the observed signal into several variational mode components by VMD, then, a new observed signal is reconsituted by the variational mode components and the original observed signal. Thus, the underdetermined blind source separation issue is converted into a positive definite blind source separation issue. Then blind source separation can be conducted on the new observed signal. The specific process is as below:

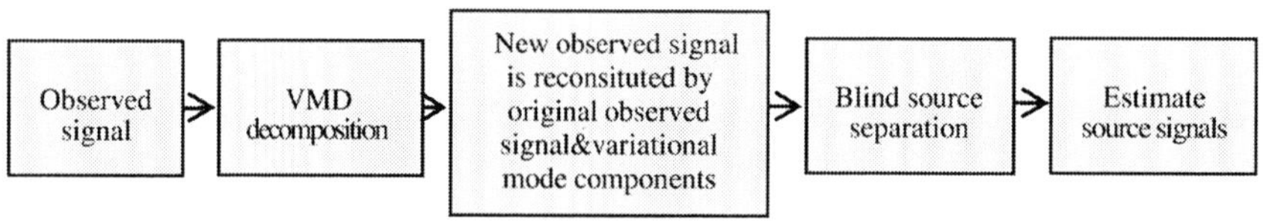

Fig. 1. Flow chart of VMD-BSS

3 Experiment

Experimental data is got from the rotating machinery fault simulation test rig of West University. The tested bearing model is 6205-2RS SKF which supports the motor axis of rotation of deep groove ball bearings.

In the test, the motor speed is 1797r/min, sampling frequency is 12000Hz, and sampling length is 1000. Single point failures were build on the bearings using electrical discharge machining technology, fault diameter is 0.1778mm. According to the structure size of the bearing and its working speed, it can calculate the inner ring fault characteristic frequency of the bearing fi=162.2Hz and the outer ring fault characteristic frequency fo=107.4Hz.

To validate the validity of the method proposed in this paper, a 2 x 3 mixing matrix A is choosen arbitrarily, and then the inner ring fault signal and the outer ring fault signal are mixed together to get the observed signal, the time domain waveform of the observed signal after mixing is shown in Fig. 2. Conducting VMD operation on the observed signal, and it gets 2 variational mode components as shown in Fig. 3. Then, a new observed signal is reconstructed by the first variation mode component and the original observed signal. At this time, the dimension of observed signal is raised. It means that the number of observed signal is equal to the number of source signal, and then the traditional blind source separation method can be applied to separate the new observed signal and get the results in Fig. 4. The envelope spectrums of Fig. 4 are shown in Fig. 5.

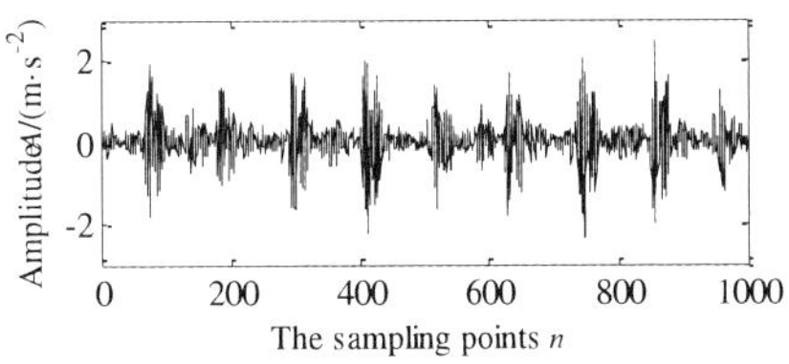

Fig. 2. The observed signal

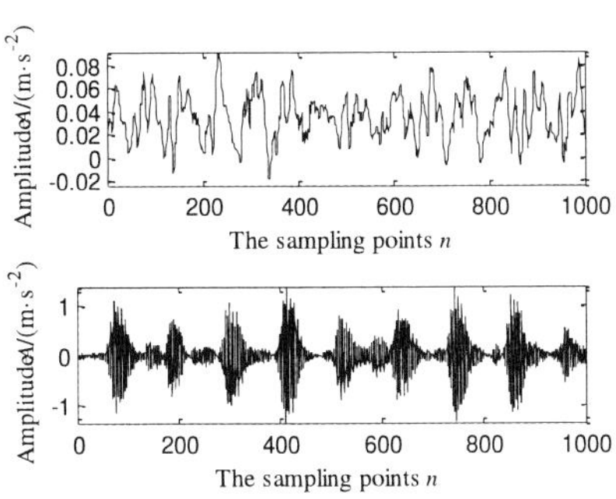

Fig. 3. Variational mode components

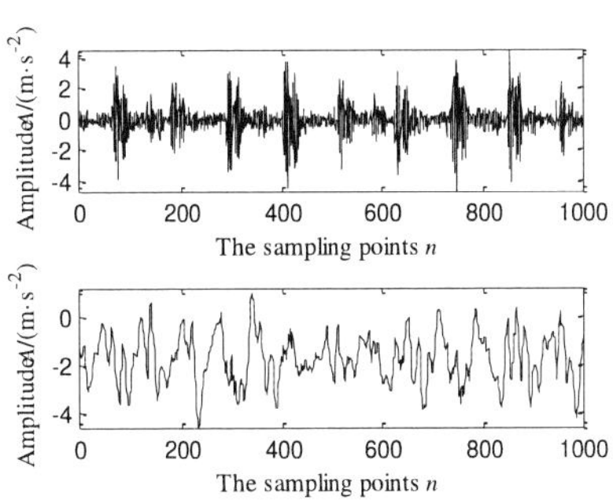

Fig. 4. VMD - BSS signals

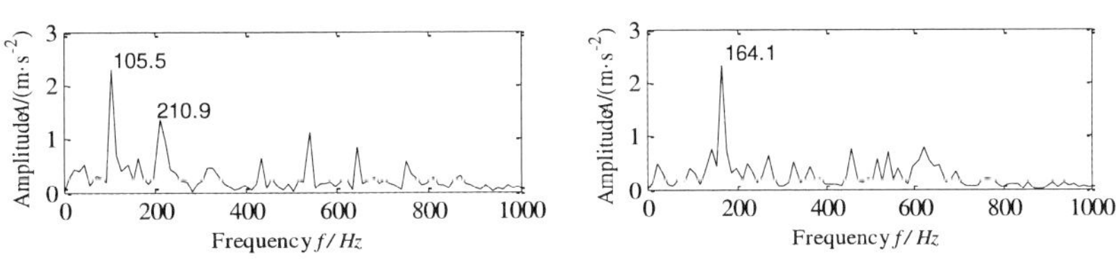

(a) Outer ring fault

(b) Inner ring fault

Fig. 5. Envelope spectrums of VMD - BSS

It can be seen from Fig. 5(a), peaks appear at 105.5Hz and 210.9Hz. Due to the influence of actual measurement, there always some small differences between the theoretical calculated values and the actual values. So 105.5Hz and 210.9Hz respectively corresponds to the outer ring fault characteristic frequency 107.4Hz and its double frequency. It nicely reflects the outer ring fault characteristic frequency. It also can be seen from Fig. 5(b) that the peak appears at 164.1 Hz, which is close to the theoretical inner ring fault characteristic frequency 162.2 Hz. It can reflect the inner ring fault characteristic frequency perfectly.

EMD method also can be used to solve the problem of underdetermined blind source separation[8], the original observed signal is decomposed by EMD at first, then new observed signal is composed of the first decomposed IMF component and the original observed signal. It gets the estimated signals shown in Fig. 6 by separating the new observed signal using traditional blind source separation. The envelop spectrums of Fig. 6 are shown in Fig. 7.

It can be seen from Fig. 7(a), the peaks appear at 105.5Hz and 210.9Hz. They are obvious reflections of the outer ring fault characteristic frequency 107.4Hz and its double frequency. But the envelope spectrum in Fig. 7(b) is some of confusing, the highest frequency is 105.5Hz, and there are also many other frequency peaks. The 164.1Hz as the red spot shown is buried in other frequencices, it has not been highlighted. Fig. 7(b) has no effective reflection on inner ring fault characteristic frequency and it is easy to cause misdiagnosis.

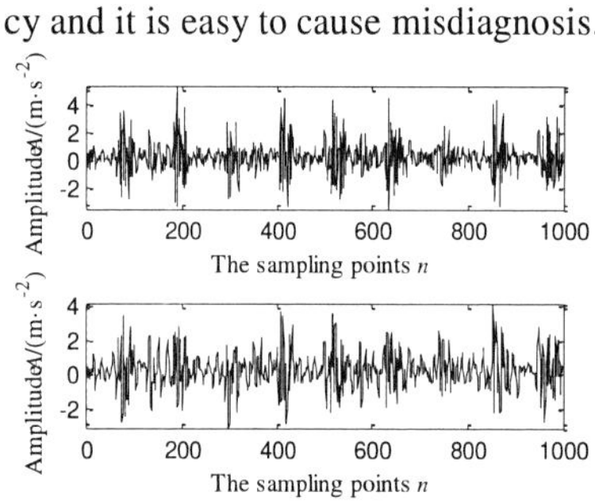

Fig. 6. EMD - BSS signals

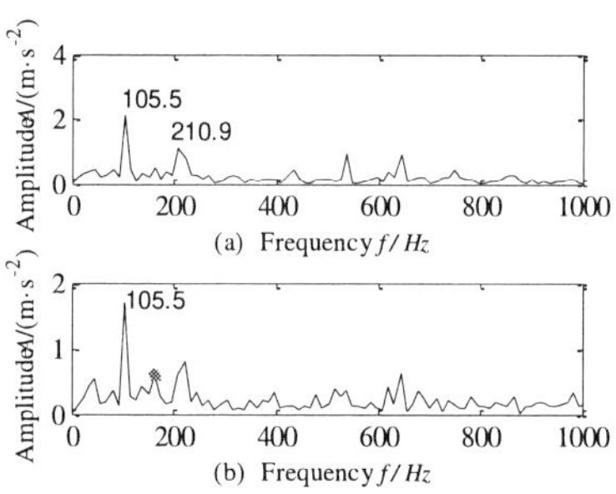

Fig. 7. Envelope spectrums of EMD - BSS

4 Conclusions

The VMD method can be used to solve the underdetermined blind source separation issue. New observed signal is reconstructed by the first variation mode components and the original observed signal. It convertes the underdetermined blind source separation issue into positive definite issue. The experimental results verified the validity and feasibility of this method, compared with the underdetermined blind source separation method based on EMD, the method based on VMD is more accurate, effective and has obvious advantages. It puts forward a new method for underdetermined blind source separation.

Acknowledgements. This work is supported by the National Natural Science Foundation of China (No.51575472), The Natural Science Foundation of Hebei Province of China (No.E2015203356), the scientific research program of Hebei Education Department(No.ZD2015049) and the scientific research program for Talents Returning from Overseas of Hebei Province(No.C2015005020).

References

1. Heng Zhang, Rongzhen Zhao.: Weighed KPCA based on fault feature selection and feature information fusion. J. Journal of Vibration and Shock. 33 (9), 89-93 (2014)
2. Jie Zhang, Hongli Gao, Chunjun Chen, et al.: Blind Source Separation Method and Application for Nonstationary Vibration Signal of High Speed Train. J. Journal of Mechanical Engineering. 50 (19), 97-104 (2014)
3. Jutten C, Herault J.: Blind separation of sources, part i: An adaptive algorithm based on neuromimetic architecture. J. Signal Processing. 24, 1–10 (1991)
4. Xiaohui Li, Pan Fu.:Application of Single-channel Blind Source Separation Based on EEMD in Bearing Fault Diagnosi. J. Journal of Mechanical Engineering. 25 (7), 924-930 (2014)
5. Qiang Li, Cong Fu, Hong Jiang, et al.: Single-channel vibration signal separation by combining empicical mode decomposition with time-frequency analysis. J. Journal of Vibration and Shock. 25 (7), 924-930 (2014)
6. Dragomiretskiy K, Zosso D.: Variational Mode Decomposition. J. Signal Processing IEEE Transactions on. 62 (3), 531 – 544 (2014)
7. Guiji Tang, Xiaolong Wang.: Parameter Optimized Variational Mode Decomposition Method with Application to Incipient Fault Diagnosis of Rolling Bearing. J. Journal of Xi'an Jiaotong University. 49 (5), 73-81 (2015)
8. Zhinong Li, Yaping Lu, Tao Fan, et al.: Underdetermined blind source separation method of machine faults based on empirical mode decomposition. J. Journal of Aerospace Power. 24 (8), 1886-1892 (2009)

MODELING AND ANALYSING ON LOCALIZED DEFECT OF TAPERED ROLLER BEARING

Yuwei Liu[1,1],Yongsheng Zhu[1],Ke Yan[1] and Dawei Gao[1]

[1]Key Laboratory of Education Ministry for Modern Design & Rotor-Bearing System , Xi'an Jiaotong University, Xi'an,710049, China
liu88246091@163.com

Abstract: To reveal the effect of localized defect on vibration characteristics of tapered roller bearing(TRB) , a simulation model was established in which the localized defect was treated as the change of the contact stiffness and damping between roller and race.The contact behaviour of roller-race was firstly analyzed based on non-Hertz line contact theory, and then the effects of stiffness and damping caused by the localized defect was obtained. A three-degrees of freedom (3-DOF)quasi-dynamic model of TRB with a localized defect was developed by considering the influence of localized defect on stiffness and damping. Simulation shows that the proposed simulation model give more reasonable results compared with the traditional fault model, and provides some valuable conclusions for faults diagnosis of TRB.

Keywords: Non-Hertz line contact theory, Tapered roller bearing, Localized defect, Contact mechanics

1 Introduction

Running state of bearing plays important role in equipment performance. The diagnostic techniques of bearing faults based on vibration signal are more applicable than other methods due to their non-invasive and high reactivity to incipient faults [1]. So far the fault detection technology for the ball bearing has been studied extensively, and the defect was always simulated as an impulse loading. For simulation model of roller element bearing[2-6], the defect was always considered as a "through damage area" in most cases, which means the length of defect is longer than that of roller, and the roller's position will suddenly change and an impulse loading is possibly induced when the roller passes the defect area, therefore the traditional defect-modeling method for ball bearing can be directly applied to roller element bearing under "through damage area" assumption. However, in most cases, only relatively small defects can be observed in local position of the contact area of roller-race in real applications [8], as shown in Fig.1, therefore more reality fault simulation model was required for element roller bearing.

In this paper, a new fault simulation model was proposed for taper roller bearing (TRB) based on three-degrees of freedom (3-DOF)quasi-dynamic model, in which the localized defect was assumed that it will change the contact stiffness and damping between roller and race at the defect location. The rest of the paper was organized as following: First, the three-DOF quasi-dynamic model of TRB was introduced. Then the effect of localized defect on contact stiffness and damping was analyzed based on non-Hertz line contact theory. Finally, a simulation model of TRB with localized defect was developed by combining the quasi-dynamic model with the change of stiffness and damping caused by localized defect. The vibration characteristics of TRB with localized defect were analyzed by the proposed model.

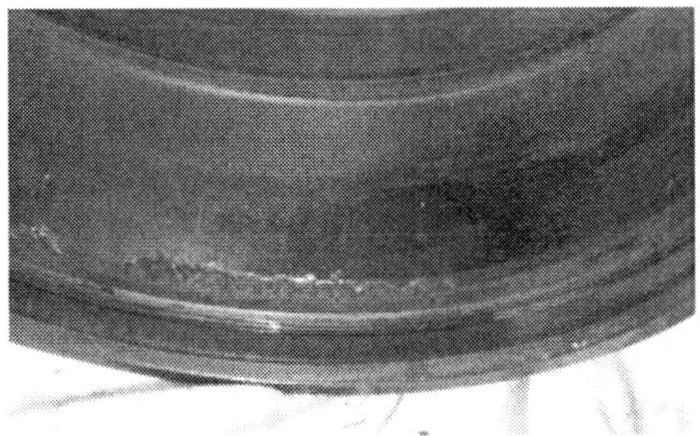

Fig.1 The realistic defect of TRB

2 Quasi-dynamic model of taper roller bearing system

The contact of roller-race can be considered as the non-linear spring-damp system, as shown in Fig.2. Based on the hypothesis of rigid ring[7], the three degree freedom dynamic model of TRB can be written as follows

$$\begin{aligned} &m\ddot{x} + \dot{x}\sum_{j=1}^{Z} c_j \cos\alpha\cos\varphi_j + x\sum_{j=1}^{Z} k_j \cos\alpha\cos\varphi_j = F_x \\ &m\ddot{y} + \dot{y}\sum_{j=1}^{Z} c_j \cos\alpha\sin\varphi_j + y\sum_{j=1}^{Z} k_j \cos\alpha\sin\varphi_j = F_y \\ &m\ddot{z} + \dot{z}\sum_{j=1}^{Z} c_j \sin\alpha + z\sum_{j=1}^{Z} k_j \sin\alpha = F_z \end{aligned} \tag{1}$$

where m is the mass of the rotor and inner race, c_j is the damping of j th roller and races(inner race and outer race), k_j are the stiffness of j roller and races(inner race and outer race), α is the contact angle between roller and race, φ_j is the position angle of j th roller.

The stiffness/damping between roller and inner-race/outer race is in series which should be considered in the calculation of k_j, c_j.

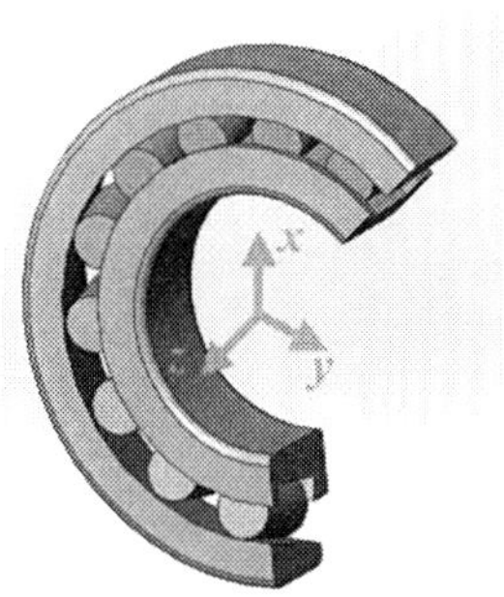

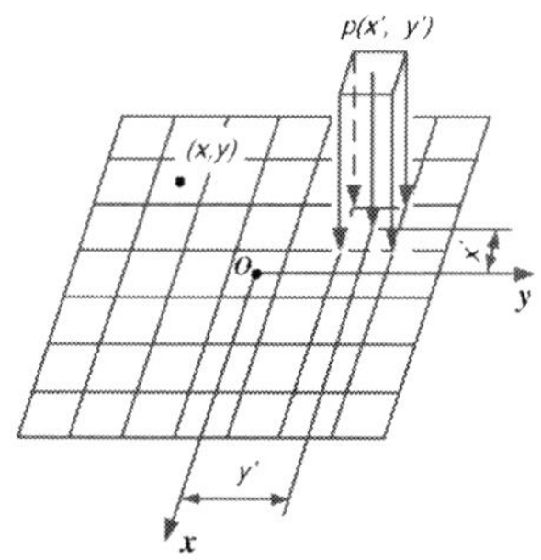

Fig.2 3-DOF quasi-dynamic model of tapered roller bearing   Fig.3 Division of contact area elements

In the traditional dynamic model, the stiffness k is based on Hertz line contact theory, which ignores the effect of the roller and race's geometrical characteristic on the contact behavior of roller-race. The contact behavior of roller-race is considered as a non-Hertz line contact problem in present paper. As shown in Fig.3, the contact behavior of roller-race based on non-Hertz contact theory can be explained as [3]

$$\iint_{S_c} p(x,y)dxdy = Q \tag{2}$$

$$\frac{1}{\pi E'}\iint_{S_c}\frac{p(x',y')dx'dy'}{\sqrt{(x-x')^2+(y-y')^2}} = \delta - Z(x,y) \tag{3}$$

where S_c is the contact area of roller-race, E' is the equivalent elastic modulus of roller-race, $p(x',y')$ is the stress of element, δ is the elastic approach of roller-race, $Z(x,y)$ is the initial distance, Q is the load between roller and race.

The roller and race's geometrical characteristic was described by initial distance $Z(x,y)$, and then the effect of the roller and race's geometrical characteristic on the contact behavior was considered. To simplify iterative computations in the dynamic model, the non-linear stiffness k of roller-race is generally described by load–deformation relationship [3]

$$Q = k\delta = K\delta^n \tag{4}$$

where K,n is the load–deflection factors of roller-race.

The load–deflection factors K,n was obtained by numerically solving Eq.(2) and Eq.(3).

Damping of bearing is composed of material damping and lubricant damping. Since the material damping factor is much less than lubricant damping factor, only the damping of lubricating film is considered in most cases. The damping of lubricating film between roller and race is described as[9]

$$c = \frac{6\pi\eta_0 R^{1.5} l_e}{2\sqrt{2}h_c^{1.5}} \tag{5}$$

Where η_0 is the lubricant viscosity, R is the equivalent curvature radius of roller-race, h_c is the central film thickness, l_e is the effective contact length between roller and race.

3 Simulation model of TRB with localized defect

Several factors, such as fatigue spalling, indentation, corrosion and so on, will cause the localized defection the contact area between roller and race. For taper roller bearing, most localized defects occur at the end of contact area. The localized defect increase initial distance $Z(x, y)$ at the defect area and the effective contact length l_e of roller-race, as shown in Fig.4, and the stiffness factors K,n and damping c is changed.

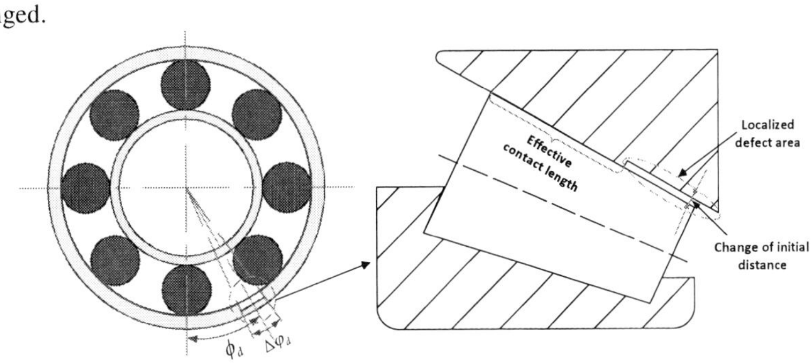

Fig.4The localized defect of outrace

The position of localized defect was described by angular position of the defect ϕ_d, angular distance of defect $\Delta\varphi_d$, as shown in Fig.4, which was used to determine whether j th roller was passing the defect area during the bearing operating. If the defect is on the surface of the outer-race, the contact stiffness factors and damping of roller-race at the angular position of the jth rolling element can be given by

$$(K_j, n_j, c_j) = \begin{cases} (K_d, n_d, c_d) & |\operatorname{mod}[\theta_j - (\phi_d + \Delta\phi_d/2), 2\pi]| < \Delta\phi_d/2 \\ (K, n, c) & \text{else} \end{cases} \tag{6}$$

where K_d, n_d, c_d is the load–deflection factors and damping of j th roller-race with the localized defect, K,n,c is the load–deflection factors and damping of j th roller-race without the localized defect.

4 Results

The simulation model of TRB (SKF30206J) is solved to obtain the radial displacements. The localized defect on outer race set to the following: the length of defect along bearing axial direction is Δl=4mm, the depth of defect is $\Delta d = 0.5mm$, the angular position of the defect is $\phi_d = 0$, angular distance of defect is $\Delta\varphi_d = 0.012$. For the shaft speed of 1000r/min, the characteristic frequency of outer race's defect f_o is 119.737Hz.

The non-linear load–deformation relation of roller-race can be obtained by the contact stress analysis, as shown in Fig.5. The elastic approach of roller-inner/outer race calculated by the non-Hertz line contact is higher than that of Palmgren function, which is in accordance with J. Zhang's analysis result[10].

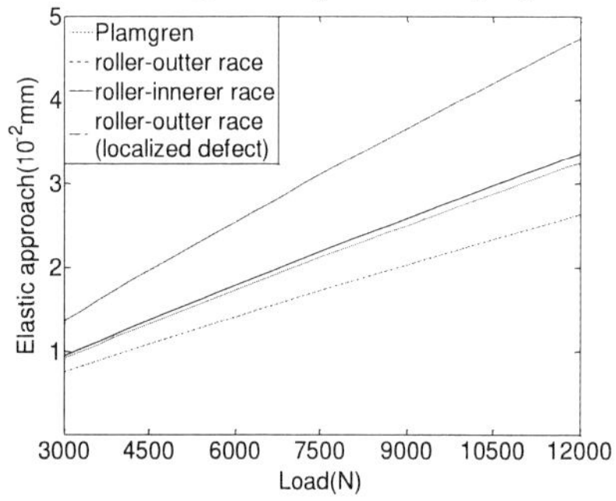

Fig.5Results of contact model and Palmgren formula

Fig.6 shows the simulation of dynamic model of TRB with localized defect on the outer-race, which includes the time domain waveform and the envelope spectrum of the vibrations. It's different with results of traditional TRB defect simulation model[1, 3], there are not obvious impulsive vibration signals in the time domain waveform, however, the characteristic frequency of outer-race's defect f_o and its frequency doubling can be reflected clearly in the envelope spectrum. The simulation shows the effectiveness and feasibility of this modeling method on fault diagnosis of TRB.

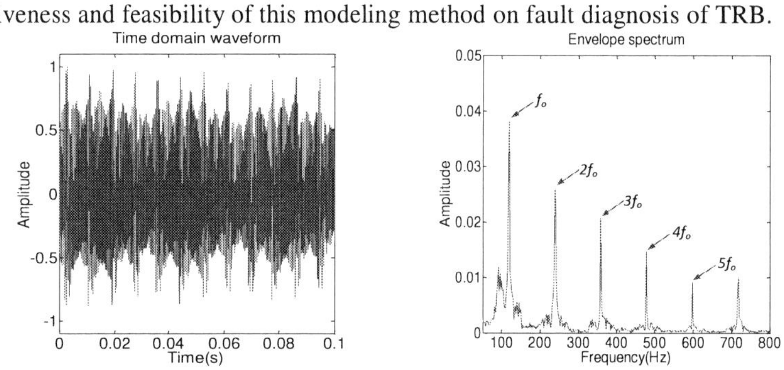

Fig.6 Simulation of defect on outer race

5 Conclusion

The influence of the roller and race's shape on contact stiffness, and the non-linear relationship of load-deformation (used to describe contact stiffness) are obtained by analyzing the contact behavior of roller-race based on non-Hertz line contact theory.

The effects of localized defect on stiffness and damping are analyzed, and the results show that the localized defect causes the decrease in contact stiffness and damping, and occurs the stress concentration at the edge of defect area. The simulation model of TRB with localized defect is developed by bringing the effect of localized defect on stiffness and damping into the three-DOF quasi-dynamic model. This simulating shows that the localized defect of TRB can be diagnosed by the characteristic frequency in the envelop spectrum, although there are not obvious impulsive vibration signals in the time domain waveform. Summarizing, the proposed simulation model of TRB can be reckoned suitable and reliable in diagnosing localized defect using vibration signal.

6 Acknowledgements

The authors would like to acknowledge the financial support by the National Natural Science Foundation of China (General Program:51275383) and the National Program on Key Basic Research Project (973 Program, No: 2012ZX04012-031).

7 References

1. Kumar R, Singh M. Outer race defect width measurement in taper roller bearing using discrete wavelet transform of vibration signal. J. Measurement. 46(1), 537--545(2013)
2. Zhang W, Han N, Wang J. Dynamic research and fault simulation on cylindrical roller bearingsbased on ANSYS/LS-DYNA. J. Coal Mine Machinery. 08, 309--311(2013)
3. Liu J, Shao M, Qing M, et al. Dynamic modeling on localized defect of cylindrical roller bearing based on non-Hertz line contact characteristics. J. Journal of Mechanical Engineering. 01, 91—97(2014)
4. Kumar R, Singh M. Outer race defect width measurement in taper roller bearing using discrete wavelet transform of vibration signal. J. Measurement. 46(1), 537--545(2013)
5. Jena D P, Panigrahi S N. Precise measurement of defect width in tapered roller bearing using vibration signal. J. Measurement. 55: 39--50(2014)
6. Renaudin L, Bonnardot F, Musy O, et al. Natural roller bearing fault detection by angular measurement of true instantaneous angular speed. J. Mechanical Systems and Signal Processing. 24(7): 1998--2011(2010)
7. Li X, Zhu Y. Han L, et al. Fault diagnosis based on nonlinear dynamic modeling in rolling element bearing systems. In: The equipment inspection diagnosis and maintenance the academic conference, national conference on equipment fault diagnosis and national equipment diagnosis engineering conference in 2014. 12--15. IEEE Press, Xi'an(2013)
8. Hoeprich M R. Rolling element bearing fatigue damage propagation. J. Journal of tribology. 114(2): 328--333(1992)
9. Y. Gong, Z. Liu, C. Yang, et al. Crowning analysis and simulation of tapered roller bearings withlogarithmic profile in skew condition. J. Mechanical Science and Technology for Aerospace Engineering. 01:37--41(2015)
10. J. Zhang. Analysis on the contact stress between tapered roller and raceway of tapered roller bearings.D .HeFei University of Technology(2003)

Research on Heating Furnace Temperature Field Distribution Object—Oriented Precision Testing Analysis Instruments*

Shi Shoujuan and Wu Guoxin

Key Laboratory of Modern Measurement and Control Technology, Beijing Information Science and Technology University, Beijing 100192, China, shoujuan16@gmail.com

Abstract. Thermal analysis instrument is a kind of precise instrument. To improve the analytic precision, change the temperature drift characteristics of the thermal analysis instrument, furnace body structure should be properly adjusted. In this paper, the heat transfer relations between furnace temperature field distribution and atmosphere flow, turbulent flow direction, the size of the furnace, direction of furnace and temperature gradient is expounded. This text reveals the heat conduction effect mechanism of the thermal analysis instrument furnace chamber.

Keywords: Thermal analysis instrument, temperature field, heat conduction

1 Introduction

- DSC (Differential Scanning Calorimeter) is a common thermal analysis instruments. Its function is exploring the relationship between temperature and the input power difference of material and reference in the process control temperature [1]. The temperature distribution in the furnace is influenced by material, size, shape, structure of furnace wall and other factors [2]. And a key to improve the thermal analysis instrument's repeat-ability, accuracy and stability is to improve the precision of instrument for temperature control. Based on the previous tests, temperature range of reheating furnace is often changed. In order to improve this situation, the mechanical structure is optimal designed. Temperature field of heating furnace is emulated by calculation software. This will provide theoretical reference for the design of the furnace hearth temperature zone.
- The distribution of uniform temperature zone in the furnace is a key factor of accurate measurement. To accurately analyze the feature of the heated sample and reference, stable temperature field in the sample heating zone is needed [3]. Different experimental conditions have an influence on the temperature field distribution. So it is significant to research temperature in different gas-

* The work is supported by Key Project of Science and Technique Development Plan Supported by Beijing Municipal Commission of Education (CIT&TCD201404120); Young talents cultivation project (2013D005007000009) and Key Laboratory of Modern Measurement & Control Technology(Beijing Information Science & Technology University), Ministry of Education.

es and heating rate. In summary, structure of the existing experimental equipment--DSC-100 is improved and an ideal experimental condition based on the improvement is provided.

2 Thermal analyzer furnace optimization solution

2.1 Introduction of traditional furnace structure

The traditional external heating furnace body structure consists of an outer cover, an inner cover, a platinum cover. The inner layer is a refractory ceramic stove core pipe around by a spiral platinum and rhodium wire winding. The sample and reference material are placed in the center of the furnace core. Absorbed and released heat is measured and outputted by the thermocouple sensor. The insulating layer and the middle layer are filled with the protective asbestos, and the insulation layer and the outer layer are filled with the continuous flowing cooling gas, which can be used to cool down to the furnace body. The furnace core needs to be filled with different heating gas to meet the protection conditions of different samples, to meet their repeatability and diversity. These functions are achieved by the gas filled pipeline. A water cooling mechanism is arranged at the bottom of the furnace body, and the cooling water at the bottom of the furnace can effectively control the sampling accuracy of the bottom of the furnace core due to the heat distortion. Furnace body structure is shown in Figure 1.

2.2 Furnace structure optimization

When heating the furnace with different heating rate, the temperature drift of the uniform temperature zone is relatively large, temperature error exists in the uniform temperature zone in the inside of the furnace, the temperature distribution alternation presents in the upper and lower parts of the uniform temperature zone under the cir cumstances that temperature-increasing blank curve with good linearity appears in traditional furnace. It will affect the instrument's accuracy analysis and increase the teat error in such situation [4, 5].

In order to improve the accuracy of the instrument and change its temperature characteristics, we can adjust the structure of the furnace and change the heat conduction mode, so that it can meet the requirements of the test. With reference to the existing experimental equipment DSC-100, structural optimization design was carried out on the heating furnace, to meet the thwart requirements. Finalize the furnace length 70mm, graphics radius of curvature 350mm, furnace diameter 11mm. The optimized structure of the furnace is shown in Figure 2.

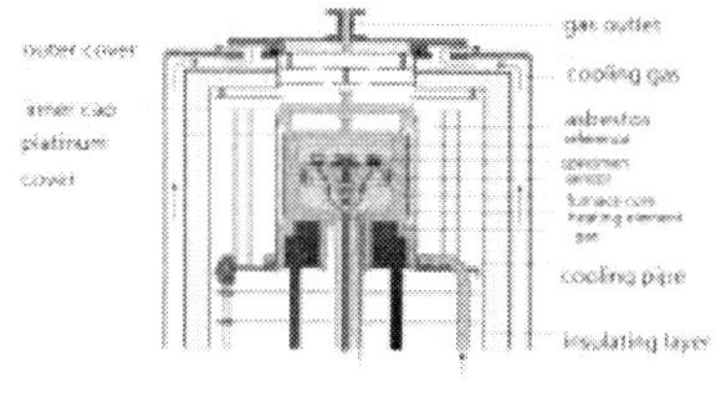

Fig. 1. Furnace body structure

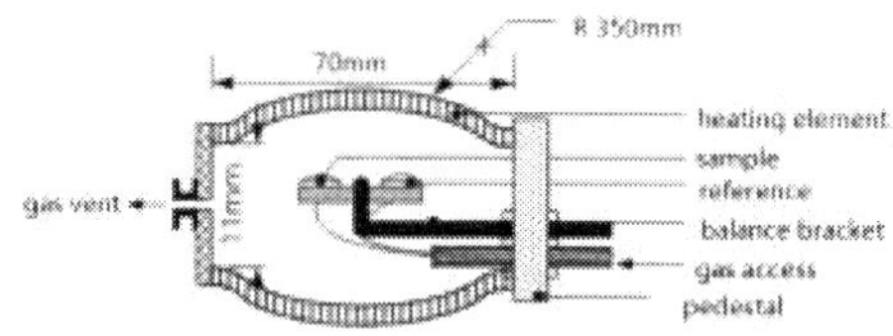

Fig. 2 Optimized furnace structure

2.3 Model of the furnace body

The temperature of the sample and reference is mainly controlled by the heat of the thermal radiation of the heating element.

The protective gas comes into turbulent in the furnace chamber. And it will affect the distribution of the temperature field inside the furnace chamber. In order to study the conjugate heat transfer process between the atmosphere and the furnace wall, the furnace model is established .Simplified model has been done based on experience, as shown in Figure 3.

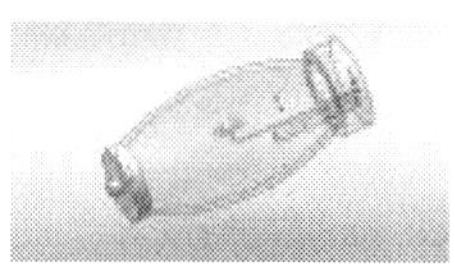

Fig. 3. Furnace model

3 Simulation of the temperature field distribution

Import the model into ANSYS Workbench, mesh it, and encrypt local structure of the model. Then precision and accuracy of turbulent analysis can be ensured. The mesh model is shown in Figure 4:

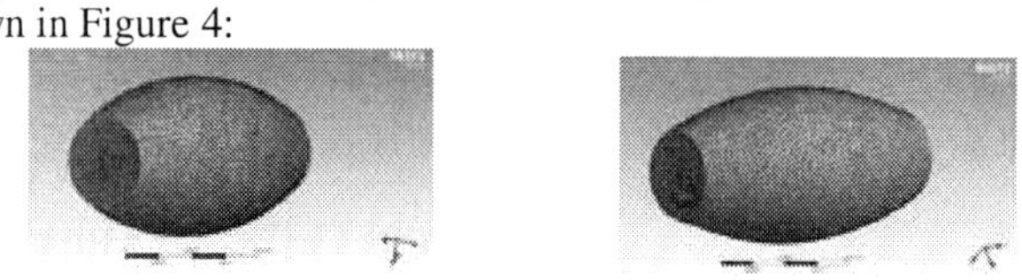

Fig. 4. Mesh model

(1) The bottom and the cover of the furnace convect with the gas in low heat transfer coefficient, and the convection heat transfer coefficient is set to $1.2w \cdot m^{-2} \cdot K^{-1}$.

- (2) Please set the wall of the furnace as heat source, heat source instantaneously transfer to the surrounding environment, the temperature increased from room temperature to 1200℃,heating rate is $20℃/\min$,analyze the temperature field in 400℃, 800℃, 1200℃.
- (3) The gas pass from the eccentric hole inlet of the bottom of the furnace into the furnace cavity. The inlet rate is set to $1mm/s$, $2mm/s$, $3mm/s$, $4mm/s$. The temperature of the inlet is 25℃. The gas is called nitrogen. The center hole of the cover is outlet. Analyze the temperature distribution of the improved model in different gas flow and controlling temperature.

In this paper, temperature field distribution in the improved furnace in different gas flow in the process of heating is researched. When the heating temperature is 400 degrees, the temperature field under four different air flow rate distribution is shown in Figure 5.As can be seen from the figure, when the temperature is 400 degrees, and the air flow rate is at $2mm/s$, $3mm/s$, the sample and reference area temperature field is relatively stable. It can be regard as uniform temperature zone. In the process of gas preheating, because of the fluid viscous forces, temperature close to the wall is higher than temperature of the center. Gas will form a vortex in the center of the furnace and finally out from the lid central hole, heat the surrounding air turbulence. But this part does not cause too much impact on the central heating zone. And entire temperature field presents a certain temperature gradient distribution.

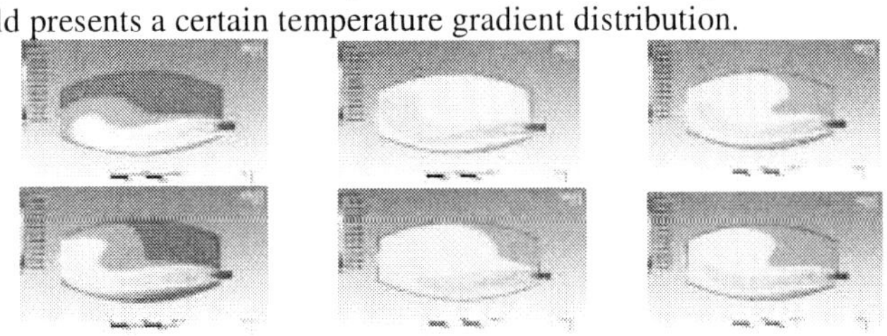

Fig. 5. Temperature field under different air flow rate

Figure 6 shows when the control temperature is 800 degrees and 1200 degrees, the distribution of temperature field under different gas flow. Combine several heating temperature, it can be seen with increasing temperature, the temperature gradient of the furnace chamber significantly increased. As shown in Table 1. A stands for the average temperature characteristic is excellent, B stands for good, and C stands for worse.

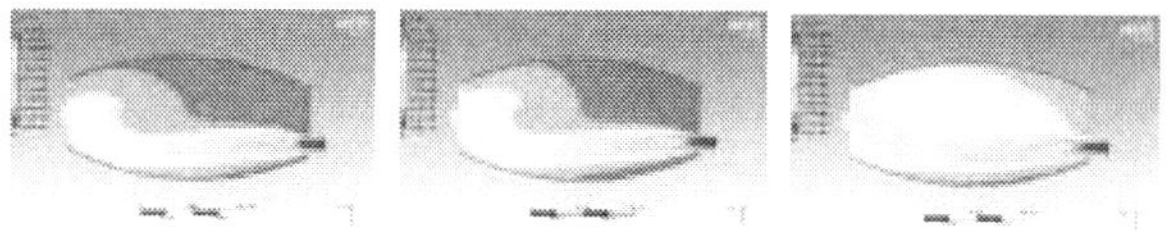

Fig. 6. Temperature field distribution

Different intake air flow will affect the cavity convection and turbulence, the heat source temperature has a greater influence on the morphology of the gas. From **the**

Table 1. Temperature and average temperature region characteristic relationship

Temperature (℃)	400				800				1200			
Gas flow rate (ml/s)	1	2	3	4	1	2	3	4	1	2	3	4
Average temperature characteristics	C	A	B	B	A	B	C	A	B	B	C	A

- analysis of the temperature distribution of the above, the heating process, when the intake air flow is $3mm/s$, the average temperature of the temperature zone becomes less good. Because the center of effective heating zone of turbulence is large, temperature field is unstable, temperature gradient is large, then the temperature difference between the sample and reference material becomes large, and comparative results are affected. Therefore, in the sample test, when heating the furnace from room temperature to 1200 degrees, the intake air flow rate can be chosen as $2mm/s$, under low control temperature .When considering the entire heating process, small intake air flow rate is not very well to provide a stable environment for the protection of the sample and the reference material and it will also affect the measurement results. Therefore, during the experiment, when the control temperature is about 400 degrees, the intake air flow rate can be changed to $4mm/s$. Thereby the uniform temperature is ensured throughout the heating process in the experimental area.

4 Conclusion

- Combined with several experiments, finding that the temperature distribution in the furnace chamber is affected by the uniformity of furnace wire, the impact of wall thickness, the distance from the sample to the furnace wall, intake air flow rate, heating rate and other factors. This article assumes that other experimental conditions are ideal, and the furnace structure has been transformed. Study on the heating temperature was raised from room temperature to 1200 degrees, theoretical analysis found that the best intake air flow in this experimental condition is at $2mm/s$ and $4mm/s$.

Reference

1. Hu Zhongquan, Gao Jinhong, Xu Xiaoli. Intellectualized Differential Scanning Calorimeter system research. Electronic Technology.2008,(4):19-22.
2. Huang Zhanghong, Zhaohui, Lu Liqiang, Gao Wenzhu. Thermal Analysis and Its Application[J].Casting·Forging·Welding, 2010,39(07):19-26
3. Wang Jialong, Luo Dongmiao, Jiang Zhucheng.The Status and Development of Chinese Thermal Analysis Instrument[J].China Instrumentation,2008,(10):28-30
4. Lu Zhihui, Cheng Yong, Liu Xiong.Various measuring instruments under conditions of constant DSC[J].Yunnan Chemical Technology, 1981,(4):34-38
5. Wang Zhangqi, An Liqiang. Adaptive finite element method analysis of temperature field[J].Journal for Engineering for Thermal Energy and Power, 2002,17(97):92-94
6. Pu Guangyi.ANSYS Workbench based tutorial and example explanation [M].Beijing: China Water Power Press, 2010.89-90.
7. Wang Xiaona.Simulation of DSC furnace temperature field [J].Computers and Applied Chemistry, 2014, 31(5):551-554.

Interaction of repair time distributions with a weather model

Helene Seyr and Michael Muskulus

NTNU, Norwegian University of Science and Technology,
Department of Civil and Transport Engineering,
N-7491 Trondheim, Norway,
Helene.Seyr@ntnu.no

Abstract. Maintenance costs account for a large part of the total cost of energy in offshore wind farms. The influence of uncertainty in the repair duration on the production losses due to maintenance is investigated as a novel approach. This is done by using a distribution for the repair times and calculating the expected delay due to sea state. The production loss is calculated based on the mean repair times and compared to the production loss with the repair time distribution. The findings are that the expected production loss is higher when taking the distribution of repair times into consideration. These findings are supported by a case study and the results for three different turbine subsystems are compared.

Keywords: Maintenance, Repair time, Uncertainty, Offshore, Wind turbines, Weather

1 Introduction and Methodology

Operations and maintenance costs account for up to 30% of the total cost of energy in an offshore wind farm [1]. To lower these costs, several approaches have been developed in the last years. Besides from condition monitoring and early failure detection [2], improving the scheduling of corrective maintenance has been in the focus of research [3]. Optimal scheduling of maintenance is influenced by several factors. Availability of transport vessels, cranes, crew and spare parts determine logistic constraints. Maintenance access is not possible in too harsh conditions, so wave height and wind speed result in weather constraints. It is well known that the weather forecasts used while scheduling are subject to uncertainties that need to be considered in the model. In this analysis the influence of information about the repair time to the maintenance costs is investigated using a simple cost model. The cost model is based on an analytical model [4] that has been validated with data from an existing site.

The model calculates the delay due to weather constraints based on the distribution of wave heights at a given location. The delays in the execution of repairs occur, because maintenance access is not possible when waves at the location exceed a given threshold. To conduct repairs the weather conditions need to be calm enough for at least as long as the repair time, otherwise the repair has to be completed at a later point in time. The interaction between the weather model and the repair time is studies, both for mean repair times and for a statistical distribution of repair times with the same mean. Data for a 2.5 MW turbine [5] is used, which is the turbine size for which data was available that is closest to the mean currently installed capacity offshore. The delay model and the distribution of repair times are presented in section 2. Next the distribution of the repair times is included in a simple maintenance cost model calculating the production losses (PL) caused by the downtime due to failures and the resulting repair actions. Here the delays in repair due to harsh weather conditions are also included. The occurrence of a failure is modeled with failure rates taken from [6]. The outputs of the cost model for the mean repair time are compared to the outputs for the distribution in section 2. The question addressed is whether using a distribution instead of a deterministic repair time significantly influences the results. The expected maintenance costs for turbine components are calculated in section 3. The results and ideas for further work are presented in section 4.

2 Maintenance Cost Model

First the model is developed without delay. A deterministic expected delay is included in a second step and finally the expected delay model is incorporated.

2.1 Repair time distribution

[5] provides the mean times to repair (MTTR) for certain turbine components. An exponential distribution of the repair times with this MTTR as the mean is assumed. The probability density function (PDF) for the repair time is then

$$f(t;\lambda) = \lambda e^{-\lambda t}\chi_0(t), \tag{1}$$

$$\text{where } \lambda = \frac{1}{\text{MTTR}} \text{ and } \chi_0(t) = \begin{cases} 1 & \text{for } t \geq 0 \\ 0 & \text{for } t < 0 \end{cases}.$$

2.2 Expected delay

For the expected delay due to sea-state, the approach presented in [4] is used. The model assumes random occurrence of faults with only one failure at a time. They further assume only single trips to conduct repair, so for multiple turbines failures, each turbine is investigated individually. The model assumes wind and wave correlation and does therefore not include a separate wind restriction limit. The correlation between wind and waves has also been investigated by [8] and the support this assumption. The expected delay is calculated with the formula presented in [4] and the weather window is demanded to be at least as long as the repair time for each turbine component. [4] describe five types of delays representing wave height threshold exceedance (type 1), duration of too short calm weather (type 2a), too late fault occurrence within the calm weather (type

2b), delay caused by a storm following a type 2 delay (type 3) and delays for a threshold exceedance following a type 1 or type 3 delay (type 4 and higher). As wave height constraint 1.5m is used as suggested by [4].

2.3 Production loss

The maintenance costs consist of the PL due to downtime, the wages of the maintenance workers, and the costs for transport and spare parts. Since the downtime varies depending on weather delays and repair times, the most uncertainty lies in the PL. It is therefore the focus of this analysis. The production loss y is the lost revenue due to downtime of a turbine caused by a failure and the subsequent repair. Therefore the production loss depends on the installed capacity, length of the downtime, the wind speeds during the downtime, power curve of the installed turbine and the current electricity price. To simplify, one can multiply the installed capacity with the capacity factor (CF) [7] instead of calculating the exact power output from the wind speeds and power curves. The losses can then be calculated by multiplying this value with the electricity price.

First, the PL is analyzed without the delay caused by sea-state. The production loss y is a function of the repair time t such that $y = g(t) = C \cdot t$, where $C = \text{installed capacity} \cdot \text{CF} \cdot \text{electricity price}$. The PDF for the PL of a single failure event becomes

$$f_Y(y) = \frac{\lambda}{C} \mathrm{e}^{-\frac{\lambda}{C} y} \chi_0(y) \qquad (2)$$

The probability p of a failure in a given year, the annual failure rate [6], enters in a Poisson model, where the probability of observing exactly k failures in a year is $f_k = \frac{p^k \mathrm{e}^{-p}}{k!}$. The total PL over one year is then

$$f(y) = \sum_{k=1}^{\infty} \frac{p^k \mathrm{e}^{-p}}{k!} f_Y^{*k}(y) \chi_0(y), \qquad (3)$$

where $f_Y^{*k}(y)$ is the $(k-1)$-fold convolution of the PDF for a single failure

$$f_y^{*k}(y) = \frac{\lambda^k}{C^k} \mathrm{e}^{-\frac{\lambda}{C} y} y^{k-1} \chi_0(y). \qquad (4)$$

The expected PL, is due to the linearity of the expectation

$$E(y|t \sim exp(\lambda)) = E(y|E(t)) = E(y|t = MTTR) = \frac{1}{\lambda} Cp.$$

For a constant delay d and $y = g(t) = C \cdot (t + d)$, similar considerations lead to

$$f(y) = \sum_{k=1}^{\infty} \frac{p^k \mathrm{e}^{-p}}{k!} \frac{\lambda^k}{C^k} \mathrm{e}^{-\frac{\lambda}{C}(y - kCd)} \frac{(y - kCd)^{k-1}}{(k-1)!} \chi_{kCd}(y), \qquad (5)$$

$$\text{with expectation } E(y|t \sim exp(\lambda)) = [E(y|t = MTTR) = \left(\frac{1}{\lambda} + d\right) Cp$$

For the delay according to [4], there exists no analytical inverse function of $y = C(t + d(t))$ and therefore no analytical expression for the PDF. The PDF of the delay is investigated numerically in the following section.

3 Case Study

In this section, the distribution of the production loss with a delay according to Feuchtwang and Infield [4] is investigated. The software R [9] is applied to model the delay due to sea state. A random sample of 10000 different repair times is generated and the resulting delay is calculated for each component. To obtain the density for the delay time, Kernel density estimation [10] is used. The PDF of repair time and delay time for a single failure for three different components, calculated based on [4,5] are plotted in Figure 1. The expectation

Fig. 1. Probability density function for the repair time and delay time distributions for the (a) Gearbox (b) Pitch/Hydraulic system (c) Power electronics

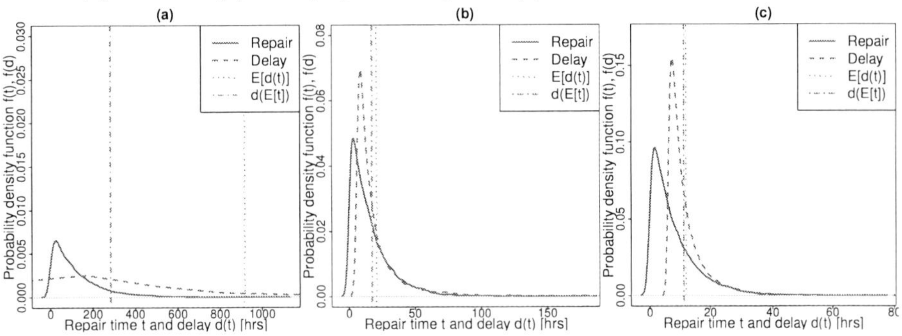

of the delay for a single failure event, calculated based on the distributed repair time, is compared to the delay based on the MTTR in Table 1 for the same three components. For longer repair times, the variations in repair times have a larger influence on the expected delay. When investigating multiple failures, linearity of the expectation can be applied and the expected delay for exactly k failures is

$E[d_k(t)] = E[\sum_{i=1}^{k} E[d(t)]] = kE[d(t)]$, which gives an expected production loss of

$$E[y] = E\left[C\sum_{k=1}^{\infty}\frac{p^k \mathrm{e}^{-p}}{k!}\left(d_k(t)+t\right)\right] =$$

$$= C\sum_{k=1}^{\infty}\frac{p^k \mathrm{e}^{-p}}{k!}\left(E\left[d_k(t)\right]+E\left[t\right]\right) = Cp\left(E\left[d(t)\right]+MTTR\right).$$

The values for the expected production loss, are presented in Table 1.The values in the last column are for a constant delay $d = d(\mathrm{MTTR})$. In a last step an analysis of the delay time distribution is conducted to see which types of delays,

Table 1. Calculation results with data from [5, 6] including the delay from [4]. Since the cost factor C is unknown, unit-free values for the production loss are presented.

Component	$E[d(t)]$	$d(E[t])$	$E[y\|d(t)]/C$	$E[y\|d(\text{MTTR})]/C$
Generator replacement	917.5	286.0	103.8	40.4
Pitch/Hydraulic system	20.9	17.4	8.5	7.4
Power electronics	11.9	11.2	11.3	10.9
Units	hrs	hrs	unit-free	unit-free

as defined in [4], had the largest influence on the delay. Figure 2 shows the distribution of the delay and the distributions of the different types of delays. An analysis of the influence of the repair time distribution on the individual delay types is presented in Table 2. The largest discrepancy exists for a delay of type 4 or higher which is consistently and substantially underestimated when using the MTTR. The delay of type 3 is consistently overestimated by a small amount. The type 1 delay does not depend on the repair time and is thus not investigated.

Fig. 2. Probability density function for the delay time and for the individual delay types for (a) Gearbox (b) Pitch/Hydraulic system (c) Powerelectronics

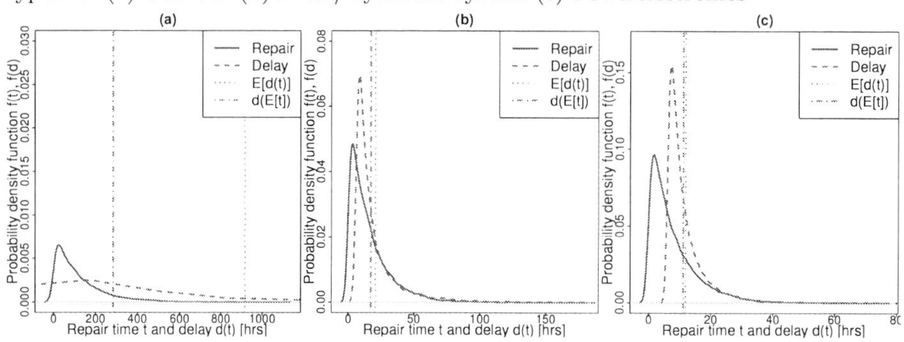

Table 2. Difference $d_i(E[t]) - E[d_i(t)]$, for each delay type $i \in \{2a, 2b, 3, 4\}$

Component	Type 2a	Type 2b	Type 3	Type 4
Generator replacement	-1.9	3.0	1.5	-636.2
Pitch/Hydraulic system	-0.28	-0.22	0.22	-3.23
Power electronics	-0.07	-0.11	0.09	-0.62
Units	hrs	hrs	hrs	hrs

4 Discussion

The analysis presented here, demonstrates a significant influence of the distribution of the repair times on the production losses. For components with short mean repair times, using the expected repair time instead of an accurate representation of the distribution has a small impact on the overall result. For components with longer repair times, as shown for the example of a gearbox replacement, the influence of the variation in repair times is amplified. The use of an accurate representation of repair times is therefore strongly suggested and further research in distributions of repair times is encouraged. For multiple turbines in a wind farm, the effects on the results of using a distribution are expected to be even larger, as the repair times influence the scheduling of maintenance actions in a nonlinear way.

Acknowledgments This work has received funding from the European Union's Horizon 2020 research and innovation programme under the Marie Skłodowska-Curie grant agreement No 642108.

References

[1] Feng, Y., Tavner, P. J., Long, H.: Early experiences with UK Round 1 offshore wind farms. Proceedings of the Institution of Civil Engineers - Energy 163, 167–181 (2010)

[2] Lu, B:, Li, Y., Wu, X., Yang, Z.: A review of recent advances in wind turbine condition monitoring and fault diagnosis. In: Power Electronics and Machines in Wind Applications (PEMWA), pp. 1-7, IEEE (2009)

[3] Dinwoodie, I., Endrerud, O.-E. V., Hofmann, M. , Martin, R., Sperstad, I. B.: Reference Cases for Verification of Operation and Maintenance Simulation Models for Offshore Wind Farms. Wind Engineering 39, 1–14 (2015)

[4] Feuchtwang, J., Infield, D.: Offshore wind turbine maintenance access: a closed-form probabilistic method for calculating delays caused by sea-state. Wind Energy 16, 1099–1824 (2013)

[5] Poore, R., Walford, C.: Development of an operations and maintenance cost model to identify cost of energy savings for low wind speed turbines. Report, National Renewable Energy Laboratory (2008)

[6] Faulstich, S., Hahn, B., Tavner, P.J.: Wind turbine downtime and its importance for offshore deployment. Wind Energy 14, 327–337 (2011)

[7] Wagner, H.J. , Mathur, J.: Introduction to wind energy systems: basics, technology and operation. Springer, Berlin Heidelberg (2012)

[8] Scheu, M., Matha, D., Hofmann, M., Muskulus, M.: Maintenance strategies for large offshore wind farms. Energy Procedia 24, 281–288 (2012)

[9] R Core Team.: R: A language and environment for statistical computing. R Foundation for Statistical Computing, Vienna, Austria. https://www.R-project.org/ (2015)

[10] Silverman, B. W.: Density estimation for statistics and data analysis. Chapmann & Hall/CRC press, London New York (1986)

Quasi-OCVT Technique for Damage Detection of Beam-like Structures

Jianfeng Zhong, Shuncong Zhong*, and Qiukun Zhang

Laboratory of Optics, Terahertz and Non-Destructive Testing, School of Mechanical Engineering and Automation, Fuzhou University, Fuzhou 350116, PR China.
E-mail: zhongshuncong@hotmail.com

Abstract. A quasi-optical coherence vibration tomography (Quasi-OCVT) measurement system for structural damage detection was investigated by taking the concept of two-dimensional OCVT technique. A printed quasi-interferogram fringe pattern (QIFP), which worked works as a sensor and is similar to the interferogram of 2D-OCVT system, was pasted on the surface of a vibrating structure. A high-speed camera as a detector captured image sequence of the QIFP. The period density of the imaged QIFP changed due to the structural vibration, from which the structural vibration information could be achieved. The quasi-OCVT system was employed to monitor the vibration of a beam-like structure with a roving auxiliary mass. The frequency shift curves obtained by the proposed quasi-OCVT method had a good agreement with the ones by finite element method (FEM). From the frequency shift curves by quasi-OCVT technique, damage location of the structures could be determined. Therefore, it could be recommended for non-contact vibration measurement or damage detection of real engineering beam-like structures.

Keywords: Quasi-optical coherence vibration tomography; Vibration measurement; Damage detection; Auxiliary mass; Frequency shift

1 Introduction

Presence of crack in a structure introduces changes in local flexibility or its dynamic behavior that can be used for damage detection. Damage identification techniques are mainly categorized as natural frequency-based methods, mode shape-based methods, curvature/strain mode shape-based methods and other methods based on modal parameters [1]. Very good review on these issues has been presented by Dimarogonas [2]. A number of studies also have been carried out to develop crack identification algorithms in different scenarios. Some of the investigators consider that a crack reduces the section modulus (*EI*) of a small segment around itself [3]. Others consider that a crack causes a local discontinuity in the slope of the deflection curve or reduces the energy content of the beam [4]. Kaushar et al. [5] and Rizos et al. [6] have represented the crack as rotational spring in modal analysis for a cantilever beam having a rectangular cross section. Zhong et al. [7] proposed an approach based on auxiliary mass spatial probing to provide a method for crack detection in beam-like structure. However, only numerical simulations were presented in the Zhong's work. Recently, Zhong et al. [8] proposed a two-dimensional optical coherence tomography

(2D-OCVT) technique offering the possibility of performing high resolution and non-intrusive vibration measurement. In the present work, inspired by the principle of the 2D-OCVT system and the features of interferogram, a quasi-optical coherence vibration tomography (Quasi-OCVT) method was investigated for vibration measurement and damage detection.

2 Theory of Quasi- Optical Coherence Vibration Tomography

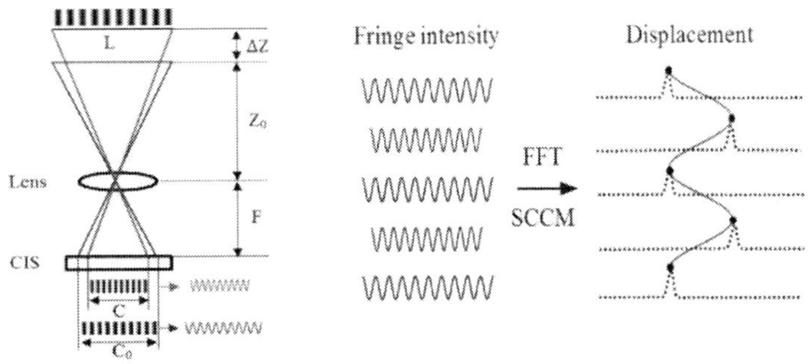

Fig.1. Principle of quasi-OCVT system and processing procedure of displacement

In the proposed quasi-OCVT system, the QIFP was pasted on the surface of a vibrating object and image sequence of the QIFP was captured by a high-speed camera. According to the imaging theory of geometrical optics, as shown in Fig. 1, the length of imaged QIFP, i.e.,C, varies with the change of the distance between QIFP and optical lens. Consequently, the period density of the imaged QIFP varies with the change of the distance between the lens and the QIFP. For each imaged QIFP, the middle row intensity $I_m(x,t)$for structural displacement calculation can be expressed as

$$I_m(x,t) = 1 + V_m \cos(\phi_m + 2\pi aNd_m(t)x) \quad (1)$$

where V_m is the local contrast, a is the width of a pixel, N is the number of pixels along the length direction, and $d_m(t)$ is the calculated period density of the imaged QIFP.

The processing procedure of the sequence of the imaged QIFP is shown in the right side of Fig. 1. The fringe intensity was used for the displacement calculation by applying Fast Fourier Transform (FFT) to it. Subsequently, the real-time vibration signals in form of QIFP fringe density variation can be obtained by the combination of the main peak of each QIFP-FFT waveform. It should be noted here that a spectral center correction method (SCCM) [9]was used to obtain almost exact period density of the imaged QIFP. According to the imaging principle shown in the left of Fig. 1, the relationships among the parameters: the initial distance between the QIFP and lens (Z_0), the initial length of the imaged QIFP (C_0), the displacement difference ($\Delta Z(t)$)and length of the imaged QIFP ($C(t)$)at time $t$can be expressed as

$$C_0 / C(t) = (Z_0 + \Delta Z(t)) / Z_0 = d(t) / d_0 \tag{2}$$

where d_0 and $d(t)$ are the computed fringe period density of imaged QIFP corresponding to C_0 and $C(t)$, respectively. The relationship between $C(t)$ and $d(t)$ can be regarded as $C(t) \propto 1/d\,(t)$ when the displacement is in millimeter scale. Therefore, the real-time displacement $\Delta Z(t)$ can be expressed as

$$\Delta Z(t) = Z_0[d(t) - d_0] / d_0 \tag{3}$$

If one wants to obtain the exact displacement amplitude, the initial object distance is required and can be expressed as

$$Z_0 = f(1 + L / C_0) \tag{4}$$

where L is the known physical length of the QIFP, C_0 is the corresponding length of imaged QIFP that could be calculated from $C_0 = aN_i$, where a is the width of the pixel whilst N_i is the number of pixels covered by the QIFP.

3 Experiment and Discussions

3.1 Experimental setup

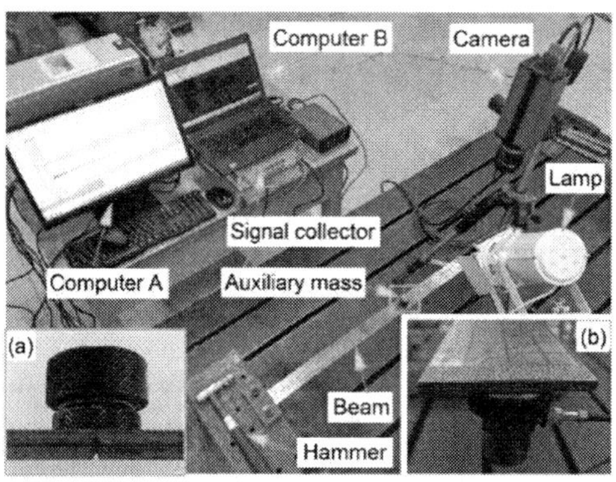

Fig.2. Photograph of the laboratory test setup

In order to evaluate the performance of the proposed method, comparison experiments both for structural dynamic characterization and for damage detection of a cracked cantilevered beam were carried out separately using the developed quasi-OCVT system and traditional accelerometer-based method. As shown in Fig. 2, a high-speed CMOS image sensor (CIS) camera (PCO.1200hs, PCO-TECH, Germany) was fixed on a tripod and placed perpendicularly to the beam surface. The QIFP was imaged in the middle of the CIS. Additional lighting lamp could be employed for better illumination. In the quasi-OCVT system controlled by Computer B, a QIFP with a period density of 20 periods per centimeter and a length of 4 cm, was pasted on the top surface of the free end of the beam, as shown in the inset (b) of Fig. 2. For comparison study, an accelerometer (Brüel & Kjær, Type 4383V, weight of 17 g) was fixed at the same position as the QIFP to measure the vibration simultaneously.

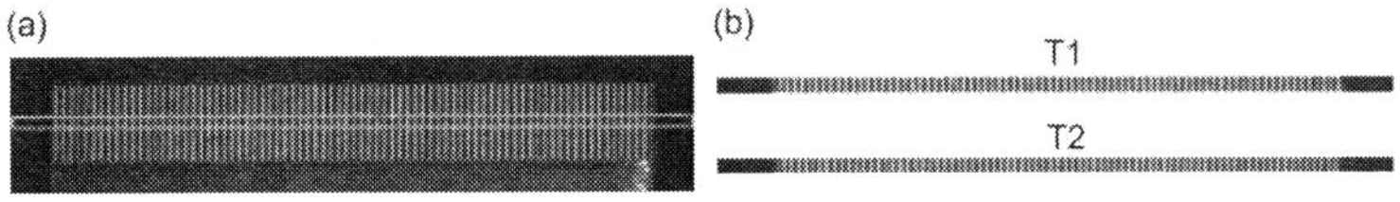

Fig.3. The images of the entire view(b)and partial view(c) of the QIFP.

Generally, there is no need to record the whole view of the QIFP (shown in Fig. 3(a)) to obtain the displacement because only one line fringe intensity of the imaged QIFP in the length direction was used for displacement calculation. In order to save memory of the camera, only part of the QIFP was recorded as shown in the green box of Fig. 3(a). Subsequently, the imaged QIFP sequence was digitized in 8-bit grey scales and streamed into the computer though a GigE cable for displacement calculation. Two images of the QIFP at time T1 and T2 during the vibration of the beam are shown in Fig. 3(b).

3.2 Experimental Results and Discussions

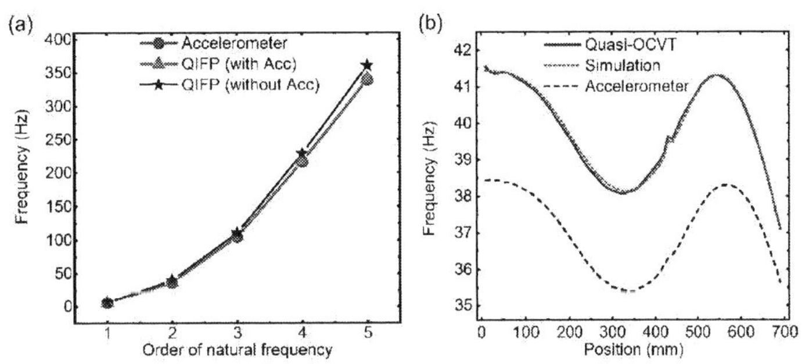

Fig.4.(a)Natural frequencies obtained by quasi-OCVT and accelerometer-based method. (b)The2nd order frequency shift curves obtained by simulation (red dotted line), quasi-OCVT method (blue solid line) and accelerometer-based method (black dashed line).

In this experiment, impulse excitation was used for the characterization the vibration modal parameters of the structure. The shock impulse was generated by a force hammer that beat slightly near the fixed end of the beam to produce a small vibration in the transverse direction. The time-domain vibration signals were recorded by the two vibration measurement systems and the natural frequencies were obtained by applying the FFT to the time-domain signals. It should be noted that spectral center correction method (SCCM) was used for accurate determination of the natural frequencies to reduce the spectral leakage in FFT [9].

Fig. 4(a) shows the first five natural frequencies obtained from the time-domain signals obtained by quasi-OCVT and accelerometer-based system in a single impulse excitation experiment. Frequencies obtained by accelerometer-based system and quasi-OCVT method are almost the same, which shows a good performance of the proposed method for dynamic measurement. Meanwhile, as could been seen from Fig.4(a), the deviations of the frequency values obtained by quasi-OCVT method with and without accelerometer on the beam become significant because the additional mass of the accelerometer has effect on natural frequencies of the beam. On the other hand, the very light QIFP nearly introduces no extra mass on the beam-like structure, making the proposed quasi-OCVT method also suitable for light structure dynamic characterization.

To illustrate the performance of the quasi-OCVT technique for damage detection, a steel beam with a transverse crack was used. The beam has the following dimensions: length of 700 mm, width of 40 mm, and height of 4 mm. One transverse saw cut (as shown in the inset (a) of Fig. 2) with a depth of 2 mm and a width of 1.5 mm, is located at430 mm from the fixed end of the beam. During the experiment, the magnetic auxiliary mass (weight of 80 g) was traversed along the beam from the fixed end to the free end. The spatial probing interval of the auxiliary mass was 10 mm, which results in a total number of 69 measuring points. The frequency shift curves obtained accordingly by the combination of the same order of natural frequencies when 69 measurements were completed using accelerometer-based system and quasi-OCVT system separately.

For comparison, a cantilevered defect beam model with an auxiliary mass was also studied using Finite Element Method (FEM).Similar with the experiment, 69 cases were simulated when the auxiliary mass was located at different positions. The 2nd order of frequency shift curves are plotted against the corresponding axial location of the auxiliary mass in Fig. 4(b). The dotted line, solid line and dashed line correspond to the simulated results, quasi-OCVT results and traditional accelerometer-based method results, respectively. Based on the theory of natural frequencies of cracked simply supported beams with a stationary roving mass [10], a discontinuity in frequency shift curve occurs when the auxiliary mass is moved closely to the crack location. From the dotted and solid lines in Fig. 4(b), the curves provide evidence of crack existence at 430 mm from the fixed end of the beam because the frequencies exhibit significant discontinuity at this position. It is evident from the dotted and solid lines that experimental and numerical results show a good agreement. It illustrated the validity, efficiency and accuracy of the proposed quasi-OCVT method. However, the frequency shift curve extracted from the acceleration data is smaller than the quasi-OCVT-based result owning to the mass effect of the accelerometer and its accessories, as shown by the dashed line in Fig. 4(b). As can been seen from Fig.4(b), the damage location cannot be identified confidently from the dashed curve, which demonstrates that the proposed quasi-OCVT technique is more accurate and crack-sensitive than the accelerometer-based one.

4 Conclusion

We proposed a novelQuasi-OCVT systemfor vibration measurement and crack detection of beam-like structure. The good agreement of simulation results with the results by quasi-OCVTmethod and the deviation of quasi-OCVT results with accelerometer results verify the efficientcy, practicability and high accuracy of the proposed quasi-OCVT method. In addition, the proposed vibration measurement system is non-contact in nature. Compared with the traditional accelerometer-based method and OCVT technique, complicated setup is not need and no mass-effect is introduced on the structure, making it attractive for the application in the characterization of structural modal parameters and damage detection of engineering beam-like structures.

Acknowledgement

We gratefully acknowledge support from the Fujian Provincial Excellent Young Scientist Fund (2014J07007), the National Natural Science Foundation of China (51005077), and the Specialised Research Fund for the Doctoral Program of Higher Education, the Ministry of Education, P. R. China (20133514110008).

References

1. Wei, F., Pizhong, Q.: Vibration-based Damage Identification Methods: A Review and Comparative Study. Struct. Health Monit. 10(1), 83-111 (2010)
2. Dimarogonas, A. D. Vibration of cracked structures: A state of the art review. Eng. Fract. Mech. 55(5), 831-857 (1996)
3. Sinha, J. K., Friswell, M. I., Edwards, S.: Simplified models for the location of cracks in beam structures using measured vibration data. J. Sound Vibrat. 251(1), 13-38 (2002)
4. Patil, D. P., Maiti, S. K.: Experimental verification of a method of detection of multiple cracks in beams based on frequency measurements. J. Sound Vib. 281(1–2), 439-451 (2005)
5. Barad, K. H., Sharma, D. S., Vyas, V.: Crack Detection in Cantilever Beam by Frequency based Method. Procedia Engineering. 51(770-775 (2013)
6. Rizos, P. F., Aspragathos, N., Dimarogonas, A. D.: Identification of crack location and magnitude in a cantilever beam from the vibration modes. J. Sound Vib. 138(3), 381-388 (1990)
7. Zhong, S., Oyadiji, S. O.: Identification of Cracks in Beams With Auxiliary Mass Spatial Probing by Stationary Wavelet Transform. J. Vib. & Acous. 130(4), 041001-041001 (2008)
8. Zhong, J., Zhong, S., Zhang, Q., Yao, L.: Two-dimensional optical coherence tomography for real-time structural dynamical characterization. Opt. Laser. Eng. 66(74-79 (2015)
9. Zhong, S., Oyadiji, S. O., Ding, K.: Response-only method for damage detection of beam-like structures using high accuracy frequencies with auxiliary mass spatial probing. J. Sound Vib. 311(3–5), 1075-1099 (2008)
10. Zhong, S., Oyadiji, S. O.: Analytical predictions of natural frequencies of cracked simply supported beams with a stationary roving mass. J. Sound Vib. 311(1–2), 328-352 (2008)

MACHINE REMAINING USEFUL LIFE PREDICTION USING A MONTE CARLO SIMULATION METHOD

Yaguo Lei[1, 2], Naipeng Li[1, 2], Jing Lin[1, 2] and Ningbo Li[1, 2]

[1] State Key Laboratory for Manufacturing Systems Engineering, Xi'an Jiaotong University, Xi'an 710049, China

[2] Shaanxi Key Laboratory of Mechanical Product Quality Assurance and Diagnostics, Xi'an Jiaotong University, Xi'an 710049, China

E-mail address: yaguolei@mail.xjtu.edu.cn (Yaguo Lei)

Abstract. This paper proposes a machine remaining useful life (RUL) prediction method based on Monte Carlo simulation. In this method, the degradation processes of machinery are described using a general stochastic process model, which assumes the degradation speed is dependent on the age and the health state. The machine RUL is predicted using a numerical approximation algorithm based on Monte Carlo simulation. The basic idea of the numerical approximation algorithm is to generate a series of simulated degradation trajectories through Monte Carlo simulation, and to calculate the RUL probability density of each trajectory using its simulated state values. A practical case study is utilized to demonstrate the effectiveness of the proposed method in the RUL prediction of machinery.

Keywords: Remaining useful life prediction, machinery, stochastic process model, Monte Carlo simulation, numerical approximation

1 Introduction

With the development of machine maintenance strategies, predictive maintenance has attracted more and more attentions in recent years [1]. One of the major tasks in predictive maintenance is remaining useful life (RUL) prediction, i.e., predicting the available operation time of machinery and providing guidance for the subsequent maintenance scheduling. In most literatures [2], the RUL is defined as the time left before the degradation processes of machinery exceed a failure threshold. So the major issue in RUL prediction is how to describe the degradation processes of machinery. Since the degradation processes of machinery are stochastic because of various uncertainty factors, they are naturally described using stochastic process models. So far, different kinds of stochastic process models have been developed and applied into the machine RUL prediction, such as geometric Brownian motion (BM) models [3], inverse Gaussian process models [4] and nonlinear stochastic process models [5]. These models assume that the degradation speed is dependent on the age of machinery. In reality, the degradation speed is not only related to the age of machinery but also influenced by the health state of machinery. For example, the degradation speed is generally low at the beginning of fault occurrence and will increase as the fault becomes severer. Therefore, the influence of both the age and the health state to the degradation speed should be considered in constructing stochastic process models.

This paper describes the degradation processes of machinery using a general stochastic process model, whose degradation speed is assumed to be dependent on the age and the health state of machinery. Based on the proposed model, the analytical solution of RUL is derived. However, the analytical solution is unable to be computed in real applications, since its computing process requires the future state values which are actually unknown. To deal with this problem, a numerical approximation algorithm based on Monte Carlo simulation is proposed, which attempts to calculate a numerical solution of RUL.

2 RUL Analytical Solution of the Stochastic Process Model

The degradation processes of machinery are described using the following stochastic process model:

$$s(t) = s_k + \int_{t_k}^{t} \mu(s(\tau),\tau)d\tau + \int_{t_k}^{t} \sigma(\tau)dB(\tau), \tag{1}$$

where s_k is the state at time t_k, $s(t)$ is the state at t with $t \geq t_k$, $\mu(s(\tau),\tau)$ is the drift coefficient function which represents the degradation speed of the stochastic process, $\sigma(\tau)$ is the diffusion coefficient function, and $\{B(\tau), \tau \geq t_k\}$ denotes a standard BM process.

Let l denote the time length after t_k and λ represent the failure threshold. The RUL L at t_k is described as

$$L = \inf\{l \mid s(l+t_k) \geq \lambda\}. \tag{2}$$

The task of RUL prediction is to estimate the probability density function (PDF) of RUL depending on the historical degradation trajectory, i.e., $f_{L|s_{0:k}}(l \mid s_{0:k})$. To derive the analytical solution of the RUL PDF based on the general stochastic process model in (1), a transformation algorithm is developed, which includes the following two transformation steps.

Translation transformation: The stochastic process $\{s(t), t \geq t_k\}$ shown in (1), which starts from t_k and s_k, can be transformed into a new stochastic process $\{\tilde{s}(l), l \geq 0\}$, which starts from $l_0 = 0$ and $\tilde{s}_0 = 0$ through the transformation:

$$\begin{cases} \tilde{s}(l) = s(t) - s_k \\ l = t - t_k \end{cases}, \tag{3}$$

where

$$\tilde{s}(l) = \int_0^l \tilde{\mu}(\tilde{s}(\tau),\tau)d\tau + \int_0^l \tilde{\sigma}(\tau)dB(\tau) \tag{4}$$

and

$$\begin{cases} \tilde{\mu}(\tilde{s}(\tau),\tau) = \mu(s(\tau + t_k), \tau + t_k) \\ \tilde{\sigma}(\tau) = \sigma(\tau + t_k) \end{cases} \tag{5}$$

Standard transformation[5, 6]: For the stochastic process in (4), if there are two time-dependent functions $c_1(l)$ and $c_2(l)$, satisfying the following equation

$$\frac{\tilde{\mu}(\tilde{s},l)}{\tilde{\sigma}(l)} = c_1(l) + c_2(l)\int_0^{\tilde{s}} \frac{dw}{\tilde{\sigma}(l)} - \int_0^{\tilde{s}} \frac{\partial}{\partial l}(\frac{1}{\tilde{\sigma}(l)})dw, \tag{6}$$

then there exists a transformation

$$\begin{cases} B(\tilde{l}) = \varphi(\tilde{s},l) = (k_1)^{1/2}\exp(-\int_{l_0}^{l} c_2(\tau)d\tau)\int_0^{\tilde{s}} \frac{dw}{\tilde{\sigma}(l)} - (k_1)^{1/2}\int_{l_1}^{l} c_1(\tau)\exp(-\int_{l_0}^{\tau} c_2(v)dv)d\tau + k_2 \\ \tilde{l} = \psi(l) = k_1\int_{l_2}^{l} \exp(-2\int_{l_0}^{\tau} c_2(v)dv)d\tau + k_3 \end{cases} \tag{7}$$

which can change the general stochastic process $\{\tilde{s}(l), l \geq 0\}$ into a standard BM process $\{B(\tilde{l}), \tilde{l} \geq 0\}$. In (7), $l_i \in [0,\infty)$, $i = 0,1,2$, $k_1 > 0$, k_2 and k_3 are arbitrary constants.

Until now, the general stochastic process in (1) has been changed into a standard BM process. Therefore, the problem of a general stochastic process crossing a constant threshold is changed into a problem of a standard BM process crossing a time-varying threshold. It has been proven that the first-passage PDF [5] of a standard BM $\{B(\tilde{l}), \tilde{l} \geq 0\}$ to a time-varying threshold $T(\tilde{l})$ is

$$f_B(\tilde{l}) \cong (\frac{T(\tilde{l})}{\tilde{l}} - \frac{dT(\tilde{l})}{d\tilde{l}})\frac{1}{\sqrt{2\pi\tilde{l}}}\exp(-\frac{T^2(\tilde{l})}{2\tilde{l}}) \tag{8}$$

Based on the above analysis, the RUL PDF of the stochastic process in (1) is

$$\begin{aligned} f_{l|s_k}(l \mid s_k) \cong & (\frac{\lambda - s_k - \int_0^l \mu(s(\tau + t_k), \tau + t_k)d\tau}{\int_0^l \sigma^2(\tau + t_k)d\tau} - \frac{\mu(s(l + t_k), l + t_k)}{\sigma^2(l + t_k)}) \\ & \times \frac{\sigma^2(l + t_k)}{\sqrt{2\pi\int_0^l \sigma^2(\tau + t_k)d\tau}}\exp(-\frac{(\lambda - s_k - \int_0^l \mu(s(\tau + t_k), \tau + t_k)d\tau)^2}{2\int_0^l \sigma^2(\tau + t_k)d\tau}) \end{aligned} . \tag{9}$$

The proof of (9) is given in the Appendix.

However, the analytical solution of the RUL PDF in (9) in inapplicable, since it requires the unknown future state values of the stochastic process from t_k to $l + t_k$. To deal with this problem, this paper proposes a numerical approximation algorithm based on Monte Carlo simulation in the following section.

3 The Numerical Approximation Algorithm

To acquire the future state values, a Monte Carlo method [7] is employed to simulate a series of possible degradation trajectories. Let $s_p(t_k) = s_k$ and $p = 1 : Ns$. The degradation trajectories of Ns samples $S_p = \{s_p(l_i + t_k)\}_{i=1:L_p}$ are generated through the following transition function:

$$s_p(l_i+t_k)=s_p(l_{i-1}+t_k)+\mu(s_p(l_{i-1}+t_k),l_{i-1}+t_k)\Delta l+V_{i-1} \tag{10}$$

where Δl is the time interval, $l_i=i\Delta l$, $V_{i-1}\sim N(0,\sigma^2(l_{i-1}+t_k)\Delta l)$, $i\in \mathrm{N}^+=\{1,2,3,...\}$, and L_p is the number of time points in the p th sample. The transition process of each sample is continued until its state values exceed the failure threshold. The RUL of the p th sample is defined as $l_p=\inf\{l_i\,|\,s_p(l_i+t_k)\geq\lambda\}$.

With inputting the p th simulated degradation trajectory into (9), the RUL probability density of the p th sample is calculated.

$$p\left(l=l_p\,|\,s_{0:k}\right)\cong\left(\frac{\lambda-s_k-\sum_{n=0}^{L_p-1}\mu(s_p(l_n+t_k),l_n+t_k)\Delta l}{\int_0^{l_p}\sigma^2(\tau+t_k)d\tau}-\frac{\mu(s_p(l_p+t_k),l_p+t_k)}{\sigma^2(l_p+t_k)}\right)$$
$$\times\frac{\sigma^2(l_p+t_k)}{\sqrt{2\pi\int_0^{l_p}\sigma^2(\tau+t_k)d\tau}}\exp\left(-\frac{(\lambda-s_k-\sum_{n=0}^{L_p-1}\mu(s_p(l_n+t_k),l_n+t_k)\Delta l)^2}{2\int_0^{l_p}\sigma^2(\tau+t_k)d\tau}\right) \tag{11}$$

Equation (11) gives the probability density of the p th trajectory crossing the failure threshold at l_p. In reality, there always exist some trajectories crossing the failure threshold at the same time. And the probability densities calculated using different trajectories are distinct. Therefore, there are several different probability densities at the same RUL. Considering this case, the mean value of the probability densities at the same RUL is calculated as the result of the RUL probability density. Let $n=\{L_d,L_d+1,...,L_u\}$, where $L_d=\min\{L_p\,|\,p=1,...,N_s\}$ and $L_u=\max\{L_p\,|\,p=1,...,N_s\}$, respectively. Find samples whose RULs equal l_n, and calculate the mean value of their probability densities.

$$p\left(l=l_n\,|\,s_{0:k}\right)=\frac{1}{M}\sum_{m=1}^{M}p\left(l=l_{p_m}\,|\,s_{0:k}\right) \tag{12}$$

4 A Practical Case Study

In this section, the proposed method is demonstrated using fatigue crack size data [8]. The data were collected to observe crack growth processes for alloys. The data includes 21 test specimens. Totally 12 specimens among the 21 test cases run to failure. To describe the accelerated degradation processes of the test specimens, three degradation models are employed in this paper. The drift coefficient functions of Model 1 (M1) and Model 2 (M2) are two age-dependent functions [5] $\mu_1(s(\tau),\tau)=ab\tau^{b-1}$ and $\mu_2(s(\tau),\tau)=a\exp(b\tau)$, respectively. In Model 3 (M3), a new drift coefficient function is employed, which is a state-dependent function $\mu_3(s(\tau),\tau)=ax^b$. All of their diffusion coefficient functions are $\sigma(\tau)=\sqrt{c+d\tau}$.

The RULs of the 12 failure cases are predicted one after another using the proposed method. Taking the first case and the last case for example, their RUL prediction results using M1-M3 are shown in Fig. 1 (a) and (b), respectively. It is seen that, the

RULs of the two cases are predicted accurately using the proposed method. In Case 1, the prediction results deviate far from the ground truth RUL at the beginning. As time goes on, the results converge to the ground truth RUL gradually. Among the three prediction results, the results of M2 and M3 converge to the ground truth RUL more quickly than that of M1. In Case 12, M3 performs a higher convergence speed and acquires a more accurate prediction result than M1 and M2. From the comparison of the prediction results it is obvious that, the state-dependent model M3 performs better than the other two age-dependent models M1 and M2 in the RUL prediction of the two cases.

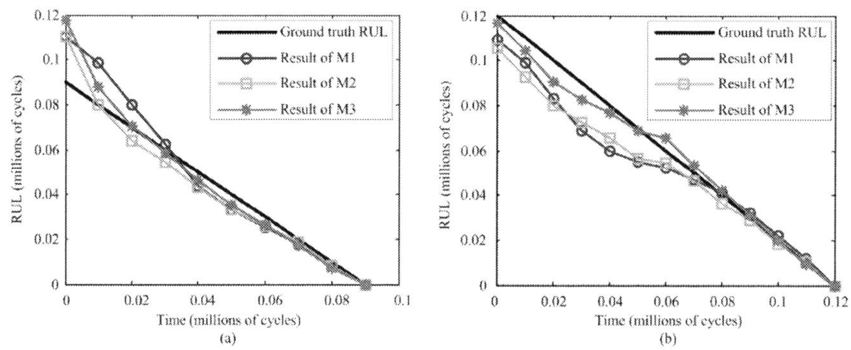

Fig. 1. Comparison of the RUL prediction results: (a) Case 1 and (b) Case 12.

5 Conclusions

A general stochastic process model is developed in this paper to describe the degradation processes of machinery. The analytical solution of the RUL PDF based on the general stochastic process model is derived. However, the analytical solution is inapplicable since it requires the future state values which are actually unknown in real application. To deal with problem, a numerical approximation algorithm based on Monte Carlo simulation is proposed, which offers a numerical solution for the RUL PDF. The effectiveness of the proposed method is demonstrated using a practical case study.

Appendix

Through the translation transformation, the stochastic process $\{s(t), t \geq t_k\}$ has been changed into $\{\tilde{s}(l), l \geq 0\}$. The failure threshold λ is changed to be $\tilde{\lambda} = \lambda - s_k$. Let $k_1 = 1$, $k_2 = k_3 = 0$ and $l_0 = l_1 = l_2 = 0$, then

$$\begin{aligned}\varphi(\tilde{s}, l) = \exp(-\int_0^l c_2(\tau)d\tau)\frac{s(l+t_k)-s_k}{\sigma(l+l_k)} - \int_0^l (\frac{\mu(s(\tau+t_k),\tau+t_k)}{\sigma(\tau+t_k)} - c_2(\tau)\frac{s(\tau+t_k)-s_k}{\sigma(\tau+t_k)} \\ + \frac{\partial}{\partial \tau}(\frac{1}{\sigma(\tau+t_k)})(s(\tau+t_k)-s_k))\exp(-\int_0^\tau c_2(v)dv)d\tau\end{aligned}.$$

To simplify the expression, let $c_2(v) = -1/\sigma(v+t_k)$. The standard transformation in (7) is changed to be

$$\begin{cases} B(\tilde{l}) = \varphi(\tilde{s}, l) = s(l+t_k) - s_k - \int_0^l \mu(s(\tau+t_k), \tau+t_k) d\tau \\ \tilde{l} = \psi(l) = \int_0^l \sigma^2(\tau+t_k) d\tau \end{cases}.$$

Correspondingly, the failure threshold is transformed into

$$\varphi(\tilde{\lambda}, l) = \lambda - s_k - \int_0^l \mu(s(\tau+t_k), \tau+t_k) d\tau \cdot$$

Since $\{B(\tilde{l}), \tilde{l} \geq 0\}$ is transformed from $\{s(t), t \geq t_k\}$ through a two-step transformation, the probability distribution of the RUL $F_{L|s_{0:k}}(l \mid s_{0:k})$ has the following relationship with the probability distribution of $F_B(\tilde{l})$.

$$F_{L|s_k}(l \mid s_k) = P_s(s(l+t_k) \leq \lambda) = P_{\tilde{s}}(\tilde{s}(l) \leq \lambda - s_k) = P_B(B(\tilde{l}) \leq \varphi(\tilde{\lambda}, l)) = F_B(\tilde{l})$$

By calculating the derivatives of $F_{L|s_k}(l \mid s_k)$ and $F_B(\tilde{l})$ with respect to l, (9) is acquired. This completes the proof of (9).

Acknowledgement

This research is supported by National Natural Science Foundation of China (51475355, 51421004), Young Talent Support Plan of Central Organization Department and Fundamental Research Funds for the Central Universities (2012jdgz01 and CXTD2014001).

Reference

1. J. Lee, F.J. Wu, W.Y. Zhao, M. Ghaffari, L.X. Liao, D. Siegel, Prognostics and health management design for rotary machinery systems—Reviews, methodology and applications, Mechanical Systems and Signal Processing, 42 (2014) 314-334.
2. N. Li, Y. Lei, J. Lin, S.X. Ding, An improved exponential model for predicting remaining useful life of rolling element bearings, IEEE Transactions on Industrial Electronics, 62 (2015) 7762-7773.
3. C. Park, W.J. Padgett, Accelerated degradation models for failure based on geometric Brownian motion and gamma processes, Lifetime Data Analysis, 11 (2005) 511-527.
4. W. Peng, Y.-F. Li, Y.-J. Yang, H.-Z. Huang, M.J. Zuo, Inverse Gaussian process models for degradation analysis: A Bayesian perspective, Reliability Engineering & System Safety, 130 (2014) 175-189.
5. X.S. Si, W.B. Wang, C.H. Hu, D.H. Zhou, M.G. Pecht, Remaining useful life estimation based on a nonlinear diffusion degradation process, IEEE Transactions on Reliability, 61 (2012) 50-67.
6. L.M. Ricciardi, On the transformation of diffusion processes into the Wiener process, Journal of Mathematical Ayalysis and Applications, 54 (1976) 185-199.
7. H.L. Anderson, Metropolis, Monte Carlo and the MANIAC, Los Alamos Science, 14 (1986) 96-108.
8. W.Q. Meeker, L.A. Escobar, Statistical Methods for Reliability Data, John Wiley & Sons:, New York, 1998.

MACHINE HEALTH MONITORING USING A NOVEL FAULT DETECTION LOGIC

Yaguo Lei[1, 2], Lang Xue[1, 2], Jing Lin[1, 2] and Shantao Niu[1, 2]

[1] State Key Laboratory for Manufacturing Systems Engineering, Xi'an Jiaotong University, Xi'an 710049, China

[2]Shaanxi Key Laboratory of Mechanical Product Quality Assurance and Diagnostics, Xi'an Jiaotong University, Xi'an 710049, China

E-mail address: yaguolei@mail.xjtu.edu.cn (Yaguo Lei)

Abstract. This paper presents a novel method for machine health monitoring. In this method, singular value decomposition (SVD) is first utilized to cancel the noise imbedded in vibration signals, thereby enhancing incipient fault characteristics. Then, root mean square (RMS) is extracted from th e de-noised signal and an adaptive threshold is established, which is constantly updated according to the new sample. Finally, a new logic of fault detection is designed and it is composed of two criteria. 1) In a sliding time window, the number of RMS values above the adaptive threshold has exceeded a safety number. 2) The RMS values in the sliding time-window present an increasing trend, which is evaluated by the Spearman coefficient. Accelerated degradation tests of rolling element bearings demonstrate the effectiveness of the proposed method.

Keywords: Machine health monitoring, fault detection logic, incipient fault, adaptive threshold, singular value decomposition

1 Introduction

Machine health monitoring is to judge whether a fault occurs on machinery based on monitoring information, which is able to ensure the safe operation of machines. One of the major tasks of machine health monitoring is to accurately detect the occurrence of incipient faults. Recently, more and more attention has been paid to the detection of incipient faults. Ginart et al. [1] presents a novel method to establish a constant alarm threshold for fault detection of machinery. Li et al. [2] detected faults of gearboxes using a probability trigger mechanism, i.e., if more than a given percentage of features are greater than a fixed fault detection threshold, then the incipient fault happens. Li et al. [3] proposed an adaptive incipient fault detection approach. It is regarded that the incipient fault has occurred when several continuous samples exceed the 3σ interval. It is named as continuous samples trigger mechanism.

These approaches have good performances in the fault detection. However, they still have two problems in them as follows: 1) since the fixed threshold is constant, which is usually determined by the historical samples in the front of full-time dataset, it cannot be adjusted online as the new sample comes. Therefore, the fixed threshold is not too accurate to judge whether subsequent samples are normal. 2) In the probability trigger mechanism, the percentage is calculated based on all of the historical measurements. The percentage of feature values above the threshold changes little when a new feature above threshold appears. So it is insensitive to the incipient fault using the probability trigger mechanism. In the continuous samples trigger mechanism, the local abnormal fluctuation in the health condition has a great influence on the detection result. Furthermore, several continuous samples above the threshold often occur when the abrupt or significant fault happens, which is not beneficial to the detection of the incipient fault.

To deal with the aforementioned problems, this paper proposes a novel method for the detection of incipient faults. The major contributions of this method are: 1) The adaptive threshold is established based on the Chebyshev's inequality. 2) To accurately detect the occurrence of incipient faults, a new logic of fault decision is proposed based on the analysis of the fault typical behaviour. For example, more abnormal fluctuation would appear in the signals and the amplitude of signals may increase gradually.

2 Adaptive Threshold Establishment

Since the distribution of random variable cannot guarantee to be a normal distribution, Chebyshev's inequality is an available tool to deal with any unknown distribution. For a random variable X, its mean and standard deviation are u and σ respectively. The Chebyshev's inequality is rewritten as:

$$P(|X-u| \geq k\sigma) \leq \frac{1}{k^2}, \forall k > 1. \tag{1}$$

where k is the threshold tolerance. For a given confidence level $1/k^2$, the threshold interval is expressed as $[u-k\sigma, u+k\sigma]$. In practice, u and σ are replaced as the mean $\overline{X}$ and standard deviation S of samples respectively.

Usually, the initial dataset $X_m = (X_1, ..., X_m)$ is selected to calculate the initial threshold:

$$T_m = \overline{X}_m \pm kS_m \tag{2}$$

where T_m is the bound of the threshold interval $[\overline{X}_m - kS_m, \overline{X}_m + kS_m]$.

As a new sample is captured, judge whether the sample belongs to the threshold interval. If the sample belongs to the threshold interval, it is added into the dataset, i.e. $X_{m+1} = (X_1, ..., X_m, X_{m+1})$, the threshold T_{m+1} is recalculated. Otherwise, the sample is regarded as the abnormal sample and the threshold T_{m+1} isn't updated.

3 Proposed Incipient Fault Detection Method

The real-time vibration signals are acquired from machinery on-line. Since mechanical fault characteristics, especially in the incipient stage, are always submerged in strong background noise, SVD is employed to cancel the noise imbedded in vibration signals [4]. Then the RMS index is extracted from the de-noised signals since RMS is able to reflect the increase of the vibration energy with the development of faults [5]. Then, an adaptive threshold is established based on Chebyshev's inequality according to the statistic characteristics of the RMS values. Since the RMS values always perform the gradually increasing trend, the upper bond $\{T_i^U\}_{i=1:N} = \overline{X}_i + kS_i$ is selected as the detection threshold.

When the upper bound $\{T_i^U\}_{i=1:N}$ has been calculated, appropriate detection logic is developed based on the analysis of the fault typical behaviour. The logic is composed of two criteria, i.e., statistical analysis and trend analysis.

3.1 Statistical Analysis

Compared with the random fluctuation, more abnormal fluctuation would appear when incipient faults occur. Therefore, a logic of fault detection is chosen, where more than a given number of feature values are greater than the fixed fault detection threshold. The given number is named as the safety number which is always set arbitrarily. In fact, it is able to be determined reasonably by the distribution characteristics of RMS values.

The inequality in Eq. (3) means that the probability of $X \notin (u - k\sigma, u + k\sigma)$ is less than $1/k^2$. If the probability is larger than $1/k^2$, it implies that the incipient fault occurs. When the probability is calculated based on all of the historical measurements, the probability changes little as a new abnormal sample appears. So the probability is calculated in a sliding time-window, instead of all of the historical measurements, which improves the sensitivity of the logic to the incipient fault. The first criterion is defined as follows: The number of RMS values above the adaptive threshold in the sliding time-window has exceeded a safety number. The safety number is determined by the threshold tolerance and the window length, which can be calculated by the following equation:

$$S_n = \frac{N}{k^2} \tag{3}$$

where S_n is the safety number, and N is the window length.

3.2 Trend Analysis

When incipient faults occur, the RMS always performs the gradually increasing trend. The Spearman coefficient between the RMS and time is an effective tool to evaluate the trend. First, RMS $\{R_i\}_{i=1:N}$ and time $\{T_i\}_{i=1:N}$ are converted to their ranks $\{r_i\}_{i=1:N}$ and $\{t_i\}_{i=1:N}$, then the Spearman coefficient ρ between them is computed using the following equality:

$$\rho = \frac{\sum_{i=1}^{N}(r_i - \bar{r})(t_i - \bar{t})}{\sqrt{\sum_{i=1}^{N}(r_i - \bar{r})^2 \sum_{i=1}^{N}(t_i - \bar{t})^2}} \tag{4}$$

where $\bar{r}$ and $\bar{t}$ are the means of $\{r_i\}_{i=1:N}$ and $\{t_i\}_{i=1:N}$ respectively. When $\{R_i\}_{i=1:N}$ is a monotone increasing or decreasing function of $\{t_i\}_{i=1:N}$, the Spearman coefficient ρ equals to $+1$ or -1. Vice versa, if $\{R_i\}_{i=1:N}$ has not a well monotonic relation with $\{t_i\}_{i=1:N}$, the ρ will close to 0.

Since the gradually increasing trend just performs after the incipient fault occurs, it is not a wise choice using the Spearman correlation coefficient in all of the historical measurements. Application of the Spearman coefficient in the sliding time-window is able to increases the sensitivity of detecting the local increasing trends. Therefore, the second criterion can be defined as follows: The Spearman coefficient ρ in the sliding time-window is higher than $1/2$.

When the above two criteria are satisfied simultaneously, it means that an incipient fault has occurred. The logic is able to make a response to the incipient fault reliably and in time.

4 Experimental Demonstration

In this section, the vibration signals acquired from accelerated degradation tests of rolling element bearings are used to demonstrate the effectiveness of the proposed method.

The experimental system is a platform called PRONOSITA [6] which is dedicated to test and validate bearings fault detection, diagnostic and prognostic approaches. It is able to provide experimental data that characterize the degradation of ball bearings along their whole operational life. The sampling frequency is 25.6 kHz and 2560 samples (i.e. 1/10 s) are recorded each 10 seconds.

SVD is employed to de-noise the vibration signals. Then the RMS is extracted from the processed vibration signals. Based on the statistic characteristics of the RMS values, the occurrence of incipient faults is determined by the adaptive threshold and the proposed fault detection logic.

To demonstrate the superiority of the proposed method, another two approaches proposed by Ginart et al. [1] and Li et al. [3], respectively, are applied for comparison. The results of these approaches are shown in Fig.2. It is seen from Fig.2 that the adaptive threshold in our approach is constantly updated as new measured sample is captured, which is trained online adaptively for more accurate and yet robust incipient fault detection. For the gradual failure case i.e. bearings 1, the incipient fault occurrence detected by Ginart's approach seriously lags behind Li's and our approach. Li's result is close to the result of our approach, but it also has a little time delay than ours. For the abrupt failure case i.e. bearing 2, the three approaches give approximate results, but our approach still performs better than Li's and Ginart's approaches. There are some abnormal samples in health condition located out of the adaptive threshold, which are marked by "*". Thanks to the help of our proposed detection logic, it ensures the accuracy detection of the incipient fault.

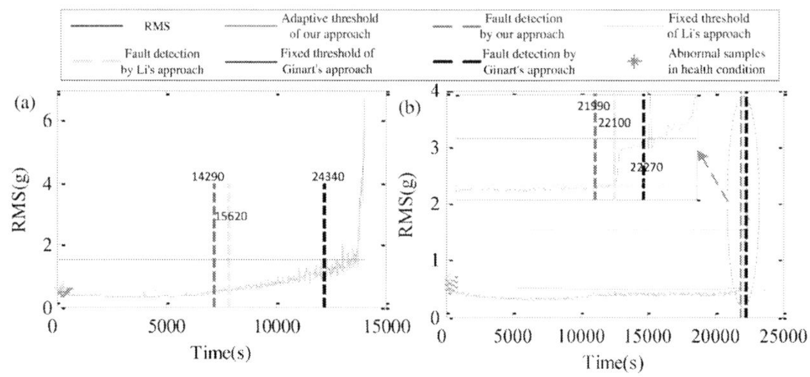

Fig. 1. Results of the incipient fault detection: (a) bearing 1, (b) bearing 2

5 Conclusions

To accurately detect the occurrence of incipient faults, a machine health monitoring method using novel fault decision logic is proposed in this paper. The noise in original vibration signal is eliminated using SVD to enhance the performance of incipient faults. RMS is extracted from the de-noised signals. Then the adaptive threshold is established. Finally, the occurrence of incipient faults is determined by the proposed logic of fault detection. The accelerated degradation tests of rolling element bearings are utilized to demonstrate the effectiveness of the proposed method. In order to validate the detection accuracy of the incipient fault, it is compared with Ginart's and Li's approach. The results show that our approach performs better than another two approaches.

Acknowledgement

This research is supported by National Natural Science Foundation of China (51475355, 51421004), Young Talent Support Plan of Central Organization Department and Fundamental Research Funds for the Central Universities (2012jdgz01 and CXTD2014001).

Reference

1. Ginart, I. Barlas, et al. Automated feature selection for embeddable prognostic and health monitoring (PHM) architectures, IEEE Autotestcon, 1 (2006) 195-201.
2. R. Li, D. He. Rotational machine health monitoring and fault detection using EMD-based acoustic emission feature quantification, IEEE Transactions on Instrumentation and Measurement, 61 (2012) 990-1001.
3. N. Li, Y. Lei, et al. An improved exponential model for predicting remaining useful life of rolling element bearings, IEEE Transaction on Industrial Electronics, 62 (2015) 7762-7773.
4. Z. Qiao, Z. Pan. SVD principle analysis and fault diagnosis for bearings based on the correlation coefficient, Measurement Science and Technology, 26 (2015) 085014.
5. Y. Lei, Z. He, et al. Fault diagnosis of rotating machinery based on multiple ANFIS combination with Gas, Mechanical Systems and Signal Processing, 21 (2007) 2280-2294.
6. P. Nectoux, R. Gouriveau, et al. PRONOSTIA: An experimental platform for bearings accelerated degradation tests, IEEE International Conference on Prognostics and Health Management, 2012 (2012) 1–8.

Remaining Useful life Estimation for Degrading Systems with Alternating Operation States

Zhengxin Zhang, Changhua Hu, Xiaosheng Si*, and Shaohua Zhou

Department of Automation Engineering, Xi'an Research Institute of High-Tech, Xi' an, 710025, Shaanxi Province, P.R. China

zhangzhengxin13@gmail.com, hch6603@263.net, sxs09@mails.tsinghua.edu.cn, 1229102037@qq.com

Abstract. This paper deals with the remaining useful life (RUL) estimation issue for systems experiencing alternating operating states which have non negligible influence on the degradation process of the system. After modeling the switching operation process through an alternating renewal process, the operation process has been incorporated into the degradation modelling of the systems, and an explicit-form approximated probability density functions (PDFs) of both lifetime and RUL defined base on the concept of first hitting time (FHT) have been derived. Such explicit PDFs with much more computing convenience compared to simulation based approaches which are quite popular in the literature, will facilitate the online updating of the RUL estimation and the sequential decision-making on health management activities including inspection, maintenance, and spare parts ordering. A numerical study is carried out to illustrate and demonstrate the proposed method.

Keywords: Reliability; Alternating renewal process; RUL; Degradation modelling; Prognostics and health management

1 Introduction

Degradation modelling based remaining useful life (RUL) estimation has been proved to be an effective alternative to the traditional lifetime based method. With lower economic costs, degradation modelling based RUL estimation lays the ground for prognostics and health management (PHM), and thus has drawn much more and more attention from both the scholars and engineers in the field of reliability all around the world [1]. Due to the fact that degradation of a system is a result of the interaction of system's inner state and its working environment [2], the influence of the system's operation process on the degradation of the system has to be considered when estimating the system's RUL.

There are various kinds of systems experiencing alternating operating states switches which affect the degradation processes of the concerned systems [3, 4]. Examples include electric vehicles (EV) which are used on workdays and parked on weekends [5], gyroscopes in an inertial navigation system (INS) which experience storage and usage operating states during their service time [6], CPUs which only process data when works are assigned. Different operating states have different influence on the degradation process of the system. As mentioned by [6], the gyroscopes in working are electrified and suffering much more additional electronic and thermal stresses which accelerate the gyroscopes degradation process. Therefore, the key performance indicators of gyroscopes degrade quite faster in working state. The Li-ion batteries in EVs undergo charge/discharge cycles which cause an increment in the internal resistance as well as a decrement in the capacity, thus degrades quicker on workdays. To achieve an accurate estimation of the RULs, the switches in the operating states have to be integrated into degradation modeling and RUL estimation.

The switches in the operating state can be caused by either changes of the working environments or the alternates of the system's task. Due to fact that the changes of environments and the alternates of tasks of a system exhibit many stochastic characteristics, a random process is required to describe such impacts. In addition, the performance of the system always shows some random dynamics even under a constant operating state. Thus, a stochastic process based method is quite suitable for the describing the degradation process. There are some related works concentrating on process degradation modelling based RUL estimation [3, 4]. However, most existing approaches, which needs either numerical approximations or Monte-Carlo simulations, are computationally costly. To meet the requirement of online RUL estimation, explicit forms of RUL distribution are urgently desired.

Based on the aforementioned discussions, this paper attempts to present a degradation modelling based RUL estimation approach for systems under alternating operation states. The attention has been focused on the derivation of explicit lifetime and RUL distributions. An alternating renewal process depicting the operating states switches is integrated into the degradation model and explicit approximates of the lifetime and RUL distribution are formulated. Numerical study is provided to validate the proposed method. The explicit lifetime and RUL distributions will facilitate the online reliability evaluation, and thus have promising practical potentials in PHM.

2 Problem Formulation

Denote by $\{X(t); t \geq 0\}$ a stochastical process depicting the performance indicator of the concerned system at time t, and the dynamics of $X(t)$ is described by the following stochastic differential equation (SDE)

$$dX(t) = \mu(t)dt + \sigma(t)dB(t) \tag{1}$$

where $\mu(t)$ is the drift coefficient function representing degradation speed of the system, $\sigma(t)$ is the diffusion coefficient function expressing the degree of uncertainty in the degradation process, and $B(t)$ is a standard Brownian motion used to characterize the randomness included in the degradation process. Equation (1) and its extensions have been widely used as degradation models in the field of reliability. See [7, 8] for more details.

To incorporate the influence of state switches in the operation on the degradation process, the drift coefficient function $\mu(t)$ is set operation state dependent. Let $\{v(t);t \geq 0\}$ be the operation process of the system, where $v(t)=1$ when the system is operating under state 1 and $v(t)=2$ if the system is in operation state 2. Therefore, $\mu(t)$ is regarded as $\mu(v(t))$. In this paper, $\mu(v(t))=\mu_1$ if $v(t)=1$, and $\mu(v(t))=\mu_2$ if $v(t)=1$, which indicating that the system degrades at a respectively constant speed under each operating state. Generally, $\{v(t);t \geq 0\}$ is also a stochastic process and a continuous time Markov chain (CTMC) has been proved to be a suitable tool to model the switches in the operation states. Furthermore, $\{v(t);t \geq 0\}$ should be fused into the degradation model and lifetime and RUL estimation. In addition, the diffusion coefficient is set to be σ_B under both operation states for simplicity.

The lifetime T of system is defined under the concept of first hitting time (FHT),

$$T = \inf\{t : X(t) \geq w\}, \tag{2}$$

where w is a preset failure threshold, cross which means an unsatisfactory performance is obtained. Accordingly, the RUL at the k-th inspection time t_k with degradation level x_k is defined as

$$L_k = \inf\{l_k : X(t_k + l_k) \geq w \mid x_k < w\}. \tag{3}$$

Obviously, both the lifetime T and the RUL at t_k are random variables. Due to its ability to model the uncertainty in the T and L_k, a probability density function (PDF) has drawn much favor of both scholars and engineers in the field of reliability. Thus, the purposed of this paper is to derive the PDF of T and L_k with the consideration of the influence of operation state switches $\{v(t);t \geq 0\}$ on the degradation process. Firstly, $\{v(t);t \geq 0\}$ should be modeled.

3 Derive the Lifetime and RUL Distributions

The distributions of lifetime and RUL defined through the concept of FHT in Section II are the foundation of decision making for lots of sequential management activities, and also the topic of this section. To accomplish the derivation, an explicit approximate of the PDF that a diffusion process crossing a preset constant threshold for the first time is quoted as Lemma 1.

Lemma 1: Assume that the probability that stochastic process $\{X(t);t \geq 0\}$ defined in (1) proceeds a preset boundary before t is negligible if $X(t)$ hits the boundary exactly at time t, and that the drift coefficient functions $\mu(t)$ in (1) is integrable, then the probability of the FHT of $\{X(t);t \geq 0\}$ crossing a preset threshold w can be approximated through

$$p_{X(t)}(w,t) \simeq \frac{1}{\sqrt{2\pi t}}\left[\frac{S_B(t)}{t}+\frac{\mu(t)}{\sigma_B}\right]\exp\left[-\frac{S_B^2(t)}{2t}\right], \tag{4}$$

where

$$S_B(t)=\frac{1}{\sigma_B}\left(w-\int_0^t \mu(\tau)d\tau\right). \tag{5}$$

The proof of Lemma 1can be found in [9] thus omitted here. Please note that the constraint that $\mu(t)$ has to be continuous has been released to integrable in lemma 1.

Based on Lemma 1, the lifetime distribution of systems with degradation process (1) can be formulated through the law of total probability. Particularly, we have the following theorem 1.

Theorem 1: The lifetime distribution $f_{T|\boldsymbol{\theta}}(t\mid\boldsymbol{\theta})$ of a system under a state-switching operation process with a degradation process (1) can be approximated by

$$f_{T|\boldsymbol{\theta}}(t\mid\boldsymbol{\theta}) \simeq \frac{1}{\sqrt{2\pi t}}\left[\frac{\tilde{S}_B(t)}{t}+\frac{\tilde{\mu}(t)}{\sigma_B}\right]\exp\left[-\frac{\tilde{S}_B^2(t)}{2t}\right], \tag{6}$$

where $\boldsymbol{\theta}$ is the collection of all the parameters in both the degradation model and operating process, and

$$\tilde{S}_B(t)=\frac{w}{\sigma_B}-\frac{\lambda\gamma^2(\mu_1-\mu_2)}{\sigma_B(\lambda+\gamma)^2}\left\{1-\exp\left[-\frac{(\lambda+\gamma)t}{\gamma\lambda}\right]\right\}-\frac{(\lambda\mu_1+\gamma\mu_2)t}{\sigma_B(\lambda+\gamma)}, \tag{7}$$

$$\tilde{\mu}(t)=\frac{(\lambda\mu_1+\gamma\mu_2)}{(\lambda+\gamma)}+\frac{\gamma(\mu_1-\mu_2)}{(\lambda+\gamma)}\exp\left[-\frac{(\lambda+\gamma)t}{\lambda\gamma}\right]. \tag{8}$$

It is observed that all the all the elements needed in (6) are of explicit forms, which is an improvement of the most popular simulation based method in the literature.

Compared to the lifetime, engineer pay much more attention on the RUL at time t_k with degradation level x_k. According to the properties of diffusion processes, the new process $\{\tilde{X}(l_k): l_k \geq 0\}$ where $\tilde{X}(l_k)=X(t_k+l_k)-x_k$ is still a diffusion process. Therefore, the RUL defined in (3) can be regarded as the FHT of $\tilde{X}(l_k)$ crossing a new threshold w_k where $w_k=w-x_k$. A similar procedure used for deriving the PDF of lifetime T is used to the formulate the PDF of L_k.

4 A Numerical Illustration

In this section, a Monte-Carlo simulation studies is carried out to demonstrate the effectiveness of the approximated lifetime distribution, as well as to illustrate the proposed model for RUL estimation. The values of the parameters are $\lambda=0.5$, $\gamma=0.25$, $\mu_1=0.6$, and $\mu_2=0.25$. Note that different values of σ_B including 0.5, and 2 have been used to test the robustness of the proposed method to approximate the PDF of FHT. The failure threshold is set as $w=5$.

After simulate an alternating renewal process to represents the operation states switches of the system, the corresponding degradation process can be generated based on Euler discretization method. The specific procedure to generate a degradation path is quite similar to that in [9], but different drift coefficient parameters should be used for different operating states. A sufficiently large number of degradation paths are generated and the corresponding FHT of each paths to w are gathered in $\boldsymbol{T}$, whose histograms under different settings of σ_B are provided below in Fig.1.

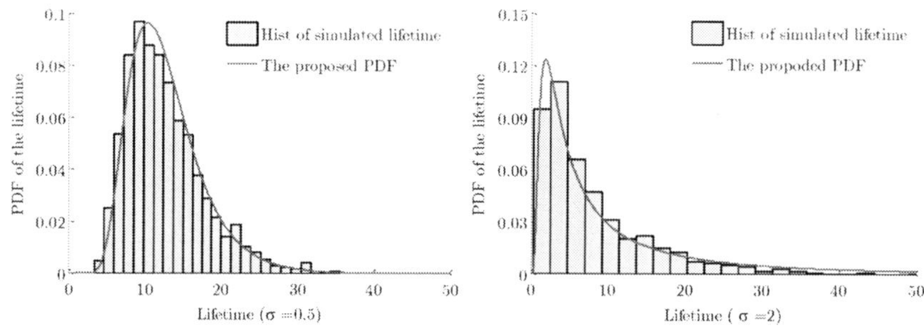

Fig. 1. The simulated life distribution and the proposed PDF

Fig.1 shows that the proposed method can provide accurate approximations to the PDFs of FHT for systems with degradation processes described by diffusion process under different scales of uncertainty. Because both results Theorem 1 for lifetime PDF and PDF of RUL are in explicit, the proposed method achieves an accurate estimation of lifetime and RUL distribution with much lower computation cost than simulation based method in [6].

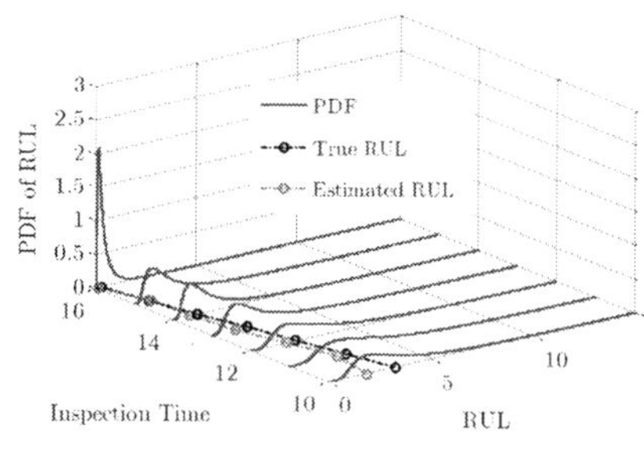

Fig. 2. PDFs of RULs at different inspection time

To further validate the presented method, a degradation path is selected to illustrate the RUL estimate result. The FHT of this degradation path crossing w is 16.29. RULs at 7 inspections times including $t_k = 10, \cdots, 16$ are estimated and the corresponding RULs are listed in Fig.2. The true RULs and the point estimations of RULs which is defined as the mode of the PDFs of RULs are also provided, indicating that the proposed method can provide an accurate point estimation of the RUL.

5 Conclusion

A degradation model incorporating the influence the operation state switches which has been depicted through an alternating renewal process has been proposed in this paper. Compared to most existing approaches where simulation based methods are employed to compute the lifetime and RUL distributions, an explicit method with much less computation cost has been proposed to approximate the PDFs of both lifetime and RUL. Results in numerical study and the results demonstrate the effectiveness of the proposed methods for degradation modeling and RUL estimation. This work can be extended in several ways. Firstly, an online parameter estimation procedure should be developed to achieve real-time updating of RUL PPFs. Secondly, Wiener process used to model the degradation of the system under each operation states can be released to nonlinear diffusion process. Last but not least, random effects could be incorporated to represent the heterogeneity in degradation among different units.

Acknowledgments

This research was partially supported by the NSFC under Grants 61573365, 61174030, 61374126, 61473094, and 61304101.

Reference

1. Sikorska, J. Z., Hodkiewicz, M., Ma L.: Prognostic modelling options for remaining useful life estimation by industry. Mech. Syst. Signal Process. 25, 1803–1836, (2011).
2. Pecht, M.: Prognostics and Health Management of Electronics. Hoboken, NJ, USA: Wiley, (2008)
3. Hawkes, A. G., Cui, L. R., Zheng, Z. H.: Modeling the evolution of system reliability performance under alternative environments. IIE Transactions. 43, 761–772, (2011)
4. Rigamonti, M., Baraldi, P., Zio, E., Astigarraga, D., Galarza, A.: Particle filter-based prognostics for an electrolytic capacitor working in variable operating conditions. IEEE Transactions on Power Electronics. DOI:10.1109/TPEL.2015.2418198, 2015.
5. Barr´e, A., Suard, F., G´erard, M., Montaru, M., Riu, D.: Statistical analysis for understanding and predicting battery degradations in real-life electric vehicle use. Journal of Power Sources. 245, 846–856, (2014)
6. Si, X. S., Hu, C. H., Kong, X. Y., Zhou, D. H.: A residual storage life prediction approach for systems with operation state switches. IEEE Transactions on Industrial Electronics. 61, 6304–6315, (2014)
7. Peng, C. Y., Tseng, S. T.: Mis-specification analysis of linear degradation models. IEEE Trans. Rel. 58, 444-455, (2009).
8. Si, X. S., Zhou, D. H.: A generalized result for degradation model-based reliability estimation. IEEE Trans. Autom. Sci. Eng. 11,632–636, (2014)
9. Si, X. S., Wang, W., Hu, C. H., Zhou, D. H., Pecht, M.: Remaining useful life estimation based on a nonlinear diffusion degradation process. IEEE Trans. Rel.61, 50–67, (2012).

Improved Empirical Wavelet Transform Based Rotor Rubbing Fault Diagnosis

Jinde Zheng[1*] Haiyang Pan[1] and Junsheng Cheng[2]

[1] School of Mechanical Engineering, Anhui University of Technology, Maanshan, Anhui, PR China, 243032.

Email: jdzheng1986@gmail.com. Tel: +86 18395585081

[2] State Key Laboratory of Advanced Design and Manufacturing for Vehicle Body, Hunan University, Changsha, Hunan, PR China, 410082. Email: jdzheng1986@gmail.com

Abstract: Empirical wavelet transform (EWT) is a novel method for analyzing multi-component signal, which is proposed based on the classical wavelet transform. To fulfill an adaptive separation of Fourier spectrum in EWT, an improved EWT (IEWT) method is introduced in this paper. The IEWT includes two parts, i.e. the adaptive segmentation of Fourier spectrum and the foundation of WT based filter group. Also the IEWT is compared with empirical mode decomposition (EMD) by analyzing the simulation signal, and the analysis results have demonstrated the effectiveness of the proposed methods. Finally, the proposed method is applied to fault diagnosis of rotor system with local rubbing and the analysis results of experiment data indicated that the proposed method could effectively fulfill rotor rubbing fault diagnosis and have a better result than the EMD method.

Keywords: empirical mode decomposition, empirical wavelet transform, rotor rubbing

1 Introduction

Since the vibration signals of most of machinery equipment are nonlinear and non-stationary, time-frequency analysis method has been widely applied to mechanical fault diagnosis for being able to simultaneously provide local information in time and frequency domains. As a classical time-frequency analysis method, wavelet transform (WT) has been applied to mechanical fault diagnosis by related scholars and a series of fault diagnosis methods have been developed [1, 2]. Nevertheless, WT also has some shortcomings, such as a specific wavelet basis should be chosen in advance, different wavelet basis will results in different conclusions.

Hilbert-Huang Transform (HHT) was a recently proposed non-stationary signal analysis method by Huang et al.[3], which mainly includes two parts, i.e. empirical mode decomposition (EMD) and Hilbert transform (HT). HHT method has been paid much attention by the related scholars of signal processing, mechanical fault diagnostics and other engineering fields [4, 5]. However, HHT also has some drawbacks, such as the fitting errors caused by cubic splines, mode mixing, end effects and the negative frequency caused by HT.

Based on EMD, other several adaptive signal methods including local mean decomposition (LMD) [6], intrinsic time-scale decomposition (ITD) [7], and local characteristic-scale decomposition (LCD) [8] are proposed, recently. However, these kinds of methods have the same disadvantages, such as mode mixing, end effects and lack of restrict mathematical foundation. Also, an EMD like tool for non-stationary data analysis, synchrosqueezed wavelet transform (SWT) was proposed by Daubechies et al. [9] by combining the classical WT analysis and time-frequency rearrangement technique. But SWT is not a fully adaptive wavelet transform in nature.

Recently, another non-stationary signal processing method called empirical wavelet transform (EWT) was proposed [10, 11]. In EWT Firstly, the Fourier frequency spectrum is normalized and separated into numbers of intervals and then the compactly supported orthonormal wavelet basis is explicitly constructed in each interval. By using EWT a multi-component signal can be adaptively decomposed into several intrinsic mode functions (IMFs). The key idea of EWT is to segment the Fourier spectrum and then apply a certain wavelet based filtering corresponding to each detected support Fourier spectrum. To fulfill an adaptive decomposition for the given signal, in this paper an improved empirical wavelet transform (IEWT) is put forward. Finally, the proposed method is compared with EMD and is employed to the fault diagnosis of rotor rubbing. The analysis results has shown its effectiveness.

2 The improved EWT method

The biggest problem of the original EWT is to segment the spectrum meaningfully and adaptively. To fulfill an adaptive decomposition of EWT, an improved EWT method (IEWT) is introduced in this paper, which includes two parts, i.e. the adaptive segmentation of Fourier spectrum and the foundation of WT based filter group. In [12], the authors introduced a scale-space representation for a given function in Fourier spectrum. And then the initial problem is transformed to a two class clustering problem. Further, the authors also introduced three different ways to automatically solve this clustering problem including Probabilist approach, Otsu's and k-Means methods. Therefore, in this paper we use the above method to the segmentation of Fourier spectrum.

We denote $f(t)$ a real value signal and its Fourier spectrum $\hat{f}(\omega)$. We restrict our discussion in the interval $[0,\pi]$ for convenience. $[0,\pi]$ can be separated N sub-intervals $\Lambda_n = [\omega_{n-1}, \omega_n]$, where $\omega_0 = 0, \omega_N = \pi$. Centered around each ω_n, we define a transition phase T_n with width $2\tau_n$ and the simplest is to choose τ_n proportional to ω_n: $\tau_n = \gamma\omega_n$, $0<\gamma<1$. The empirical wavelets are defined as band-pass filters on each Λ_n. Based on Λ_n and inspired from Meyer's and Littlewood-Paley wavelets, a wavelet tight frame $\mathrm{B} = \left\{ \{\phi_n(t)\}_{n=1}^{N-1}, \{\psi_n(t)\}_{n=1}^{N-1} \right\}$ are defined. For $\forall n>0$, their Fourier transforms, i.e. the empirical scaling function $\hat{\phi}_n(\omega)$ and the empirical wavelets $\hat{\phi}_n(\omega)$ are given by expressions of equations (1) and (2), respectively

$$\hat{\phi}_n(\omega) \begin{cases} 1, & |\omega| < (1-\gamma)\omega_n \\ \cos\left[\frac{\pi}{2}\beta\left(\frac{1}{2\tau_n}|\omega| - (1-\gamma)\omega_n\right)\right], & (1-\gamma)\omega_n \le |\omega| \le (1+\gamma)\omega_n \\ 0, & otherwise \end{cases} \tag{1}$$

$$\hat{\psi}_n(\omega)\begin{cases} 1, & (1+\gamma)\omega_n \le |\omega| \le (1-\gamma)\omega_{n+1} \\ \cos\left[\frac{\pi}{2}\beta\left(\frac{1}{2\tau_{n+1}}|\omega|-(1-\gamma)\omega_{n+1}\right)\right], & (1-\gamma)\omega_{n+1} \le |\omega| \le (1+\gamma)\omega_{n+1} \\ \sin\left[\frac{\pi}{2}\beta\left(\frac{1}{2\tau_n}|\omega|-(1-\gamma)\omega_n\right)\right], & (1-\gamma)\omega_n \le |\omega| \le (1+\gamma)\omega_n \\ 0, & otherwise \end{cases} \quad (2)$$

where the function $\beta(x)$ most often used is $\beta(x) = x^4(35-84x+70x^2-20x^3)$.

Similar to the classic wavelet transform, the detail coefficients $W_f^{\varepsilon}(n,t)$ are given by the inner products between the signal with the empirical wavelets and the approximation coefficients $W_f^{\varepsilon}(0,t)$ are defining by the inner product between the signal with the scaling function. Then the empirical mode $f_k(t)$ can be given as $f_0(t) = W_f^{\varepsilon}(0,t) * \phi_1(t)$ and $f_k(t) = W_f^{\varepsilon}(n,t) * \psi_n(t)$, where k represents the mode number and "$*$" represents the convolution. A detailed description can be found in literature [10]. In nature, for an arbitrary multi-component signal, the IEWT method tries to establish a set of adaptive band-pass filter and then decompose the analyzed signal into a number of mono-component signals with compact supported spectrum.

3 Simulation signal analysis

To verify the validity of the proposed IEWT, we consider the mixed signal $s(t)$ superposed by three modes and a trend shown in Eq.(5)

$$s(t) = s_1(t) + s_2(t) + s_3(t) + s_4(t), \quad t = [0,1] \quad (5)$$

where $s_1(t)$ is an AM-FM signal and the frequency of carrier signal is 200 Hz. $s_2(t) = (1+0.3\cos 2\pi\cdot 5t)\sin(2\pi\cdot 80t + 2\pi 5t^2)$, $s_3(t) = 2(t^2+1)\cos(2\pi\cdot 40t)$ and $s_4(t) = 2t^2$.

The time domain waveforms of the four components are shown in Fig. 1.

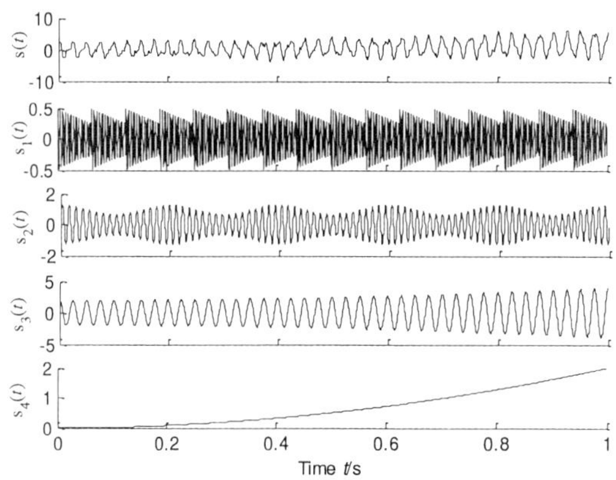

Fig. 1 the time domain waveforms of mixed signal $s(t)$ and its components

Firstly, the IEWT is used to decompose $s(t)$ and the results are shown in Fig. 2. The obtained four component signals C_1 ,C_2, C_3 and C_4 are respectively corresponding to the $s_1(t)$, $s_2(t)$, $s_3(t)$ and $s_4(t)$ and the correlation coefficients (CCs) are 0.9929, 0.9935, 0.9987 and 0.9973. Hence, the components obtained by IEWT have very high consistencies with the true components. For comparison, the EMD is used to decompose $s(t)$ and the results of EMD are given in Fig.3, from which it can be found that the IMFs C_2 and C_3 obtained by EMD have serious mode mixing and the IMFs C_4, C_5 and C_6 are false-components obtained by fitting errors and thus the IMFs obtained by EMD have large differences from the true components. Therefore, the comparison results show that IEWT has better decomposition effects than that of EMD method.

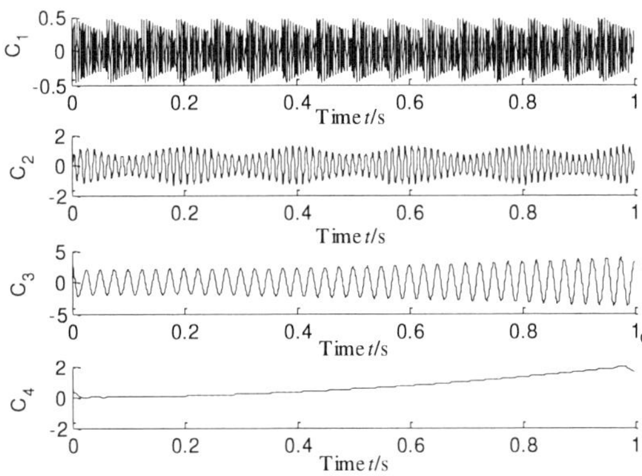

Fig. 2 the decomposed results of $s(t)$ by using IEWT

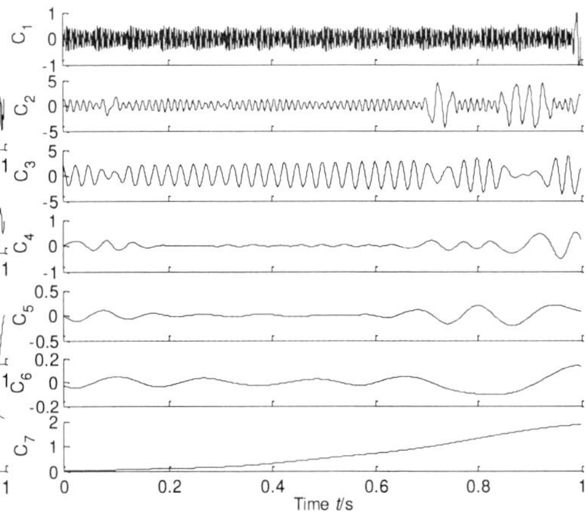

Fig.3 the decomposed results of $s(t)$ by using EMD

4 Application of the IEWT in rotor rubbing

To verify the effectiveness of IEWT, it is employed to analyze the experiment data of rotor system with local rubbing fault. The Z-T3 fault simulation test rig of rotor system is used to generate the data set and a detailed description of the test system is shown in literature [13]. The sample frequency is 2048 Hz and the rotating frequency of shaft is 50 Hz. Under these conditions, the radial vibration displacement signals of rotor system with local rubbing are collected and one of them and its spectrum are shown in Fig. 4, from which it can be found that only the component of rotating frequency is clear and the fault information related with local rubbing is covered by the background signal and noises.

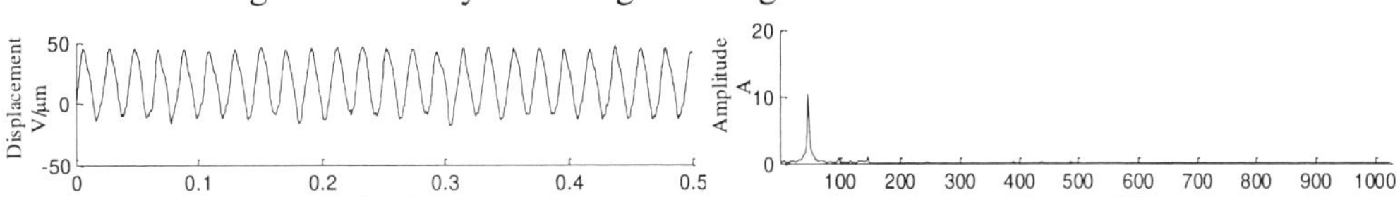

(a) time domain waveform (b) frequency spectrum

Fig. 4 time domain waveform and spectrum of the radial displacement signal of rotor with local rubbing

The IEWT method is employed to decompose the radial vibration displacement signal of rotor system for extracting the fault information and the decomposing result is shown in Fig. 5(a). The envelope spectrum of the first IMF C_1 is shown in Fig. 5(b), respectively. From Fig. 5, it can be found that the first frequency component has an apparent amplitude-modulated characteristic and is modulated by a lower frequency component and from its envelope spectrum it can be seen that the frequency of AM signal is 50Hz equals the rotating frequency. This can be explained that the rotating part will rub the fixed part once when the rotor rotates per round. Therefore, by using IEWT, the AM signal containing main fault information can be extracted from the strong background signal. This example shows the effectiveness of IEWT in extracting fault information.

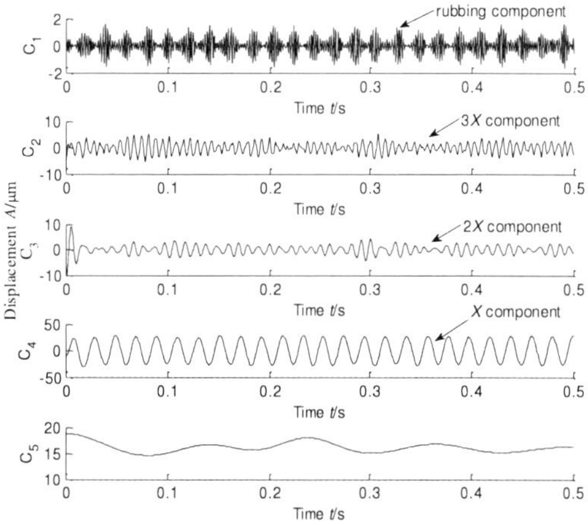

(a) the decomposed results by IEWT

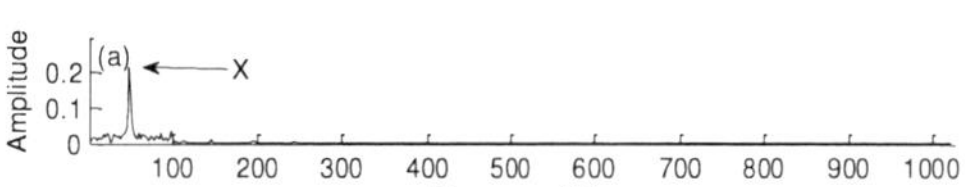

(b) envelope spectrum of the first IMF of IEWT

Fig. 5 IEWT results

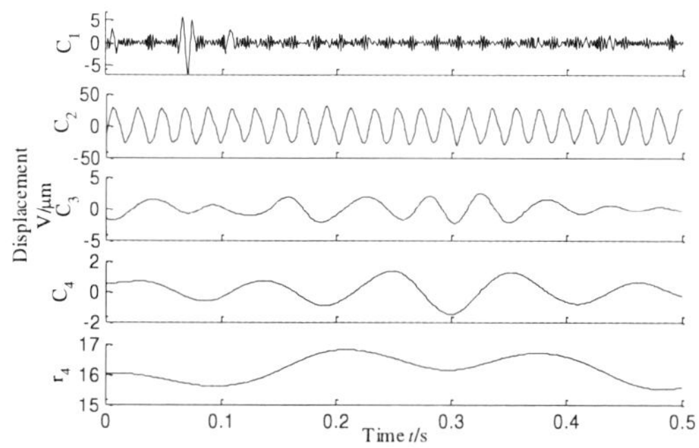

(a) the decomposed results by EMD

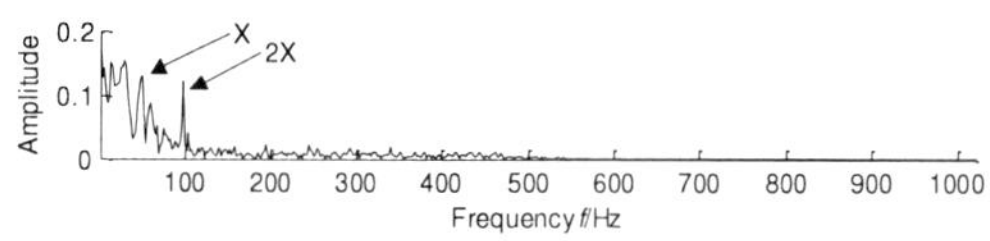

(b) envelope spectrum of the first IMF of EMD

Fig. 6 EMD results

For the purpose of comparison, the same analyzed signal is decomposed by EMD. The decomposing result is given in Fig. 6(a) and the envelope spectrum of the first IMF component is shown in Fig. 6(b), from which it can be found that though the first component holds an AM characteristics, the mode mixing arises in some place. the frequency of AM part contains the rotating frequency and its double frequency, as well as the other lower frequency, which will mislead the process of fault diagnosis. Therefore, the comparison results indicate that the effect of IEWT is superior to that of EMD in fault diagnosis of rotor rubbing.

5 Conclusions

An improved empirical wavelet transform (IEWT) is introduced and is compared with EMD by analyzing simulation signal and the vibration signal of rotor system with local rubbing. The analysis results indicate that IEWT can effectively diagnose the local rubbing fault of rotor and get a much better diagnosis effect than that of EMD. There are also some shortcomings of IEWT, such as the end effect and the dependence on spectrum separation. The authors will focus on these issues in future work.

Acknowledgments

This work was supported by the National Natural Science Foundation of China (Grant No.51505002).

References

[1] Peng, Z. K., P. W. Tse, and F. L. Chu.: A comparison study of improved Hilbert–Huang transform and

wavelet transform: Application to fault diagnosis for rolling bearing. Mechanical Systems and Signal Processing. 19 (5), 974--988 (2005)

[2] Yan, Ruqiang, R. X. Gao, and X. Chen.: Wavelets for fault diagnosis of rotary machines: A review with applications. Signal Processing. 96, 1--15 (2014)

[3] Huang, N. E., Shen, Z., Long, S. R, et al.: The empirical mode decomposition and the Hilbert spectrum for nonlinear and non-stationary time series analysis. Royal Society of London Proceedings A. 454(1971), 903--995 (1998)

[4] Yan, J., and Lu, L.: Improved Hilbert–Huang transform based weak signal detection methodology and its application on incipient fault diagnosis and ECG signal analysis. Signal Processing. 98(5), 74--87 (2014)

[5] Zheng, J., Cheng, J., and Yang, Y.: Generalized empirical mode decomposition and its applications to rolling element bearing fault diagnosis. Mechanical Systems and Signal Processing. 40 (1), 136--153 (2013)

[6] Smith, J. S.: The local mean decomposition and its application to EEG perception data. Journal of the Royal Society Interface. 2 (5), 443--454 (2005)

[7] Frei, M. G.: Intrinsic time-scale decomposition: time-frequency-energy analysis and real-time filtering of non-stationary signals. Royal Society of London Proceedings. 463(2078), 321--342 (2007)

[8] Zheng, J., Cheng, J., Yang, Y.: A rolling bearing fault diagnosis approach based on LCD and fuzzy entropy. Mechanism and Machine Theory. 70, 441--453 (2013)

[9] Daubechies, I., Lu, J., and Wu, H. T.: Synchrosqueezed wavelet transforms: an empirical mode decomposition-like tool. Applied & Computational Harmonic Analysis. 30(2), 243--261 (2011)

[10] Gilles, J.: Empirical wavelet transform. IEEE Transactions on Signal Processing. 61(16), 3999--4010 (2013)

[11] Gilles, J., Tran, G., Osher, S.: 2D Empirical Transforms: Wavelets, Ridgelets and Curvelets revisited. SIAM Journal on Imaging Sciences. 7(1), 157--186 (2014)

[12] Gilles, J., Heal, K.: A parameterless scale-space approach to find meaningful modes in histograms-Application to image and spectrum segmentation. International Journal of Wavelets Multiresolution and Information Processing. 12 (6), 1450044 (1--17) (2014)

[13] Zheng, J., Cheng, J., Yang Y.: Partly ensemble empirical mode decomposition: An improved noise-assisted method for eliminating mode mixing. Signal Processing. 96, 362--374 (2014)

A new fault diagnosis method of planetary gear set based on sparse decomposition

Yongzhuo Li, Kang Ding, Guolin He

School of Mechanical and Automotive Engineering, South China University of Technology, Guangzhou 510640, PR China

liyongzhuo123@foxmail.com

Abstract: There are modulation sidebands in the vibration signal measured from a planetary gear set in the healthy state, so it is difficult for traditional methods to diagnose correctly when a local fault appears. A new fault diagnosis method of the planetary gear set based on the sparse decomposition is proposed. Both the steady modulation dictionary and the impact modulation dictionary are designed according to the operational parameters of the gearbox and the features of the vibration signal. The steady modulation dictionary is utilized for separating the steady component caused by the changes of the meshing forces' direction and the signal transmission path. The impact modulation dictionary are optimized by correlation filtering adaptively, and the signal is divided into many segments before performing the matching pursuit. Experimental analysis shows that the proposed method can extract the faulty feature of a planetary gearbox effectively.

Keywords: Planetary gear, Fault diagnosis, Sparse decomposition

1 Introduction

Planetary gear sets are typically used as crucial transmissions in industrial products such as wind turbine, helicopter and heavy machinery. Their vibration signals in healthy state have abundant modulation sidebands [1], which are comparable to those of fixed shaft gearboxes with distributed damage. When a failure occurs, the signal becomes more and more complicated [2]. Some traditional fault diagnosis methods of gearboxes do not work well any more. Sparse decomposition has a particular advantage on signal separation and feature extraction, and has been successfully applied to the fault diagnosis of the rotating machinery [3-4].

This paper proposes a new fault diagnosis method of the planetary gear set based on the sparse decomposition. Two dictionaries with definite physical significance are constructed according to the signals' features. The steady component and the impact component are extracted successfully in the experiment.

2 A new fault diagnosis method

A signal $\boldsymbol{x} \in \boldsymbol{R}^{n\times 1}$ can be represented by the linear combination of a redundant dictionary $\boldsymbol{D} \in \boldsymbol{R}^{n\times m}$ ($m \gg n$). That is $\boldsymbol{x} = \boldsymbol{D\alpha}$, where $\boldsymbol{\alpha} \in \boldsymbol{R}^{m\times 1}$ is a sparse coefficient vector which includes nonzero numbers.

The matching pursuit algorithm is a generally used method for sparse decomposition. After approximate solved by matching pursuit, the signal will be decomposed into the linear combination of atoms $\boldsymbol{d}_\beta$ ($\|\boldsymbol{d}_\beta\| = 1$), and be represented by Eq.(1). In order to obtain the minimum residue $\|\boldsymbol{Rx}_N\|$, each atom $\boldsymbol{d}_{\beta n}$ must ensure that the inner product $|\langle \boldsymbol{Rx}_n, \boldsymbol{d}_{\beta n} \rangle|$ is maximum.

$$x = \sum_{n=0}^{N-1} |\langle \boldsymbol{Rx}_n, \boldsymbol{d}_{\beta n} \rangle| \boldsymbol{d}_{\beta n} + \boldsymbol{Rx}_N \tag{1}$$

2.1 Constructing steady modulation dictionary

Because of the periodic change of the mesh forces' direction and the signal transmission path, the vibration signal measured from the planetary gear set in the healthy state has abundant modulation sidebands.

If each planet suffers equal load-share, there are sidebands around the mesh frequency and its harmonics whose modulation frequencies are $N_p f_c$ and its harmonics, as shown in Fig.1 (a). Otherwise, the modulation frequencies of the sidebands become f_c and its harmonics, as shown in Fig.1 (b). Where f_c and N_p represent the rotational frequency of the carrier and the number of planets respectively.

The atom d_{std}^j of the steady modulation dictionary $\boldsymbol{D}_{std}$ can be constructed according to the feature of the vibration signal.

$$\begin{cases} d_{std}^j(t) = \cos(2\pi f_j t + \varphi_j),\ j = 1,2,\ldots,J,\ \|d_{std}^j(t)\|_2 = 1 \\ f_j = i f_c,\ i \in \{mz, \pm k \mid m = 1,2,\ldots,M; k = 0,1,\ldots,K\} \end{cases} \tag{2}$$

The parameters M and K should be designed to ensure that the range of all the atoms covers the whole mesh frequencies and sidebands of the processed signal.

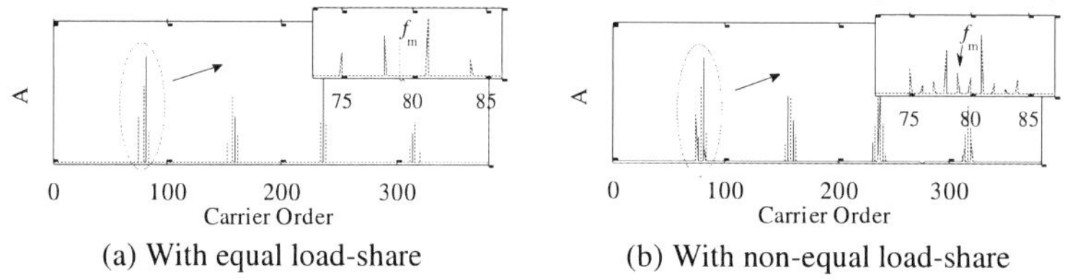

Fig. 1. Feature frequency of the planetary gearbox in the healthy state

2.2 Constructing impact modulation dictionary

When there is a local fault such as pitting corrosion, spalling damage and broken tooth, periodic impacts occur just when the fault point goes into meshing. The vibration signal will contain a series of impact responses whose period is the feature period of faulty gear. Therefore, the atom $\boldsymbol{d}_{imp}^{j}$ of impact modulation dictionary $\boldsymbol{D}_{imp}$ can be constructed by Eq.(3).

$$d_{imp}^{j}(t) = \exp[-2\pi\zeta/\sqrt{1-\zeta^2} f_d(t-\tau)]\sin[2\pi f_d(t-\tau)],\ \ t \ge \tau, \left\|d_{imp}^{j}\right\|_2 = 1 \tag{3}$$

For reducing the redundancy of dictionary $\boldsymbol{D}_{imp}$, natural frequencies and damping ratios $(\bar{f}_d, \bar{\zeta})$ are identified by the correlated filtering method [5].

2.3 A new fault diagnosis method

This algorithm mainly includes two steps. Firstly, the steady component is separated by using the matching pursuit algorithm. The signal-noise ratio of the impact component in the residue can be enhanced. Secondly, the optimized impact modulation dictionary is constructed, then the matching pursuit is undertaken by segments. The details of this algorithm are summarized as follows:

(1) Obtain the acceleration signal $x(t)$ of the planetary gear set, and get its amplitude spectrum $A(f)$ and phase spectrum $\varphi(f)$ by fast Fourier transform (FFT).

(2) Build the steady modulation atoms $\boldsymbol{d}_{std}^{j}$ according to Eq.(2). The phase of the ith atom is $\varphi(f_j)$.

(3) Separate the steady component $x_{std}(t)$ by the matching pursuit:

(A) Construct the steady modulation dictionary $\boldsymbol{D}_{std}$, and set the threshold as ε_{std}.

(B) Apply the matching pursuit to the signal $x(t)$ based on Eq.(1). The atom d_{std}^{n} will be selected during the nth decomposition. And the amplitude recovery coefficient

μ_n is $A(f_n)/c_n$, where c_n is the inner product between the residue after the $(n-1)^{th}$ decomposition and the atom d^n_{std}, and f_n is the frequency of atom d^n_{std}.

(C) If the residue satisfies the terminal condition of Eq.(4), or the number of decomposition is more than the length of atoms, the steady separation is completed. The steady component $x_{std}(t)$ can be reconstructed by the linear combination of each $\mu_n d^n_{std}$. The residuary component $Rs(t)$ can be obtained.

$$\left|\sqrt{\sum_t [Rx_n(t)]^2} - \sqrt{\sum_t [Rx_{n-1}(t)]^2}\right| \le \varepsilon \tag{4}$$

(4) Identify the natural frequencies and damping ratios $(\bar{f}_d, \bar{\zeta})$ of the gearbox from the residue $Rs(t)$ by the correlation filtering method.

(5) Extract the impact component $x_{imp}(t)$ by the matching pursuit:

(A) Note the minimum feature period of the components as T_n. The range τ of the impulse response in each period is $[0:\Delta\tau':T_n]$. The optimized impact modulation dictionary $\boldsymbol{D}_{imp}$ is constructed according to the parameters $(\bar{f}_d, \bar{\zeta}, \tau)$ and the Eq. (3).

(B) Divide the residuary signal $Rs(t)$ into W segments with the length of T_n. Set the iteration threshold as ε_{imp}.

(C) Apply the matching pursuit to the each segment. The impact modulation component $x_{wimp}(t)$ can be reconstructed by the linear combination of $c_m d^m_{imp}$, where c_m is the sparse coefficient and d^m_{imp} is the best-matched atom of the mth decomposition. The ultimate impact component $x_{imp}(t)$ is the connection of each $x_{wimp}(t)$.

(6) The signal $x_{imp}(t)$ is demodulated after applying the bandpass filter. The Hilbert envelop spectrum will show the faulty feature frequency.

3 Experimental analysis

A broken tooth of the sun gear is set on the planetary gear set. The structure of the gearbox and the fault location are shown in Fig.2. The gearbox with fixed ring has three planets. The tooth number of the ring gear, the sun gear and the planet gear are 79, 20 and 28, respectively. Let the input speed of the sun gear be 1500r/min and the load of the carrier be 250N.m.

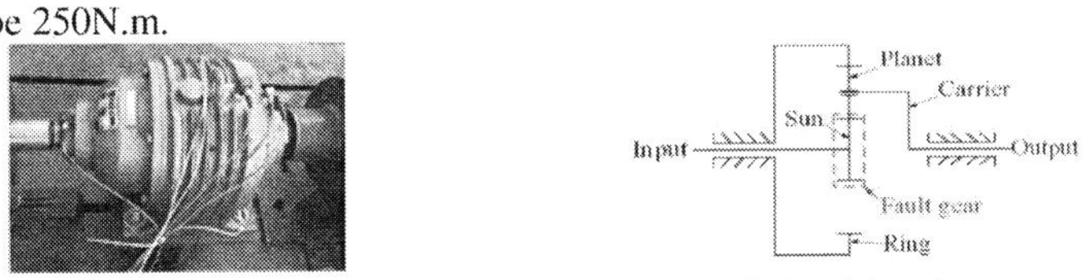

(a) Planetary gear set (b) Fault location

Fig. 2. The planetary gear set and fault location

Fig.3 shows an example of acceleration signal collected from the bearing house of the input shaft with 24000Hz sampling frequency. It can be seen that there is no visible impact in the time domain wave. The signal is further demodulated in the resonance region between 1500Hz and 2500Hz. Although there are feature frequencies of the sun gear with local fault (59.8Hz) and its harmonics, too much interference frequencies exist.

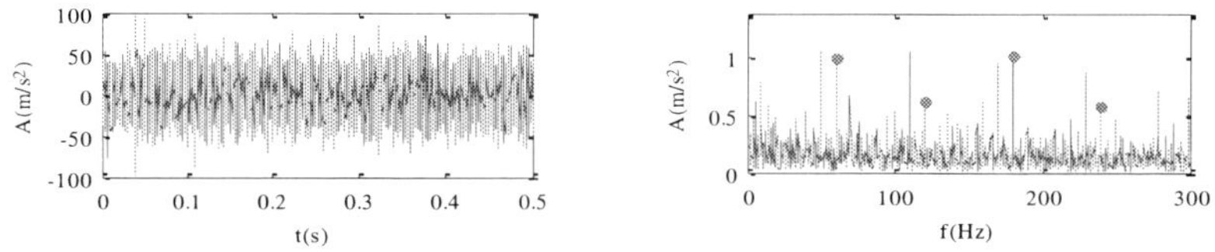

(a) Time domain (b) Hilbert envelope spectrum

Fig. 3. The acceleration signal of the planetary gearbox

(A) Separating the steady component

Let the maximum harmonic of mesh frequency M be 12, and the width of side-bands K be 10. The range of frequencies covers almost all the steady component. Let the threshold ε_{std} be 0.05, and apply the sparse decomposition according to the third step in Section 2.3. Fig.4 shows the reconstruction result of the steady component. The frequency $3f_c$ and its harmonics are extracted mainly.

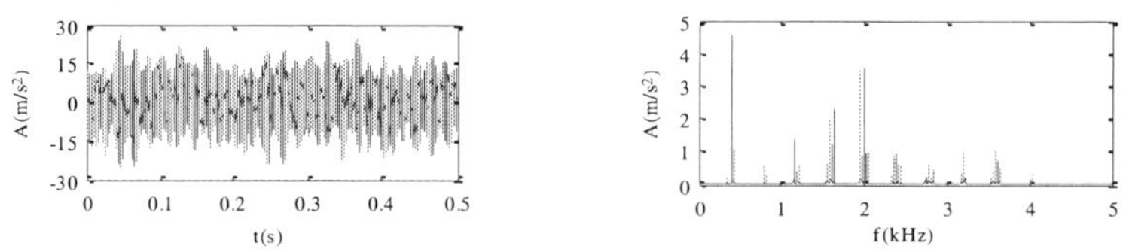

(a) Time domain (b) Frequency domain

Fig. 4. The reconstruction result of the steady component

(B) Extracting the impact component

The system parameters $(\bar{f}_d, \bar{\zeta})$ are identified by the correlated filtering method and shown in Fig.5. The red line represents the natural frequencies and the green one represents the damping ratios, respectively.

The residue $Rs(t)$ component is divided into several segments by the minimum feature period T_n after separating the steady. In this case, T_n is the feature period of the sun gear with local fault T_{s_loc} equaling 16.71ms. Set the scope of the moment τ when the impact occurs in each segment as $\tau \in [0:0.001:T_n]$. Let the threshold ε_{imp} be 0.5. The reconstruction result of the impact component after the sparse decomposition is shown in Fig.6. The average period of impact responses is 16.61ms which is close to

T_{s_loc}. The reconstructed signal is demodulated in the same region as the original signal. Fig.6 (b) shows that the modulation frequency is the feature frequency of the faulty sun gear (60Hz). Therefore, it can be confirmed that a localized fault occurs on the sun gear.

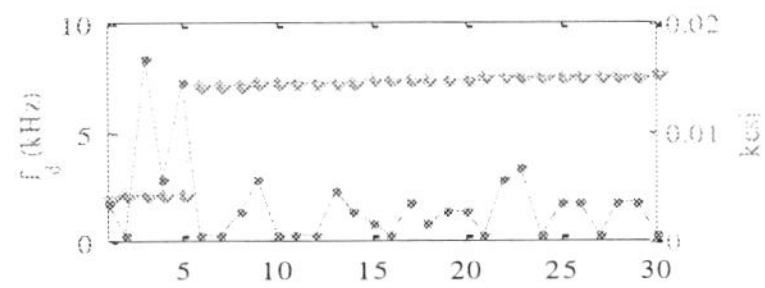

Fig. 5. Natural frequencies and damping ratios

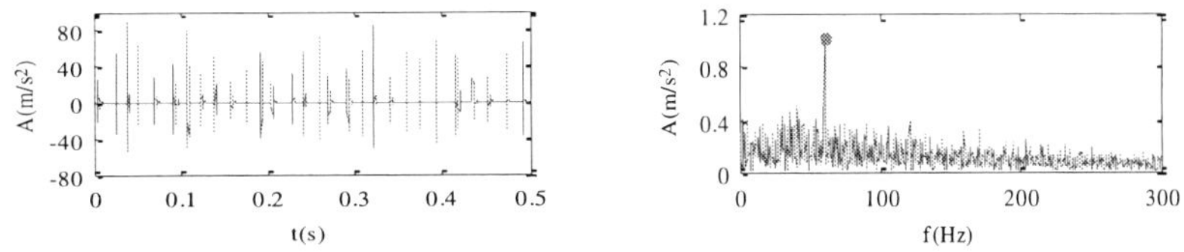

(a) Time domain (b) Hilbert envelop spectrum

Fig. 6. The reconstruction result of the impact component

4 Conclusion

A new fault diagnosis method of the planetary gear set based on the sparse decomposition is proposed. The steady modulation dictionary is designed considering the feature frequency of the vibration signal. And the impact modulation dictionary is optimized by the correlation filtering method. Then the residue is decomposed by segments. Finally, experiment shows that the proposed method can extract the faulty feature effectively.

Reference

1. Inalpolat, M., Kahraman, A.: A theoretical and experimental investigation of modulation sidebands of planetary gear sets. Journal of Sound and Vibration, 323, 677--696 (2009)
2. Feng, Z., Zuo, M.: Vibration signal models for fault diagnosis of planetary gearboxes. Journal of Sound Vibration, 331, 4919--4939 (2012)
3. Cai, G.G., Chen, X.F., He, Z.J.: Sparsity-enabled signal decomposition using tunable Q-factor wavelet transform for fault feature extraction of gearbox. Mech. Syst. Signal Process. 26, 34--53 (2012)
4. He, G.L., Ding, K., Lin, H.B.: Gearbox coupling modulation separation method based on match pursuit and correlation filtering. Mech. Syst. Signal Process. 66, 597--611 (2016)
5. Wang, S., Huang, W., Zhu, Z.K.: Transient modeling and parameter identification based on wavelet and correlation filtering for rotating machine fault diagnosis. Mech. Syst. Signal Process. 25, 1299--1320 (2011)

A Novel Approach for TV-L1-L2 Compressive Sensing Reconstruction

Yanxue Wang, Shuilong He, and Zhansi Jiang

Guilin University of Electronic Technology,
No.1, Jinji Road, Guilin, China, wyx1999140@126.com

Abstract. Compressive sensing (CS) is a framework that enables the recovery of a sparse signal from a few of its measurements by exploiting the sparsity as the prior knowledge of the natural signal. In this paper, we propose a novel parallel proxi-mal splitting algorithm for compressive reconstruction of natural image from limited noisy observations. The FISTA (fast iterative shrinkage thresholding algorithm) is embedded into the parallel proximal decomposition algorithm which solves the TV-L1-L2 CS image reconstruction. Simulations of the compressive sensing for natural images verify that it has better reconstruction SNR than other conventional techniques with less substantial time consuming. The proposed algorithm provides an inspiration to solve the issues of sharing, storage, transfer and visualization of big data such as so large 2D time-frequency data used in long-term condition monitoring of machine.

Keywords: Compressive sensing, big data, sparse, fast iterative shrinkage-thresholding algorithm

1. Introduction

The increasing demand for both high resolution and high frame rate cameras leads to a large amount of image data. Image compression algorithms convert high-resolution images into a relatively small bit streams (while keeping the essential features intact), in effect turning a large digital data set into a substantially smaller one.

Compressive sensing or compressed sampling (CS) is a novel framework that enables sampling below Shannon-Nyquist rate without (or with little) sacrificing reconstruction quality. CS was initiated by Donoho [1] and Candès etc. [2] in 2006. In the compressive sensing method, instead of sensing the entire image and then subsequently removing redundant information during the compressive step, only the required or non-redundant information is sensed. The CS paradigm has attracted increased interest in the past decade because it intrinsically avoids sensing redundant information that commonly exists in imaging or video data. The fields of image processing, such as compressive imaging [3] and medical imaging [4, 5], are the main beneficiaries from the recent progress made in the theory and practice of CS.

CS scheme performs acquisition and compression simultaneously, and shifts almost all computation burdens to the decoder (nonlinear reconstruction), resulting in a low-complexity encoder (linear projection). The key operation of CS is to recover a sparse signal by convex optimization from very few non-adaptive and linear measurements. Hence, a fast and accurate image compressive reconstruction algorithm is required only using a very limited number of CS measurements. A fast and robust algorithm is proposed for TV-L1-L2 signal reconstruction via FISTA in this work, which has better reconstruction SNR than other conventional techniques with less substantial time consuming and very limited the noisy observations in natural image compressive sensing.

The remainder of the paper is organized as follows. In section 2, the reconstruction TV-L1-L2 model is briefly introduced. The proposed parallel FISTA-like proximal decomposition algorithm for the above model is given in section 3. Experimental evaluation with comparisons to many of the state-of-the-art CS algorithms is presented in section 4. Conclusions and discussions are provided in section 5.

2. CS Reconstruction TV-L1-L2 Model

Models for CS recovery in nature are the linear inverse problems with sparsity constraints. There are at least five major classes of commutating techniques for solving those sparse approximation problems: greedy pursuit, convex relaxation, Bayesian framework, nonconvex optimization and brute force [6]. When the TV, L1 and L2 are all considered, thus the reconstruction TV-L1-L2 model is then formulated as follows

$$\arg\min_{\mathbf{x}\in\mathbb{R}^N}\left\{\alpha\mathrm{TV}(\Psi\mathbf{x})+\beta\|\mathbf{x}\|_1+\frac{1}{2}\|A\mathbf{x}-\mathbf{b}\|_2^2\right\} \tag{1}$$

where α and β are two positive parameters. To simplify the model, the $g_\beta(\mathbf{x})=\beta\|\mathbf{x}\|_1$, $g_\alpha(\mathbf{x})=\alpha\,\mathrm{TV}(\Psi\mathbf{x})$ and $f_1(\mathbf{x})=\frac{1}{2}\|A\mathbf{x}-\mathbf{b}\|_2^2$ are used in the following sections. Eq. 1, it is much more difficult to solve, which is also the bottleneck in practical applications. Aiming at the recovery of signals and operators from a very limited number of measurements, an efficient algorithm which combines the fast iterative shrinkage-thresholding algorithm and parallel proximal splitting algorithm is developed in this communication.

3. Parallel FISTA-like Proximal Decomposition Algorithm for CS Reconstruction

3.1 Fast iterative shrinkage-thresholding algorithm

Fast iterative shrinkage-thresholding algorithm (FISTA) was proposed to solve a general non-smooth convex optimization model of two functions [7]:

$$\arg\min_{\mathbf{x}\in\mathbb{R}^N}\{F(\mathbf{x}):=g(\mathbf{x})+f(\mathbf{x})\} \tag{2}$$

The proximal map is often used for convex function optimization. Given any scalar $\rho>0$, the proximal map associated with a closed proper convex function g is defined by

$$prox_\rho(g)(\mathbf{y}):=\arg\min_{\mathbf{x}\in\mathbb{R}^N}\left\{g(\mathbf{x})+\frac{1}{2\rho}\|\mathbf{x}-\mathbf{y}\|^2\right\} \tag{3}$$

Based on gradient model algorithm, the solver of Eq. 2 is equal to the following statement hold:

$$\mathbf{x}^*=(I+t\partial g)^{-1}(I-t\nabla f)(\mathbf{x}^*) \tag{4}$$

Call for the fixed point iterative scheme, the above Eq. 4 is rewritten as

$$\mathbf{x}_k=(I+t_k\partial g)^{-1}(I-t_k\nabla f)(\mathbf{x}_{k-1}) \tag{5}$$

where $t_k > 0$ is the suitable stepsize. Based on the important properties of proximal map given in *Lemma 3.1* in [7], Eq. 5 can be written as

$$\mathbf{x}_k = prox_{t_k}(g)\left(\mathbf{x}_{k-1} - t_k \nabla f(\mathbf{x}_{k-1})\right)$$
$$= \arg\min_{\mathbf{x}\in\mathbb{R}^N}\left\{ g(\mathbf{x}) + \frac{t_k}{2}\left\| \mathbf{x} - \left(\mathbf{x}_{k-1} - \frac{1}{t_k}\nabla f(\mathbf{x}_{k-1})\right)\right\|_2^2 \right\} \quad (6)$$

When $f(\mathbf{x}) = f_1(\mathbf{x})$ and $g(\mathbf{x}) = g_\beta(\mathbf{x})$ are designated, Eq. 6 can be reduced to the popular iterative shrinkage-thresholding algorithms (IST) [8],

$$\mathbf{x}_k = \mathrm{T}_{\beta t}\left(\mathbf{x}_{k-1} - 2tA^T\left(A\mathbf{x}_{k-1} - \mathbf{b}\right)\right) \quad (7)$$

where t is an appropriate constant stepsize is the shrinkage operator defined as

$$\mathrm{T}_\sigma(\mathbf{x})_i = \left(|x_i| - \sigma\right) + \mathrm{sgn}(x_i) \quad (8)$$

Actually, we can construct the following quadratic approximation of $F(\mathbf{x})$ at given $\mathbf{y}$ for any Lipschitz constant $L_f > 0$,

$$Q_{L_f}(\mathbf{x},\mathbf{y}) := f(\mathbf{y}) + \langle \mathbf{x}-\mathbf{y}, \nabla f(\mathbf{y})\rangle + \frac{L_f}{2}\|\mathbf{x}-\mathbf{y}\|_2^2 + g(\mathbf{x}) \quad (9)$$

admitting a unique solution which is denoted by map $p_{L_f,g}(\mathbf{y})$,

$$p_{L_f,g}(\mathbf{y}) := \arg\min_{\mathbf{x}\in\mathbb{R}^N}\left\{Q_{L_f}(\mathbf{x},\mathbf{y})\right\} \quad (10)$$

where L_f is often defined as the smallest Lipschitz constant of $\nabla f(\mathbf{x})$. After ignoring constant terms, this equation can then be written as

$$p_{L_f,g}(\mathbf{y}) = prox_{1/L_f}(g)\left(\mathbf{y} - \frac{1}{L_f}\nabla f(\mathbf{y})\right) \quad (11)$$

Thus Eq.11 leads to

$$\mathbf{x}_k = p_{L_f,g}(\mathbf{x}_{k-1}) \quad (12)$$

3.2 The parallel FISTA-like proximal decomposition algorithm

The combination of the FISTA with parallel proximal decomposition algorithm, thus parallel FISTA-like proximal decomposition algorithm (PFPDA) is proposed for the CS reconstruction problem. The main idea of the novel algorithm is that FISTA is embedded into the parallel proximal decomposition algorithm. The routine of the proposed PFPDA algorithm is given as follows:

Input: $L_f \geq L(f)$, an upper bound on the Lipschitz constant of ∇f

Step 0: Take $\mathbf{y}_1 = \mathbf{x}_0 \in \Omega,\ t_1 = 1,\ \sum_i \omega_i = 1$

Step k: ($k \geq 1$) compute

$$\mathbf{z}_k^\alpha = p_{L_f,g_\alpha}(\mathbf{y}_k) \quad (13)$$

$$\mathbf{z}_k^\beta = p_{L_f,g_\beta}(\mathbf{y}_k) \quad (14)$$

Due to $g_\beta(\mathbf{x}) = \beta\|\mathbf{x}\|_1$, Eq. 17 is the special case and can be computed using Eqs. 7 and 8. While $g_\alpha(\mathbf{x}) = \alpha\mathrm{TV}(\Psi\mathbf{x})$, Eq. 16 is computed using FISTA algorithm. The core of the parallel algorithm is applied to obtain $\mathbf{z}_k$.

$$\mathbf{z}_k = \omega_1 \mathbf{z}_k^{\alpha} + \omega_2 \mathbf{z}_k^{\beta} \tag{15}$$

The traditional FISTA is not a monotone algorithm which will not guarantee to be nonincreasing. A monotone version FISTA (MFISTA) was developed in [7] in which the convergence was also verified theoretically (*Theorem* 5.1 in [7]). Then MFISTA algorithm is applied instead of the traditional FISTA,

$$\mathbf{x}_k = \arg\min\left\{F(\mathbf{x}):\ \mathbf{x} = \mathbf{z}_k, \mathbf{x}_{k-1}\right\} \tag{16}$$

$$\mathbf{y}_{k+1} = \mathbf{x}_k + \left(\frac{t_k - 1}{t_{k+1}}\right)(\mathbf{x}_k - \mathbf{x}_{k-1}) + \left(\frac{t_k}{t_{k+1}}\right)(\mathbf{z}_k - \mathbf{x}_k) \tag{17}$$

This implementation is used to solve recover model in Eq. 1, and it may well preserve the advantages of the two techniques. The proposed algorithm also provides an inspiration to solve the non-smooth convex optimization model of more than two terms as well as to develop a new CS method using TV and L1 regularization. Here, we give the algorithm in detail when the three functions are designated as in the model of Eq. 1. This algorithm is only a simplified version of the general model. In effect, this algorithm provides an inspiration to solve the non-smooth convex optimization model of more than two terms based on the framework of the combination of the FISTA with the parallel proximal decomposition algorithm. The application of this new method in compressive sensing of natural image is evaluated in the following section.

4. Experimental Evaluation

Since many CS reconstruction algorithms have been developed until now, which can be found from the introduction in this work, it is impossible to compare them all. We only compare the performance of PFPDA to five recently developed CS reconstruction algorithm: RecPF [9], TwIST [3] , fast-CS [10], TVCMRI [4] and CSA [11]. The RecPF, fast-CS, TVCMRI and CSA methods among them are selected as competitors in this work, because they were all proposed to solve Eq. 7. In addition, most of these methods have fast algorithms and have also been compared with other CS methods in their respective works. All softwares are written in MATLAB and run on PCs with 2.5 GHz CPU and 4 GB memory. In addition, in any practical application devices of measurement do not have infinite precision. Noise is inevitable in the measured data. It is therefore small perturbations is used in the experiments. CS reconstruction is known to be stable in the presence of noise [12]. The measurement and sparsity decomposition are the same with those in [4].

We considered a subset of images from Microsoft Research in Cambridge [13]. Six different classes of natural image are selected: cows, trees, buildings, planes, flowers and boats; eight images are selected at random from each class. These images altogether are depicted in Fig. 1. Each row in Fig. 1 represents one class of the image. The comparison is performed for all algorithms discussed above. Some important parameters used in these algorithms are given as follows. Variable density random under-sampling is used in this research. The parameters α and λ mentioned in Eq. 7 are set to 0.001 and 0.035, respectively. These two parameters are empirically selected referring to former published corresponding works. Sigma of the noise added in the sampling data is 0.02. Orthogonal Daubechies wavelet (Db 4) is used in the sparse representation of the images except for the fast-BCS method, while biorthogonal CDF 9/7 wavelet is used in the fast-BCS method. Results of SNR and CPU times using the motioned recover methods are shown in Fig. 2. Those randomly selected three different classes of nature are illustrated in Fig. 2 (1)(a), (2)(a) and (3)(a). As we can see in Fig.2 (1)(a), (2)(b) and (3)(b), the SNRs also rise with the increase of the sampling rate while the CPU time illustrated in Fig. 2 (1)(c), (2)(c) and (3)(c) do not raise obviously. The proposed method always outperforms the other algorithms in aspects of SNR for all the tests.

Fig. 1 Natural images from six different classes. Each row represents one class, with eight images selected randomly from each class. From top to bottom, the six classes are successively cows, trees, buildings, planes, flowers and boats.

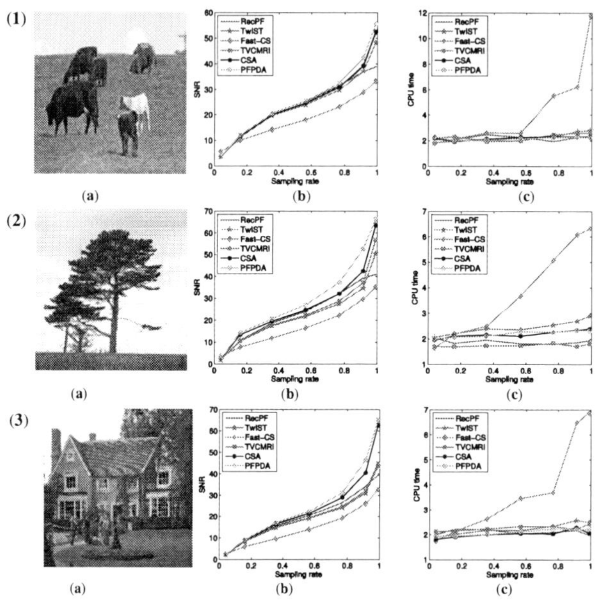

Fig. 2 Comparisons of SNR and CPU times using the RecPF, TwIST, Fast-CS, TVCMRI, CSA, PFPDA methods for different images: (**1**) cows, (**2**) tree, (**3**) house. (**a**) The used image, (**b**) The results of SNR with six recovery methods, (**c**) The results of CPU time with six recovery methods. (The results of Plane, Flower and boats are similar to those shown in Fig. 2)

5. Conclusions

A new algorithm for reconstruction of natural image in CS is presented in this work. This proposed PFPDA algorithm combines the parallel proximal decomposition and FISTA algorithm to solve the TV-L1-L2 compressive sensing reconstruction. The comparisons of the reconstruction performance between RecPF, TwIST, Fast-CS, TVCMRI, CSA and the PFPDA are investigated. The results show that significant improvement in SNR can be obtained using the proposed algorithm with little computational time consuming. It is worth mentioning that the proposed algorithm provides an inspiration to solve the issues of sharing, storage, transfer and visualization of big data such as so large 2D time-frequency data used in long-term condition monitoring of machine.

Acknowledgements

The financial sponsorship from the project of National Natural Science Foundation of China (51105085, 51475098 and 61463010) and Guangxi Natural Science Foundation (Grant No. 2013GXNSFBA019238), Guangxi key laboratory of manufacturing system & advanced manufacturing technology (15-140-30-001Z) are gratefully acknowledged. The project is also sponsored by Guangxi Experiment Centre of Information Science (20130312).

References

[1] D. L. Donoho, "Compressed sensing," *IEEE Transactions on Information Theory,* vol. 52, pp. 1289-1306, 2006.

[2] E. J. Candes, J. Romberg, and T. Tao, "Robust uncertainty principles: exact signal reconstruction from highly incomplete frequency information," *IEEE Transactions on Information Theory,* vol. 52, pp. 489-509, 2006.

[3] J. M. Bioucas-Dias and M. A. T. Figueiredo, "A new TwIST: two-step iterative shrinkage/thresholding algorithms for image restoration," *IEEE Transactions on Image Processing,* vol. 16, pp. 2992-3004, 2007.

[4] S. Ma, W. Yin, Y. Zhang, and A. Chakraborty, "An efficient algorithm for compressed MR imaging using total variation and wavelets," *2008 IEEE Conference on Computer Vision and Pattern Recognition (CVPR),* p. 8 pp., June 2008.

[5] M. Lustig, D. Donoho, and J. M. Pauly, "Sparse MRI: The application of compressed sensing for rapid MR imaging," *Magnetic Resonance in Medicine,* vol. 58, pp. 1182-1195, 2007.

[6] J. A. Tropp and S. J. Wright, "Computational methods for sparse solution of linear inverse problems," *Proceedings of the IEEE,* vol. 98, pp. 948-958, Jun 2010.

[7] A. Beck and M. Teboulle, "Fast gradient-based algorithms for constrained total variation image denoising and deblurring problems," *IEEE Transactions on Image Processing,* vol. 18, pp. 2419-2434, Nov 2009.

[8] M. A. T. Figueiredo and R. D. Nowak, "An EM algorithm for wavelet-based image restoration," *IEEE Transactions on Image Processing,* vol. 12, pp. 906-916, Aug 2003.

[9] J. Yang, Y. Zhang, and W. Yin, "A fast alternating direction method for TVL1-L2 signal reconstruction from partial Fourier data," *IEEE Journal of Selected Topics in Signal Processing,* vol. 4, pp. 288-297, Apr 2010.

[10] T. T. Do, L. Gan, N. H. Nguyen, and T. D. Tran, "Fast and efficient compressive sensing using structurally random matrices," *IEEE Transactions on Signal Processing,* vol. 60, pp. 139-154, Jan 2012.

[11] J. Huang, S. Zhang, and D. Metaxas, "Efficient MR image reconstruction for compressed MR imaging," *Medical Image Analysis,* vol. 15, pp. 670-679, 2011.

[12] E. J. Candes, J. K. Romberg, and T. Tao, "Stable signal recovery from incomplete and inaccurate measurements," *Communications on Pure and Applied Mathematics,* vol. 59, pp. 1207-1223, Aug 2006.

[13] *Microsoft Research in Cambridge.* Available: http://research.microsoft.com/en-us/projects/objectclassrecognition/

Adaptive multiwavelet for detecting transient impulse responses of wind turbine

Shuilong He[1,2,3,*], Yanxue Wang[1], Enyong Xu[3], Zhansi Jiang[1], Rongjiang Tang[1], and Yanyang Zi[2]

[1] Guilin University of Electronic Technology, No.1, Jinji Road, Guilin, China
[2] Xi'an Jiaotong University, No.28, Xianning Road, Xi'an, Shaanxi, China
[3] Dongfeng Liuzhou Motor Co., Ltd, Liuzhou, Guangxi, China
[*] Corresponding author, Email: xiaofeilonghe@163.com

Abstract. Due to the characteristic of spectral diversity and complexity as well as the amplitude modulation and frequency modulation (AMFM) nature of the vibration signals of wind turbine, early-stage fault detection turns into a challenging task. Multiwavelet is a powerful tool to extract the non-stationary fault features. However, the existing predetermined multiwavelet bases are independent of the vibration measurement and the sensor-based dynamic response signals, which make the result not really perfect. To overcome this drawback, this paper proposed an adaptive multiwavelet method by taking a new index which combines minimum quantizing error (MQE) with spectral entropy as target in the construction process. MQE show the minimum distance between definition and input vector, and spectral entropy reflects the uniformity of the probability distribution. Based on the lifting scheme and taking maximum index as the optimization objective, optimal multiwavelet can be constructed for best matching the rolling bearing fault. Moreover, benefiting by the valuable features of the Hilbert transform demodulation analysis of the multiwavelet decomposition results, we extract the fault features. And the effectiveness of this proposed technique is validated via numerical simulation.

Keywords: adaptive multiwavelet, minimum quantizing error, spectral entropy, feature extraction, wind turbine

1. Introduction

As a kind of regenerative clean energy, wind energy has attracted increasing attention of many investigators recently. According to the report of the world wind energy association, the global wind power cumulative installed capacity will reach 700 GW in 2020 [1]. However, due to the harsh working environments, such as the variability of wind, shifting loads, severe corrosion and inadequate lubrication make the key components, such as rolling bearings, susceptible to damage [2]. Accidents in key parts may result in huge economic losses and safety problems. Therefore, it is significant to detect the fault feature of the key equipment as early as possible to avoid the occurrence of abnormal accidents.

Several methods have been developed for the fault diagnosis in the rolling bearing. Li [3] selected an appropriate base wavelet for analyzing vibration signals. Kankar [4] focused on fault diagnosis of ball rolling bearings with localized defects on the various rolling bearing components by the method of wavelet-based feature extraction. He [5] proposed the manifold learning technique to explore the wavelet packet energy flow characteristics of vibration signals. Chen [6] detects a fault of generator bearing successfully based on empirical wavelet transform. One of the most difficult issues on fault diagnosis is how to effectively extract the weak characteristic of incipient fault from noisy signals. Therefore, an advanced analytical tool is needed urgently.

Multiwavelet transform, as a newer development of the wavelet theory with the excellent properties of orthogonality, compact support and high vanishing moments simultaneously [7], has been proved as an effective tool in extracting weak characteristics. In this paper, an adaptive multiwavelet constructing technique is proposed for enhancing fault symptom extracting performance. With the acquired data and a known multiwavelet basis, this technique can derive an optimal basis that perfectly matches the critical non-stationary and transient fault signatures via genetic algorithm. In this technique, a new index is constructed by minimum quantizing error (MQE) and spectral entropy. Then, the index is adopted as the optimization objective in the proposed method to obtain the best multwavelet basis functions. Moreover, the resolutions in the time and frequency domain have been enhanced via the redundant multiwavelet decomposition. Finally, benefiting by the valuable features of the Hilbert transform demodulation analysis of the redundant multiwavelet decomposition results, we extract the fault features from the high frequency modulation signal. The effectiveness of the proposed technique is validated via numerical simulation signals analysis.

2. Redundant multiwavelet transform

2.1 Summary of multiwavelet theory

A multiwavelet is generated by two or more wavelets, and the two refinement equations between the vector-valued multiscaling function $\Phi(t)$ and multiwavelet function $\Psi(t)$ are expressed as:

$$\begin{aligned} \Phi(t) &= \sqrt{2}\sum_k H_k \Phi(2t-k) \\ \Psi(t) &= \sqrt{2}\sum_k G_k \Phi(2t-k) \end{aligned}, \quad k \in Z \tag{1}$$

Difference to scalar wavelet, multiwavelet decomposition is a multi-input and multi-output system.

2.2 Redundant multiwavelet transform

Discrete wavelet transform is time-varying process and down-sampling, which is adverse for extracting the fault characteristics. In order to eliminate the defect, the redundant transform is used. The details are presented as follows:

Let T be the operator that pads zeros terms alternatively between the coefficients of a series. That is, for any integer i,

$$(Tx)_{2i} = x_i \,,\ (Tx)_{2i+1} = 0 \tag{3}$$

The redundant multiwavelet' recursion coefficient matrices are defined as

$$\begin{cases} H_k^l = G_k^l = 0 & \text{if } k \text{ is not a multiple of } 2^l, \\ H_{2^l i}^l = (T^l H)_{2^l i} = H_i, G_{2^l i}^l = (T^l G)_{2^l i} = G_i & \text{otherwise} \end{cases} \tag{4}$$

3. Adaptive multiwavelet construction

3.1 Lifting scheme

Sweldens proposed the lifting scheme to construct new wavelets.The multiwavelet has more designing freedom than scalar ones, therefore the lifting scheme is promising for utilizing such advantage to produce new multiwavelet with certain properties [8]. Analogous to the lifting scheme in the scalar case, the lifting scheme for multiwavelet, with a multiplicity of 2 and a dilation factor 2, is defined as in Lemma 3.1.

Lemma 3.1 (Strela [9]) Suppose we have two compactly supported multiwavelet families that satisfy the conditions of biorthogonality. Denote these two families by $\{\Phi_{old}, \Psi_{old}\}$ and $\{\Phi_{new}, \Psi_{new}\}$. Let $\{H_{old}(z), G_{old}(z)\}$ and $\{H_{new}(z), G_{new}(z)\}$ be the symbols for the recurrence relations for these families. Then they are related as

$$\begin{aligned} H_{new}(z) &= H_{old}(z) \\ G_{new}(z) &= T(z^2)(G(z) + S(z^2)H_{old}(z)) \end{aligned} \tag{5}$$

where $T(z)$ and $S(z)$ are of finite degree and the determinant of $T(z)$ is monomial.

It has proven that the matrix $T(z)$ has little effect on the vanishing moment of the multiwavelet. Let $T(z) = I$ in Eq (5). The algorithm for constructing $S(z)$ is referenced in Ref [10]. According to the engineering experience, 5 order vanishing moments is enough to capture the hidden fault features effectively. Choosing GHM as an initial multiwavelet, then obtain the lifting scheme matrix as

$$S = \begin{bmatrix} 0.626+0.463a & -1.039-0.374a & 0.137-0.975a & -0.409+5.977a & 0.336-0.275a & -1.034 & -0.206-0.303a & -0.209+0.454a & 0.021+1.312a & -0.039-7.068a \\ -0.041+0.084b & 0.075-0.187b & -0.069-0.203b & -0.177+0.29b & 0.053+1.353b & 0.117 & 0.042-0.193b & -0.032b & -0.137+0.076b & -0.013-2.173b \end{bmatrix} \tag{6}$$

where a, b are nonzero variables. Substituting the S in Eq (5), a pair of multiwavelet $\{\Phi_{new}, \Psi_{new}\}$, providing vanishing moment 5 can be obtained via Lemma 3.1. They are influenced by factors a and b together.

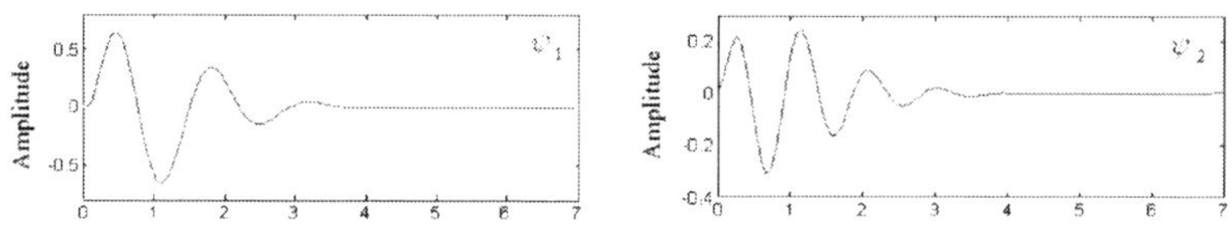

Fig. 1. An example of new multiwavelet: (a) ψ_1; (b) ψ_2.

An example is shown in Fig. 1 with a=1 and b=1. It shows that the improved multiwavelet have a unilateral oscillation decay which is more similar to impulse of local failure of rolling bearings. So the improved multiwavelet are suitable for wind turbines feature extraction.

3.2 Objective optimization

3.2.1 Minimum quantization error

Self Organizing Maps (SOM) is an unsupervised learning feed forward network. With a chosen input vector the weight vectors of the best matching unit (BMU) whose weight vector is closest to input vector are updated, and the distance between it and all the weight vectors of the SOM is calculated using some distance measure. At the end of the learning process, the weight vectors are grouped into clusters depending on their distance in the input space.

The distance between the data collected in fault condition and the weight of BMU trained by data normal condition can be used as a performance indicator to quantitatively measure the degradation status. This distance is defined as minimum quantization error (MQE).

$$MQE=\| n-m_{BMU} \| \tag{7}$$

where n is the input data vector and m_{BMU} stands for the weight vector of the BMU.

A higher value of MQE represents the larger deviation from the current bearing state to the normal state, indicative of greater damage severity.

3.2.2 Spectral entropy

Let $\{ F_i \}$ be the envelope spectrum, and $\{ p_i \}$ describes the proportion of $\{ F_i \}$. According to the formula of entropy, the spectral entropy can be defined as Eq. (8).

$$E_f = -\sum_{i=1}^{k} p_i \log_k p_i \tag{8}$$

Spectral entropy reflects the uniformity of the probability distribution and can provide useful information about the dynamic process. The smaller value of E_f represents the more obvious periodic signal.

3.2.2 Optimization index

To correctly extract the fault feature, MQE which reflects the degree of deviating from the normal state, is combined with the spectral entropy, forming a new index K_c.

$$K_c = \frac{MQE}{E_f} \tag{9}$$

K_c combines the comprehensive advantages of MQE and spectral entropy, and reflects periodicity of the signal effectively. A higher value of K_c represent the greater damage severity of rolling bearing.

3.2.3 Objective optimization

Recalling the lifting scheme for multiwavelet in Eq. (6), there are two free parameters *a, b* for designing new multiwavelet.

Genetic algorithm (GA) is recommended as a tool to optimize multiwavelet with the two free parameters. In this paper, the ranges of these parameters are uniformly searched within the interval $[-50,0) \cup (0,50]$. To accelerate the optimization of GA, the population scale is set to 100 and the number of iteration to 50, the probability of crossover to 0.6 and the probability of mutation to 0.05.

The procedure of the method for the detection of the rolling bearing faults of wind turbine can be presented as follows:

(1) Chosen five indicators (waveform index, peak index, kurtosis index, impulsion index, tolerance index) under the normal state as input vector **X** of SOM.

(2) Obtain the best matching unit (BMU) whose weight vector is closest to **X**.

(3) Compute the minimum quantization error (MQE) based on the data collected in fault condition and the weight of BMU trained by data normal condition.

(4) Compute the spectral entropy, and obtain the optimization index K_c.

(5) Generate the optimal adaptive multiwavelet by GA based on the maximum K_c.

(6) Conduct redundant multiwavelet transform with the adaptive multiwavelet functions.

(7) Output the corresponding envelope spectrum of the multiwavelet coefficients for

extracting fault features.

4. Experimental Evaluation

To simulate the inner fault-related transient vibration responses of rolling bearing in wind turbine, an AMFM signal of nonstationarity is constructed as:

$$f(t) = A\sin(2\pi f_a t)\sum_{k=1}^{L} h(t-kT) + n(t)$$
$$h_k(t)=\mathrm{e}^{-\beta t}\sin(2\pi f_r t+\theta_k) \tag{10}$$

where $A\sin(2\pi f_a t)$ represents the amplitude modulation, β is the damping characteristic of the mechanical system, T is the period associated with the occurrence rate of the periodic impulses, and f_r is the excited resonance frequency. What's more the terms θ_k is introduced to represent the inconsistency inherent in the periodic impacts due to a variety of factors such as slip, varying load angle and transition path effect of engineering mechanical systems. Specifically, we assign the above parameters as $A=1$, $\beta=200$, $f_a=10$, $f_r=100$, $T=0.0835$, and θ_k are random variables ranged in $(-\pi,\pi]$. The noisy term $n(t)=0.5N(0,1)$, where $N(0,1)$ denotes white Gaussian noise subjected to normal distribution.

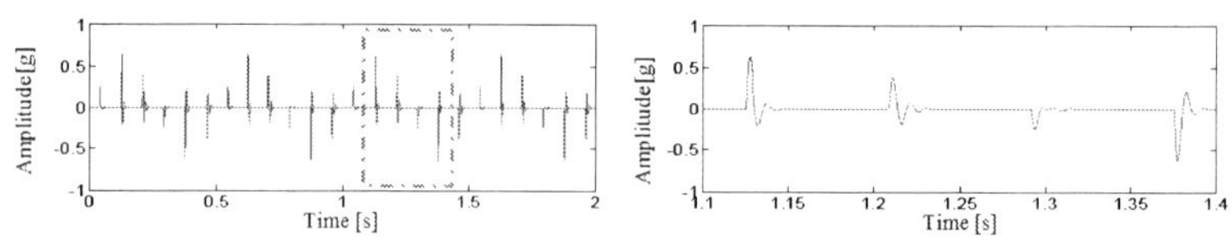

Fig. 2. **(a)** The simulated signal without noise and **(*b*)** the zoom-in plot of the midpoint square.

Let the sampling rate and sampling length be 2048Hz and 4096 respectively. The simulated signal without noise and the zoom in pluses are shown in Fig. 2(a) and (b). The noise contaminated signal is shown in Fig. 3(a). For the noise contained case, the signal-to-noise ratio (SNR) is computed to be -11 dB according to the following definition of SNR. Masked by such severe noise, the original periodic impact signal is completely unrecognizable. And the corresponding Hilbert envelope spectrum of the simulated signal is shown in Fig. 3(b), from which it is difficult to recognize the frequency peak of 12 Hz. Thus, this method is not applicable to extract the weak fault feature of the signal.

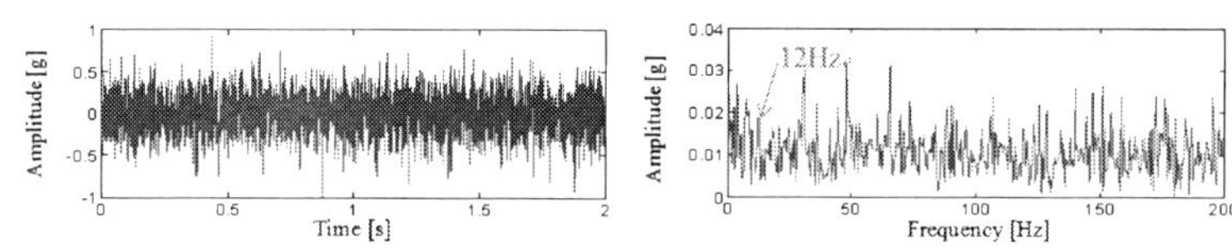

Fig. 3. (a) The simulated signal with added noise and (*b*) the envelop spectrum of the simulated signal.

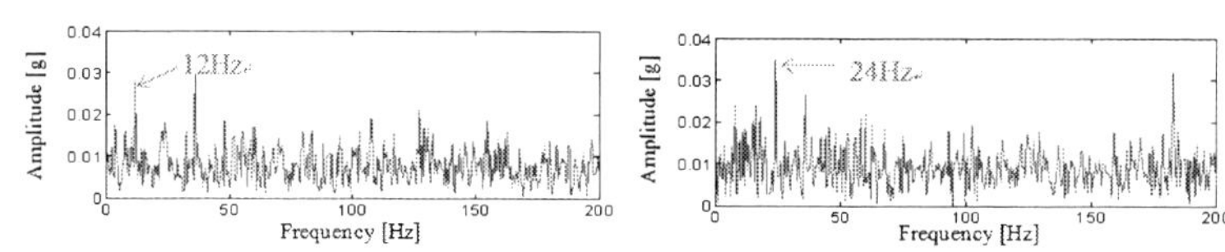

Fig. 4. The envelop spectrum of purified signal via the proposed method: (a) ψ_1; (b) ψ_2.

We apply the proposed method in this paper to analyze the simulated signal. The GHM is used as the initial multiwavelet. The optimal multiwavelet factors *a, b* were determined at [12.698, 7.262] by the process in GA. The Hilbert envelope spectrums of decomposition coefficients performing by proposed method to the simulated signal are shown in Fig. 4. From Fig.4, the characteristic frequency 12 HZ and its frequency doubling 24 Hz are quite dominant respectively. Obviously the weaken features of original simulated signal are extracted effectively with this method.

5. Conclusions

Considering the spectral complexity and the AMFM nature of the vibration signal from rolling bearings of wind turbine, an adaptive multiwavelet analysis method, which can enhance fault symptom extracting performance, is proposed for detecting transient vibration responses caused by wind turbine faults. This technique combines the advantages of MQE and spectral entropy which form a new index to guide the adaptive optimization of the multiwavelet to signals' hidden transient vibration responses via genetic algorithm. And a simulated signal is applied to verify the validity of this proposed method.

Additionally, the signals of wind turbine are more complicated compared with simulated signal. The authors are conducting experiments on the test bench to verify the proposed method effectiveness in their future work.

Acknowledgements

The financial sponsorship from the project of National Natural Science Foundation of China (51105085, 51475098 and 61463010) and Guangxi Natural Science Foundation (Grant No. 2013GXNSFBA019238), Guangxi key laboratory of manufacturing system & advanced manufacturing technology (15-140-30-001Z) are gratefully acknowledged. The project is also sponsored by Guangxi Experiment Centre of Information Science (20130312).

References

[1] WWEA. World wind energy report 2015, http://www.wwindea.org; 2015.
[2] Andrew K, Anoop V. Analyzing bearing faults in wind turbines: A data-mining approach. Renewable Energy 2012; 8:110-116.
[3] Li K, Chen P, Wang HQ. Intelligent diagnosis method for rotating machinery using wavelet transform and ant colony optimization. IEEE Sensors Journal 2012; 12: 2474-2484.
[4] Kankar PK, Sharma SC, Harsha SP. Rolling bearing fault diagnosis using wavelet transform. Neurocomputing 2011; 74: 638–1645.
[5] He QB. Vibration signal classification by wavelet packet energy flow manifold learning. Journal of Sound and Vibration 2013; 332: 1881–1894.
[6] Chen J, Pan J, Li Z, et al. Generator bearing fault diagnosis for wind turbine via empirical wavelet transform using measured vibration signals[J].Renewable Energy, 2016, 89: 80-92.
[7] Keinert F Wavelets and multiwavelets, Studies in advanced mathematics, Chapman & Hall/CRC Press, Boca Raton, FL 2004.
[8] Keinert F. Wavelets and multiwavelets. Chapman and Hall/CRC, 2004ISBN: 1-58488-304-9.
[9] Davis GH, Strela V, Turcajova R. Multiwavelet Construction via the Lifting Scheme, in Wavelet Analysis and Multiresolution Methods Lecture Notes in Pure and Applied Mathematics Marcel Dekker 1999; 57–79.
[10] Keinert F. Raising multiwavelet approximation order through lifting. Siam Journal on Mathematical Analysis, 2001, 32 (5): 1032-1049.

Novel Condition Monitoring Method Using a Parabolic Sound Reflector Microphone and Separation of Composed Sound Waves

H. Ohta [1], S. Fukunaga [2], M. Nakamura [1], T. Matsuyama [2], T. Watanabe [1]

[1] Department of Ocean Mechanical Engineering, National Fisheries University, Shimonoseki, Japan
[2] Fisheries Science and Mechanical Engineering, Graduate Student of National Fisheries University, Shimonoseki, Japan

ABSTRACT

This paper proposes a novel method for monitoring the condition of machinery by using a parabolic sound reflector microphone and signal separation processing technique concerning composed sound generated by the monitored equipment. Discrimination of the operating frequency components of several machines has been performed by estimating the intensity difference of the frequency components in the frequency domain. These intensity differences of the operating frequencies are estimated using the directivity of a parabolic microphone that has high directional sensitivity. The individual operating frequency components are extracted by signal processing with attention to the amplitude and phase values in the frequency domain. Finally, the individual frequency components are converted into a time series by using an inverse Fast Fourier Transform. The time series data generated by a specific piece of equipment can be differentiated from several other machines in the environment by using the proposed method.

Keywords: Dynamic facility, Condition monitoring, Diagnosis, Parabolic reflector type microphone, Separation of composed sound wave.

1. INTRODUCTION

Many production facilities are arranged in a confined space such as a chemical plant or foundry. Moreover, many pieces of equipment are compactly placed in a vehicle, such as a ship, that has a small engine room. Vibration and acoustic signals generated by the equipment are often used for monitoring the condition and evaluating the condition of the equipment. Vibration and acoustic waves produced by many machines produce composed waves in the field. The composed wave signals are transmitted to vibration and acoustic sensors for condition monitoring and diagnosis. The accuracy of the monitoring and diagnosis is much decreased by the reason of the compound waveforms. In order to perform accurate monitoring and diagnosis, the vibration and acoustic signals generated by only the target facility must be extracted from the sum of all the waveforms. An improvement in the signal to noise ratio is also necessary [1,2]. The simplest way to improve the signal to noise ratio is to increase the number of sensors that the measure vibration and acoustic signal in the composed field. However, the cost of sensors, amplifiers, and analogue to digital converters necessary for condition monitoring increases. However, a high signal to noise ratio and can be realized with a small number of sensors. This paper proposes a novel, very-efficient, precise monitoring method for dynamic plural facilities using a parabolic reflector microphone [3] and signal separation processing. First, we identify the propagation direction of driven frequency components generated by a target machine by estimating intensity differences of the envelope spectrum on the driven frequency components in the frequency domain using the parabolic reflector microphone. Second, the separation of the individual driven frequency components is performed by signal processing operation with respect to their amplitude and phase. Finally, the extracted individual frequency components are converted into time series by inverse FFT. The time series data generated by only the targeted piece of equipment from all pieces of equipment can be extracted using the proposed method. In this research, this method has been used to analyze the frequency components of three dynamic machines: two rotary machines and a vacuum pump. With the rotary machines, one had an outer race fault of the rolling element bearing, and the other had an excessive wear of a gear. During the experiment, the extraction of time series data of each driving frequency component associated with the above faults could be effectively implemented. Consequently, the high performance, high precision, and practical use of the proposed method has been demonstrated, and it is a practicable solution for condition surveillance.

2. SEPARATING THE COMPOUND SOUND WAVES

Figure 1 shows the experimental setup used for monitoring the condition of three machines using a parabolic reflector microphone (parabolic mic.) and the proposed separation method for compound sound waves.

Only sound waves that come directly from the front of the parabolic mic are amplified by the parabolic reflector. For example, when the parabolic mic is directed at target machine 1, the sound waves caused by rotational and pass frequency elements generated by the machine can be detected. Conversely, the sound waves generated by machine 2 and machine 3 are attenuated. The origin of sounds generated by only the target machine can be determined in this way. Specifically, identification of the location of each target machine can be performed on the basis of the intensity difference in the frequency domain associated with the rotational and pass frequency component. After identifying the source of each sound

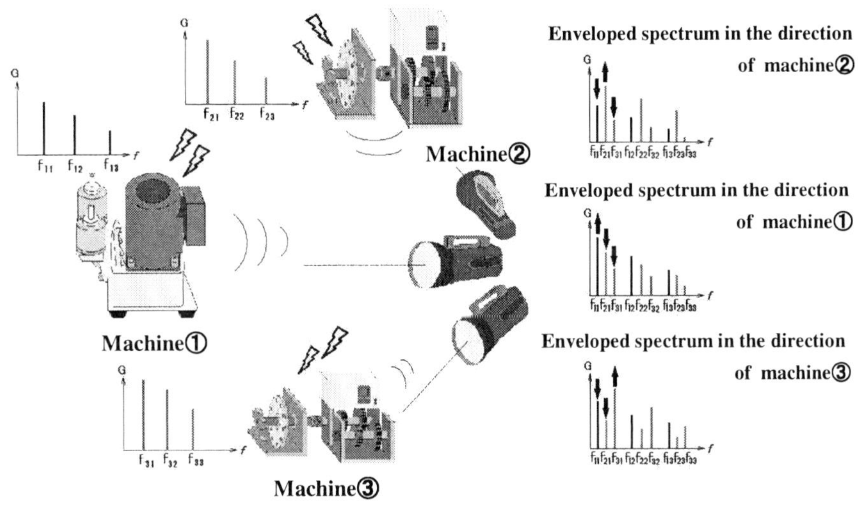

Figure1 Conceptual scheme of condition monitoring using a parabolic reflector type microphone

in the monitored machine, only the rotational and pass frequency components that are of concern relating to the condition of the machine need to be effectively extracted. Estimation of the intensity difference of the rotational and pass frequency components can be done using the enveloped spectrum [4] in the frequency domain. The method of separation using signal processing is based on direct FFT processing value. The direct FFT processing value implies only the FFT process and not the power spectrum. Initially, the extraction of only rotational and pass frequency components in the frequency domain is required to improve the signal to noise ratio. The amplitude and phase values of the rotational and pass frequency components identified by the parabolic mic are replaced by 0 in the direct FFT calculation. In the next step, the difference between the direct FFT value replaced for value 0 and not replaced can be recommended. The direct FFT processing values of difference includes only the rotational and pass frequency components of the machine being monitored, effectively excluding the other two machines. The obtained rotational and pass frequency components of only the monitored machine are transformed using the inverse FFT. We can obtain the necessary time series data associated with only the rotational and pass frequency components of the target machine.

3. EXPERIMENTAL LAYOUT

The machines used in this study were two rotary machines and a vacuum pump. Figure 2 shows their positions relative to the parabolic mic used for recording. These machines and the mic are approximately in a straight line, the distances and the angles are shown in the figure. The parabolic mic reflector has a diameter of 96 mm and a depth of 79 mm. The mic is a ¼-inch condenser mic, and its amplifier is a CCLD. The target rotary machines are known as machine 1 and machine 2 in this study. A rolling bearing that has a faulty outer race is mounted on the machine1, the dimensions of the faulty outer race

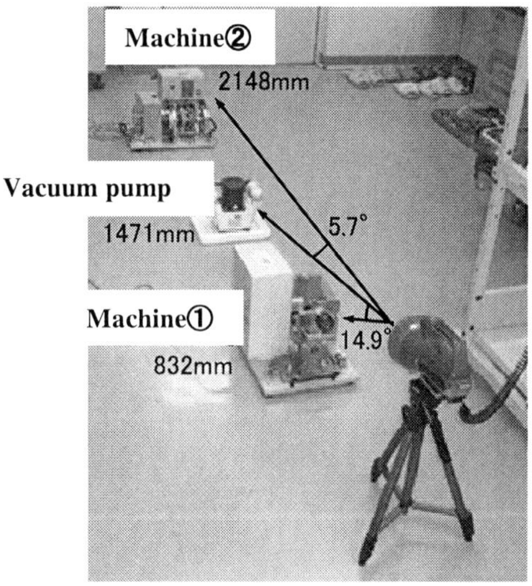

Figure2 Positional relation between the three facilities and a parabolic mic

Table1 Rotational and pass frequency components of monitored facilities

Monitored facility	Machine①	Vacuum pump	Machine②
Kind of fault	Auter race fault	Normal	Local wear of gear
Fundamental frequency[Hz]	104.6	28.5	34.2
Second	209.2	57.0	68.5
Third	314.9	85.5	102.6
Fourth	417.5	114.1	137.0
Fifth	522.3	142.6	
Sixth		171.1	

Table2 Measurement condition on the signal processing

Sampling frequency	2048 [Hz]
Cut off frequency	600 [Hz]
Number of line	16384
Frequency resolution	0.0125 [Hz/line]
Measurement time	8.0 [s/point]

are as follows: 0.5 × 0.3 × 0.3 mm. On machine 2, a spur gear that has a local wear on two teeth has been installed. The vacuum pump is in a working state. Table 1 shows the rotational and pass frequencies generated from the machines. When the parabolic mic is directed at the machines in turn, these frequency components can be measured and visually confirmed by the enveloped spectrum in the frequency domain. The pass frequency components from the fundamental at 104.6 Hz to the fifth harmonic at 522.3 Hz for the outer race fault can be estimated from the calculated value of the pass frequency of machine 1. The drive frequency components of the vacuum pump from the fundamental at 28.5 Hz to the sixth harmonic at 171.1 Hz can be seen. In addition, the rotational frequency components of machine 2 associated with the wear on the spur gear have the fundamental at 34.2 Hz and continue to the fourth harmonic at 137.0 Hz. Table2 shows the measurement conditions on the signal processing of the sound wave.

4. RESULTS AND DISCUSSION

4.1 Sound Source Identification in the Monitored Machines

Table 3 shows the mean intensity values of 150 times about enveloped spectrum associated with the pass frequency component and its harmonics for machine 1, the vacuum pump, and the machine 2. Entries (1), (4), and (7) in this table indicate the mean intensity values of machine 1, the vacuum pump and machine 2, when the parabolic mic is directed towards machine 1. Entries (2), (5), and (8) indicate the mean intensity values of each machine, when the parabolic mic is directed towards the vacuum pump. Similarly, entries (3), (6), and (9) refer to the mean intensity values when the mic is directed towards machine 2. Rows (1), (2), and (3) show the changes in the mean intensity of the pass frequency components in machine 1, when the parabolic mic is directed towards each machine. Rows (4), (5), and (6) show the change of the mean intensity values of the drive frequency components of the vacuum pump, when the parabolic mic is directed towards each of the machines. Similarly, rows (7), (8), and (9) show the change of the mean intensity values of the rotational frequency components associated with the wear of the spur gear of machine 2, when the parabolic mic is directed towards each of the machines.

Table 3 Intensity values of the spectrum envelope of rotational and pass frequency components of the monitored machines

Pass frequency of outer race fault of the machine ①		Intensity value of enveloped spectrum obtained by the parabolic mic			Comparison of the intensity values
		In the direction of the machine①	In the direction of the vacuum pump	In the direction of the machine②	
Fundamental [Hz]	104.6	7.44×10^{16}	8.00×10^{15}	7.59×10^{15}	(1)>(2)>(3)
Tecond	209.2	1.83×10^{16}	2.77×10^{15}	1.37×10^{15}	(1)>(2)>(3)
Third	314.9	(1) 4.60×10^{15}	(2) 2.30×10^{14}	(3) 2.19×10^{14}	(1)>(2)>(3)
Fourth	417.5	2.14×10^{15}	3.09×10^{14}	6.86×10^{13}	(1)>(2)>(3)
Fifth	522.3	1.37×10^{15}	1.17×10^{14}	7.93×10^{13}	(1)>(2)>(3)
Driving frequency of the vaccum pump					Comparison of the intensity values
Fundamental [Hz]	28.5	8.71×10^{14}	1.44×10^{17}	4.85×10^{16}	(5)>(6)>(4)
Second	57.0	8.64×10^{14}	5.10×10^{16}	1.40×10^{16}	(5)>(6)>(4)
Third	85.5	3.99×10^{14}	6.34×10^{15}	2.21×10^{15}	(5)>(6)>(4)
Fourth	114.1	(4) 2.18×10^{14}	(5) 5.93×10^{14}	(6) 2.42×10^{14}	(5)>(6)>(4)
Fifth	142.6	1.02×10^{14}	1.87×10^{15}	1.11×10^{15}	(5)>(6)>(4)
Sixth	171.1	1.48×10^{14}	8.18×10^{14}	1.08×10^{15}	(5)>(6)>(4)
Rotational frequency of the local wear of the gear					Comparison of the intensity values
Fundamental [Hz]	34.2	2.27×10^{14}	2.90×10^{15}	5.12×10^{15}	(9)>(8)>(7)
Second	68.5	2.58×10^{14}	8.69×10^{14}	1.37×10^{15}	(9)>(8)>(7)
Third	102.6	(7) 3.14×10^{14}	(8) 3.56×10^{14}	(9) 6.37×10^{14}	(9)>(8)>(7)
Fourth	137.0	5.40×10^{13}	1.54×10^{14}	2.34×10^{14}	(9)>(8)>(7)

The sound source of the pass frequency components generated from machine 1 can be identified by comparison with the above intensities in (1), (2), and (3) of the enveloped spectrum. The order of intensities is (1) > (2) > (3). Therefore, the sound source of the pass frequency components is clearly machine 1. Similarly, the sound source of the drive frequency components generated by the vacuum pump can be identified by comparison with the above intensity values of (4), (5), and (6). The order of intensities is (5) > (6) > (4). Therefore, the sound source of drive frequency components is the vacuum pump. Further, the sound source of the rotational frequency components associated with the wear of the spur gear generated from machine 2 can be identified by comparison with the above intensity values of (7), (8), and (9). The order of intensities is (9) > (8) > (7). Therefore, the sound source of the wear of the spur gear comes from machine 2. From the above results, the identification of each sound source in the monitored machines can be effectively and conveniently estimated using a parabolic mic.

4.2 Separation of the Compound Sound Wave

Figure 4 shows the signal processing of the proposed separation of the compound wave in the faulty outer race of the rolling bearing mounted in machine 1. Figure 4a shows the direct FFT processing values

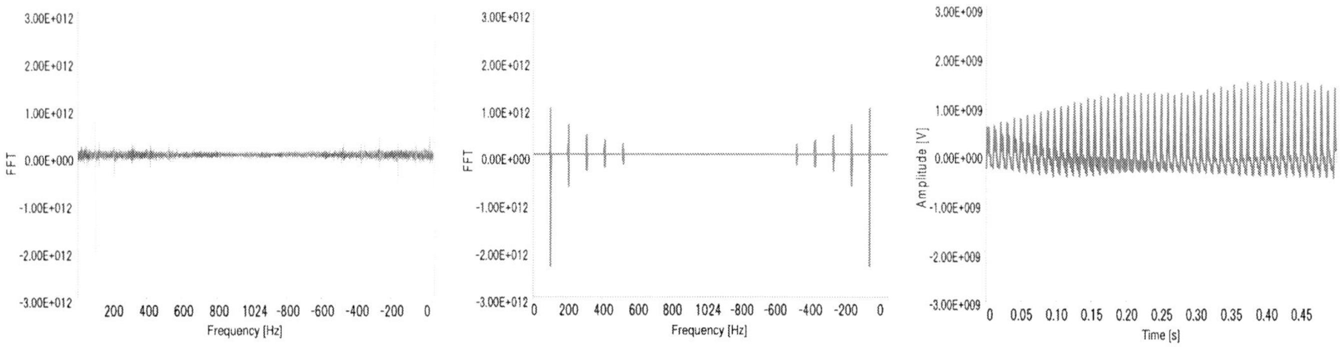

(a) Direct FFT processing value (b) Direct FFT processing value of only the pass frequency component (c) Times series of only the pass frequency

Figure4 Separation of composed sound wave in the outer race fault of monitored machine①

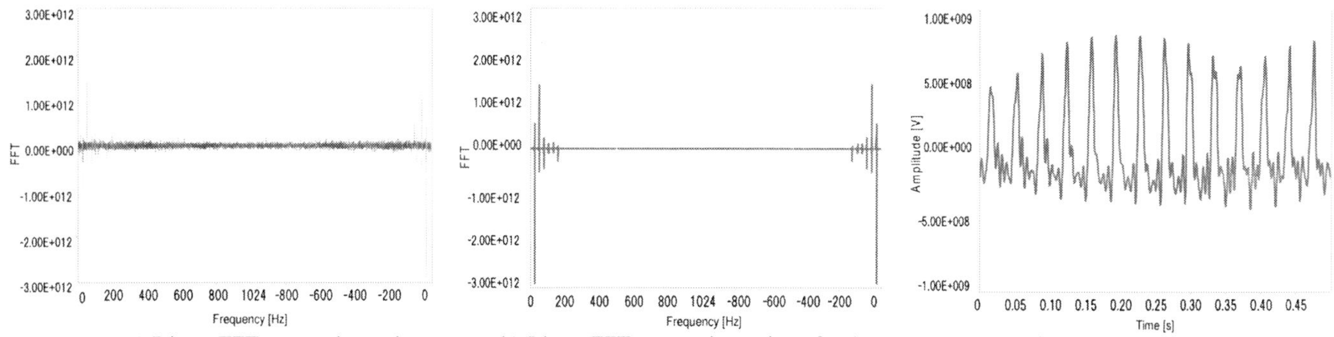

(a) Direct FFT processing value (b) Direct FFT processing value of only the rotational frequency component (c) Times series of only the rotational component

Figure5 Separation of composed sound wave in the rotational frequency of the monitored vacuum pump

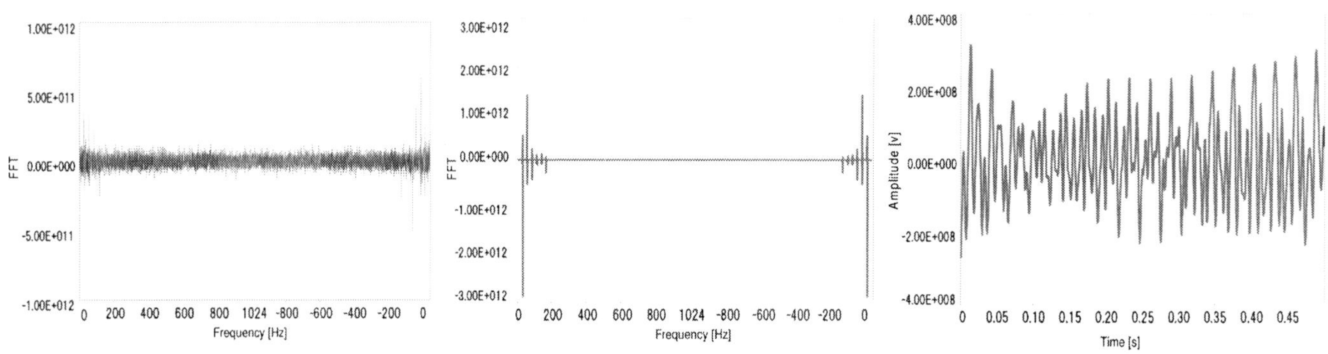

(a) Direct FFT processing value (b) Direct FFT processing value of only the local wear frequency component (c) Times series of only the local wear component

Figure 6 Separation of composed sound wave in the local wear of the monitored machine② which are measured when the parabolic mic is directed towards machine 1. Figure 4b shows the extracted difference value between the direct FFT processing values replaced for value 0 and not replaced in the intensity and its phase associated with only the pass frequency of outer race fault. Finally, the time series data shown in figure 4c, which is associated with only the outer race fault in machine 1, can be obtained by inverse FFT of the previous frequency domain data. Hence, diagnostically important time series data can be extracted effectively and accurately by means of the proposed separation of the compound sound wave recorded by a parabolic mic. Figure 5 shows the signal processing of the drive frequency of the monitored vacuum pump. Figure 5a indicates direct FFT processing, and figure 5b extracts the difference value between the direct FFT processing values replaced for value 0 and not replaced in the intensity and its phase associated with only the driving frequency components of the vacuum pump in frequency domain. In addition, the time series data expressed in figure 5c, which is associated with only the drive frequency of the monitored vacuum pump, can be obtained by inverse FFT of the previous frequency domain data. Hence, diagnostically important time series data can be extracted effectively and accurately by means of the proposed separation of compound sound wave recorded by a parabolic mic. Figure 6 shows the case of the worn gear mounted in machine 2. Again, diagnostically important time series data can be extracted effectively and accurately by the proposed method.

5. CONCLUSION

In this paper, an efficient and accurate method of monitoring the condition of machines' moving parts is proposed, based on recordings of sound using a parabolic mic and then doing the required signal processing on the compound sound waves. Diagnostically important time series data can be extracted effectively and accurately using this method. The method thus has applications in condition surveillance technology.

ACKNOWLEDGMENTS

This work was supported by Japan Society for Promotion of Science (JSPS), Grant-in-Aid for Science Research (C) (No.26420839), and Fundamental Research Developing Association for Shipbuilding and Offshore (REDAS) of the shipbuilders' Association of Japan. Therefore, the authors wish to express our deep gratitude to all JSPS and REDAS partners.

REFERENCES

[1] L. Wang and R.Gao, "Condition Monitoring and Control for Intelligent Manufacturing", Springer Press, (2005), p.109-136.

[2] R.B. Randall, Jerome Antoni, "Rolling element bearing diagnostics -A tutorial", Mechanical Systems and Signal Processing, Vol.25, No.2, (2011), p.485-520.

[3] Yoshihito Tamanoe et al, "Machine Diagnosis Using Acoustical Signal Processing Techniques and Special Sound Collecting Hood", IEICE Trans. FUNDAMENTALS, Vol. E78-A, No.12, (1995), p.1627-1633.

[4] Iron and steel institute of Japan, "Handbook of Condition Diagnosis Technology", Maruzen publications (1986), p.47-48. (in Japanese)

Convolutive Blind Source Separation Based on Reference Signals and Its Application

Zhifa Liu[1], Zhousuo Zhang[1,*], Hongqing Huo[2], Jie Zhang[1], Teng Gong[1]

[1]State Key Laboratory for Manufacturing Systems Engineering, School of Mechanical Engineering, Xi'an Jiaotong University, Xi'an 710049, China, zzs@mail.xjtu.edu.cn
[2]Lanzhou Institute of Physics, Lanzhou 730000, China, huohq@sina.com

Abstract. Micro-vibration caused by disturbance sources on board spacecraft can severely degrade the working environment of sensitive payloads. Extraction of effective source information from measured vibration signals is a fundamental task for the satellite micro-vibration reduction. BSS is a powerful tool to recover main vibration source from observed signals. In this paper, a new higher-order contrast function based on cross-statistics between the estimated output and the reference signal is introduced. By means of the fixed point iteration and the optimal step size, the contrast function is optimized in the classical deflation scenario. In the meanwhile, the methods to choose the reference signals and the characteristic of reference signals are discussed. The effectiveness of the proposed methods is validated by experiment.

Keywords: Micro-vibration, MIMO convolutive mixtures, blind source separation (BSS), reference signal, contrast function

1 Introduction

The micro-vibration produced by the mechanical systems of satellite can degrade the pointing precision and stability performance of instruments [1]. Micro-vibration is primarily generated by the onboard spinning devices such as crycoolers, solar array drive mechanisms, antenna pointing mechanisms and reaction wheel assemblies [2]. The reduction of micro-vibration is a key issue for satellites which require high pointing accuracy and stability. Due to the limitation of test condition, the vibration signal of micro-vibration source always can't be obtained. The vibration signals of sensitive payload points, which are away from vibration sources, contain a wealth of information about the mechanical running conditions. The vibration signals of sensitive payload points are easily measured. Extraction of effective source information from measured vibration signals is a fundamental task for the satellite micro-vibration reduction. Blind source separation (BSS) is a powerful tool to recover the source signals from the mixed signals without the mixing mode information and source signals in-

* Corresponding author .Tel .: 029-82663689
E-mail address: zzs@mail.xjtu.edu.cn

formation. However, the large number of micro-vibration and noise sources, the complexity of transmission paths make the BSS in satellites have a poor performance.

Independent component analysis (ICA) has been proved to be an effective method to solve the source separation and source extraction problem in which the source signals and mixing mode are unknown. However, the separation performances of ICA algorithms are affected by random sampling and random initial value of the variable. Further researches indicate that if the prior information are introduced, the blind source separation can be improved. The methods are firstly proposed by Lu and Rajapakse. They proposed some methods which are called independent component analysis with reference (ICA-R) or constraint independent component analysis (cICA) [3]. When these algorithms are applied in our research, we find that the Lu's method is computationally expensive and not stable in some cases [4]. Recently, contrast functions referred to as "reference -based" have been proposed [5]. These contrast function are particularly efficient because they are quadratic with respect to the search parameters. The contrast functions based on reference signals is generalized and a family of reference based contrasts are introduced [6]. The contrast functions are available for both i.i.d. and non i.i.d. source signals. Furthermore, these new contrasts require no constraint on the reference signals, unlike other referenced approaches. A maximum method based on SVD has been proposed for quadratic contrast function, but the SVD optimization is very sensitive to a rank estimation and thus it is not appropriate to use in a deflation process [7].

The reference-based contrast function of BSS provides an effective way to use prior knowledge about the sources and the transmission paths to improve separation performances. Those prior knowledges include the number of micro-vibration sources, the characteristic of the micro-vibration sources, the vibration transfer paths, the rotational speed of momentum wheel assemblies and reaction wheel assemblies and so on.

This paper deals with the blind source separation problem in the framework of MIMO convolutive mixture. A new higher-order contrast function which based on the reference signals is introduced firstly. Then, the algorithm which based on the optimal step size and which does not require any rank estimation is given to optimize the contrast function. The separation of micro-vibration source signals in a box test bench verifies the effectiveness of the proposed methods in the end.

2 Model And Assumptions

The linear and time-invariant (LTI) multiple input and multiple output (MIMO) convolutive mixing model can be described by

$$\mathbf{x}(n)=\sum_{k\in Z}\mathbf{M}(k)\mathbf{s}(n-k)\triangleq\{\mathbf{M}*\mathbf{s}\}(n), \tag{1}$$

where, $\mathbf{M}(n)$ represents the $Q\times N$ matrix corresponding to the impulse response of the LTI mixing system. $\mathbf{x}(n)$ denotes the Q dimensional discrete-time observation signals. $\mathbf{s}(n)$ denotes the N dimensional discrete-time source signals. The objection of BSS is to find an inverse source separating system to restore the sources blindly. Considering that the source signals are extracted one by one, the BSS is reduced to the estimation of a $(1,Q)$ row vector filter $\mathbf{w}(n)$, such that output scalar signal

$$\mathrm{y}(n)=\sum_{k\in Z}\mathbf{w}(k)\mathbf{x}(n-k), \tag{2}$$

which restores one of the sources $\mathrm{s}_i(n), i\in\{1,\ldots,N\}$, in other words, the independent components (ICs).

In order to carry out the estimation blindly and successfully, some fundamental assumptions on the source are made [8]. These assumptions are shown as follows:

A1: For all i, the source signals $s_i(n), i\in\{1,\ldots,N\}$ is stationary, zero-mean and with unit variance, the fourth order cumulant $\kappa \triangleq \left|\mathrm{Cum}\{s(\mathrm{n}),s^*(\mathrm{n}),s(\mathrm{n}),s^*(\mathrm{n})\}\right|$ exists and is assumed to be nonzero.

A2: The source signals $s_i(n)$ are statistically mutually independent

3 Contrast Function and Reference Signal

3.1 Contrast Function

Many separation criteria provide good results which rely on higher order statistics. It is proved that the criterion $\kappa_{\mathrm{R,r}}\{\mathrm{y(n)}\}=\mathrm{Cum}\{\mathrm{y(n)},\mathrm{y}^*(\mathrm{n}),\mathrm{r}_1(\mathrm{n})\ldots\ldots\mathrm{r}_{\mathrm{R}-2}(\mathrm{n})\}$ $(\mathrm{R}\geq 3)$ is a contrast function which leads to the source separation when maximized with respect to the separating filter [5].

In this paper, we define the following criterion:

$$J(\mathrm{w})=\left|\mathrm{Cum}\{\mathrm{y}(n),\mathrm{y}^*(n),\mathrm{r}(n),\mathrm{r}^*(n)\}\right|, \tag{3}$$

where $\mathrm{Cum}\{\bullet\}$ denotes the cumulant, $\mathrm{r(n)}$ is a given signal called reference signal.

3.2 Reference Signal

How to choose or construct the reference signals is one of the most important issues in the S-BSS. James and Gibson's work implies that the shape of the reference signals should be closely matched to that of the desired source signals [9]. There are several ways to construct the reference signals when S-BSS is applied to recover the micro-vibration signal in satellites. On the condition that the source signals can be measured, the measured source signals or their signs can be reference signals. Otherwise, we construct reference signals which correlate with source signals in frequency domain by means of analyzing observation signals. At least, the reference signals can be periodic signals which have the same frequency with rotational frequency of micro-vibration. Furthermore, if we have a first estimation of source signals, then this can easily be used as a reference signal.

4 Separation Method

It is assumed that mixing filter admits a MIMO-FIR left inverse filter of length L, which can be assumed to be casual because of the delay indetermination. The row vectors, which define the impulse response, can be stacked in the following row vector

$$\underline{\mathbf{w}} \triangleq (\mathbf{w}(0)\ \ \mathbf{w}(1)\ \cdots\ \mathbf{w}(D-1)) \tag{4}$$

We also define the $(QD,1)$ column vector

$$\underline{\mathbf{x}}(n) \triangleq \left(\mathbf{x}(n)^T\ \ \mathbf{x}(n-1)^T\ \cdots\ \mathbf{x}(n-D+1)^T\right)^T \tag{5}$$

So, the equation (2) can be written as

$$\mathrm{y}(n) = \underline{\mathbf{w}}\underline{\mathbf{x}}(n) \tag{6}$$

Then, the power of the output is

$$\mathrm{E}\left\{|\mathrm{y}(n)|^2\right\} = \underline{\mathbf{w}}\mathrm{R}_{\underline{\mathbf{x}}}\underline{\mathbf{w}}^H, \tag{7}$$

where , $\mathbf{R}_{\underline{\mathbf{x}}} \triangleq \mathrm{E}\left\{\underline{\mathbf{x}}(n)\underline{\mathbf{x}}(n)^T\right\}$ is the variance of $\underline{\mathbf{x}}(n)$.

Finally, using cumulant multilinearity, one can easily obtain that

$$J(\mathbf{w}) = \underline{\mathbf{w}}\mathbf{C}_{\underline{\mathbf{x}},\mathbf{r}}\underline{\mathbf{w}}^H \tag{8}$$

where, $\mathbf{C}_{\underline{\mathbf{x}},\mathbf{r}}$ is defined as $\left(\mathbf{C}_{\underline{\mathbf{x}},\mathbf{r}}\right)_{i,j} = \mathrm{Cum}\left\{\underline{\mathbf{x}}_i(n), \underline{\mathbf{x}}_j^*(n), \mathbf{r}(n), \mathbf{r}^*(n)\right\}$.

The source separation problem is reduced to an optimization process:

$$\max\left|\underline{\mathbf{w}}\mathbf{C}_{\underline{\mathbf{x}},\mathbf{r}}\underline{\mathbf{w}}^H\right|, \text{ under the constraint: } \underline{\mathbf{w}}\,\mathrm{R}_{\underline{\mathbf{x}},\mathbf{r}}\,\underline{\mathbf{w}}^H = 1. \tag{9}$$

It has been noticed that the optimization is a quadratic problem with respect to the parameters. A method based on SVD (Singular Value Decomposition) has been proposed [7]. Because of the unknown rank of $\mathbf{R}_{\underline{\mathbf{x}}}$, the performance of SVD method seriously degrades as more sources are separated. In order to avoid this drawback, the optimization based on optimal step size gradient is a good choice to optimize the above separation problem. ∇ denotes a gradient operate. $k_{\max}$ denotes the iteration number of fixed point method. μ_{opt} denotes the optimal step size which has been introduced in[10]. The new algorithms are given as follows:

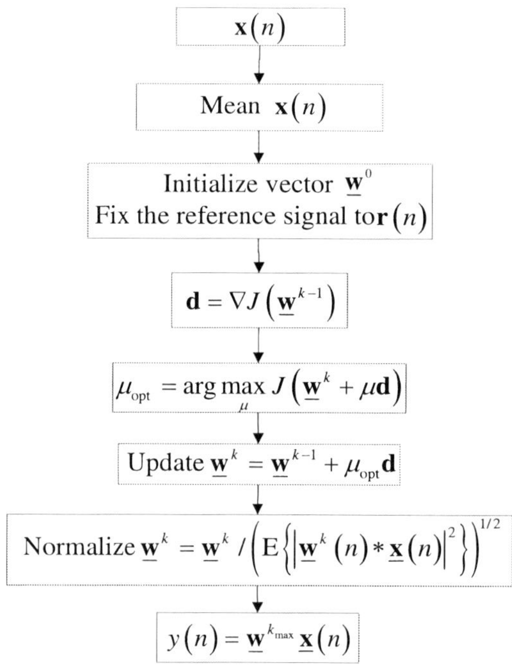

Fig. 1. The algorithm based on the optimal step size

5 Experiment

Box structures are one of the most common structures in satellites. The proposed methods are applied in the separation of the source signals through the micro-vibration signals collected from a thin box test bench which is shown in Fig.2. There are two exciters to simulate the micro-vibration source signals on the test bench. The excitation frequencies of two sinusoidal signals are 94Hz and 31Hz. The observed signals are displayed in Fig.3. (a).

In order to obtain the ICs, the proposed algorithm is applied to analyze the observed signals. The reference signals which are generated for the proposed method in this case are shown in Fig.3. (a). The waveforms of the ICs are shown in Fig. 4(a). The correlation coefficients of waveform ρ between source signals and ICs are calculated. The ρ of the proposed method is $\rho = [0.970\ 0.981]$.

In order to compare with the separation method without reference signals, the classical kurtosis optimization is used to analysis the same observed signals. And the ICs are displayed in Fig. 4(b). It is obvious that the source signals are not separated successfully.

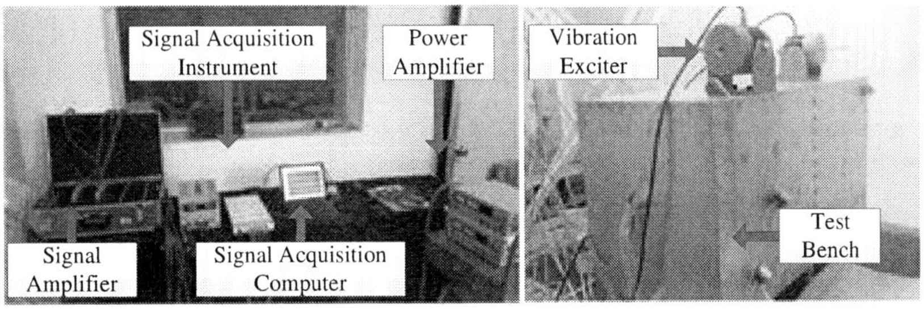

Fig. 2. The photo of test bench

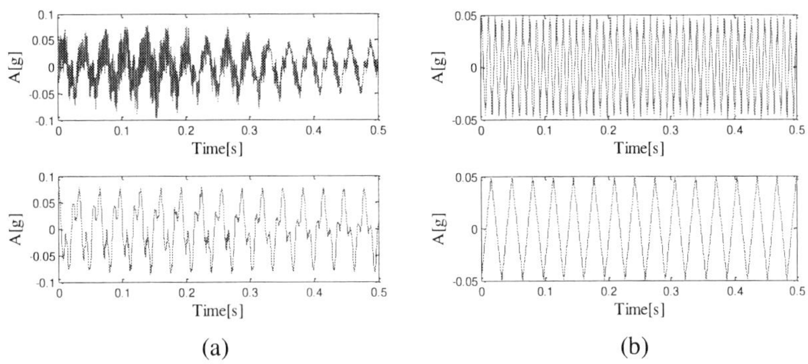

Fig. 3.(a) Observation signals (b) Reference signals

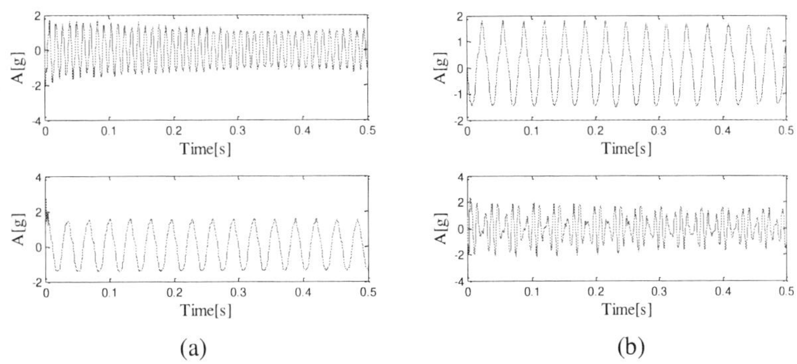

Fig. 4. (a)The ICs by the proposed algorithm (b) The ICs by classical kurtosis optimization

6 Conclusion

In order to recognize the main micro-vibration source in satellite, a new contrast function is introduced in this paper. The new contrast function which introduces the reference signal makes full use of the prior information in satellites. Due to the use of reference signals, the proposed method gives a more accurate separation. The experiment in a box test bench verifies the effectiveness of the new contrast function and optimization method.

Reference

1. Liu, C., Jing, X., Daley, S., et al.: Recent advances in micro-vibration isolation. J. Mechanical Systems and Signal Processing. 56, 55-80 (2015)
2. Kamesh, D., Pandiyan, R., Ghosal, A.: Modeling, design and analysis of low frequency platform for attenuating micro-vibration in spacecraft. J. Journal of sound and vibration. 329(17), 3431-3450 (2010)
3. Lu, W., Rajapakse, J. C.: Constrained independent component analysis. In in Advances in Neural Information Processing Systems 13 (NIPS2000 (2000)
4. Lin, Q. H., Liu, J., Zheng, Y. R., Liang, H., Calhoun, V. D.: Semiblind spatial ICA of fMRI using spatial constraints. Human brain mapping, 31(7), 1076-1088 (2010)
5. Castella, M., Rhioui, S., Moreau, E., et al.: Quadratic higher order criteria for iterative blind separation of a MIMO convolutive mixture of sources. J. Signal Processing, IEEE Transactions on, 55(1), 218-232 (2007)
6. Dubroca, R., De Luigi, C., Castella, M., Moreau, E.: A general algebraic algorithm for blind extraction of one source in a MIMO convolutive mixture. Signal Processing, IEEE Transactions on, 58(5), 2484-2493 (2010)
7. Castella, M., Moreau, E. (2009, April). : A new optimization method for reference-based quadratic contrast functions in a deflation scenario. In Acoustics, Speech and Signal Processing, 2009. ICASSP. IEEE International Conference on (pp. 3161-3164). IEEE(2009)
8. Castella, M., Moreau, E.: New kurtosis optimization schemes for MISO equalization. J. Signal Processing, IEEE Transactions on. 60(3), 1319-1330 (2012)
9. James, C. J., Gibson, O. J.: Temporally constrained ICA: an application to artifact rejection in electromagnetic brain signal analysis. J. Biomedical Engineering, IEEE Transactions on. 50(9), 1108-1116 (2003)
10. Castella, M., Moreau, E., Zarzoso, V.: A survey of kurtosis optimization schemes for MISO source separation and equalization. M. Advances in heuristic signal processing and applications. Springer Berlin Heidelberg, 183-217 (2013)

Multi-modulation Feature Separation via Modified Iterative Hilbert Transform for Planetary Gearboxes Fault Diagnosis

Biao Wang[1], Yanyang Zi[1*], Jinglong Chen[1]

[1]State Key Laboratory for Manufacturing and Systems Engineering, School of Mechanical Engineering, Xi'an Jiaotong University, Xi'an 710049, China, ziyy@mail.xjtu.edu.cn
[*]corresponding author

Abstract. The vibration signal of planetary gearboxes is a typical multi-modulated AM-FM signal due to its time-varying vibration transfer path and the inevitable manufacturing defects. Aiming at excavating the detailed multi-modulation feature related with the state information, the modified iterative Hilbert transform (MIHT) is proposed for the planetary gearboxes fault diagnosis. By utilizing the empirical mode decomposition (EMD) based low-pass filtering, the signal can be demodulated adaptively by the proposed MIHT method and therefore have a better effect on revealing the fault-related multi-modulation feature. The experimental signal is analysed by the proposed method and the results verify its effectiveness and good performance.

Keywords: multi-modulation, adaptive filtering, Hilbert transform, planetary gearbox, fault diagnosis

1 Introduction

Working under the condition of heavy load and corrosion, planetary transmission systems are more likely to be damaged after being operated for a long time. Therefore, it is of great importance to implement fault diagnosis and condition monitoring for planetary gearboxes to ensure its safety and operational reliability [1].

The vibration signal of planetary gearboxes is modulated by the slow-varying signal resulting from the revolution of planet gears [2]. In addition, the inevitable manufacturing defects may cause transmission accuracy reduction and eventually result in the more complicated modulation [3]. As a result, the vibration signal of planetary gearboxes is a typical multi-modulated AM-FM signal, which is quite difficult to extract its fault features. Aiming at solving this problem, Feng et al. applied Fourier transform on the amplitude envelope and instantaneous frequency of the IMF or PF to reveal the fault features in the FM components [4-5]. The existing demodulation analysis methods worked reasonably well but neglected the multi-modulated feature of the planetary gearbox vibration signal and thus remain to be improved.

The iterative Hilbert transform (IHT) is an effective method to demodulate nonstationary signals and extract the multi-modulation feature. IHT decompose the signal by iteratively applying the Hilbert transform on the residue signal that is acquired by high-pass filtering from the envelope signal [6]. One of the critical point of IHT is the way the low-pass filter designed. Qin et al. adopted the same high-pass filter designed with FDA tool and for each iteration [7]. The filter does act better than the fixed one do, but is complicated and not data-driven. In this paper, an adaptive filter based on EMD is introduced to the original IHT method and the modified iterative Hilbert transform (MIHT) is presented.

This paper is organized as follows: In section 2, a brief introduction of IHT is given. The MIHT is presented in section 3. In section 4, an experimental signal is analyzed to prove the effectiveness of the MIHT. Finally, section 5 concludes this paper.

2 Iterative Hilbert transform

For a given signal $x(t)$, by virtue of Gabor analytical representation, it can be calculated by Hilbert transform as [8]:

$$z(t) = x(t) + jH[x(t)] = x(t) + j\hat{x}(t) \tag{1}$$

Therefore, by virtue of (1), $x(t)$ can be written as

$$x(t) = a_0(t)\cos[\alpha_0(t)] \tag{2}$$

where $a_0(t) = |z(t)|$ and $\alpha_0(t) = \arg[z(t)]$.

The core concept of the IHT method is applying the Hilbert transform iteratively to derive a multi-component representation of a signal. Let $a_j(t)$ denote the envelope signal after j iteration. It is obvious that the envelope signal $a_j(t)$ is oscillating but always nonnegative, and thus unsuitable for further Hilbert transform. Therefore, an adaptive low-pass filter is needed to separate the trend $\bar{a}_j(t)$ from $a_j(t)$, and the residual $\tilde{a}_j(t)$, can be processed iteratively as in (2). Set $j = N$, we can obtain [6]

$$x(t) = \sum_{j=0}^{N} \frac{\bar{a}_j(t)}{2^j} \sum_{i=1}^{2^j} \cos[\phi_j^i(t)] + \frac{\tilde{a}_N(t)}{2^N} \sum_{i=1}^{2^N} \cos[\phi_N^i(t)] \tag{3}$$

where

$$\phi_j^{2l-1}(t) = \alpha_j(t) - \phi_{j-1}^l(t), \phi_j^{2l}(t) = \alpha_j(t) + \phi_{j-1}^l(t) \tag{4}$$

for $l = 1, ..., 2^{j-1}$ and $j = 1, ..., N$. This is a asymptotically exact decomposition of the signal $x(t)$ in which the residual(second) term in Eq. (3) rapidly vanishes as N increases if a fraction $k < 1/2$ of the signal energy is kept in the oscillating component $\tilde{a}_j(t)$ after the aforementioned low-pass filtering.

3 The modified iterative Hilbert transform.

This section presents the modified iterative Hilbert transform (MIHT) for multi-modulation feature separation by introducing the EMD based adaptive filter.

The Empirical mode decomposition (EMD) is a classical data-driven decomposition method for non-stationary and nonlinear signals, which decomposes the multi-component signal into several intrinsic mode functions (IMF) and the residual adaptively [9]. Patrick et al. have proved that EMD acts as a dyadic filter bank [10]. The inherent adaptivity of the EMD make it possible to develop better adaptive filter based on itself. This paper introduces the adaptive filter based on EMD to better satisfy the requirement of the filter. First, $x(t)$ is decomposed into $c_i(t), i=1,\dots n$, and a residual $r_n(t)$ by EMD. Then, calculate the defined energy ratio of the energy sum of the first $m(m=1,\dots n-1)$ IMFs and the signal $x(t)$ as

$$R_m = \sum_{i=1}^{m}\sum_{j=1}^{N}[c_i(t_j)]^2 / \sum_{j=1}^{N}[x(t_j)]^2 (m=1,\dots,n-1) \quad (5)$$

where N is the length of the signal (also IMF). The energy ratio represents the energy percentage kept in the oscillating component, which should be less than 0.5 to guarantee the convergence of the IHT method. In practice, the criteria of the energy ratio λ should be set between 0.15-0.25 according to experiment experience. If $R_m > \lambda$, stop calculating the next energy ratio and construct the signal trend $T(t)$ and oscillating component $X(t)$ as

$$T(t) = \sum_{j=m}^{n} c_j(t) \quad (6)$$

$$X(t) = x(t) - T(t) \quad (7)$$

Otherwise, calculating the energy ratio successively until the energy ratio exceeds the criteria. In this way, the trend of the envelope signal is acquired and a more accurate oscillating component is demodulated by Hilbert transform in every iteration, which eventually improves the demodulation effect of the IHT.

The procedure of the MIHT method is summarized as follows:

1. Apply Hilbert transform to the signal to get the envelope and phase signal.
2. The instantaneous frequency signal is obtained by derivation.
3. Decompose the envelope signal into n IMFs and set $m=1$.
4. Calculate the energy ratio R_m by virtue of Eq. (5).
5. If $R_m < \lambda$, set $m=m+1$ and go to step 4. Otherwise, construct the signal trend $T(t)$ by virtue of Eq. (6).
6. Calculate the oscillating component $X(t)$ by virtue of Eq. (7) and go to step 2.
7. Demodulate the signal iteratively until the iterations reach the pre-set number k.

4 Experimental signal analysis

In this section, the vibration signal captured from a test rig is analyzed to validate the effectiveness of the proposed MIHT. The structure of the test rig is shown in Fig.1 and the parameters of the 1st stage gears are listed in Table 1. To simulate the fault signal, a 1st stage sun gear with crack of which the depth is about 1mm is processed, which is shown in Fig.2. The vibration data is collected with a sample frequency of 25600 Hz under the speed of 2376 rpm. Therefore, the rotating frequency of sun gear and carrier is 39.6 Hz and 6.6 Hz, respectively. The fault characteristic frequency of sun gear crack is calculated to be 99 Hz according to Ref [3].

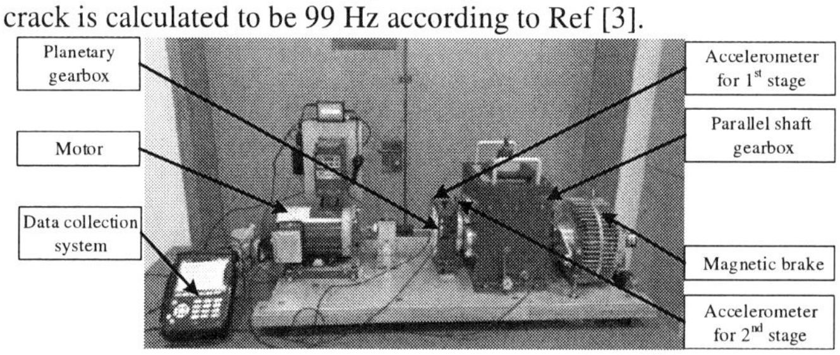

Fig. 1. Test rig and sensor location

Table 1. parameters of the 1st stage Gears

Gear	Sun	Planet	Ring
Number of teeth	20	40	100
Number of planet gears		3	

Fig. 2. 1st stage sun gear with crack

Fig.3 shows the analysis result of the fault signal by MIHT. Fig.3 (a) and (b) shows the original vibration and its Fourier spectrum. Before applying MIHT, the signal whose frequency floating around mesh frequency need to be filtered. Fig.3 (c) and (e) show the envelopes and phase signals acquired by applying MIHT to the filtered signal. And the Fourier spectrum of the envelopes is shown in Fig.3 (d), respectively. It is clear that the first envelope contains frequency component of 6.6 Hz that is corresponding to the carrier rotating frequency, which indicates that MIHT can demodulate the AM components result from the carrier rotation and make contribution to further modulation of the multi-modulated vibration signal. The unwrapped phase signals shown in Fig.3 (e) have a mean slope of 652.2 Hz, 78 Hz, 99.2 Hz and 118.2 Hz, respectively. The first and third frequency approximately correspond to the difference

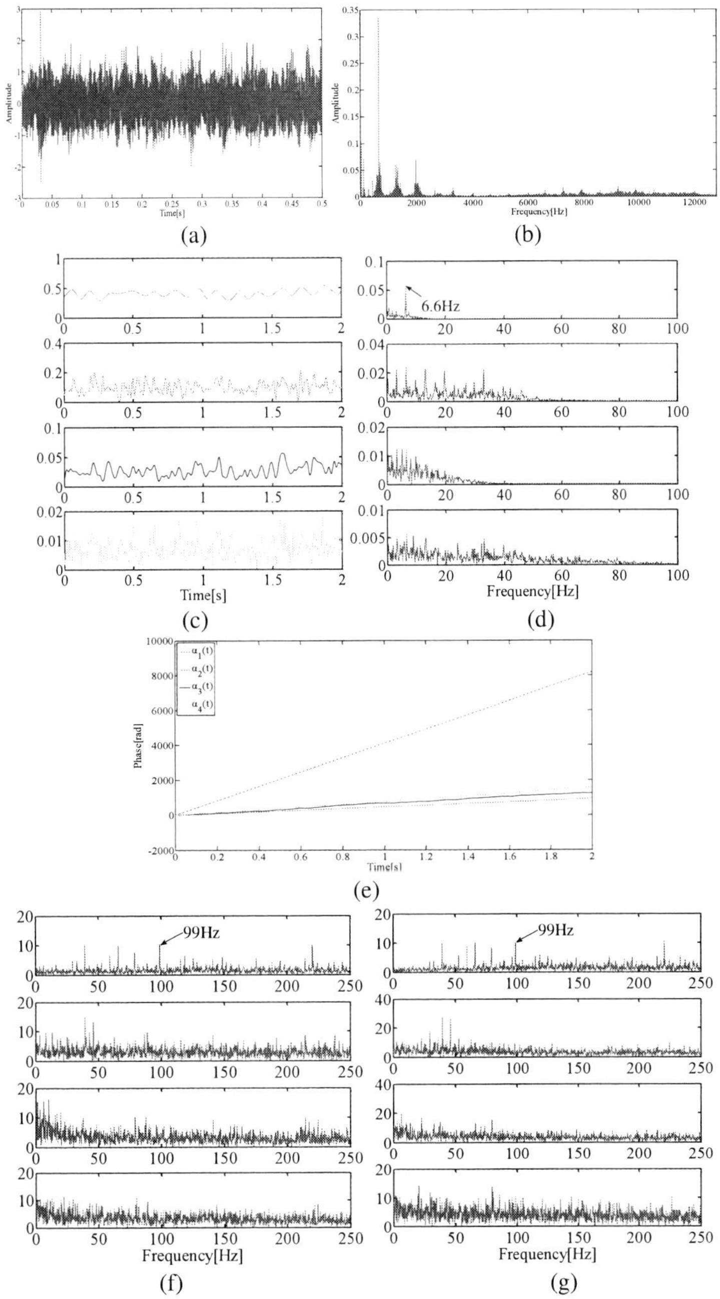

Fig. 3. (a) Vibration signal, (b) Fourier transform of (a), (c) The envelope signals, (d) Fourier transform of (c), (e) The phase signal, (f) Fourier transform of IF, (g) Envelope spectrum of IF

of the mesh frequency and the carrier rotating frequency and the characteristic frequency of the sun gear, respectively. Fig.3 (f) and (g) show the corresponding Fourier spectrum and envelope spectrum of the IF (instantaneous frequency) calculated from phase signal, in which the characteristic frequency of the sun gear is exhibited.

Having analyzed the fault signal of the planetary gearbox test rig, the fault feature of the sun gear with crack is successfully extracted by MIHT. Besides, some of the multi-modulated components of the signal can be extracted correctly, which makes the fault feature clearer and more credible.

5 conclusion

This paper introduces the EMD based adaptive filter to the original iterative Hilbert transform and presents the modified iterative Hilbert transform (MIHT) method. The adaptive filter utilizes the adpativity of EMD, and a simple and effective criterion is defined to ensure the effect of the trend extraction. With these improvements, the MIHT is particularly suitable for multi-modulation feature extraction and its effectiveness is validated by experimental signal.

References

1. Lei Y, Lin J, Zuo M J, et al. Condition monitoring and fault diagnosis of planetary gearboxes: a review[J]. Measurement, 2014, 48: 292-305.
2. Koch J P T, Vicuña C M. Dynamic and phenomenological vibration models for failure prediction on planet gears of planetary gearboxes[J]. Journal of the Brazilian Society of Mechanical Sciences and Engineering, 2014, 36(3): 533-545.
3. Feng Z, Zuo M J. Vibration signal models for fault diagnosis of planetary gearboxes[J]. Journal of Sound and Vibration, 2012, 331(22): 4919-4939.
4. Feng Z, Liang M, Zhang Y, et al. Fault diagnosis for wind turbine planetary gearboxes via demodulation analysis based on ensemble empirical mode decomposition and energy separation[J]. Renewable Energy, 2012, 47: 112-126.
5. Feng Z, Zuo M J, Qu J, et al. Joint amplitude and frequency demodulation analysis based on local mean decomposition for fault diagnosis of planetary gearboxes[J]. Mechanical Systems and Signal Processing, 2013, 40(1): 56-75.
6. Gianfelici F, Biagetti G, Crippa P, et al. Multicomponent AM–FM representations: an asymptotically exact approach[J]. Audio, Speech, and Language Processing, IEEE Transactions on, 2007, 15(3): 823-837.
7. Qin Y, Qin S, Mao Y. Research on iterated Hilbert transform and its application in mechanical fault diagnosis[J]. Mechanical Systems and Signal Processing, 2008, 22(8): 1967-1980.
8. Hahn S L. Hilbert transforms in signal processing[M]. Artech House on Demand, 1996.
9. Huang N E, Shen Z, Long S R, et al. The empirical mode decomposition and the Hilbert spectrum for nonlinear and non-stationary time series analysis[C]//Proceedings of the Royal Society of London A: Mathematical, Physical and Engineering Sciences. The Royal Society, 1998, 454(1971): 903-995.
10. Flandrin P, Rilling G, Goncalves P. Empirical mode decomposition as a filter bank[J]. Signal Processing Letters, IEEE, 2004, 11(2): 112-114.

Direct Driven Motor Sensor Fault Diagnosis Based on Optimal Unknown Input Observer for Electric Vehicle

Xing Xu[1,2], Te Chen[1], Hongtao Xue[1], Long Chen[1,2]

(1. School of Automotive and Traffic Engineering, Jiangsu University, Zhenjiang 212013, Jiangsu, China;

2. Automotive Engineering Research Institute, Jiangsu University, Zhenjiang 212013, Jiangsu, China)

Abstract: Using direct driven in-wheel motor as driving system for distributed driven electric vehicle can quickly and accurately realize the torque distribution of every driving wheel. Aiming at the matter of current sensor fault and speed sensor fault of distributed driven electric vehicle, the robust fault detection and location method is studied. Considering the unknown input and noise in motor model, decoupling the unknown input via system order reduction method, using Kalman filter (KF) to filter the white noise of the decoupled subsystem, then the optimal unknown input observer (UIO) is designed to realize the system state estimation, and a robust residual generator is obtained. At the same time, the generalized likelihood ratio (GLR) method is used to evaluate the residual signals and determine the threshold, a sensor fault location method is proposed. Bench experiment results show that the sensor fault diagnosis method can realize the sensor fault diagnosis and location for direct driven motor system.

Keywords: sensor fault; unknown input observer; diagnosis; direct driven motor; electric vehicle

1 Introduction

Owing to the impact of environmental factors and components aging, the current and rotational speed sensors of distributed driven electric vehicle are prone to performance failure. Consequently, it is necessary to study the sensor fault diagnosis for distributed driven electric vehicle[1].

In fault diagnosis methods based on model analysis, the state estimation method depends on the accuracy of mathematical model of the controlled objects. At present, the research of state estimation methods are mainly focused on improving the sensitivity to fault and the robustness of the detection system for modeling errors, disturbances, noise[2-5]. Fault diagnosis method based on unknown input observer is one of the research directions in recent years, there have been many research achievements[6-10].

The research on sensor fault diagnosis with the characteristics of unknown input and noise for direct drive motor system of electric vehicle is not common at present. In this paper, a robust residual generator based on optimal UIO is proposed combining with Kalman filter algorithm. The current and rotational speed sensor fault diagnosis and location of the direct driven motor system are realized.

2 Design of robust residual generator based on optimal UIO

Assume that the state vector is $x=[i\ \omega]^T$, input u is the line voltage U, unknown input d is the load torque T_L, y is the sensor measurement output, w and v are the uncorrelated zero mean white noise series, the direct driven in-wheel motor model can be obtained as

$$\dot{x} = Ax + Bu + Dd + Ew \tag{1}$$

$$y = Cx + Fv \tag{2}$$

where, $A=\begin{bmatrix} -\frac{R}{L} & -\frac{K_a}{L} \\ \frac{K_t}{J} & -\frac{b}{J} \end{bmatrix}$, $B=\begin{bmatrix} \frac{1}{L} \\ 0 \end{bmatrix}$, $C=\begin{bmatrix} 1 & 0 \\ 0 & 1 \end{bmatrix}$, $D=\begin{bmatrix} 0 \\ -\frac{1}{J} \end{bmatrix}$, $E=F=\begin{bmatrix} 1 \\ 1 \end{bmatrix}$。$J$ is the rotary inertia of motor rotor, b is the damping coefficient, K_t is the motor torque constant, T_L is the motor load torque, i is the line current, U is the line voltage, R is the equivalent winding resistance, L is the equivalent winding inductance, K_a is the inverse electromotive force coefficient. According to equation (1) and (2), it can be deduced that rank (CD) =rank (D), the system satisfies the observer matching condition. It is possible to construct a nonsingular matrix T, then, the system given by (1) and (2) can be converted into the standard type of observation such that

$$\dot{\bar{x}} = \bar{A}\bar{x} + \bar{B}u + \bar{D}d + \bar{E}w \tag{3}$$

$$y = \bar{C}\bar{x} + \bar{F}v \tag{4}$$

where, $\overline{A}=T^{-1}AT=\begin{bmatrix}\overline{A}_{11} & \overline{A}_{12}\\ \overline{A}_{21} & \overline{A}_{22}\end{bmatrix}$, $\overline{B}=T^{-1}B=\begin{bmatrix}\overline{B}_1\\ \overline{B}_2\end{bmatrix}$, $\overline{C}=CT=\begin{bmatrix}CN & CD\end{bmatrix}$,

$\overline{D}=T^{-1}D=\begin{bmatrix}0\\ \overline{D}_2\end{bmatrix}$, $\overline{E}=T^{-1}E=\begin{bmatrix}E_1\\ E_2\end{bmatrix}$, $\overline{F}=F$, $x=T\overline{x}=T\begin{bmatrix}\overline{x}_1\\ \overline{x}_2\end{bmatrix}$。

In equation (3) and (4), it is able to decouple the model into two parts: former system that is related to the unknown input, and latter system that does not see the unknown input. The latter system can be defined as

$$\begin{bmatrix}I_{n-1} & 0\end{bmatrix}\dot{\overline{x}}=\begin{bmatrix}\overline{A}_{11} & \overline{A}_{12}\end{bmatrix}\overline{x}+\overline{B}_1u+\overline{E}_1w \quad (5)$$

$$y=\begin{bmatrix}CN & CD\end{bmatrix}\overline{x}+\overline{F}v \quad (6)$$

A nonsingular matrix U can be constructed, where $U^{-1}=\begin{bmatrix}U_1\\ U_2\end{bmatrix}=\begin{bmatrix}U_{11} & U_{12}\\ U_{21} & U_{22}\end{bmatrix}$, multiplying both side of equation (6) by U^{-1}, then the state space equation of the subsystem can be expressed as

$$\dot{\overline{x}}_1=\tilde{A}_1\overline{x}_1+\overline{B}_1u+H_1y-\overline{A_{12}U_1}\overline{F}v+\overline{E}_1w \quad (7)$$

where, $\tilde{A}_1=\overline{A}_{11}-\overline{A}_{12}U_1CN$, $H_1=\overline{A}_{12}U_1$. In the same way, the state space equation of the other subsystem can be expressed as

$$\dot{\overline{x}}_2=\tilde{A}_2\overline{x}_2+H_2\overline{x}_1+H_3u+U_1\dot{y}-U_1CN\overline{E}_1w \quad (8)$$

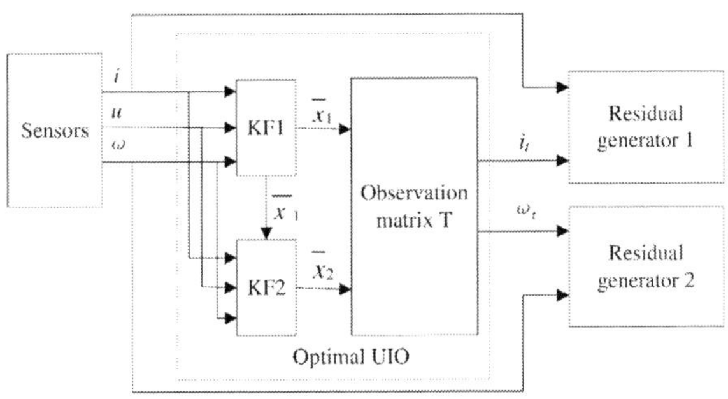

Fig. 1. Design flow of the residual generator

where, $\tilde{A}_2=-U_1CN\overline{A}_{12}$, $H_2=-U_1CN\overline{A}_{11}$, $H_3=-U_1CN\overline{B}_1$. At this time the system does not contain unknown input. It is able to design two KF that can be used to filter out the noise and estimate $\bar{x}_1$ and $\bar{x}_2$. Assuming that the theoretical values of current and rotational speed is i_t and ω_t respectively, the optimal UIO for the direct driven in-wheel motor system can be expressed as $y_t=[i_t \ \omega_t]^T$. The residuals are given by $r_1=i-i_t$, $r_2=\omega-\omega_t$. Design flow of the residual generator is shown in Fig.1.

As shown in Fig. 1, in the system with unknown input and noise interference, the residual generators combine the advantages of KF and UIO, which helps to improve the robustness of residual error identification.

3 Sensor fault diagnosis and location

In order to reduce the influence of system uncertainty and environmental factors which are difficult to be modeled, the method of GLR is used for residual signals evaluation. Residual GLR can be represented as $J_i=r_i^2/\sigma_i^2$, where, r_i is the actual output residual signal of residual generator, σ_i^2 is the variance of residual signal under the condition of no fault. Here, Ji compliance the χ^2 distribution with 1 degree of freedom. According to the false alarm rate of the actual system requirements, the threshold T_r can be set in the form of $P\{J_i > T_r|s=0\}=P_f$, where, s=0 represents that GLR does not exceed the threshold, s=1 represents that GLR exceed the threshold. Set the false alarm rate as P_f, then the threshold T_r can be obtained by finding the χ^2 distribution table. The sensor fault diagnosis logic can be expressed as

$$s=\begin{cases}0, J_i < T_r \\ 1, J_i > T_r\end{cases} \tag{9}$$

By calculation, it can be obtained as $U_{11}=0$. It means that, in the sensor measurements of the input y, only the rotational speed measured by speed sensor is the effective input. That is to say, the designed optimal UIO is not sensitive to the current sensor fault, but is sensitive to the speed sensor fault. The sensor fault location logic table is shown in Table 1.

Table 1. Sensor fault location logic

Residual treatment results		Diagnostic results
$J_1 \geqslant T_r$	$J_2 \leq T_r$	Current sensor fault
$J_1 \geqslant T_r$	$J_2 \geqslant T_r$	Speed sensor fault
$J_1 \leq T_r$	$J_2 \leq T_r$	No sensor fault

4 Bench experiment and verification

Fig. 2 shows the experimental bench. In the experiment, Digatron battery test system provided stable voltage to the motor. Magnetic powder brake is used as a loading device. Rotational speed&torque sensor is used to measure the rotational speed and driving torque of the motor under different control signals. The sensor fault is simulated at t=400s.

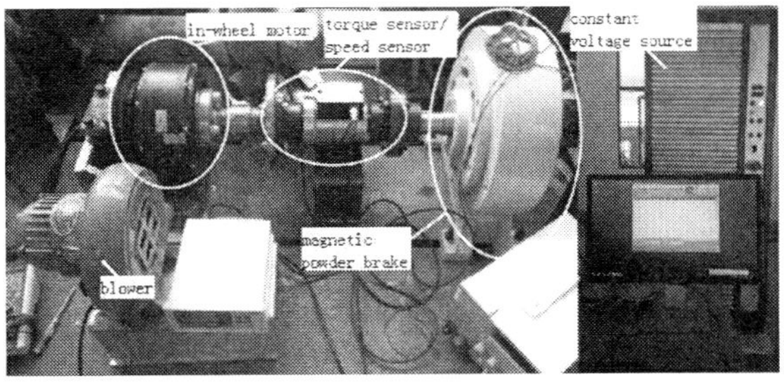

Fig. 2. Bench of in-wheel BLDC motor characteristic experiment

When the sensor has no fault, the result is shown in Fig. 3. When the sensor has no fault, the residual signal based on the common UIO is obviously larger than that of the optimal UIO, this is easy to cause false alarm and is not conducive to the correct fault diagnosis. On the contrary, the residual generator based on the optimal UIO has an inhibitory effect on the system noise, which improves the robustness of the residual error.

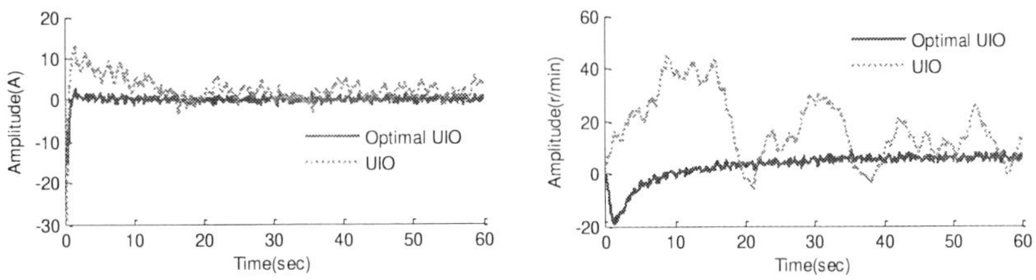

Fig. 3. Current/speed residual signal without fault

The residual signal is evaluated by the method of GLR.The false alarm rate is selected as P_f=0.0001. The threshold T_r=15.137 is obtained according to the χ^2 distribution table. When the current sensor fails, the experimental results are shown in Fig. 4. The current residual GLR estimation has always been in the range of the threshold, but it significantly exceed the threshold at t=400~1000s. The variation of rotational speed residual GLR estimation is relatively small and has been kept in the range of threshold all the time. Comparing the two figures and referencing the fault location logic table, current sensor fault can be inferred.

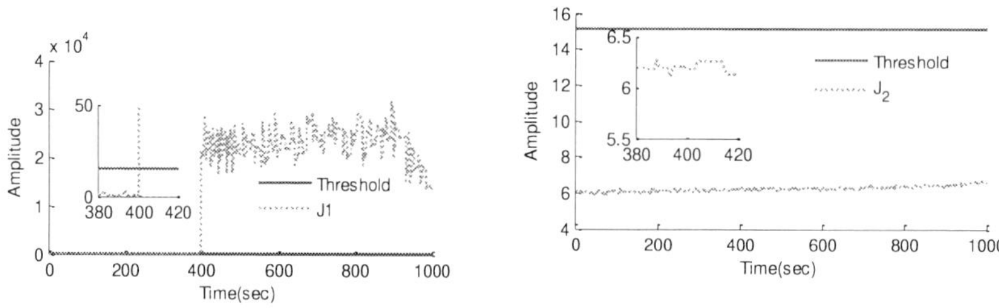

Fig. 4. Comparison of current/speed residual GLR estimation and threshold

5 Conclusion and discussion

This paper proposes a sensor diagnosis and location method for direct driven motor system of distributed driven electric vehicle. The optimal UIO can decouple the unknown input, and the KF has a better inhibitory effect on the noise, which makes the residual generator has a strong robustness. Moreover, the GLR method improves the sensitivity of residual signal to the fault and reduces the false alarm rate. Consequently, the sensor fault location method can be used to locate the fault of the current and rotational speed sensor for direct driven motor system.

ACKNOWLEDGMENTS

The authors are grateful for the support provided by China Postdoctoral Science Foundation (No. 2015M571680), "Qing Lan" Project of Jiangsu Province, and "Six Major Talent" Project of Jiangsu Province (No. 2014-JXQC-004).

References

1. Yozo Hori. Future Vehicle Driven by Electricity and Control-research on Four-wheel-motored"UOT Electric March II"[J]. *Vehicle System Dynamics*, 2012, 50(9): 1473-1494.
2. Zhou D H, Liu Y, He X. Review on Fault Diagnosis Techniques for Closed-loop Systems[J]. *Acta Automatica Sinica*, 2013, 39(11): 1933-1943.
3. Hwang I, Kim S, Kim Y. A survey of fault detection, isolation and reconfiguration methods [J]. *IEEE Transactions on Control Systems Technology*, 2010, 18(3): 636-653.
4. Zhong M Y, Ding S X. An LMI Approach to design robust fault detection filter for uncertain LTI systems [J]. *Automatica*, 2003, 39:543-550.
5. Zhang K, Jiang B. Fault Diagnosis Observer-based Output Feedback Fault Tolerant Control Design[J]. *Acta Automatica Sinica*, 2010, 36(2): 274-281.
6. Sedighi T, Koshkouei A J, Burnham K J. Nonlinear unknown input observer design for nonlinear systems: A newmethod[C]//*Preceedings of the 18th IFAC World Congress*. New York, NY, USA: Elsevier, 2011: 11067-11072.

7. Mondal S, Chakraborty G, Bhattacharyya K. LMI approach to robust unknown input observer design for continuous systems with noise and uncertainties [J]. *International Journal of Control Automation and Systems*, 2010, 8(2): 210-219.
8. Hu Z K, Sun Y, Jiang B, et al. An Optimal Unknown Input Observer Based Fault Diagnosis Method[J]. *Acta Automatica Sinica*, 2013, 39(8): 1225-1230.
9. Qian H M, Fu Z D, Ning X L, et al. Fault diagnosis algorithm for GPS/SINS with unknown perturbation distribution matrix[J]. *Chinese Journal of Scientific Instrument*, 2013, 34(1): 208-214.
10. Chen W T, Saif M. Fuzzy nonlinear unknown input observer design with fault diagnosis applications [J]. *Journal of Vibration and Control*, 2010, 16(3): 377-401.

A Method of Determining the Scale of Source Signals

Teng Gong[1], Zhousuo Zhang[1*], Hongqing Huo[2], Jie Zhang[1] and Zhifa Liu[1]

[1]State Key Laboratory for Manufacturing Systems Engineering, School of Mechanical Engineering, Xi'an Jiaotong University, Xi'an 710049, China, zzs@mail.xjtu.edu.cn
[2]Lanzhou Institute of Physics, Lanzhou 730000, China, huohq@sina.com

Abstract. The micro-vibration of the observation points is an important factor affecting the accuracy of observation satellites. It's of great engineering significance to analyze and evaluate the vibration sources. Blind source separation (BSS) is a new technique used to recover source signals from their mixture. However, due to the scale ambiguities of BSS algorithm, the amplitude of the vibration sources can't be obtained directly. In this paper, integrating BSS algorithm with the transmission characteristic of the structure which can be got in advance we propose a method to determine the scale of the vibration source signals. This method mainly includes sources extraction by BSS algorithm and determining the scale of the vibration source signals. We consider only the case that each source signal is a single frequency signal. It can be proved that the convolution mixture can be simplified as the instantaneous mixture model in that case. So we can use the ICA algorithm to deal with the observation signals, which greatly simplifies the process of calculation. Firstly, by ICA algorithm, the sources signals are extracted and the mixing matrix is also obtained. Then, the observation signals while only one source is operating can be further calculated. Finally, by using the transmission characteristic of the structure from the source to the observation points, the scale of the source signal can be determined. At this point, the source signals are completely reconstructed. Besides, the reconstruction coefficient is presented to assess the performance. The effectiveness of the proposed method is verified by experiment.

Keywords: Micro-vibration, blind source separation (BSS), scale ambiguities, transmission characteristic, sources extraction, the reconstruction coefficient

1 Introduction

Observation satellites are widely used in weather forecasting, resource exploration, environmental monitoring and other applications. The micro-vibration from internal sources are regarded as an important factor influencing the resolution of observation satellites. Vibration source estimation and analysis can provide the basis for monitoringthe running state of the vibration sources and reducing the vibration of the satellite systems, eventually improving the precision of observation satellites.

[*] Corresponding author.Tel.:029-82663689

E-mail address:zzs@mail.xjtu.edu.cn

Vibration signals contain a wealth of information about the mechanical running conditions. At present, the desired information is obtained mainly through sensors measurement and signal processing. In terms of sensors measurement, the improvement of the ability to obtain the source signals is mainly achieved through the improvement of the hardware performance and the reduction of the testing distance. However, it is often difficult to install the sensors inside an equipment to obtain the clean vibration information, which causes that the measured vibration signals are the mixture of various vibration source signals and noises [1].

The task of signal processing is to extract the useful information from the measured signals. Signal processing methods have been widely applied to mechanical engineering[2-5]. But for the complex mechanical systems, frequency ranges of different vibration sources often overlap, so the traditional signal processing methods can't effectively achieve the source separation.

The theory of independent component analysis (ICA) has been treated as a new method for the separation and identification of mechanical vibration sources. ICA, which can extract and separate the source signals from the received observation signals without knowing any prior knowledge or only a small amount of prior knowledge, is widely used in the field of biomedical signal processing [6], communication systems [7], image processing [8], and so on. Comon first proposed the term "independent component analysis" and gave a rigorous mathematical definition of it[9]. Hyvarinen and Oja attained a robust ICA algorithms called FastICA by using fixed-point iteration to optimize classic contrast functions [10-12]. Because of its high efficiency and high precision, FastICA has been widely used in engineering field. However, ICA algorithm has the characteristics of scale ambiguities and the signals without amplitude information can't play the role of monitoring the running states of vibration sources, therefore, it is significant to reconstruct the source signals with amplitude information.

In this paper, a method of reconstructing the source signals with amplitude information is proposed. We consider only the case that each source signal is a single frequency signal. In this case the convolution mixture can be simplified as the instantaneous mixture model and the proof is given, demonstrating the feasibility of dealing with the problem by ICA algorithm. Then the theory of FastICA is briefly introduced. After that the method of determining the scale of source signals based on the results of the ICA algorithm is given. In addition, reconstruction coefficient is defined to evaluate the performance of signal reconstruction. Finally, the proposed method is verified by experiment, proving its validity.

2 ICA Algorithm

2.1 Mixing Model

Assume that the observation signals $\mathbf{X}(t)=[x_1(t),\cdots,x_M(t)]^T$ are convolutive mixtures of the source signals $\mathbf{S}(t)=[s_1(t),\cdots,s_N(t)]^T$, which can be expressed as

$$x_j(t)=\sum_{i=1}^{N}\sum_{p=0}^{P-1}h_{ij}(p)s_i(t-p), j=1,2,\cdots,M \tag{1}$$

where $x_j(t)$ is the j th mixed signals measured by the j th sensor; $s_i(t)$ is the i th source signal; $h_{ij}(p)$ denotes the impulse response coefficient.

The mixing model in frequency domain can be expressed as

$$\mathbf{X}(f,\tau)=\mathbf{H}(f)\mathbf{S}(f,\tau) \tag{2}$$

where $\mathbf{S}(f,\tau)=[\mathbf{S}_1(f,\tau),\cdots,\mathbf{S}_N(f,\tau)]^T$ and $\mathbf{X}(f,\tau)=[\mathbf{X}_1(f,\tau),\cdots,\mathbf{X}_M(f,\tau)]^T$ respectively denote the source signals and the observation signals in frequency f and time τ ; $\mathbf{H}(f)$ represents the mixing matrix in frequency f .

In this paper, we consider only the case that each source signal is a single frequency signal, that is, $s_i(t)$ is a single frequency signal whose frequency is f_i, which means that $\mathbf{S}_i(f_j,\tau)=0, i\neq j$. Then $\mathbf{X}(f_i,\tau)=\mathbf{H}(f_i)\mathbf{S}(f_i,\tau)$ can be simplified as

$$\mathbf{X}(f_i,\tau)=\mathbf{H}_i(f_i)\mathbf{S}_i(f_i,\tau), i=1,2,\cdots,N \tag{3}$$

where $\mathbf{H}_i(f_i)=[h_{1i}(f_i),h_{2i}(f_i),\cdots,h_{Mi}(f_i)]^T$ denotes the i th column of $\mathbf{H}(f_i)$.

Owing to that $\mathbf{S}_i(f_i,\tau)$ is equivalent to $s_i(t)$. Formula (3) can be transformed into

$$\mathbf{X}(f_i,\tau)=\mathbf{H}_i(f_i)s_i(t) \tag{4}$$

then,

$$\mathbf{X}(t)=\sum_{i=1}^{N}\mathbf{X}(f_i,\tau)=\mathbf{H}^*\mathbf{S}(t) \tag{5}$$

where

$$\mathbf{H}^*=\begin{bmatrix} h_{11}(f_1) & h_{12}(f_2) & \cdots & h_{1N}(f_N) \\ h_{21}(f_1) & h_{22}(f_2) & \cdots & h_{1N}(f_N) \\ \vdots & \vdots & \ddots & \vdots \\ h_{M1}(f_1) & h_{M2}(f_2) & \ldots & h_{MN}(f_N) \end{bmatrix}$$

Thus it can be seen that the convolution mixture model of source signals can be simplified into instantaneous mixture model in this case. Therefore, the source signals can be effectively separated by ICA algorithm. The source signals can be extracted by

$$\mathbf{Y}(t)=[y_1(t),\cdots,y_N(t)]^T=\mathbf{W}\mathbf{X}(t) \tag{6}$$

where $y_i(t)$ is the estimation of i th source signal; $\mathbf{W}$ denotes the separating matrix.

2.2 FastICA Algorithm

Blind source separation is actually an optimization problem, and the optimization goal is the maximization of non-Gaussianity. Negentropy is widely used as the measurement of non-Gaussianity for its good robustness. In the practical application, it is simplified as

$$\mathbf{J}(y) \propto \left[E\{G(y)\} - E\{G(v)\}\right]^2 \quad (7)$$

where v is the normalized variable; y has been normalized; and function F can be chosen by experience.

By means of fixed-point iteration, the following iterative formula is obtained:

$$\mathbf{w} = E\{\mathbf{z}g(\mathbf{w}^T\mathbf{z})\} - E\{g'(\mathbf{w}^T\mathbf{z})\}\mathbf{w} \quad (8)$$

where $\mathbf{z}$ is the whitened mixed signals; and function g is the derivative of function G in (7).

By the FastICA algorithm, source signals and the mixing matrix can be obtained.

3 Determining the Scale of Source Signals

3.1 Method of Determining the Scale of Source Signals

Because of the scale ambiguities of BSS algorithm, the components extracted by the FastICA algorithm don't have the amplitude information.If the complete source signals are obtained, we can monitor the running state of the vibration sources through them, and can also analyze and evaluate the source signals more intuitively. We give out the method of determining the scale of source signals based on the results of the FastICA algorithm.The true source signals S and the mixed matrix A can be expressed as

$$\mathbf{S} = [k_1 \cdot y_1, k_2 \cdot y_2, \cdots, k_n \cdot y_n]^T \quad (9)$$

$$\mathbf{A} = \begin{bmatrix} \tilde{a}_{11}/k_1 & \tilde{a}_{12}/k_2 & \dots & \tilde{a}_{1N}/k_N \\ \tilde{a}_{21}/k_1 & \tilde{a}_{22}/k_2 & \dots & \tilde{a}_{2N}/k_N \\ \vdots & \vdots & \ddots & \vdots \\ \tilde{a}_{M1}/k_1 & \tilde{a}_{M2}/k_2 & \dots & \tilde{a}_{MN}/k_N \end{bmatrix} \quad (10)$$

where y_i denotes i th separated source signal; $\tilde{a}_{ji}$ is the element of mixing $\tilde{\mathbf{A}}$ obtained by ICA algorithm; k_i is the unknown scale factor; M and N respectively represent the number of observations and the number of sources.

The observation signals while only 1st source is operating can be calculated by

$$\tilde{\mathbf{X}}_1 = \mathbf{A} \cdot \begin{bmatrix} k_1 \cdot y_1 \\ \mathbf{O} \end{bmatrix} \quad (11)$$

where $\mathbf{O}$ is a zero matrix with $N-1$ rows.

Although k_i cannot be obtained, it is eliminated during the operation. And then $\tilde{\mathbf{X}}_1$ can be worked out.

$$\tilde{\mathbf{X}}_1 = \left[\tilde{a}_{11} \cdot y_1, \tilde{a}_{21} \cdot y_1, \cdots \tilde{a}_{m1} \cdot y_1\right]^T \quad (12)$$

Next, we need to use the transmission characteristics of the structure. By harmonic analysis of the model we get the amplitude frequency curves, which can be used as the prior knowledge. From the curves we can get the amplification factors of each observation point to 1st vibration source, which are respectively represented by β_{11}, $\beta_{21}, \cdots, \beta_{m1}$. The observation signals divided by the amplification factors respectively, i.e. $\tilde{a}_{11} \cdot y_1 / \beta_{11}$, $\tilde{a}_{21} \cdot y_1 / \beta_{21}$, $\cdots$, $\tilde{a}_{m1} \cdot y_1 / \beta_{m1}$, all can be regarded as the estimations on the 1st source signal. Then the reconstructed signal of 1st source whose amplitude is certain can be obtained by taking the average.

The reconstructed signals of other sources can be calculated by the same method.

3.2 Evaluation Index of the Reconstruction

Waveform similarity of two signals is usually evaluated by means of cross correlation analysis. However, there are phase differences between the source signals and the observation signals in the mechanical system, and the correlation analysis also can't reflect the accuracy of the determined scale, leading to the limitation of the traditional cross correlation analysis method. Besides, using the amplitude to evaluate the results is not comprehensive. Therefore, it is necessary to construct an indicator that can reflect both the similarity of the waveforms and the accuracy of amplitude.

Based on the normalized cross-correlation function, the reconstruction coefficient ρ_S is defined to evaluate the reconstruction performance.

$$\rho_S = \frac{\max R_{sy}(k)}{\max\{\|\mathbf{s}\|^2, \|\mathbf{y}\|^2\}} \qquad 0 \le k \le N \quad (13)$$

where N is the sampling length; $R_{sy}(k) = \sum_{i=0}^{N-k-1} s(i+k)y(i)$, $\|\mathbf{s}\|^2 = \sum_{i=0}^{N-1} s(i)^2$, $\|\mathbf{y}\|^2 = \sum_{i=0}^{N-1} y(i)^2$. The introduction of Variable k eliminates the effect of phase difference, and the change of denominator introduces the effect of amplitude.

The range of the indicator is [0,1]. And the closer to 1, the more accurate the result is.

4 Experimental Verification

The shell structure is one of the most important structures in the satellite. In order to verify the proposed method, a shell structure test system mainly consisting of the experimental shell, the support, 2 vibration sources and the signal measurement system is set up. Fig. 1 displays the photos of the test system.

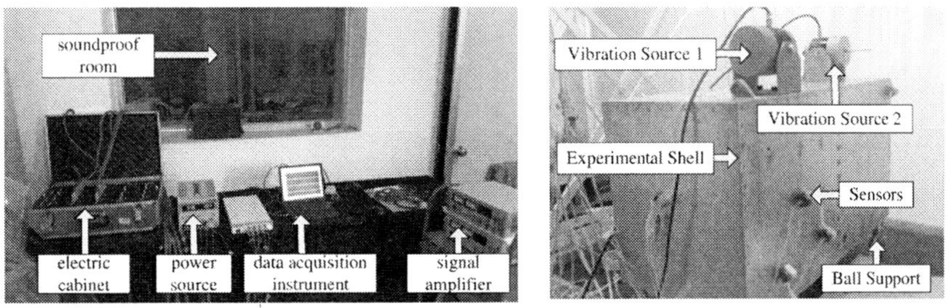

(a) (b)

Fig. 1. Photos of the shell structure test system. (a) signal measurement system. (b) experimental shell and vibration sources and so on.

Two kinds of experiments have been performed, including experiment in single source case and that in two sources case. The former is used to obtain the transfer characteristics and the true source signals. The latter is used to obtain the experimental data to verify the feasibility of the method. The proposed method is used to process the experimental data, and the reconstructed signals is compared with the true source signals measured under the experiment in single source case to verify the effectiveness of the proposed method.

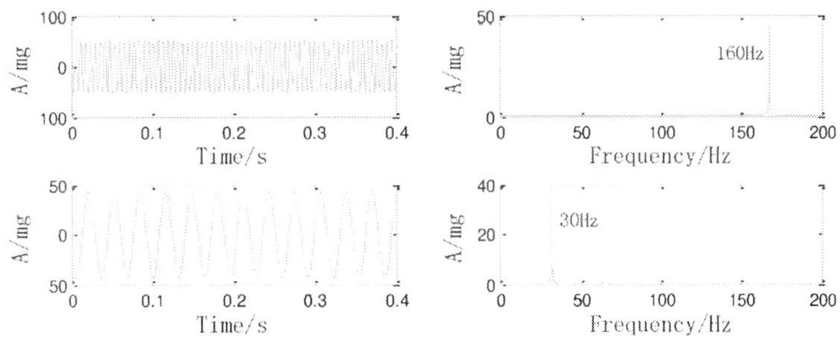

(a) (b)

Fig. 2. The source signals. (a) Waveforms. (b) Fourier spectrums.

The true source signals are displayed in Fig. 2. Two observation signals are selected, which are displayed in Fig.3.By the proposed method, the mixed signals are processed to obtain the reconstructed signals. The reconstructed signals are displayed in

Fig.4. It is clear that the reconstructed signals are sine waves with some interference. The frequencies of these signals are 160 and 30 Hz, respectively, which are in accordance with the vibration source signals displayed in Fig. 2. The reconstruction results are shown in Table 1. It is found that, the amplitude errorsare5.81% and 1.51% respectively. The reconstruction coefficients are 0.948 and 0.971 respectively, validating the effectiveness of the method. And the two evaluation results are consistent, which proves the feasibility of the indicator.

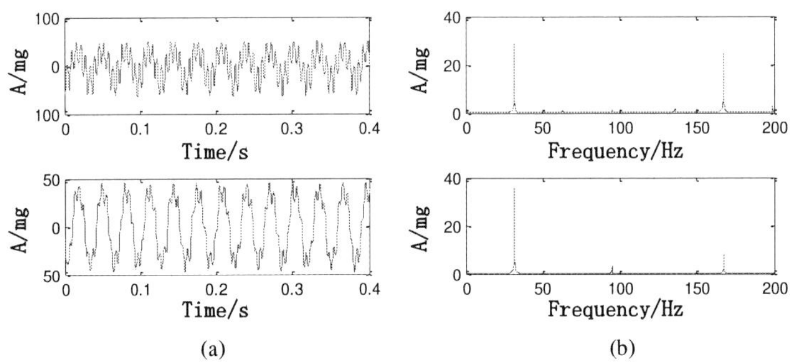

Fig. 3. The observation signals. (a) Waveforms. (b) Fourier spectrums.

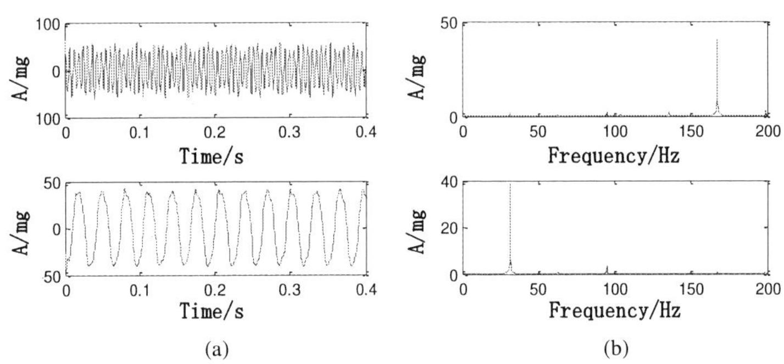

Fig. 4. The reconstructed signals. (a) Waveforms. (b) Fourier spectrums.

Table 1. Reconstruction results evaluation

Source signal	Measured amplitude (mg)	Calculate amplitude (mg)	Relative error	Reconstruction coefficient
Source signal 1	45.35	42.72	5.81%	0.948
Source signal 2	39.47	38.87	1.51%	0.971

5 Conclusion

In order to analyze the vibration sources and monitor the running state of each micro-vibration source in observation satellites, a method for determining the scale of source signals is proposed in this paper. Combining the ICA algorithm and amplitude-frequency response curves from the sources to the observation points, the vibration waveform and the scale of the vibration sources can both be obtained. The proposed method is applied to reconstructing the vibration source signals from the observation data of a shell structure test system, demonstrating the effectiveness of this method. In addition, this method can be further extended to the reconstruction of complex source signals, which will be carried out in future work.

References

1. Forrester, B.D.: Advanced Vibration Analysis Techniques for Fault Detection and Diagnosis in Geared Transmission Systems. Swinburne University of Technology, Melbourne(1996)
2. Wang, X., Liu, C., Bi, F.: Fault Diagnosis of Diesel Engine Based on Adaptive Wavelet Packets and EEMD-fractal Dimension. Mechanical Systems and Signal Processing. 41, 581--597 (2013)
3. Li, R., Yu, D., Chen, X., Liu, J.: A Compound Fault Diagnosis Method for Gearboxs Based on Chirplet Path Pursuit and EEMD. Journal of Vibration & Shock. 33,51--6 (2014)
4. Wang, Y.X., Markert, R., Xiang, J.W.: Research on Variational Mode Decomposition and Its Application in Detecting Rub-impact Fault of the Rotor System. Mechanical Systems and Signal Processing. 60, 243--251 (2015)
5. Miesowicz, K., Staszewski, W.J., Korbiel, T.: Analysis of Barkhausen Noise Using Wavelet-Based Fractal Signal Processing for Fatigue Crack Detection. International Journal of Fatigue. 83, 109--116 (2016)
6. Foresta, F., Mammone, N., Morabito, F.: PCA-ICA for Automatic Identification of Critical Events in Continuous Coma-EEG Monitoring. Biomedical Signal Processing and Control. 4, 229--235 (2009)
7. Gao, J., Zhu, X., Nandi, A.: Independent Component Analysis for Multiple-Input Multiple-Output Wireless Communication Systems. Signal Processing. 91, 607--623 (2011)
8. Drew, M., Bergner, S.: Spatio-Chromatic Decorrelation for Color Image Compression. Signal Processing: Image Communication. 23, 599--609 (2008)
9. Comon, P.: Independent Component Analysis, A new concept. Signal Process. 36, 287--314 (1994)
10. Hyvarinen, A., Oja, E.: A Fast Fixed-Point Algorithm for Independent Component Analysis. Neural Comput. 9, 1483--1492 (1997)
11. Hyvarinen, A.: Fast and Robust Fixed-Point Algorithms for Independent Component Analysis. IEEE Trans. Neural Netw. 10, 626--634 (1999)
12. Hyvarinen, A., Karhunen, J., Oja, E.: Independent Component Analysis. John Wiley and Sons, New York (2001)

A Non-linear STFT with Application to Estimation of Instantaneous Frequency

Gang Yu[1] Yiqi Zhou[2] and Mingxiao Dong[1]

[1] College of Mechanical and Electronic Engineering, Shandong Jianzhu University, Jinan 250014, PR China.

[2] Key Laboratory of High-efficiency and Clean Mechanical Manufacture, Shandong University, Jinan 250061, PR China.

yugang2010@163.com

Abstract. Time-frequency (TF) analysis (TFA) method is a common tool to characterize the non-linear behaviors in the rotating machines. However, the conventional TFA method is restricted to provide a precise result, because of their inherent shortcomings, such as poor energy concentration, low TF resolution and cross-terms .etc. In order to obtain TF representation with more satisfactory energy concentration and higher TF resolution, we propose a non-linear short time Fourier-transform method. Finally, we take the advantage of synchrosqueezing transform to enhance the TF resolution further more. The numerical signal and experimental signal are utilized to validate the effectiveness of the proposed method.

Keywords. Rotating machines, time-frequency analysis, synchrosqueezing transform

1 Introduction

Rotating machines play an important role in modern industrial applications [1-2]. To prevent economic loss and catastrophic failure, the development of fault diagnosis technology is essential. Under the influence of fault, the rotating machines will show many non-linear behaviors which can be recorded by vibration signals. The classical

Fourier transform (FT) is the commonly signal-processing tool to obtain the underlying information about the machines.

By assuming the considered signal is stationary, FT has the ability of revealing the overall frequency content. However, the assumption of stationary cannot always hold in practice, such as the vibration response of rotary machinery under non-stationary events. The concept of instantaneous frequency (IF) is more suitable for describing these non-stationary features than overall frequency. The short-time Fourier transform (STFT) provides a powerful solution to effectively characterize the IF features of non-stationary signals. By assuming the signal is piece-wisely stationary in a short time window, the energy distribution of time-frequency (TF) representation (TFR) generated by STFT is concentrating at and around the IF of analyzed signal in the TF domain. However, according to the Heisenberg uncertainty principle, the STFT cannot achieve an arbitrarily high TF resolution simultaneously. Moreover, when strongly demodulated frequency signal encountered, the IF of signal in the analyzed window is time-varying, which will lead to poor TF resolution. Herein, given a numerical non-stationary signal to illustrate the above analysis, the signal model is as

$$s(t)=\begin{cases}\sin(2\pi\cdot(25\cdot t+10\cdot\sin(t))), & 0<t\le 6\\ \sin(2\pi\cdot 34.2\cdot \mathrm{t}) & , 6<t\le 14\end{cases} \tag{1}$$

The IF of the signal is shown in Fig. 1(a), which can be seen that in time 0s-6s it has a non-linear IF and in time 6s-14s the IF is constant. The TFR generated by STFT is shown in Fig. 1(b). It can be seen that, STFT can provide a satisfactory energy concentration for the part of constant frequency, but the energy of modulated frequency part smear heavily, which leads to a low TF resolution. In order to obtain TFR with high TF resolution, we propose an effective TFA method, which can characterize the time-varying IF feature more precisely.

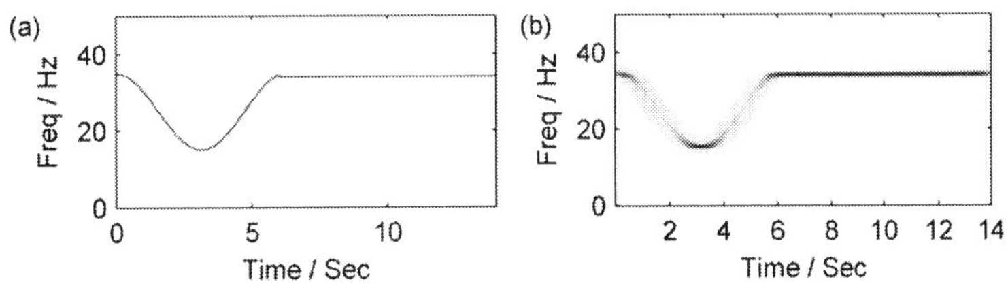

Fig. 1. (a) IF and (b) TFR by STFT.

2 Non-linear STFT

The well-known definition of IF was proposed by Ville and it describes how fast a signal oscillates locally at a time instant [3]. According to the theory of Ville, the IF of a signal can be calculated from the derivative of the phase of its analytic signal. Given an analytic signal of mono-component, suppose its IF is $\varphi(t)$, then this analytic signal can be written as

$$s(t) = A \cdot e^{i\int \varphi(t)dt} \tag{2}$$

In a short time u, the IF of signal $s(t)$ can be regarded as linear equation approximately, which can be expanded by Taylor expansion,

$$\varphi(t)|_{t\in u} = \varphi(u) = \varphi(t') + \varphi'(t')(u-t') \tag{3}$$

where the remainder item is ignored, $\varphi(t')$ is the IF at time t', $\varphi'(t')$ is the first order derivative of IF at time t'. Given a window $w(u-t')$ to truncate this signal, then FT amplitude of $w(u-t')\bullet s(u)$ in the IF $\varphi(t')$ can be calculated as

$$\begin{aligned}
&|S(t',\varphi(t'))| \\
&= \left|\int_{-\infty}^{+\infty} w(u-t')\cdot s(u)\cdot e^{-i\varphi(t')u}du\right| \\
&= \left|\int_{-\infty}^{+\infty} w(u-t')\cdot A\cdot e^{i\int\varphi(u)du}\cdot e^{-i\varphi(t')u}du\right| \\
&= \left|\int_{-\infty}^{+\infty} w(u-t')\cdot A\cdot e^{i\varphi(t')u+i\varphi'(t')(u-t')^2/2}\cdot e^{-i\varphi(t')u}du\right| \\
&= \left|\int_{-\infty}^{+\infty} e^{i\varphi'(t')(u-t')^2/2}\cdot w(u-t')\cdot A\cdot e^{i\varphi(t')u}\cdot e^{-i\varphi(t')u}du\right| \\
&\le \left|\int_{-\infty}^{+\infty} w(u-t')\cdot A\cdot e^{i\varphi(t')u}\cdot e^{-i\varphi(t')u}du\right| \\
&= A\cdot\hat{w}(0)
\end{aligned} \tag{4}$$

where $\hat{w}(0)$ is the FT amplitude of window function in zero frequency bin. The Eq. (4) denotes that, due to the presence of modulated element $e^{i\varphi'(t')(u-t')^2/2}$, the FT amplitude of $w(u-t')\bullet s(u)$ in its IF is lower than the harmonic component. Actually, the calculation by Eq. (4) is the STFT amplitude of analytic signal in its IF. And this is the reason why the energy of TFR of modulated part will sprout out around the IF in a large area, but the energy of TFR of harmonic part can concentrate on the IF with higher amplitude. Given the standard STFT formulation as

$$S(t,\omega)=\int_{-\infty}^{+\infty} w(u-t)\cdot s(u)\cdot e^{-i\omega u}du \tag{5}$$

In order to eliminate the influence of modulated element, it is necessary to introduce the demodulated operator. Consider that the linear modulated ratio $\phi'(t')$ is the function of time, i.e. time-variant, the demodulated operator should be time-variant as well. Then the STFT of signal *s*(*t*) considering the time-variant demodulated operator, can be written as

$$NS(t,\omega)=\int_{-\infty}^{+\infty} w(u-t)\cdot s(u)\cdot e^{-i\omega u}\cdot e^{-ic(t)(u-t)^2/2}du \tag{6}$$

Due to the introduction of time-variant demodulated operator, the Eq. (6) is called as non-linear STFT (NLSTFT). To enhance the concentrated energy further more, the post-processing tool synchrosqueezing transform (SST) is introduced [4], which is based on NLSTFT. The proposed method is to squeeze the TF energy into the IF region, which can be written as

$$Ts(t,\omega')=\int_{-\infty}^{+\infty} NS(t,\omega)\cdot\delta(\omega'-\varphi(t,\omega))d\omega \tag{7}$$

where the parameter $\phi(t,\omega)$ can be estimated by

$$\varphi(t,\omega)=-i\cdot\frac{\partial_t NS(t,\omega)}{NS(t,\omega)} \tag{8}$$

The TFR generated by NLSTFT and SST are listed in Fig. 2, which show more energy-concentrated than STFT. And the IF can be characterized by SST precisely.

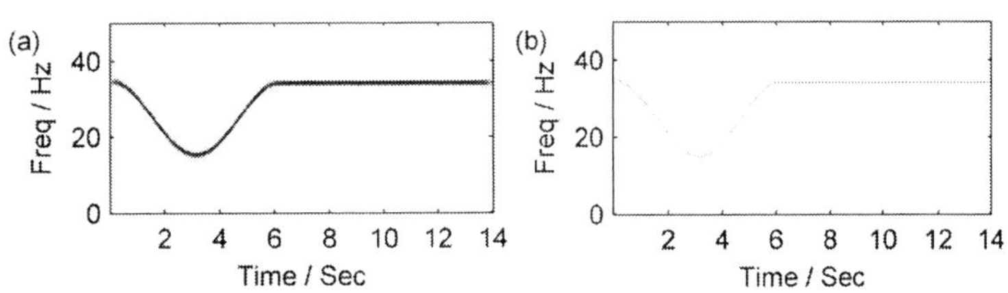

Fig. 2. TFR generated by (a) NLSTFT and (b) SST.

3 Experimental validation

Herein, we focus on an abnormal vibration of a heavy oil catalytic machine set [5]. The rotating speed of the gas turbo is about 5381 rpm (about 90 Hz). In running condition, the vibration in bearing is larger than the alarm limitation. So we take the recorded signal to analyze. The vibration signal and its spectrum are shown in Fig. 3. It can be seen that, the vibration signal consists of first order (1X) component of rotating frequency and its higher order components. And the 1X component has the largest energy, which should be the main reason of causing abnormal vibration.

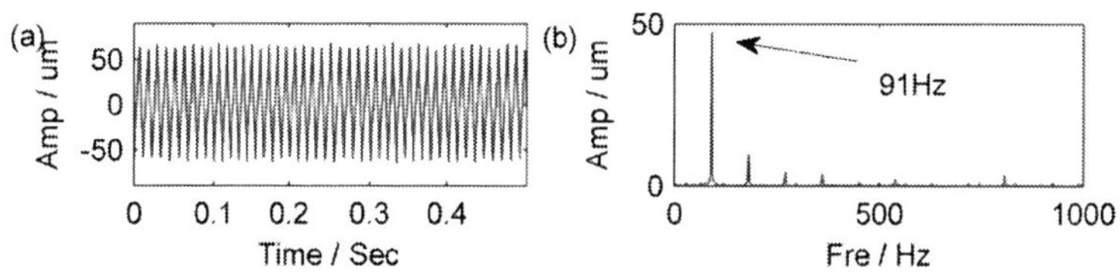

Fig. 3. Vibration signal

To characterize the fault features more precisely, we utilize the proposed method to analyze the vibration signal and list the TFR of 1X component, as shown in Fig. 4(a). It can be found that the 1X component has a clearly periodic oscillatory IF. By detecting the peak data of the TFR, we redraw this oscillatory IF in Fig. 4(b), and calculate its spectrum (see Fig. 5), which shows the same frequency with rotating frequency. It should be noted, the IF of 1X component is corresponding to the instantaneous speed of gas turbine shaft. And it means the shaft running in an unstable condition. If a rub-impact fault exists between rotor and the static element, then during each rotation, it will cause a local speed-down and speed-up effect on shaft. And the oscillatory frequency of instantaneous speed should be equal to the rotating frequency. Therefore, we can conclude that a rub-impact fault does exist.

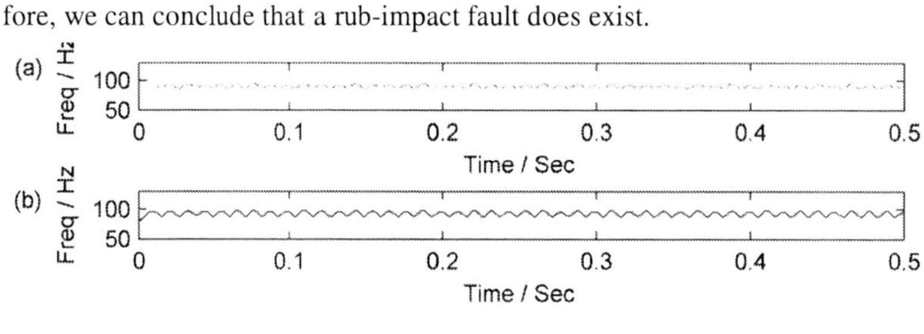

Fig. 4. (a) The TFR and (b) the detected IF

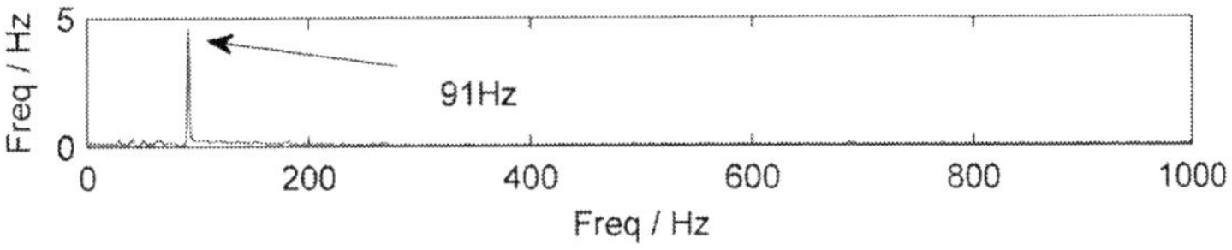

Fig. 5. The FT of IF

4 Conclusion

A NLSTFT and its SST are proposed. By introducing a demodulated operator whose demodulated ratio is equal to the first derivative of IF, the NLSTFT and SST have the ability of generating superior TFR over conventional TFA method. A numerical signal and an experimental signal are utilized to illustrate the effectiveness of the proposed method.

Acknowledgements

This work was financially supported by National Key Technology Research and Development Program of china (Grant No.2015BAF07B04), the National Natural Science Foundation of China (Grant No. 51475277) and Foreign Expert Project in Shandong Province.

References

1. K.C. Gryllias, I.A. Antoniadis: Estimation of the instantaneous rotation speed using complex shifted Morlet wavelets, Mech. Syst. Signal Process. 38, 78 95 (2013)
2. G. Yu, Y. Zhou: General linear chirplet transform, Mech. Syst. Signal Process. 70-71 958-973 (2016)
3. J. Ville: Theorie et application de la notion de signal analytic, Cables Transm. 93 61-74 (1948)
4. I. Daubechies, J. Lu, H.T. Wu: Synchrosqueezed wavelet transforms: an empirical mode decomposition-like tool, Appl. Comput. Harmonic Anal. 30 243-261 (2011)
5. S. Wang, X. Chen, G. Li, et al: Matching demodulation transform with application to feature extraction of rotor rub-impact fault, Instrum. Meas. IEEE Trans. 63 1372-1383 (2014)

GENERATOR SET FAULT DIAGNOSIS USING COMPLETE INTRINSIC MODE DECOMPOSITION*

Duan Chendong[1], Yuan Ye[2], Song Suchen[3]

School of Electronic & Control Engineering, Chang'an University, Xi'an, China
1 cduan@chd.edu.cn; 2 yy_chd@foxmail.com; 3 ssc_chd@foxmail.com

Abstract. In order to extract some effective characteristic information from monitoring signals, a time-frequency approach is proposed by using the complete ensemble empirical mode decomposition (CEEMD). On each decomposition scale, correlation coefficient between the residues and the original signal is applied as a terminal condition. If the coefficient is less than the preset threshold, the decomposition algorithm is set to stop. Finally, a series of effective mode components are obtained. By making further processing, features on time and frequency domain that contain in these components are extracted and separated. The results of numerical experiment and engineering application indicate that the proposed method is represented the original signal much more efficiently. Because some insignificant components are excluded, interference on diagnosis and analysis is decreased greatly. It effectively reveals the looseness fault features of a bearing-bush on a generator set in a power plant.

Keywords. complete decomposition, intrinsic mode component, time-frequency analysis, feature extraction

1 Introduction

The Empirical mode decomposition (EMD) is thought as an adaptive time-frequency signal processing approach which decomposes a signal into a series of subcomponents called intrinsic mode functions (IMF)[1], and in the past decade it was used as a helpful method in fault diagnosis[2] But the problem of mode aliasing is a fatal disadvantage that makes some IMFs difficult to explain. To eliminate the mode aliasing phenomenon, Wu developed an improved approach called ensemble empirical mode decomposition (EEMD)[3]. The EEMD does the EMD over an ensemble of a group of composite signals which are built by adding white noises to the

* Supported by the Natural Science Foundation of Shaanxi (No. 2014JM7246)

analyzed signal. The EEMD indeed overcomes the mode aliasing between two adjacent IMFs. However new bugs are introduced. The reconstructed signal of the EEMD includes residual noise, and different realizations of composite signals may produce different number of modes. In order to overcome these problems, Torres developed a new version EEMD called complete EEMD (CEEMD)[4]. Different from the EEMD, on every decomposition scale all the decomposed components must be constructed again. On each scale the IMFs of white noises added to the analyzed signal are used to built the decomposed components. The CEEMD provides an perfect reconstruction and gives a better decomposition for the signal. In this paper, In order to extract some effective information from signals, a modified approach based on the CEEMD is proposed which applies a correlation coefficient index as a terminal condition. Experiment and engineering application indicate that the proposed method can represent the original signal much more efficiently. Because some insignificant IMFs are excluded, interference on diagnosis and analysis is decreased greatly. It is effectively revealed features of bearing-bush looseness fault in a generator set.

The paper is organized as follows. In Section.2 the CEEMD algorithm is introduced. In Section 3 numerical experiment is studied and new terminal condition index for the CEEMD by using correlation coefficient as terminal index is stated. An engineering case is described in Section 5. Finally, Section 6 is the conclusions.

2 Complete ensemble empirical mode decomposition

The CEEMD algorithm is stated as following [4]

(1) Add white noise to the original signal $x(t)$ to build a group of composite signals for the 1st decomposition scale:

$$x_i(t) = x(t) + \beta_0 n_i(t)\,,\quad i = 1, 2, \ldots, N \tag{1}$$

here $\beta_0 \geq 0$, $n_i(t)$ is white noise with $N(0,1)$.

(2) Decompose the signals $\{x_i(t), i = 1, 2, \ldots, N\}$ with the EEMD, and then calculate the component $\tilde{c}_1(t)$ of the 1st scale as

$$\tilde{c}_1(t) = \frac{1}{N}\sum_{i=1}^{N} c_i(t) \tag{2}$$

where $c_i(t)$ is the 1st IMF of the signal $x_i(t)$ which is obtained by the EMD.

(3) Compute the residual of the 1st scale

$$r_1(t) = x(t) - \tilde{c}_1(t) \tag{3}$$

(4) Do the EMD to a set of white noises { $w_i(t)$, $i = 1, 2, \ldots, N$ }, $w_i(t)$ with $N(0,1)$, the length of $w_i(t)$ is the same as the original signal $x(t)$, we obtain

$$\{\varepsilon_{i,j}(t), i = 1, \ldots, N, j = 1, \ldots, M\}$$

$\varepsilon_{i,j}(t)$ denoted as a IMF of $w_i(t)$ on jth decomposition level.

(5) For a decomposition scale $k = 2,...,M$, the residual of kth scale is

$$r_k(t) = r_{k-1}(t) - \tilde{c}_k(t) \tag{4}$$

Then the composition signal on kth scale is defined as

$$x_i^k(t) = r_k(t) + \beta_k \varepsilon_{i,k}(t) \tag{5}$$

where $\beta_k \geq 0$. Thereafter, make the EMD only one level to $x_i^k(t)$, and gain its IMF denoted as $c_i^k(t)$. Then we get component of the kth scale

$$\tilde{c}_k(t) = \frac{1}{N}\sum_{i=1}^{N} c_i^k(t) \tag{6}$$

The CEEMD process is ended when extreme point number of $r_k(t)$ is less than 2, or $r_k(t)$ is a monotonic function. Obviously, the above process satisfies

$$x(t) = \sum_{j=1}^{k} \tilde{c}_j(t) + r_k(t) \tag{7}$$

Therefore, the CEEMD is perfect reconstruction mathematically.

3 Numerical experiment

Experimental signal is $s(t)$ designed as follows which has 3 components

$$s(t) = s_1(t) + s_2(t) + \varepsilon n(t) \tag{8}$$

here $s_1(t) = 4\sin(2\pi f_1 t)e^{-200t_1}$, $s_2(t) = 2\sin(2\pi f_2 t)$ and white noise $n(t)$ with $N(1,0)$. $t_1 = \text{mod}(t, 1/f_1)$, ε is set to 0.2, f_1 =300Hz and f_2 =50Hz. The signal is shown in Fig.1 with the length of 1024, the sample frequency is 2000 Hz.

Fig.2 displays the results obtained by the CEEMD and the EEMD respectively. Obviously, Fig. 2a gives much better filtering property than that of Fig. 2b. Table 1 lists correlation coefficient (CR) values of decomposition components with respect to $s(t)$, $s_1(t)$ and $s_2(t)$ separately. In Table 1, the CR value between c5 and $s_2(t)$ is close to 1 (CR = 0.99), and the average absolute error between them is 0.19, we have reason to believe that c5 may represent $s_2(t)$ perfectly. The components c1, c2and c3 have higher correlation with $s_1(t)$, they are the main elements of $s_1(t)$. Computing the CR between $s_1(t)$ and the composition of c1, c2, c3 and c4, the value is 0.96, and the average absolute error between these 2 signals is 0.24, so we sure that the composition signal is a recover copy of $s_2(t)$. Moreover, as we seen in Table 1, the others c7~c9 and r9 are not the significant components of $s(t)$ because their CR values to $s(t)$

are too small (CR<< 0.3). There is no need for further processing. As the reason the CEEMD should be ended at the Scale 5th.

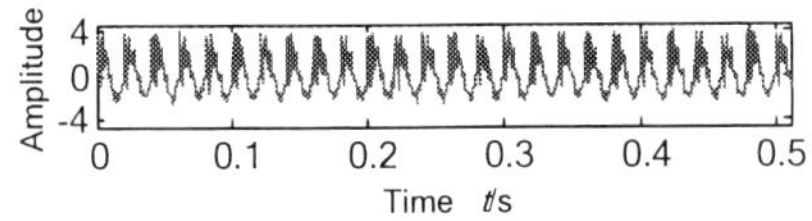

Fig. 1. Experimental signal

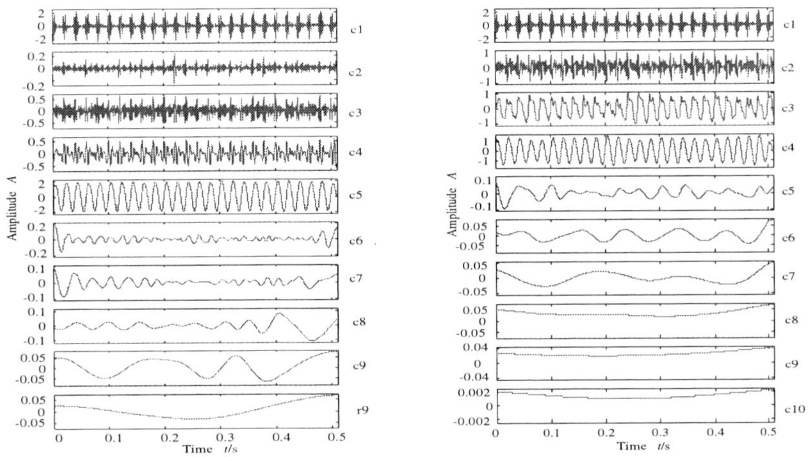

(a) Decomposed by the CEEMD (b) Decomposed by the EEMD

Fig. 2. Components obtained by the CEEMD and the EEMD respectively

As we investigated above, a modified CEEMD with a new termination condition is proposed. On each decomposition scale, the CR value between the residue and the original signal is calculated. If the value is less than ε (ε>0), then the CEEMD is over. Otherwise, continue decomposition until $\mathrm{CR} < \varepsilon$.

Table 1. CR of CEEMD components and $s(t)$, $s_1(t)$, $s_2(t)$

Component	c1	c2	c3	c4	c5	c6	c7	c8	c9	r9
$s(t)$	0.51	0.41	0.44	0.26	0.82	0.52	0.25	0.01	0.03	0.02
$s_1(t)$	**0.87**	**0.68**	**0.74**	0.27	0.04	0.033	0.010	0.0001	0.0028	0.001
$s_2(t)$	0.01	0.001	0.0004	0.131	**0.99**	0.62	0.29	0.01	0.02	0.02

4 Engineering case

As a 50MW generator unit booting up and loading after a heavy maintenance, it wa s found that vibration on Bush No.4 was abnormal. When the operation speed reached 740 r/min, the peak-to-peak value (PPV) of vibration was 24.7 μm, up to 3000 r/min, the PPV became 63.2μm. At the speed of 3000 r/min, the loading process began to ap ply. When load was applied up to 6MW, the PPV was about 74μm, tuned to 16MW, the PPV was 104 μm and when added to 20MW, the PPV rose to 132μm. Along with load increased, the PPV was enlarged much more. It did not suit to loading. Then the unit unloaded. Unloaded down to 6MW, the PPV was fall to 75μm which was still greater than the alarm value 50 μm. But in the process of loading, the PPVs of all the other bearing bushes were less than the alarm value.

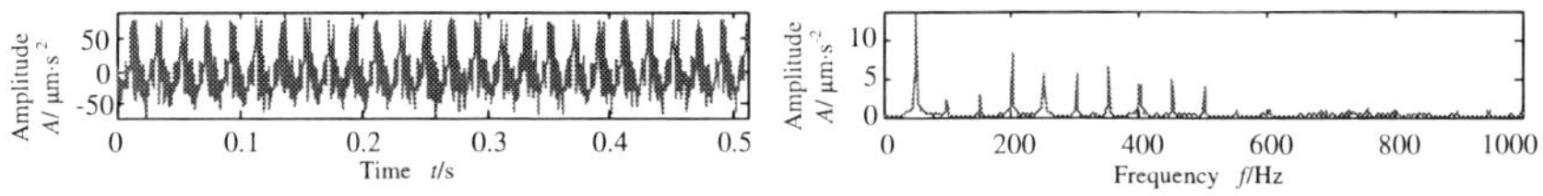

(a) vibration signal (b) spectrum of the vibration signal

Fig. 3. Vibration signal and its spectrum

An acceleration signal from Bush No. 4 and its spectrum is shown in Fig. 3 with the sample frequency 2000 Hz, the signal length is 1024. As we observed, in Fig. 3b the peak value of the operation frequency (50Hz) is the highest one, and 2~12 harmonics of the operation frequency also exist in the spectrum.

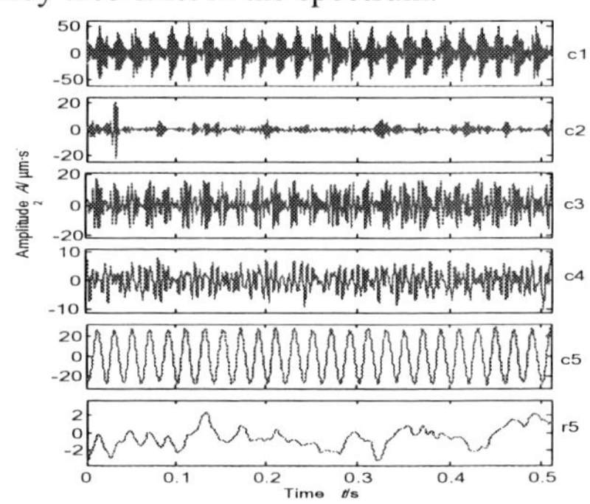

Fig.4. CEEMD components of the signal

Processing the signal with the modified CEEMD, we obtain components c1~c5 and a residue r5 as seen in Fig.4. It is clearly that c1 reveals features of impact response, the impact interval time is about 20 ms, and c3 emerges the amplitude modulation signatures. There is no doubt that c5 represents the operation frequency component.

Time-frequency analysis is performed for c1, c3 and c5 respectively, and the maps are displayed in Fig.5. In Fig.5a, the main energy is concentrated over the band 200-600 Hz, it is demonstrated that c1 includes those main subcomponents with the center frequencies as 2-10 harmonics of the operation frequency. These subcomponent's frequencies fluctuate within a certain range, the signatures show that all these frequencies are time variant. From Fig.4 and Fig.6a it is sure that c1 is an impulse response component with unilateral attenuation and time variant in amplitude and frequency.

In Fig.5b, the main energy of c5 is spread over the band 100-500Hz. The map reveals that the subcomponents with the center frequencies as 2-8 harmonics of the operation frequency, their frequencies also fluctuate yet. In Fig.5c the operation frequency (5

0Hz) component is dominant, its color change over time, this means that the envelop of c5 is time variant even though we can hardly find any the samc signatures in Fig. 4.

As investigating above, we could get the information as follows: (i) Vibration beca me stronger when the load was increased; the same situation was also existence as the unit operation speed is raised. (ii) In the spectrum, the peak of the operation frequenc y is dominant, and the frequency time variant. (iii) Some high harmonics components, 2-12 harmonics, are found both in spectrum and time-frequency map. (iv) Amplitude and frequency of the signal are time variant. (v) Powerful impact response characterist ics are discovered in the signal.

According to the above analysis, it is inferred that excessive vibration on Bush No. 4 was caused by the bush looseness. Stopping the unit and making inspection for the b ush, it is found that the pre-tightening force of the bush and the reserved clearance bet ween the bush and its shim did not meet the installation process requirement.

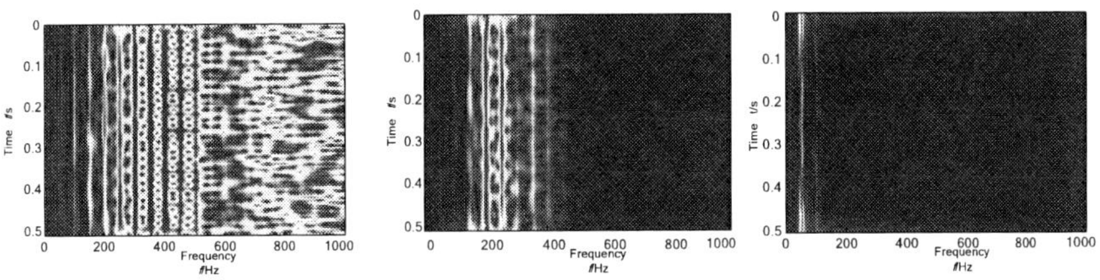

(a) The component c1 (2) The component c3 (3) The component c5

Fig. 5. The time-frequency map of the components

5 Conclusion

The CEEMD is an improved version of the EEMD. It has much better properties than the EEMD on filtering, component separation, perfectly reconstruction and so on. Applying a correlation coefficient of CEEMD residue and the original signal as a terminal condition, it may greatly decrease the decomposition timc and cxclude the insignificant components of the signal effectively, it represents the original signal much more efficient. In some extent, interferences in feature extraction and analysis are decreased, so this is helpful to fault diagnosis.

6 Reference

1. N E Huang, Z Wu, An Adaptive Data Analysis Method for nonlinear and Nonstationary Time Series: The Empirical Mode Decomposition and Hilbert Spectral Analysis, Wavelet Analysis and Applications, 2007,363-376
2. Y Lei, J Lin, Z He, et al. A review on empirical mode decomposition in fault diagnosis of rotating machinery, Mechanical Systems and Signal Processing, 2013, 35(1–2):108–126.
3. Z Wu, N E Huang, Ensemble empirical mode decomposition: a noise-assisted data analysis method. Advances in Adaptive Data Analysis, 2009, 1(1): 1–41
4. M E Torres, M A Colominas, G Schlotthauer, et al. A complete ensemble empirical mode decomposition with adaptive noise, Preceding of 2011 IEEE International Conference on Acoustics, Speech, and Signal Processing, Prague, 2011, 4144 – 4147

Collision Detection in Virtual Operation System of Cranes

Mingxiao Dong, Songjun Han, Mengmeng Lu

School of Mechanical and Electrical Engineering, Shandong Jianzhu University, Jinan, 250101, China. mxdong@sdjzu.edu.cn

Abstract. Collision detection is the key technique of virtual operation system for cranes. First, based on the LOS intersect vector method, we designed the sling as the collision target of the container in Isector module on Lynx interface and the Isector class vector masks of the container are the same as ones of the sling. Then, a BUMP collision detection instance of the sling is created by calling vgIsector functions in C++ Software and the BUMP detection method of the sling is bound with the role object of the sling. Finally, the sling can move only if the sling does not collide with other objects by programming. Therefore, the virtual scene around the sling is detected in real time to avoid the sling traversing other objects. The authenticity of the virtual scene is improved and the user's sense of immersion is enhanced in the virtual operation system for cranes.

Keywords. Crane; Virtual Operation; Collision Detection; Intersect Vector

1 Introduction

The main purpose of the virtual operation system for cranes is simulating the operation process of downloading and uploading, providing a realistic visual scene during crane drivers take part in examination [1]. However, when the port crane uploads and downloads the containers, it is inevitable that the collision occurs between the spreader and the crane beams, the spreader and container, and the spreader and other objects around the spreader. If the wrong traversing phenomena occur, it will reduce the verisimilitude of the virtual scene and trainees' immersive sense [2]. Therefore, according to the requirement of collision detection between the spreader and the containers in the virtual operation system of the port crane, the intersection vector technique in the Vega software is used to set the collision detection between the spreader and the virtual objects based on the principle of the collision detection in the Vega software [3]. That will prevent the spreader from colliding with the cranes and the virtual objects and ensure the authenticity of the virtual scene and users' immersive sense [4, 5].

2 Principle of Collision Detection

The Vega software provides intersection vector technique that can meet the collision detection in real time. The intersection vector technique realizes intersection test function by mainly packaging the collision detection as Isector class through the two modes of Isector and Volume. When the Isector class carries out the intersection test, a volume, a target and an intersection vector method are necessary. Namely, the intersection test is complete between the corresponding volumes of the goal designated by the Isector instance and the method set by the Isector instance. Vega software supports the Volumes including cubes, spheres, cylinders, frustums, dots, lines, and surfaces.

Before carrying out the collision detection, first, the Isector instance was bound with the role object. After bounding, the Isector instance can position and move following the bounding target in the scene and performs various types of real time computing of intersection test. Then the role object can respond in real time based on all kinds of information resulted from intersection test and make the appropriate condition adjustment in the virtual scene to simulate the complex interaction effects[6].

3 Intersect Vector Technique

3.1 Intersection Vector Masks

Because there are different types of intersection test targets in the scene, the Vega software provides a 32 bit Isector mask to distinguish different types of targets. When two objects intersect and collide, only if ensure that the definition of Target Isector Class is consistent with one of the Object Isector Class of the test object of the Isector instance, then the Isector instance can accurately position the corresponding targets and collision detection can be carried out correctly. Here the intersection vector masks are divided into four large groups including terrain, static objects, dynamic objects and special objects. Taking into account the intersection test occur between the spreader and the ground or the port crane in the virtual operation system, we set two types of terrain and static objects.

3.2 Intersection Vector Methods

The Vega software provides seven types of intersection testing methods whose volumes are lines. When carrying out intersect detection of two objects, the red vector line becomes green to reminder that a collision occurs and the system need to respond to the collision [3].

The corresponding volume of Z method is a segment that is perpendicular to and through the current Isector position. The corresponding volume of Los method is a segment that starts from the current Isector position along the axis of the Y body and whose length is Range. The volume of TRIPOD method consists of three segments which are respectively vertical segment respectively through (x-Width/2, y - Length / 2), (x + Width / 2, y + Length / 2), (x, y + Length / 2). The ZPR method, HAT method and XYZPR method include only one vertical line segment. BUMP collision detection method defines a Segment class internal body includes six vertical line segments which are perpendicular to each other that extend along positive and negative directions of X, Y, and Z and the internal body is determined by three parameters of the width, length, height. If one of the lines collides with other objects in the scene, the results of the collision detection are stored in an array of result [24]. Then we can extract the collision information about the coordinates on the collision axes and the collision object. The motion of the virtual objects is restricted by C++ program [2-5]. The collision information in the array of result [24] is shown as in Table 1.

Table 1. Collision information in the array of result [24]

Result[24]	Collision or not in axis	Result[24]	Collision or not in axis
Result[0]<0	Non-collision in positive X	Result[8]>=0	Collision in positive Y
Result[4]<0	Non-collision in negative X	Result[12] >=0	Collision in negative Y
Result[0]>=0	Collision in positive X	Result[16]<0	Non-collision in negative Z
Result[4] >=0	Collision in negative X	Result[20]<0	Non-collision in positive Z
Result[8]<0	Non-collision in positive Y	Result[16] >=0	Collision in negative Z
Result[12]<0	Non-collision in negative Y	Result[20] >=0	Collision in positive Z

4 Realization of Collision Detection

When the container and the spreader move, they may collide with all surrounding objects in six directions. When the port crane is downloading, the unloaded spreader may collide with the containers on the trucks and crane beams, while the loaded spreader may collide with the container ship and crane beams. When the port crane is unloading, the unloaded spreader may collide with the containers, crane beams, while the loaded spreader may collide with the containers and the crane beams. Therefore BUMP method is used to set collision detection of the spreader.

The BUMP intersect vector method is more precise detection method. It has been calculating whether the six line segments collide with the object in the scene with a great number of calculations. In order to improve the system speed and reduce the CPU load, a more efficient LOS vector intersect method is first set on the container. When the Los method detect that the collision occurs between the spreader and the container, the BUMP method is used to carry out precise detection. Therefore it is necessary first enable the LOS collision detection method on the container.

4.1 Realization of Collision Detection of Container

First the test target of the Isector instant, intersect vector method and intersect vector class of the container are set in the Isector model on lynx interface. Then, the same intersect vector class of container test object is designed in the Object model. Because Los intersect vector line segment is horizontal and should be used to detect the spreader in vertical direction, we should program to add vgRotMat (matStac k_container, 90.0f, 'x') in the C++ program that has the function of rotating the LOS collision detection method on the container.

4.2 Realization of Collision Detection of Spreader

Under condition of the spreader without load, the intersect vector is put on the geometric center position of the unloaded spreader. When detecting that the vector line segment collides with the objects in the scene, the position of the spreader is needed update in time. After the spreader grasps the container, the spreader and the container are regarded as a whole and the intersect vector is put on the geometric center of the whole. When collision between the intersect vector line and the objects is detected in the scene, collision information should be fed back to the system which can respond to collision in time.

Creating Spreader Collision Detection Instance.

First, the intersect vector masks is set in the Object model on Lynx interface. The BUMP collision detection instance of the spreader is created by calling VegaAPI functions in C++. Then BUMP method of the spreader is bounded with the role object of the spreader to detect the target moving with spreader in real time.

Intersection Test of BUMP Method of Spreader.

The BUMP intersection test instance created can only detect whether the collision occurs, but cannot respond to the collision. Therefore the spreader can move in a direction where there exists no collision by programming in C++ based on the collision information stored in an array of result [24]. BUMP collision detection of the spreader without load is shown in Fig. 1 and BUMP collision detection of the spreader with load is shown in Fig. 2.

Fig. 1. Collision detection of the spreader without load

5 Conclusion

According to the requirements of the virtual operation system of cranes, the problems of the collision detection between the spreader and virtual objects or the containers during uploading and downloading in the virtual scene is solved based on the intersection test methods provided by the Vega software. Therefore, the authenticity of the virtual scene, user's immersive sense, and interactive sense is ensured in the virtual operation system of cranes.

Fig. 2. Collision detection of the spreader with load

Acknowledgements

This work was financially supported by the National Natural Science Foundation of China (Grant No. 51475277) and Foreign Expert Project in Shandong Province.

References

1. Xie Shushuang, Li Fuwu, Zhu Jiehong: Research status and feasibility analysis of port gate and loading and unloading simulator. Light Industry Science and Technology. Vol. 1, pp 44-154 (2013)
2. Zhang Rui, Zhang Xien, Xie Jianhua: Collision detection applications operating in Training Simulation System. Advanced Manufacturing Technology. Vol. 24, pp 22-23 (2005)
3. Wang Cheng, Li Lijun, Zhou Jun Qing: Vega real-time three-dimensional visual simulation technology. Wuhan: Huazhong University of Science and Technology Printer, 2005
4. Wang cheng, Li Lijun, Zhou Junqing: Vega collision detection technology and research. Information Technology. Vol. 11, pp78-80 (2007)
5. T. Teixidó, E. G. Artigot, J. A. Peña, F. Molina, T. Nájera, F. Carrión: Geoarchaeological Context of the Motilla de la Vega Site (Spain) Based on Electrical Resistivity Tomography. Archaeological Prospection. Vol. 20, pp11-22 (2013)
6. Bakker Adam, Moseley Jessica, Friedrich Jeffrey: Vehicle factors and outcomes associated with hand-out-window motor vehicle collisions. Trauma and Acute Care Surgery. Vol. 74, pp 87-91(2013)

Risk-State Identification for a Stephenson II Six-Bar Linkage Based on Kinematic and Kinetic Analysis

Jiajia Gao[1], Xuefeng Chen[1,*], Zhibo Yang[1]

1State Key Laboratory for Manufacturing Systems Engineering, School of Mechanical Engineering, Xi'an Jiaotong University, Xi'an 710049, China, chenxf@mail.xjtu.edu.cn

Abstract. The Stephenson II six-bar linkage has been widely used in airplane retraction system, high-lift systems and automatic machine, etc. The analysis towards its kinematic and kinetic performance is the key for fault tracing and prognosis. In this paper, the kinematic and kinetic analysis was performed to identify the risk-state of a Stephenson II six-bar linkage. Firstly, a Stephenson II six-bar linkage's operating range is determined by analysing the mechanism's angular displacements. After that, the kinematic curves and hinge forces during a certain operating range are obtained via further analysing the mechanism's kinematic and kinetic performance. According to the hinge forces, the risk-state of the linkage during this operating range is identified.

Keywords: Stephenson II six-bar linkage, risk-state identification, kinematic and kinetic analysis

1 Introduction

The Stephenson II six-bar linkage and the derived mechanism have been widely applied in airplane retraction system, high-lift systems and automatic machine, etc [1, 2]. The analysis towards its kinematic and kinetic performance is the key for fault tracing and prognosis. For a linkage mechanism, most failures occur on the hinges where the reaction forces could be obtained by establishing and solving quasi-static equations. The greater the hinge force, the riskier the mechanism becomes. Therefore, a

* Corresponding author. Tel.: +86 13689253672, E-mail address: chenxf@mail.xjtu.edu.cn

linkage in the risk-state could be identified when the hinge force increases sharply or reaches the maximum. Before calculating hinge forces, the operating and stagnant range for a specific mechanism should be distinguished firstly, and only the analysis towards the operating range, that is, the mechanism's normal working range is meaningful. Katsumi and Hiroaki [3] proposed a method to investigate the kinematic performance of Stephenson six-bar linkages in early years. Analytical method was used to establish a six-order equation and the displacements of the mechanism could be obtained by solving this equation. Chao-Hwa et al. [4] used both analytical and numerical methods to investigate the kinematic performance of a flapping-wing MAV with Watt six-bar linkage. Dewen et al. [5] utilized Watt six-bar mechanism in prosthetic knees and analysed the knee joint's kinetic performance during swing phase through quasi-static procedure. Lluisa et al. [6] proposed a new method to determine a mechanism's driving and constraint forces in the quasi-static analysis by replacing actuator or using auxiliary mechanisms.

According to the current literatures, it appears that the forward kinematic and kinetic analysis of the Stephenson II six-bar linkage and the derived mechanism is crucial both in theoretical exploration and engineering application. In this paper, the kinematic and kinetic analysis was performed to identify the risk-state of a Stephenson II six-bar linkage. The high-lift mechanism authorized by the Boeing Company [1] was used as a benchmark to demonstrate the risk-state identification procedure. Firstly, the Stephenson II six-bar linkage's operating and stagnant ranges were distinguished by analysing the mechanism's angular displacements. Then, the kinematic curves and hinge forces during a certain operating range were calculated, according to which, the Stephenson II six-bar linkage's risk-state during this operating range was identified. The research work in this paper could address a theoretical foundation for the fault diagnosis of Stephenson II six-bar linkage.

2 Operating and Stagnant Ranges Distinguishment

Displacement analysis towards the input link's entire rotation period of a Stephenson II six-bar linkage was performed, from which, the rotation period was classified as operating and stagnation ranges. The mechanism's geometry configuration was shown in Fig. 1(a). The Stephenson II six-bar linkage was composed of two coupled vector loops, that was, loop consisted of links①②③④⑤ and loop consisted of links ①②③⑥⑤, the vector loop equations could be written as

$$\begin{cases} \overrightarrow{EA} + \overrightarrow{AB} + \overrightarrow{BC_1} + \overrightarrow{C_1D_1} = \overrightarrow{ED_1} \\ \overrightarrow{EA} + \overrightarrow{AB} + \overrightarrow{BC_2} + \overrightarrow{C_2D_2} = \overrightarrow{ED_2} \end{cases} \tag{1}$$

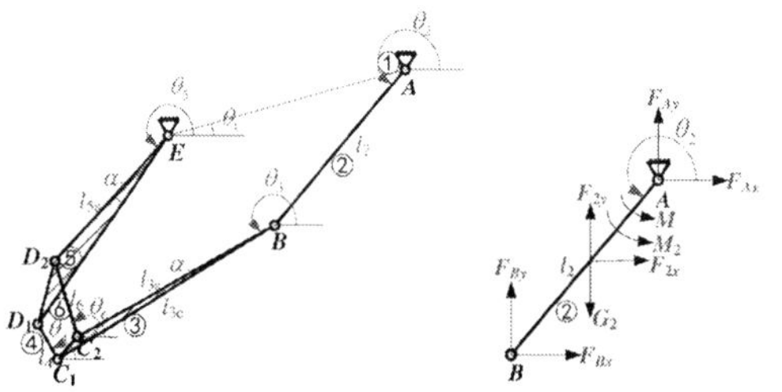

(a) Geometry configuration (b) Force diagram of link②

Fig. 1. The geometry configuration and force diagram of a Stephenson six-bar linkage

After page-long formula derivation[3], a six-order algebraic equation was obtained

$$h_0t^6 + h_1t^5 + h_2t^4 + h_3t^3 + h_4t^2 + h_5t^1 + h_6 = 0 \tag{2}$$

In this equation, coefficients $h_i(i=0,1,...,6)$ contained the information of input angle θ_2 and link lengths; the unknown t contained the information of link③'s angular displacement θ_3. Once θ_3 was obtained, the other angular displacements could be calculated by simple derivation[3]. Specific to the Stephenson II six-bar linkage in Ref. [1], the input link② was driven by a rotatory actuator. Substituting the pre-defined data into Eq. (2), and solving the equation, one could obtain the angular displacements corresponding to the entire rotation of the input link, as shown in Fig. 2.

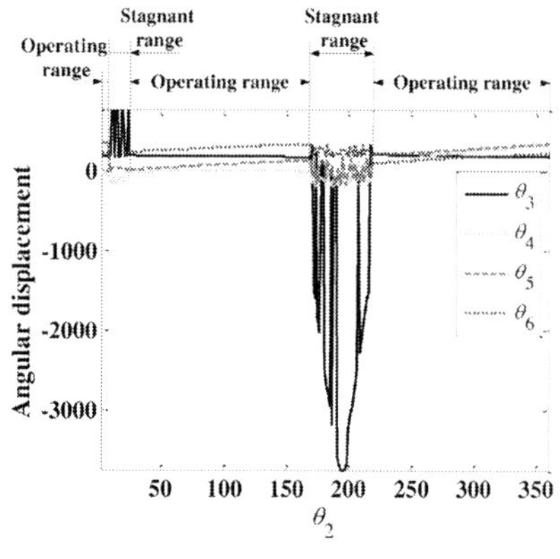

Fig. 2. Angular displacements corresponding to the entire rotation of the input link

It could be observed that during some ranges the angular displacements of links③④⑤⑥ fluctuated continuously while during other ranges the displacement values oscillated drastically. Reading the calculated data, one could find that the solutions of

Eq. (2) were meaningless complex numbers during the oscillation range. That implied a spatial rotation was needed for the mechanism to cross the oscillation range. Naturally, the planar Stephenson II six-bar linkage was clamped. The continuously operating ranges corresponding to 0~7°, 24~170°and 219~360°of the input link would be referred as operating ranges hereafter; the clamping ranges corresponding to 8~23°, 171~218°of the input link would be referred as stagnant ranges hereafter.

3 Kinematic and Kinetic Analysis

An operating range 219~360° of the input link was selected to perform further kinematic and kinetic analysis since this range generated the best match with the extending-retracting range of the flaps in Ref. [1].

3.1 Kinematic Analysis

Extract the calculated angular data in the range 219~360° from section 2, assuming the input link was driven with a speed of 20°/s, and then other components' displacements corresponding to time could be seen in Fig. 3. The mechanism's velocities and accelerations during this range could be calculated by differentiating and double differentiating displacements respectively.The kinematic data would be used to calculate the mechanism's inertial force in section 3.2.

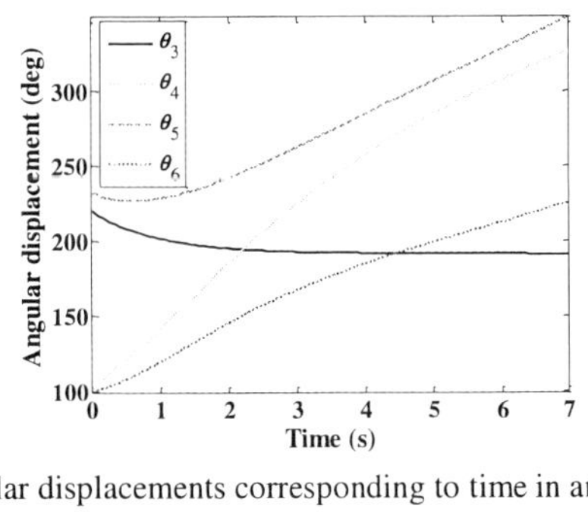

Fig. 3. Angular displacements corresponding to time in an operating range

3.2 Kinetic Analysis and Risk-State Identification

To get the Stephenson II six-bar linkage's hinge forces, the component quasi-static equations should be established at first, as in Eq. (3). Taking link② as an illustration, the equations were as follows: F_{2x}, F_{2y}, M_1 denoted horizontal inertia force, vertical inertia force, and inertia moment respectively (corresponding to Fig. 1(b)). The other components' quasi-static equations could be established in a similar way. By synthesizing the component equations, one could establish the system quasi-static matrix equation, as shown in Eq. (4). In this equation, matrix $F_{(15\times1)}$ were composed of unknown hinge forces, matrix $A_{(15\times15)}$ were composed of links' lengths and derived kinematic parameters in section 3.1, matrix $B_{(15\times1)}$ were composed of the known forces including gravity, external forces, inertia forces and moments. By performing matrix inversion, the hinge forces in matrix $F_{(15\times1)}$ could be calculated.

$$\begin{cases} M + M_2 - F_{2x}\dfrac{l_2}{2}\sin\theta_2 - F_{Bx}l_2\sin\theta_2 + (F_{2y} - G_2)\dfrac{l_2}{2}\cos\theta_2 + F_{By}l_2\cos\theta_2 = 0 \\ F_{Ax} + F_{2x} + F_{Bx} = 0 \\ F_{Ay} + F_{2y} + F_{By} - G_2 = 0 \end{cases} \tag{3}$$

$$A_{(15\times15)}F_{(15\times1)} = B_{(15\times1)} \tag{4}$$

As for a two-force link, either hinge point's resultant force should be same to keep the link in balance. In this Stephenson II six-bar linkage, links②④⑥were two-force components, hence $F_A=F_B$, $F_{C1}=F_{E1}$, $F_{C2}=F_{E2}$, besides $F_A=F_D$ in the two-hinged-joint frame. Therefore, the mechanism's hinge force behaviour could be expressed by plotting the reaction forces of hinge points A, C_1, C_2, which were shown in Fig. 4. The external force was applied on link⑤, perpendicular to the edge D_1D_2.

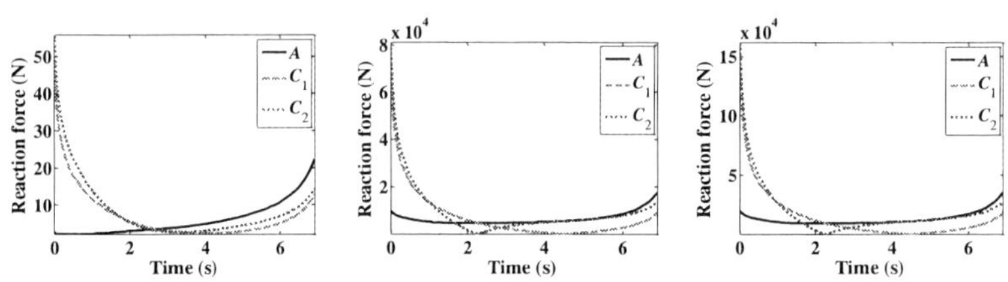

(a) Without external force (b) With External force of 5000N (c) With external force of10000N

Fig. 4. Reaction forces of some hinge points corresponding to time in an operating range

It could be found that the hinge forces increased with the increase of external force, but the force curves fluctuated basically the same no matter external force was applied on or not. In Fig. 4(a-c), the reaction forces of hinge point C_1, C_2 reached the maximum when the mechanism was in initial state, the reaction forces of hinge point A reached the maximum when the mechanism was in end state. This phenomenon was induced by the variation of transmission angles which were much less in initial and end states than that in the intermediate state. Since the maximum of reaction forces had appeared in initial or end state, these inspected states corresponding to 219°and 360°of the input angle could be identified as risk-states.

4 Conclusions

The kinematic and kinetic performances of a Stephenson II six-bar linkage have been investigated analytically. The results show that the displacement curve fluctuates continuously during normal operating ranges but oscillates drastically during clamping stagnant ranges. An operating range 219°~360°of the input angle is selected to perform quasi-static analysis, the results show that the hinge forces attain to the maximum in the initial or the end state no matter external forces exist or not. Therefore, the initial and end states corresponding to 219°and 360°of the input angle are identified as risk-states of this Stephenson II six-bar linkage.

Acknowledgement.

This work is supported by National Natural Science Foundation of China (Nos. 51225501 & 51405369), and of Shaanxi Province (No. 2016JQ5049).

References

1. K. W. Beyer and S. Sakurai, "Link mechanisms, including Stephenson II link mechanisms for multi-position flaps and associated systems and methods," ed: US, (2012).
2. S. Hernandez, S. Bai, and J. Angeles, "The Design of a Chain of Spherical Stephenson Mechanisms for a Gearless Robotic Pitch-Roll Wrist," Journal of Mechanical Design, vol. 128, pp. 422-429, (2006).
3. K. Watanabe and H. Funabashi, "Kinematic Analysis of Stephenson Six-Link Mechanisms : 1st Report, Discrimination of Composition Loops," Lecture Notes in Computer Science, vol. 27, pp. 2863-2870, (1984).
4. C.-H. Liu and C.-K. Chen, "Kinematic Analysis of a Flapping-wing Micro-aerial-vehicle with Watt Straight-line Linkage," Journal of Applied Science and Engineering, vol. 18, p. 355r362, (2015).
5. D. Jin, R. Zhang, H. O. Dimo, R. Wang, and J. Zhang, "Kinematic and dynamic performance of prosthetic knee joint using six-bar mechanism," Journal of Rehabilitation Research & Development, vol. 40, pp. 39-48, (2003).
6. L. s. Jordi-Nebot, J. Puig-Ortiz, and S. Cardona-Foix, "Determination of the force of the actuators and constraint forces in the inverse dynamic analysis of mechanisms," Journal of Mechanical Science & Technology, vol. 24, pp. 1595-1601, (2010).

Rolling Bearing Fault Diagnosis Based on Laplacian Score and Adaptive Fuzzy C-Means Clustering

Sheng Fu, Lei Cheng, Kun Liu, Hao Zheng, Dianwei Xue, and Hongmo Wang

College of Mechanical Engineering and Applied Electronics Technology, Beijing University of Technology, No.100, Pingleyuan, Chaoyang District, 100124 Beijing, China.
{Lei Cheng: chenglei@emails.bjut.edu.cn}

Abstract. Rolling element bearings are key components of rotating machinery. For the non-stationary features and the problem of feature selection for the high dimensional data of the vibration signals of a bearing with faults, a fault diagnosis method based on Laplacian score and adaptive Fuzzy C-Means (FCM) clustering is proposed. Firstly, nine typical time domain parameters of vibration signals are calculated, from which an initial feature vector is formed. Secondly, by applying Laplacian score method to the initial feature vector, the optimal fault feature vector is obtained. Thirdly, by using adaptive FCM clustering to the fault feature vector, the optimal clustering number and centers are obtained and then different fault patterns can be classified. Finally, the fault pattern can be also identified by calculating the Hamming nearness degree between the unknown-fault samples and known-fault ones. Experiment results showed the effectiveness of this method in diagnosing rolling bearing faults.

Keywords: rolling bearing, fault diagnosis, time domain parameters, adaptive Fuzzy C-Means (FCM) clustering, Laplacian score, Hamming nearness degree

1 Introduction

Rolling element bearings are widely used in rotating machinery. The faults associated with rolling bearings are one of the major causes that lead to the breakdown of machines, for example, over 40 percent of failures in induction machines resulted from rolling bearings [1]. Therefore, to prevent these failures, bearing fault diagnosis should be conducted as early as possible.

In fact, the fault diagnosis of rolling bearings is a matter of pattern recognition [2]. Selecting appropriate fault features and finding out appropriate pattern recognition method are the two key points to realize the fault diagnosis of rolling bearings. In many pattern recognition problems, we need to calculate a huge number of parameters based on vibration signals, but, in reality, the fault-related and useful ones are limited. The high dimensional data often makes the pattern classification more difficult and less accurate, even lead to "dimensional curse". Therefore, it is fairly important to select the most representative features based on an effective feature selection algorithm to avoid the high dimensional data. In this paper, we introduce the Laplacian score method [3]. The Laplacian score selects the features based on the capability of locality preserving and data variance.

Apart from feature selection techniques, pattern recognition is another important part of rolling bearing fault diagnosis [4]. Considering the fuzziness of operating conditions of rolling bearings from the normal to the fault, fuzzy C-means (FCM) clustering is considered in this paper. FCM algorithm as an effective tool for classification has shown successful application in fault diagnosis [5, 6]. However, the number of clusters for data sets is needed to be given in advance for most of clustering algorithms. We will not get the right classifications if the clustering number is not given properly. So the adaptive FCM clustering is introduced, which is exactly the adaptive process of the clustering number based on FCM clustering.

In this paper, the fault feature vectors are firstly extracted from nine typical time domain parameters which form the initial vector. To lower the dimensional number of the feature vector, the Laplacian score is applied to the initial vector, and we can get the optimal and low dimensional feature vector. And then the operation condition of the rolling bearing can be classified by adaptive FCM clustering. Finally, the fault patterns can also be recognized by Hamming nearness degree.

2 Related works

2.1 Laplacian score for feature selection

The basic idea of Laplacian Score (LS) is to evaluate the features according to their locality preserving power. LS can extract the intrinsic structure information by learning directly the feature sets, and select those features which have relatively small scores, thus converting the complicated and high dimensional feature space into simple and low dimensional feature space. LS reflects the importance of the feature. The lower the LS is, the more important the feature is. So we can choose the first few features with lowest scores as the final results. LS algorithm can be found in [7].

2.2 Adaptive Fuzzy C-Means (FCM) clustering

Adaptive FCM clustering is based on maximization of the following objective function:

$$L(c)=\frac{\sum_{i=1}^{c}\left(\sum_{j=1}^{n}u_{ij}^{m}\right)\|v_i-\bar{x}\|^2/(c-1)}{\sum_{i=1}^{c}\left(\sum_{j=1}^{n}u_{ij}^{m}\right)\left\|x_j-v_i\right\|^2/(n-c)} \tag{1}$$

where n and c denotes the number of data points and clusters respectively. m is the weighting exponent. x_j is the j-th data point. $\|*\|$ represents the Euclidean distance. $\bar{x}$ denotes the central vector of population sample. u_{ij} is the membership matrix. v_i is the center of the cluster. The relevant mathematical formulas and algorithm can be found in [8].

For the function $L(c)$, its numerator indicates the distance between different classes, while the denominator shows the distance between the data point and the center in the same class. Therefore, a more reasonable clustering can be realized if the $L(c)$ is larger, and the optimal c will be the one which makes the $L(c)$ largest.

2.3 Hamming nearest degree

Hamming nearest principle is given as follows:

$$N(A,B) = 1 - \frac{1}{n}\sum_{k=1}^{n}|A(x_k) - B(x_k)| \tag{2}$$

where A and B denotes standard mode and identifying mode respectively. B is more similar to A if we get the larger $N(A,B)$. In fuzzy diagnosis, we can firstly gain the cluster centers of the known samples by adaptive FCM clustering, and then calculate the nearest degrees between the feature vectors of testing samples and cluster centers to recognize the pattern of the unknown samples.

3 Fault feature extraction of rolling bearings

Bearing vibration signals usually present non-stationary characteristics because bearings are inherently dynamic, therefore it seems that extracting fault feature in time-frequency domain is the most suitable analysis method [9]. In this paper, we choose nine typical time domain feature parameters. They are Average value (X_{ave}), Root mean square (X_{rms}), Maximum value (X_{max}), Kurtosis (X_{kur}), Standard deviation (X_{std}), Skewness (X_{ske}), Form factor ($S_f = X_{rms}/X_{ave}$), Peak factor ($C_f = X_{max}/X_{rms}$), Impulse factor ($I = X_{max}/X_{ave}$).

4 Experimental verification and results

To verify the effectiveness of the proposed method, we apply it to analyze the actual data measured from rolling bearings. The data is from the Case Western Reserve University Bearings Data Center [10]. The testing bearings are 6205-2RS JEM SKF deep groove ball bearings. Single point faults ranging in diameter from 0.007 to 0.021 inches (0.1778 mm, 0.3556 mm, 0.5332 mm) were introduced to the bearings using electro-discharge machining (EDM). The sampling frequency is 12 kHz.

In this experiment, 6 conditions including the normal (N), inner raceway fault with three severities (0.1778 mm (I-1), 0.3556 mm (I-2), 0.5332 mm (I-3)), outer raceway fault (O) and ball fault (B) both with 0.1778 mm at 1730 rpm are considered. 7 groups of data is prepared for every condition. Choose randomly 5 groups as the training samples, 30 groups in total, and the rest as the testing samples, 12 groups in total. Number 1~5 is the normal, 6~10 for the ball fault, 11~15 for the inner raceway fault, 16~20 for the outer raceway fault all with 0.1778 mm, 21~25 for the inner raceway fault with 0.3556 mm, and number 26~30 is the inner raceway fault with 0.5332 mm. Firstly, nine typical time domain feature parameters of 30 groups of data and their LS values are shown in Table 1. Because we are not able to visualize directly the nine feature parameters in low dimensional space, two most important feature parameters, I and X_{max} which has the lowest LS values are selected to visualize the clustering results. Next, applying adaptive FCM clustering ($m = 2, \varepsilon = 0.001$) to the feature vector composed by I and X_{max}, and the results are shown in Fig.1,Table 2, and Table 3. Obviously, according to Fig.1, 30 groups of data are clearly classified into 6 clusters which indicate the 6 conditions of the bearing. As shown in Table 2, $L(c)$ is largest when c is 6, that is, the optimal clustering number is 6, which is exactly the number of conditions.

In order to compare with the results with the assistance of LS, we select randomly two feature parameters, such as X_{kur} and X_{ske}, to form the feature vector which is also clustered by adaptive FCM clustering method, and the results are shown in Fig.2 and Table 4. It can be seen from Fig.2 that there are only 4 clustering centers while other data points are scattered. Table 4 also show that the optimal clustering number is 4. However, the actual conditions is 6. Therefore, we may not get the right results if the feature parameters are not chosen properly, and it also indicates that the fault diagnosis method based on adaptive FCM clustering and LS is more superior.

Both Fig.1 and Table 2 show the effectiveness of the proposed method in classifying different bearing conditions including different fault types and the fault with different severities at early stage. To further verify its effectiveness, we normalize the feature vectors of 12 testing samples and cluster centers in Table 3, and then calculate the nearest degrees between them. The diagnosis results are shown in Table 5.

Table 1. Nine feature parameters and their LS values

No.	X_{ave}	X_{rms}	X_{max}	X_{kur}	X_{ske}	X_{std}	S_f	C_f	I
1	0.0124	0.0655	0.2253	2.8715	-0.1103	0.0643	5.2843	3.4393	18.1743
2	0.0125	0.0657	0.2422	2.9569	-0.0795	0.0646	5.2680	3.6838	19.4064
3	0.0132	0.0658	0.2422	2.9592	-0.0738	0.0644	4.9886	3.6833	18.3743
4	0.0140	0.0655	0.2512	2.9560	-0.1419	0.0640	4.6668	3.8347	17.8958
5	0.0141	0.0668	0.2512	3.0970	-0.1753	0.0653	4.7252	3.7577	17.7559
6	0.0033	0.1547	0.6239	2.7916	0.0342	0.1547	46.9548	4.0323	189.3366
7	0.0036	0.1549	0.5593	2.8220	0.0246	0.1549	43.0160	3.6098	155.2808
8	0.0039	0.1548	0.5593	2.9054	0.0031	0.1547	39.9022	3.6131	144.1718
9	0.0039	0.1537	0.5157	2.8385	0.0018	0.1537	39.1585	3.3554	131.3908
10	0.0039	0.1549	0.5487	2.7943	0.0277	0.1549	39.2349	3.5424	138.9848
11	0.0046	0.3159	1.5016	5.2264	-0.0125	0.3159	69.3078	4.7527	329.4004
12	0.0044	0.3077	1.3312	5.1455	-0.0211	0.3077	69.2983	4.3260	299.7818
13	0.0046	0.3048	1.3695	5.2196	-0.0257	0.3048	66.8453	4.4928	300.3216
14	0.0046	0.3049	1.3989	5.3184	-0.0135	0.3049	66.3116	4.5885	304.2724
15	0.0044	0.3063	1.3989	5.2356	-0.0242	0.3063	69.1449	4.5670	315.7849
16	0.0045	0.5681	2.9705	8.0323	-0.0115	0.5681	125.6421	5.2291	656.9946
17	0.0047	0.5723	3.0863	7.9781	0.0015	0.5724	123.0166	5.3923	663.3481
18	0.0048	0.5733	3.0863	8.0556	-0.0144	0.5733	119.7168	5.3832	644.4585
19	0.0048	0.5705	3.1630	8.0654	-0.0013	0.5705	119.5785	5.5440	662.9487
20	0.0049	0.5727	3.1841	7.9736	0.0340	0.5727	115.8553	5.5599	644.1448
21	0.0030	0.1740	1.6050	16.3100	-0.0972	0.1740	58.4775	9.2237	539.3790
22	0.0029	0.1755	1.6869	17.0643	-0.0345	0.1755	60.8605	9.6109	584.9256
23	0.0030	0.1817	1.6890	17.2303	0.1104	0.1817	60.8987	9.2931	565.9377
24	0.0030	0.1866	1.6911	17.9761	0.0991	0.1865	61.3842	9.0645	556.4198
25	0.0030	0.1877	1.6911	18.1911	0.1300	0.1877	62.7632	9.0090	565.4318
26	0.0032	0.4567	2.9933	8.1962	0.3634	0.4567	144.4351	6.5542	946.6534
27	0.0029	0.4472	2.8824	7.8584	0.3643	0.4472	152.0265	6.4460	979.9585
28	0.0029	0.4380	2.8097	8.1576	0.3481	0.4380	150.5264	6.4151	965.6447
29	0.0030	0.4389	2.8097	8.2049	0.3379	0.4389	146.6230	6.4024	938.7444
30	0.0032	0.4372	2.9648	8.9251	0.3247	0.4372	137.1424	6.7812	929.9943
LS	0.3358	0.2928	0.1135	0.1912	0.2973	0.2922	0.1214	0.2243	0.0412

Table 2. Adaptive FCM-LS clustering objective function

c	1	2	3	4	5	6	7	8
$L(c)$	0	107.32	185.97	401.42	734.52	2595.32	2236.89	0

Table 3. Adaptive FCM-LS clustering center

Name	N	B	O	I-1	I-2	1-3
X_{max}	0.2426	0.5584	3.0942	1.3979	1.6736	2.8919
I	18.4091	149.9710	654.2827	309.5551	561.9243	952.1439

Table 4. Adaptive FCM clustering objective function

c	1	2	3	4	5	6
$L(c)$	0	124.40	539.28	1795.59	1602.15	0

Table 5. Diagnosis results of testing samples

No.	X_{max}	I	N	B	O	I-1	I-2	I-3	Corresponding with reality? (Y/N)
			Hamming nearness degree						
1	0.2660	19.2860	**1**	0.93	0.72	0.87	0.83	0.67	Y
2	0.5487	139.3579	0.92	**1**	0.76	0.91	0.86	0.71	Y
3	3.1841	617.8791	0.72	0.76	**1**	0.85	0.90	0.93	Y
4	1.4830	328.4874	0.87	0.91	0.85	**1**	0.93	0.80	Y
5	1.6596	537.3295	0.83	0.86	0.90	0.93	**1**	0.84	Y
6	3.0708	946.5933	0.67	0.71	0.93	0.80	0.84	**1**	Y
7	0.2660	18.3378	**1**	0.93	0.74	0.87	0.82	0.67	Y
8	0.5237	136.3529	0.92	**1**	0.78	0.91	0.85	0.70	Y
9	2.9782	572.5407	0.74	0.78	**1**	0.87	0.92	0.92	Y
10	1.4830	331.0331	0.87	0.91	0.87	**1**	0.94	0.79	Y
11	1.7538	562.8175	0.82	0.85	0.92	0.94	**1**	0.85	Y
12	3.0708	969.9669	0.67	0.70	0.92	0.79	0.85	**1**	Y

Table 5 show clearly that the largest Hamming nearness degree between the optimal clustering centers of 6 conditions and the feature vector of each testing sample is 1. The results showed that all the testing samples are identified correctly, which verify the feasibility and effectiveness of the proposed method again.

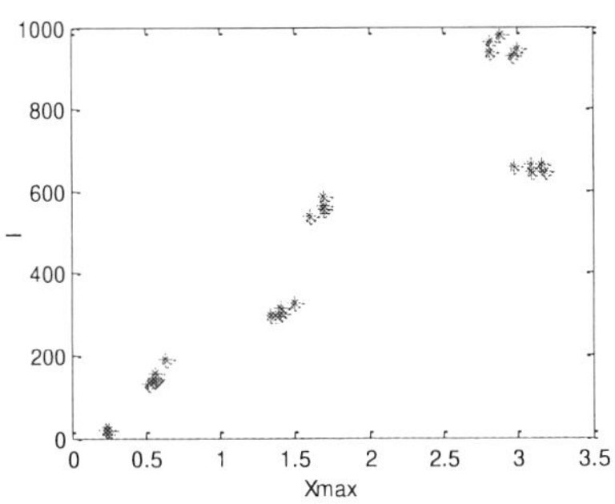

Fig. 1. The clustering result with Laplacian score

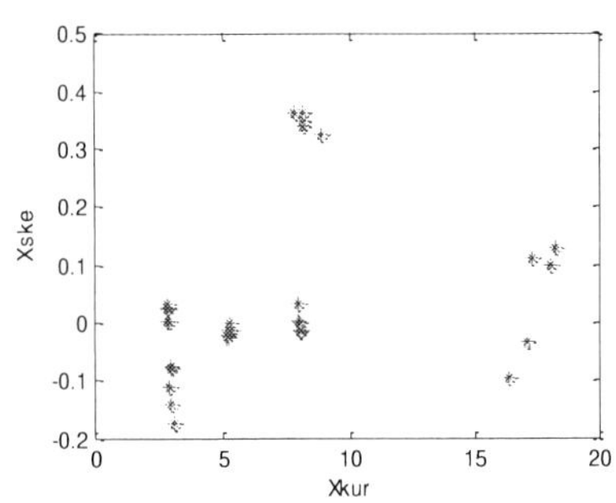

Fig. 2. The clustering result without Laplacian score

5 Conclusions

In fuzzy diagnosis, selecting appropriate fault features and finding out appropriate pattern recognition method are two key points to realize the fault diagnosis. In this paper, we proposed an effective method based on LS and adaptive FCM clustering. Two most important feature parameters were selected from nine typical time domain feature parameters based on LS. The optimal clustering number could be gained through the adaptive algorithm of FCM. The experiments demonstrated that this method could classified the different conditions of rolling bearings at early stage, even the fault with different micro severities. The Hamming nearness degree experiments verified its effectiveness again.

Acknowledgments

This work is supported from the National Science and Technology Major Project of China (Grant No. 2010ZX04007-051). This work is also supported in part by the National Natural Science Foundation of China (Grant No. 51375020), which is very much appreciated. The authors would like to thank the Case Western Reserve University Bearings Data Center for kindly providing the experimental data.

References

1. Calis, H., Cakir, A., Dandil, E.: Artificial Immunity-Based Induction Motor Bearing Fault Diagnosis. Turkish Journal of Electrical Engineering &Computer Sciences. 21, (2013)

2. Qiu, H., Lee, J., Lin, J., Yu, G.: Wavelet Filter-based Weak Signature Detection Method and its Application on Rolling Element Bearing Prognostics. Journal of Sound and Vibration. 289(4-5), 1066--1090 (2006)

3. He, X., Cai, D., Niyogi, P.: Laplacian Score for Feature Selection. In: NIPS, pp. 507--514(2005)

4. Sugumaran, V., Ramachandran, K.I.: Fault Diagnosis of Roller Bearing Using Fuzzy Classifier and Histogram Features with Focus on Automatic Rule Learning. Expert Systems with Applications. 38 (5), 4901--4907 (2011)

5. Zogg, D., Shafai, E., Geering, H.P.: Fault Diagnosis for Heat Pumps with Parameter Identification and Clustering. Control Engineering Practice. 14(12), 1435--1444 (2006)

6. Pan, Y.N., Han, J., Li, Z.N.: Vector Power Spectrum-Fuzzy C Means Clustering Method for The Fault Diagnosis of Rotating Machine. Turbine Technology. 48(3), 212--214 (2006)

7. He, X.F., Cai, D., Niyogi, P: Laplacian Score for Feature Selection. Advances in Neural Information Processing System. Cambridge. MA: MIT Press, 2005

8. Fu, S., Liu, K., Xu, Y.G., Liu, Y.: Rolling Bearing Diagnosing Method Based on Time Domain Analysis and Adaptive Fuzzy C-Means Clustering. Shock and Vibration. http://dx.doi.org/10.1155/2016/9412787

9. Yang, Y., Wang, H.H, Cheng, J.S., Zhang, K.: A Fault Diagnosis Approach for Roller Bearing Based on VPMCD under Variable Speed Condition. Measurement. 46, 2306--2312 (2013)

10. Case Western Reserve University Bearings Data Central. https: //csegroups. case. edu/ bearingsdatacenter/pages/12k-drive-end-bearings-fault-data

The axle AE signal feature extraction based on the LMD and demodulated resonance technique

Lin Li[1],Xu Sijia[1],Zhou Yong[2],Fulei Chu[3],Zhou Yue[4]

[1]Dalian JiaoTong University ,China
[2]Dalian University of Technology ,China
[3]Tsinghua University, China
[4]The Ohio State University, America
Correspondence should be addressed to Linli: 173670324@qq.com

Abstract. A new approach of combining the method of resonance demodulation and local mean decomposition is presented to characterize the acoustic emission (AE) signals. Firstly, LMD is used to decompose the acoustic emission signal into several PF components and a residual. Then the energy of each PF components is calculated to draft the percentages of each components on total energy. Finally, based on the technology of resonance demodulation, PF component which occupies the most energy is drawn by power spectrum and gets the feature of signal. According to the experimental data analysis, and taking it compare with knock signal and crack signal, it shows that this method can effectively extract the characteristics of acoustic emission and it is a new way of acoustic emission feature extraction.

Keywords: Local mean decomposition; Resonance demodulate; AE signal; Feature extraction

0 Introduction

Local resource in material releases its energy quickly can produce transient elastic wave, we refer to this phenomenon as acoustic emission (AE). AE technology means that we use instrument to detect, record, analysis and deduce AE source by AE signal[1]. AE source signals own many features, such as diversity, sudden, uncertainty and so on, which makes AE signal embody non-stationary and time-variability. So, the only way to recognize the nature of AE source is to process it.

In the process of extracting AE signal's fault characteristics, The time-frequency analysis method based on LMD decomposes original signal into instantaneous amplitude and frequency on the basis of local characteristic scale, then combines time-frequency spectrum according to time-frequency to show the signals' characteristics in time and frequency. This ensures higher resolution. Hence, LMD has been adopted widely in recent years. J. Qi analyzed the wind farms vibration signals with a combination of LMD and order analysis, and proved the method is valid[2]. J. S. Cheng, Y. Yang proposed a method and applied it to gear fault detection based on kurtosis and LMD, it can extract the fault features effectively by verification[3]. Y. Yang, L. X. Yang putted forward a fault diagnose method in rotor system based on LMD and AR model, they illustrated the advantages of this method through comparison[4]. In the research of rail vehicle AE signals, LMD was used to detect axle faults by F. C. Li and achieved substantive progress[5]. Through researching and analyzing, X. Tan confirmed that the AE signals of mechanical failure can be decomposed by LMD effectively, and their fault information will be present exactly[6].

Resonance demodulation technology is a part of the vibration monitor technology, developing and extending on this basis. Academia normally uses the Hilbert transmit to describe resonance demodulation. Better than traditional monitor technology, resonance demodulation can make the distinction between normal vibration and fault impact.

The remainder of this paper is organized as follows. In Section 1, we introduce the reason of signal preprocessing. The principle and characteristics of LMD and resonance demodulation were recommended in Section 2 and 3. And then, we report experimental results and give discussion in Section 4, before concluding our work in Section 5.

1 Signal Preprocessing

While the signal transmitted over sensors amplifier or convertor and data acquisition unit. Usually, the measured vibration will be influenced by various factors of test system. In the process of the output, the signal will contain many other ingredients. So, the preprocessing of signals must be done to make sure the preliminary processing of the results restore the true face as real as possible. The effect involve that modifying the distorted waveform, removing the mixed noise and interference of the signal, weakening the excess content of signal and reinforcing the required parts.These interference would lead signal data deviate from the baseline. This process is called signal trend. The most common way to eliminate the trend is method of least square by using least square method of polymonial[7]. Therefore, in this paper, we use the method to process the vibration signal.

2 The theory of LMD

2.1 The Principle of LMD

Most often, the signals of locomotive axle fault are all nonstationary signals and have multi-component AM-FM nature. Local Mean Decomposition is very applicable to complex multi-component signals. Vibrate signals are broken down into several mono-component signals. Each of them is shown by a product function which is made up of an envelope signal (it can be sure the amplitude characteristics of mono-component signal) and a pure frequency modulation signal (it determines the instantaneous frequency characteristics of mono-component).

2.2 The LMD Algorithm

LMD separates mono-component from original signals adaptively. PF component is an AM-FM signal contains amplitude and frequency, In other words, PF component is a carrier for some important features which can be easily extracted and get the identifying information of original signal[8]. Therefore, for any signal $x(t)$, to get PF component we can use LMD deal with it. Here is how it worked:

Determine all the local extreme value and time value of original $x(t)$, and calculate an average m_i of extreme value n_i and n_{i+1}.

$$m_i = \frac{n_i + n_{i+1}}{2} \quad (1)$$

The use of cubic spline interpolation function method can work out the local mean value function $m_{11}(t)$.

By using the local peak, we can get the local amplitude a_i.

$$a_i = \frac{|n_i - n_{i+1}|}{2} \quad (2)$$

In the same way, according to the envelope estimation function $a_{11}(t)$.

To separate out the local mean value function $m_{11}(t)$ from original signal $x(t)$.

$$h_{11}(t) = x(t) - m_{11}(t) \quad (3)$$

Next, divide envelope estimation function $a_{11}(t)$ into $h_{11}(t)$ for demodulation.

$$s_{11}(t) = h_{11}(t) / a_{11}(t) \quad (4)$$

Calculate the envelope estimation function $a_{12}(t)$ of $s_{11}(t)$, If $a_{12}(t)$ and 1 do not equal each other, It shows that $s_{11}(t)$ is not a pure FM signal, So it is need to repeat the iterative process, until it gets a pure FM signal $s_{1n}(t)$ ($-1 \le s_{1n}(t) \le 1$), and satisfy that $a_{1(n+1)}(t) = 1$.

$$\begin{cases} h_{11}(t)=x(t)-m_{11}(t) \\ h_{12}(t)=s_{11}(t)-m_{12}(t) \\ \vdots \\ h_{1n}(t)=s_{1(n-1)}(t)-m_{1n}(t) \end{cases} \tag{5}$$

Among them:

$$\begin{cases} s_{11}(t)=h_{11}(t)/a_{11}(t) \\ s_{12}(t)=h_{12}(t)/a_{12}(t) \\ \vdots \\ s_{1n}(t)=h_{1n}(t)/a_{1n}(t) \end{cases} \tag{6}$$

Multiply all of envelope estimation function which produced in the process of iteration to get the envelope signal $a_1(t)$.

$$a_1(t)=a_{11}(t)a_{12}(t)...a_{1n}(t)=\prod_{q=1}^{n} a_{1q}(t) \tag{7}$$

The first PF component multiplied by envelope signal $a_1(t)$ and pure FM signal $s_{1n}(t)$.

$$\mathrm{PF}_1(t)=a_1(t)s_{1n}(t) \tag{8}$$

By separating component $\mathrm{PF}_1(t)$ from original signal $x(t)$ to obtain a new signal $u_1(t)$, and regard it as original data to repeat these steps. Through k times circulation,until $u_k(t)$ becomes a monotone function.

$$\begin{cases} u_1(t)=x(t)-\mathrm{PF}_1(t) \\ u_2(t)=u_1(t)-\mathrm{PF}_2(t) \\ \vdots \\ u_k(t)=u_{k-1}(t)-\mathrm{PF}_k(t) \end{cases} \tag{9}$$

So far, original signal $x(t)$ is broken down into the following form

$$x(t)=\sum_{p=1}^{k} \mathrm{PE}_p(t)+u_k(t) \tag{10}$$

We know through the algorithm that, LMD is an adaptive decomposition method which is based on the local extreme points of signal. The decomposition process can be seen as multiple iterations. Its algorithm flow chart shown below.

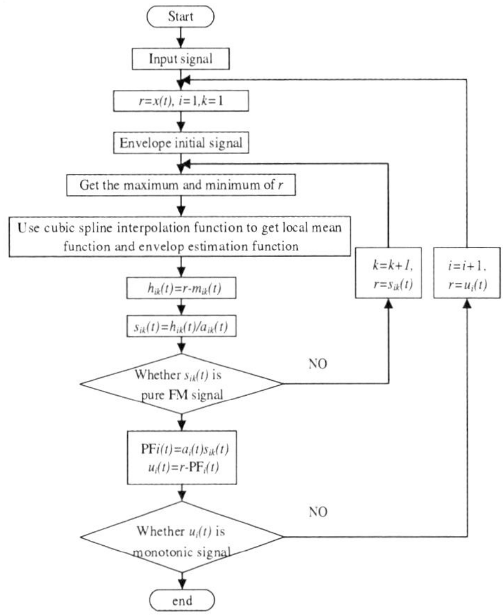

Fig. 1 The diagram of LMD algorithm

3 The Principle and Feature of Resonance-demodulation

3.1The Feature of Resonance-demodulation

Resonance demodulation bases on demodulation theory of waveform and demodulation spectrum to be sure. The traditional fault diagnosis still have uncertainty--there is no inevitable relation between spectral line and fault. But the signal has characteristic means it is a fault in technology of resonance demodulation, in a sense, it enhances the reliability of diagnosis.

Resonance demodulation technology has the following characteristics:

(1) Parallelism: Each shock corresponds to excite a resonance demodulation wave, it does not miss any impact caused by a fault.

(2) Selectivity: Only fault impact can excite a resonance demodulation wave.

(3) Magnification: A tiny fault impact can excite a great resonance demodulation wave.

(4) Proportionality: Resonance demodulation wave has a variation of direct proportion with the range of fault impact.

(5) Malleability: After the original fault wave broadened by resonance demodulation, the level of broadening is irrelevant to resonant and impacting frequency.

(6) Low frequency: The high frequency energy of impact fault transform into low-frequency signal by resonance demodulation.

(7) Multi-Level: Resonance demodulation wave's spectrum is an evenly spaced comb line.

3.2 The Principle of Resonance-demodulation

There are many ways that the most common is the use of the Hilbert transform to achieve resonance demodulation processing aimed at original time-domain or frequency domain signal. The resonance demodulation software process the raw signal can be divided into four steps[9]:

(1) The purpose of time domain signal analyzed FFT is to find resonant frequency of mechanical resonator, and this frequency becomes the band-pass filter's center frequency;

(2) Hilbert transform of filtered signal;

(3) Draw an envelope curve of shock signal

(4) Resample the envelope curve and analyze spectrum.

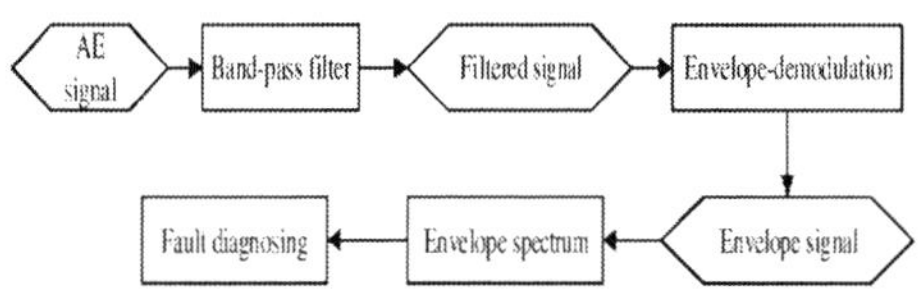

Fig. 2 The principle of resonance demodulation

4 The Means of Fault Diagnosis and Analysis of Test Results

4.1 Diagnosis Method for Failures

LMD is able to reduce the low frequency interference and highlight high frequency sections, thus played an instrumental role in noise cancellation. The diagnose method of LMD-Resonance demodulation has main components as following: (1) Collect signal and preprocess it; (2) Signal is divided into appropriate PF components and margin through the using of LMD; (3) The energy of each PF components is calculated to draw the percentages of each components on total energy; (4) Analysis the PF component which suffers the most energy (the component takes on the main features) and demodulated it; (5) By drawing the power spectrum to extract fault signatures. The flow chart shown below.

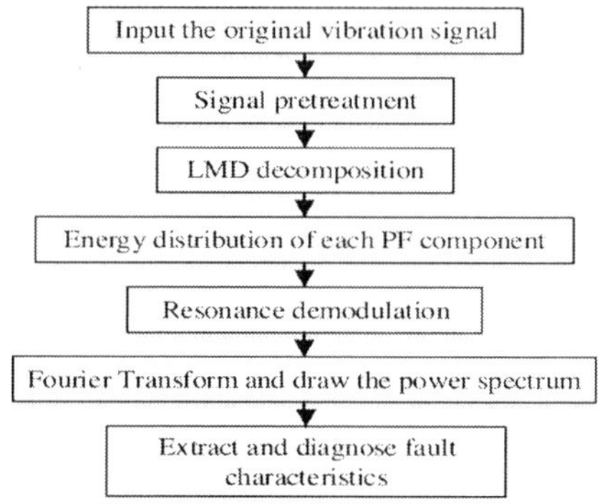

Fig. 3. Method flow chart

4.2 Verification of Experiment Results

For this article, we use the experimental data which were once measured by our project group for validation and judge the feasibility of this method[10,11]. That is knock and the crack signal of axle. The sampling frequency of the signal is 10^6Hz, their sampling point are 1024. The result of time domain and frequency domain of two signals are shown in the figure 4. We can see from the time domain graph that knock signal is hybrid AE signal, that contains not only onset AE signal, but also continuous AE signal. Crack signal's amplitude significant increases at 0.27ms and decays quickly, It belongs in the typical onset AE signal, failure mechanism is crack formation. In the frequency domain graph, the main frequency ingredients of both signals are in the range of 50KHZ-200KHZ. Due to the influence of noise and interference, we can't tell the specific situation of AE signals, thus can't extract the characteristic effectively.

Upon further analysis, in order to restore the vibration condition as real as possibility, pretreatment of AE signals are completed to removal of the trends. The result is shown in Figure 5. The aberrations of signals get fixed after least square method(LSM) processing. The trend decreased with time delay, knock signal decreased significantly more than crack signal, it indicates that knock signal is seriously affected by the outside.

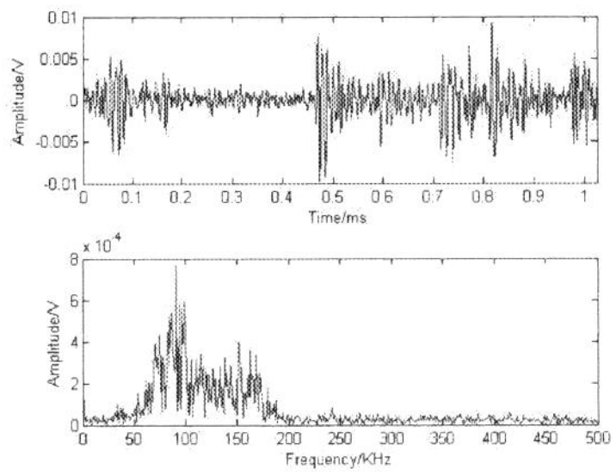

(a) Knock signal

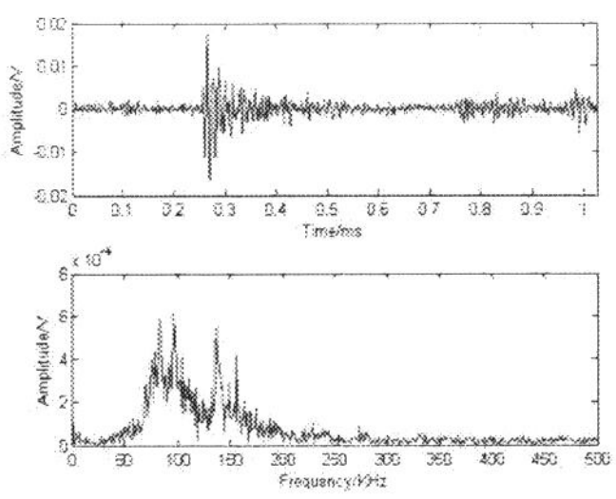

(b) Crack signal

Fig. 4. The time domain and the frequency range chart of AE signal

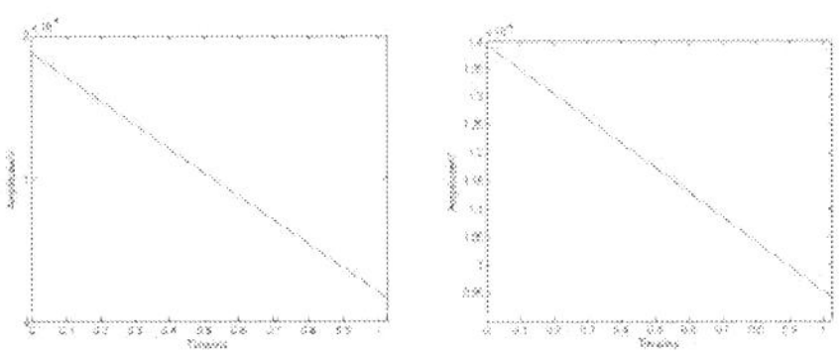

(a) (b)

Fig. 5. Signal preprocessing (a) knock signal (b) Crack signal

Then knock signal and crack AE signal are resolved into several PF components and residual (see chart6) and make the energy diagram distribution of PF components(see figure 7). We are told from the energy diagram that because of low in the proportion of PF4 and RES components, they could be negligible. Inversely, the PF1 component has the most energy, that illustrates PF1 contains wealthy information of the original signal. So, choose the PF1 as an effective component for the following demodulated resonance. The same to the crack AE signal. Most often, the main information of AE signals are in high frequency part. So we use high-pass filter instead of low-pass filter when envelope detection. The result of demodulated resonance is shown in figure 8.

Comparing power spectrums of the two signals, we see that knock signal's frequency focuses mostly around 75KHZ and the peak value appears at 73.24KHZ. We can judge the characteristics of knock signal roughly accordance with the figure 8(a). Crack AE signal is quite different from the knock signal. There is an obvious spike appear at 95.7KHZ. Therefore, we can be sure that the fault frequency of crack AE signal is 95.7KHZ, this characteristic can be used to judge equipment faults.

5 Conclusions.

This paper comes up with a method that combines LMD with resonance demodulation. After verification, it proves the feasibility of this method, which opens up a new way for fault diagnosis research.

Compared with traditional methods, LMD can decomposes AE signals availably and realize the separation of non-stable signal.

By withdrawing the signal characteristics successfully. This indicated the proposed method is feasible and effective.

Acknowledgements

This work was funded by the National Science Foundation of China under Grant 51275066.

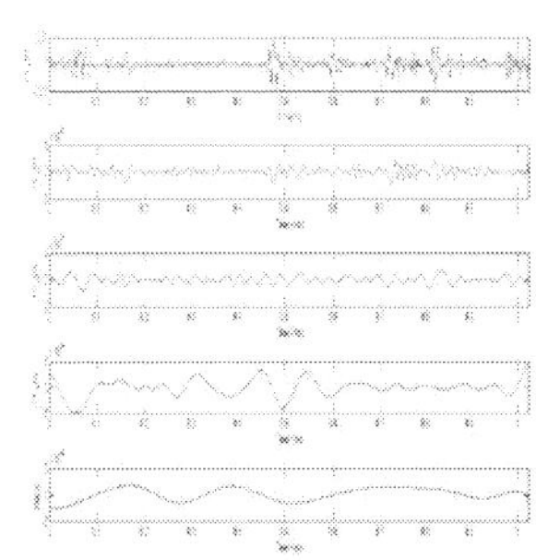

(a) LMD decomposition of knock signal

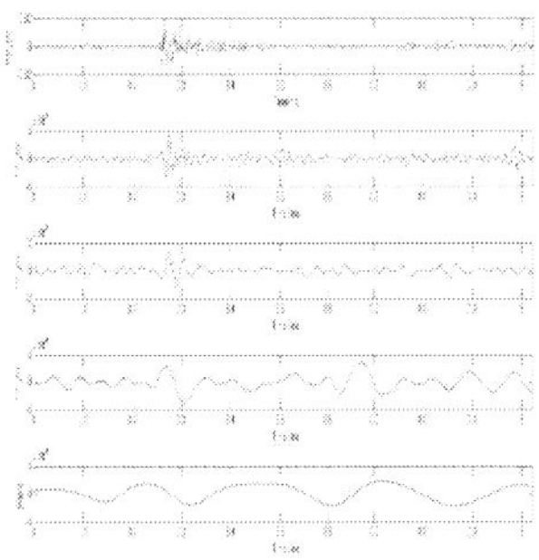

(b) LMD decomposition of crack signal

Fig. 6. LMD decomposition of AE signal

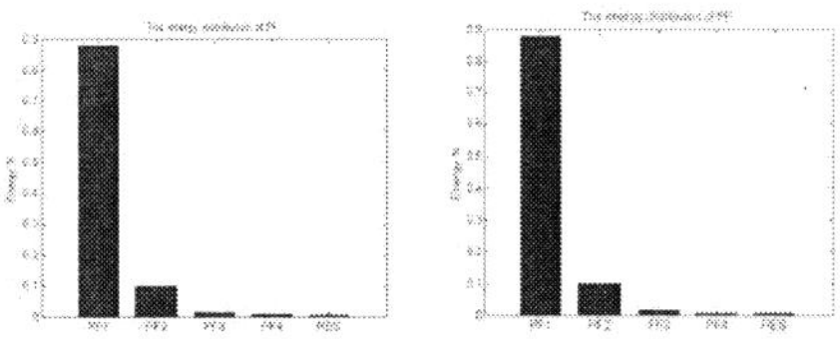

(a) (b)

Fig. 7. The energy distribution of AE signal(a) knock signal (b) Crack signal

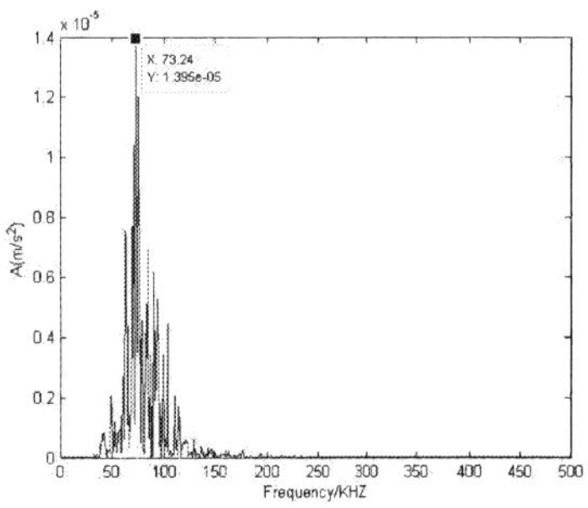

(a) The power spectrum of knock signal

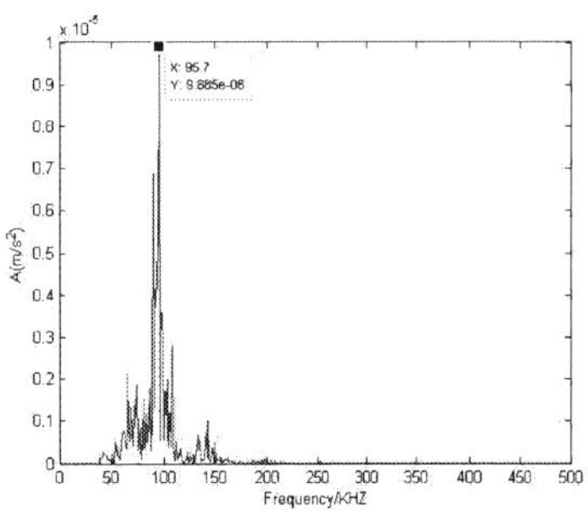

(b) The power spectrum of crack signal

Fig. 8. The power spectrum of AE signal

Reference

1. G. T. Shen, G. Dai, S. F. Liu. The developments of Chinese acoustic emission testing [M].Non-Destructively, 2003， 25 (6)： 302-307.
2. J. Qi, The vibration signal analysis of wind farms based on LMD [D]. Harbin University of Science and Engineering, 2014, 03
3. J. S. Cheng, Y. Yang, Y. Yang. The method of kurtosis and LMD and its application to gear fault detection [J]. Vibration and Shock, 2012,31 (18)： 20-23
4. Y. Yang, L. X. Yang, J. S. Cheng. Fault diagnosis in rotor system based on LMD and AR model [J]. Hunan University Journal, 2013： 15
5. F. C. Li The acoustic emission LMD research of railway vehicle axle [D]. Dalian Jiaotong University. 2014(6)
6. X. Tan The analysis research of mechanical AE fault signals based on LMD [D]. Shenyang University of Technology. 2014(2)
7. J. Hu, X. Wang. The use of MATLAB in vibration signal processing [M]. Beijing: China Water Resources and Hydropower Press, 2005: 49
8. J. S. Cheng, K. Zhang, Y. Yang, The local mean decomposition of machine vibration signal, [M] Hunan: Hunan University Press, 2013: 15
9. L. X. Hou, P. S. Yang, Fault diagnose of the motor lorry bearing based on demodulated resonance technique [J].Shijiazhuang Tiedao University Journal, 2013, 26 (4): 47-51
10. L. Lin, D. Y. Zhao, Jacket platform AE signal recognition system [J]. Non-desstructively, 2009， 31(1)： 42-47
11. Li Lin, Fulei Chu HHT-based AE characteristics of natural fatigue cracks in rotating shafts. Mechanical Systems and Signal Processing, 26 (2012) 181–189
12. Jun Ma, Jiande Wu, Yuguang Fan, Xiaodong Wang The Rolling Bearing Fault Feature Extraction Based on the LMD and Envelope Demodulation. Mathematical Problems in Engineering, 2015 1-13.

Investigation of the synchronous response of a back-back centrifugal compressor due to unbalance and excessive seal clearance – Case study

Mohamed Al Yahyai[1], Fang Duan[2*], Mohamed Al Nabhani[3], Suresh Sampath[4], David Mba[2]

[1] Petroleum Development Oman, Rotating Equipment Department, Postal Office Box 81, Postal Code 100, Oman.
[2] School of Engineering, London South Bank University, London, SE1 0AA
[3] Petroleum Development Oman, Condition Monitoring Head, Postal Office Box 81, Postal Code 100, Oman.
[4] Gas Turbine Systems & Operations, Centre for Propulsion, Cranfield University, MK43 0AL

Abstract This paper presents a case study on the effect of fouling and worn long-span labyrinth seals on the rotor dynamics of a high-speed gas centrifugal compressor. The back to back, two stage, nine-impeller unit is used to compressor natural gas associated with oil production in a particular field within the Sultanate of Oman. The highly pressurised natural gas is injected into the oil reservoir to enhance and maintain the oil flow as well as compensate for any pressure depletion in the wells. In this study, vibration problems of an operational gas compressor were replicated on a rotor-dynamic model so as to understand the root cause and investigate the influence of seal wear and hydrocarbon fouling on the compressor synchronous vibration. In addition, the paper demonstrates the dominant influence of inlet stages on rotor vibration in the presence of fouling.

Key words: Gas compressor, rotor-dynamics, fouling, seal wear, vibration, condition monitoring

1. Introduction

Mature and depleted oil fields require unconventional technical solutions for Enhanced Oil Recovery (EOR). One such solution requires the use of centrifugal compressors. The increasingly difficult production environments due to deep oil wells, reservoir formation, gas properties, viscous crude oil and high Hydrogen Sulfide (H_2S) contents all influence the need for more powerful and reliable turbomachinery that can handle extreme head, flow rates and unpredictable fluid properties. The cost associated with the unplanned downtime of gas lift compressors is extremely high [1] and causes of such downtime include high vibration levels.

Generally centrifugal compressor rotor dynamic problems are caused by two main sources, forced response and self-induced vibration. The former source is driven by the residual unbalance and process fouling while the latter is driven by the cross coupling forces generated by high rotational speed, flexible rotors, the number of impeller stages and the high process pressure on the impellers and labyrinth seals. Identifying rotor critical speeds plays an important role in mapping the safest, optimum and reliable operating range, which in turn enhances the efficiency and life cycle of the compressor. Moreover vibration levels at critical speeds provide an insight on clearance margins at impellers and seals. To ensure rotor integrity an assessment of the likelihood for self-induced vibration is vital. Predicting self-induced vibration, commonly known as rotor instability is achieved by considering impellers, seals and bearings cross coupling forces. A detailed explanation of how to undertake forced response and self-induced vibration analysis in centrifugal compressor is provided by the American Petroleum Institute [2]. The open literature reveals many case studies where forced response and self-induced vibration (rotor instability) are reported due to a variety of causes such as poor balancing, fouling, seal design degradation and ineffective or worn bearings [3,4,5].

2. Compressor train description

The gas lift centrifugal compressor train investigated in this paper consisted of two back-to-back Low Pressure (LP) and High Pressure (HP) compressors driven by a 9.24 MW brushless synchronous electric motor via a speed increasing fluid coupling gearbox (figure 1). The compressor train has a rated power of 9MW with a Rated Continuous Operating Speed (RCOS) of approximately 8800rpm. The

Maximum Continuous Operating Speed (MCOS) was approximately 9500rpm with an overall pressure ratio of 45 (LP ratio of 5 and HP ratio of 9). The LP compressor is divided into two stages (1st and 2nd) with four closed impellers at each stage, acting in opposite directions (back-to-back configuration). The HP Compressor is also divided into two opposed stages (3rd & 4th) having four impellers on the 3rd stage and five impellers on the 4th stage. As both compressors utilise the back-to-back rotor configuration they employ an interlocked mid-centre labyrinth seal between the two stages. In addition two long span abradable Tooth-on-Rotor (TOR) labyrinth seals located at the rotor's Drive End (DE) and Non Drive End (NDE) were employed. Dry Gas Seals (DGS) are used to seal the compressed gas from leaking into the atmosphere, or to the bearings, at both rotor ends. The DGS is located in-between the TOR seal and the journal bearing. For both compressors the impeller shroud uses Tooth on Stator (TOS) labyrinth seals with no swirl brakes. The LP and HP compressors both use 4-pad tilting journal bearings with a Load-Between-Pivot (LBP) configuration. The thrust bearings employed were tilting pad (8-pads) journal bearing with LBP. Finally, three flexible couplings were used between the prime mover (Motor) and Gearbox, Gearbox and LP compressor, and, LP and HP compressor see figure 1.

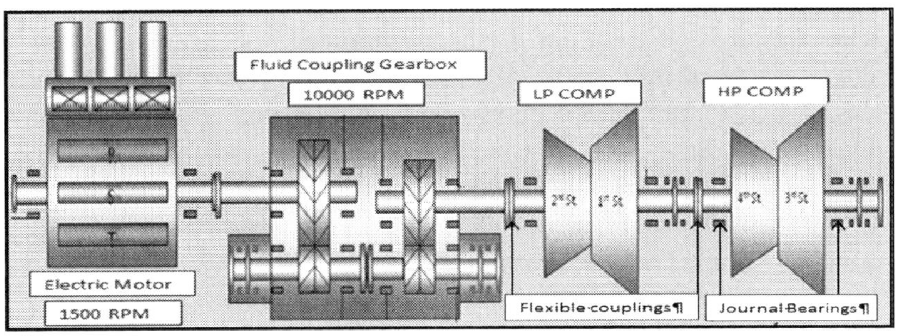

Fig. 1. Gas lift Compressor Train layout

3. Problem description and field observation

After commissioning, the unit operated for a year without any major performance or vibration problems. At 12-months of operation the compressor performance degraded to a point where it entered intermittent surge with high vibration levels noted at HP NDE (Table 1 and figure 3). Table 1 shows the compressor thermodynamic performance at optimum and degraded conditions. The optimum data was taken directly after a major overhaul (Dec. 2011) where most critical parts such as compressor rotor, dry gas seals and bearings were replaced while the degraded data was taken prior to the unscheduled major overhaul (Nov. 2012). Pressure, temperature and overall efficiency of each stage were used to monitor the compressor performance on a daily basis. The data in Table 1 shows the efficiencies of the various stages under optimum and degraded conditions.

Table 1: Compressor thermodynamic performance degradation

Stage	Efficiency at (Optimum condition (Dec. 2011))	Efficiency at Degraded condition (Nov. 2012)
LP	84%	79%
LP2	82%	75%
LP3	70%	50%
HP	66%	46%

The compression train had several vibration problems with the HP compressor throughout the degradation period (Dec. 11 to Dec. 12) whilst the LP compressor maintained relative stable vibration behaviour during this period. For the HP compressor, vibration problems were associated with start-ups and shutdowns such that whilst traversing the HP compressor critical speed, a temporary trip multiplier was used to ensure continuous operation irrespective of the high vibration amplitude.

Figures 2 and 3 show corresponding bode plot of the HP DE and NDE respectively taken from proximity probes in Jan. 2012, at the onset of the gradual reduction in efficiency and prior to the overhaul. These figures show direct and 1X components of vibration displacement. From the plots, it can be seen that the amplitude at critical speed at the HP DE side was 30μm, below the specified alert setting of 40μm whilst at the HP NDE the vibration amplitude reached a value of 52μm close to the trip setting of 55μm. Moreover, figure 2 shows that the vibration amplitude at MCOS at HP DE side was 13μm whilst the corresponding amplitude at the HP NDE side was relatively high at 30μm. Figure 3 shows a similar bode plot in Nov 2012, one month before shutdown for overhaul (Nov. 2012). By this period at the MCOS, the vibration amplitude at the NDE had reached 48μm. Essentially the run-up vibration data in Nov. 2012 showed vibration levels at MCOS exceeded that measured at critical speed (NDE), reaching 50μm, see figure 3.

The vibration amplitude under this condition of operation is intolerable for long-term operation with high probability of rotor damage (impellers, seals and bearings) as well as other sealing components such as the dry gas seal. Also noted in Nov. 2012 was that the vibration level at the critical speed (45μm) was lower than that measured in Jan. 2012 (52μm).

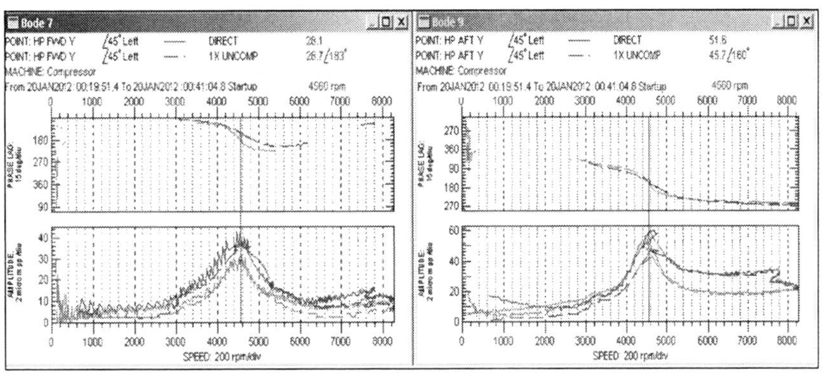

Fig. 2. Bode Plot for HP Compressor DE (Direct and 1X vibration components)

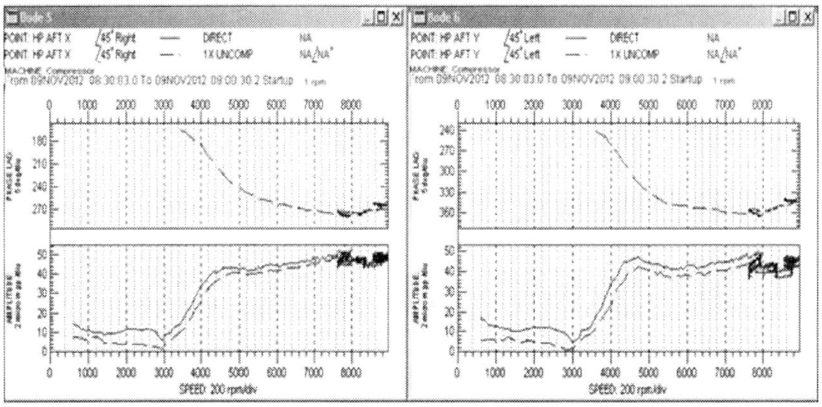

Fig. 3. Bode Plot for HP Compressor NDE (Direct and 1X vibration components)

4. Rotor-dynamic model

A rotor–bearing model was generated to understand the root cause of the high vibration levels from the HP compressor. The rotor model was built and analysed using 'XLRotor' which employs a Polynomial transfer matrix method to solve for Undamped Critical Speeds (UCS), Eigenvalue and Eigenvector, Imbalance Response... etc [6]. The Transfer Matrix method is widely used by turbomachinery designers and manufacturers for rotor-dynamics analyses. Hozler [7], initially developed the Linear Transfer Method then Myklestad improved it in 1944 for airplane wings [8] and Prohl [9] in 1945 adapted it for steam turbines. In1974 Lund applied it with adaptation for rotor bearing system analyses. Finally, Murphy improved the transfer matrix method using the Polynomial method overcoming some limitation in application [10].

The aerodynamic excitation forces from the impellers were calculated using the dimensional form of Wachel's equation [11]. The HP rotor was modelled using 69 stations (see figure 4) comprising the two stages, all nine impellers, the seals, and mass and inertias of each impeller. Bearing characteristics were analysed using 'XLHydroDyn' which provides a full film solution of Reynolds equation [12]. The bearing stiffness and damping terms were calculated for different operating conditions that were to be considered in the rotor dynamic analysis. Labyrinth seal forces were calculated based on analyses of the compressible flow in annular seals employing the 'one control volume' model presented by Childs [13]. In the model seal clearances were also varied from a minimum of 0.2mm to a maximum of 0.9mm, the operating range recommended by the compressor manufacturer.

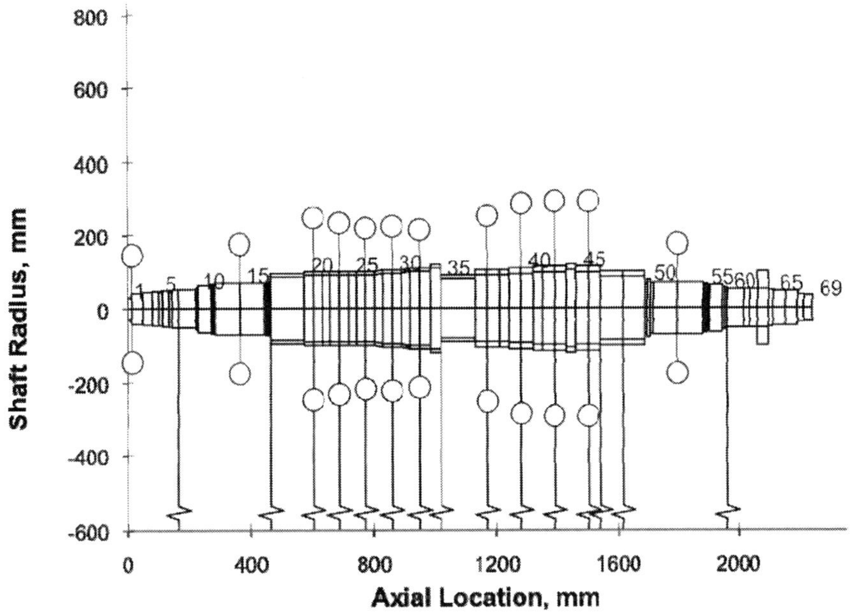

Fig. 4. HP Rotor Model

5. Rotor dynamic simulation

An unbalance response analysis was conducted in accordance with the American Petroleum Institute API 617, 7th Edition [3]. The used unbalance amount for the rotor weight (W) of 683 kg and speed (N) of 8600 RPM was determined to be 504 g-mm, as recommended by API specification [3]. This unbalance mass was placed at an anti-node location so that the first mode was excited. The anti-node location is the mid span of the rotor, which corresponds to station 38. The unbalance response analysis was carried without considering the effect of forces from the impellers and labyrinth seals so as to compare the results against the Factory Acceptance Test (FAT) (referred to as 'no-load'). The results show a peak displacement amplitude of approximately 11μm, 10.7μm and 30μm at the DE, NDE and rotor mid-span respectively. Table **2** shows that both critical speed and Amplification Factor (AF) compared favourably with the Factory test results and within the API recommended numerical model variance of 5%. The calculated first critical speed (first bending mode) was estimated to be 4833rpm with an Amplification Factor (AF), based on half power point of 3.92.

Table 2: FAT Test against Numerical Model Comparison

	Factory Acceptance Test (FAT)	**Numerical Model**
1st Critical Speed	4802 (RPM)	4833 (RPM)
Amplification Factor	3.75	3.92

To simulate the condition of the HP compressor that represents the inoperable state (trip) as noted in Nov. 2012, different scenarios were considered which included simulations for varying bearing and seal clearance conditions. The bearing clearances simulated ranged from 0.15mm to 0.18mm and seal clearances ranged from 0.2mm to 0.9mm. In both cases this was the range recommended by the OEM. A sensitivity analyses was conducted to explore the effect of impeller and seal cross coupling forces on the rotor synchronous response while maintaining the same initial unbalance amount (504g-mm). Figure 5 shows the synchronous response of the HP DE for a variety of simulation conditions that included the rotor response without inclusion of the impellers and seal forces. A maximum displacement of 33 μm was noted at the critical speed for the HP DE under rotor model conditions that included the seals and impellers. The inclusion of the long span labyrinth seals and impeller forces reduced the critical speed to 4416 rpm, which compares well with the field data critical speed of 4570rpm. In addition at MCOS the displacement noted was approximately 6.5μm at the HP DE. It was noted that the labyrinth seals and impeller forces increased the response amplitude prior to and whilst traversing critical speed but had no effect at MCOS, see figure 5.

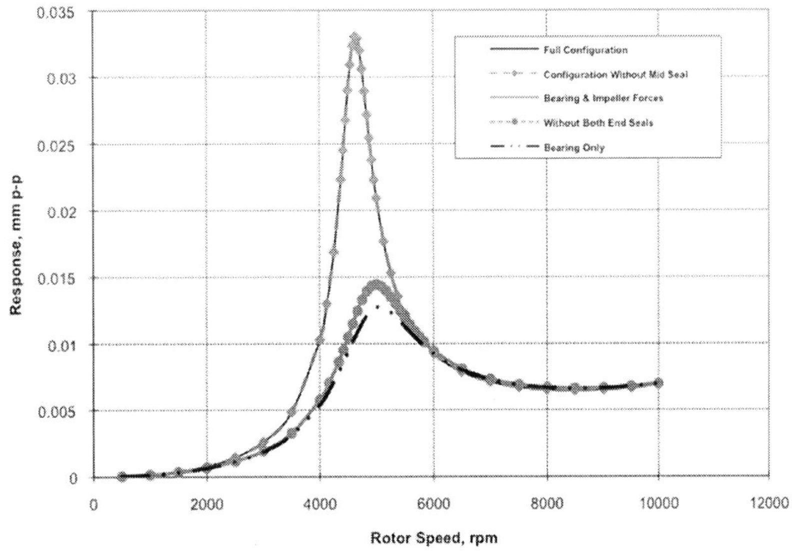

Fig. 5. Synchronous Response Analysis at varying conditions (station 9, HP DE bearing with a clearance of 0.15μm)

The next phase of the analysis involved investigating the effect of minimum and maximum labyrinth seals clearance for the same amount of unbalance and bearing clearance (0.12μm) used in the previous analysis. From the analysis, it was observed that higher vibration amplitude levels were noted at the critical speed for minimum seal clearances compared to that of maximum clearance while the same amplitude was observed after traversing the critical speed for both cases. An increase in amplitude of up to 100% at the critical speed was noted for the case with minimum seal clearance (0.2mm) as opposed to maximum seal clearance (0.9 mm) as it can be shown in figure 6.

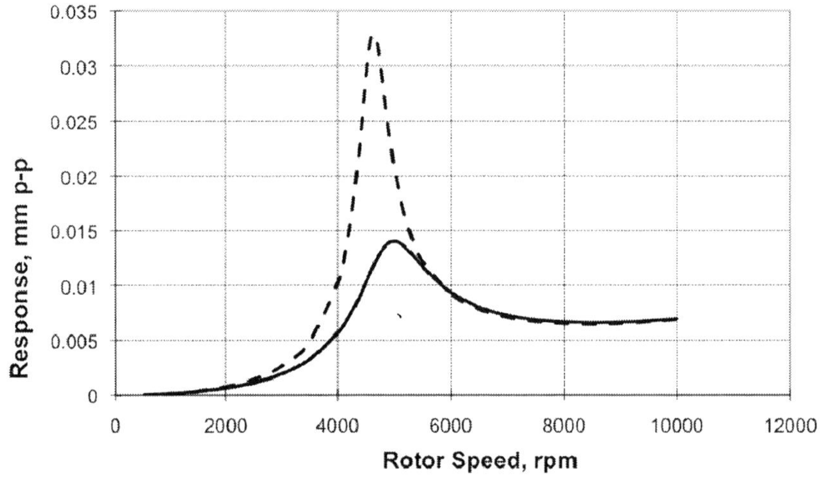

Fig. 6. The effect of worn seals on the synchronous response (- - min clearance; HP DE, bearing 0.12μm).

Finally to simulate the rotor dynamic conditions of the degraded compressor the DE bearing was kept at its minimum clearance (0.12μm) while the NDE bearing was kept at its maximum clearance (0.15μm). Simultaneously all seals clearances were set to an out-of-tolerance value of 0.9mm whilst varying the amount of unbalance along the rotor. Different unbalance values and associated locations were considered starting with an equal unbalance distribution (total 504g-mm) along the rotor, concentrated unbalance at mid-span position, single sided unbalance locations, etc. The simulation which provided the closest match of the compressor rotor-dynamic behaviour (target condition) prior to the inoperable condition was achieved by placing a total unbalance amount of 6900 g-mm distributed between the 3rd stage first impeller and 4th stage first impeller, with 90% of the unbalance placed on the 3rd stage (1st impeller) and 10% on the 4th stage 1st impeller. Comparative results of the distributed and targeted unbalance analysis are presented in figures 7 and 8. This shows that the equally distributed unbalance produces higher amplitude at critical speed than the targeted condition. Also the equally distributed unbalance condition showed decreasing vibration amplitude after critical speed whilst the targeted unbalance showed an increasing vibration level away from the critical speed as was noted on the operational data (see figure 3), particularly at the NDE.

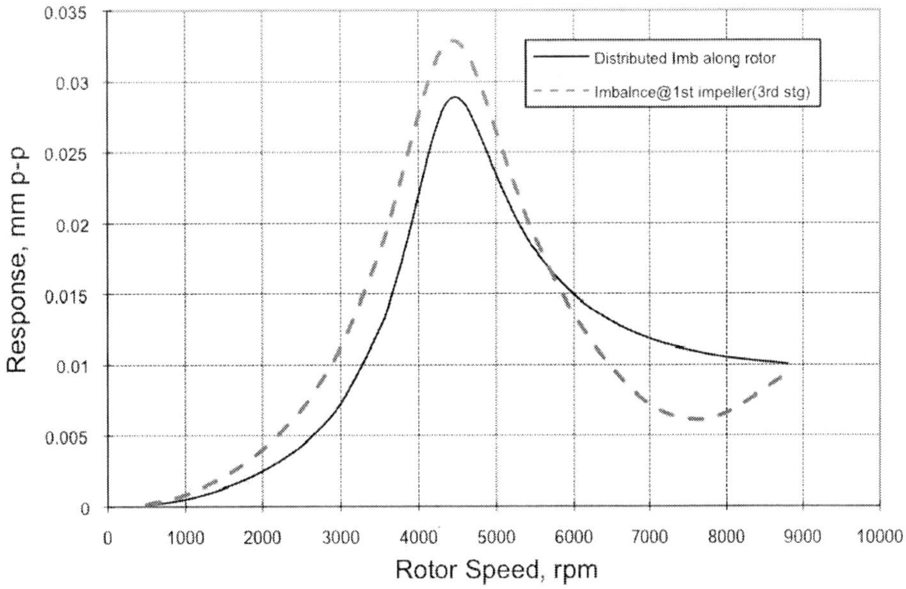

Fig. 7. Equal and sided distributed Imbalance response plot at DE

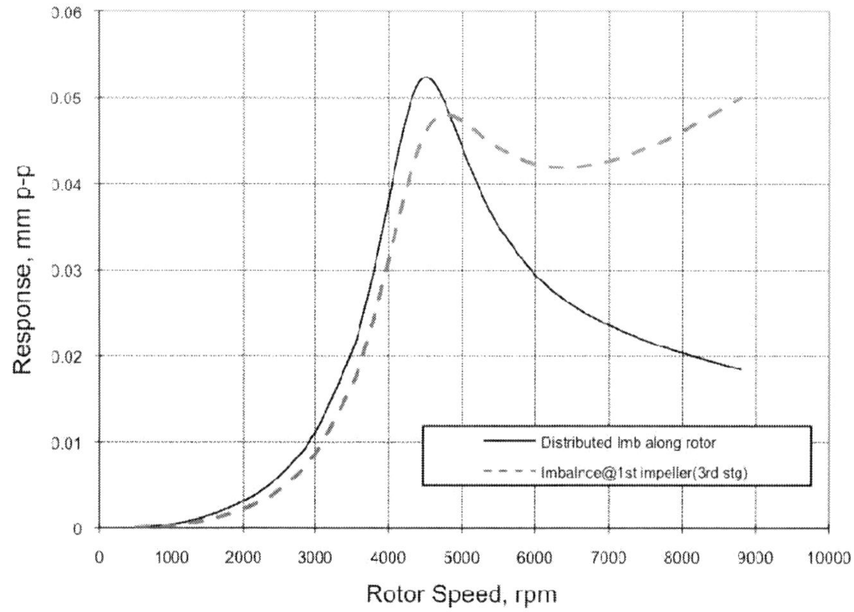

Fig. 8. Equal and sided distributed Imbalance response plot at NDE

6. Discussion and field observations

A major overhaul was undertaken in Jan. 2013 and several observations were noted such as carbon deposits on the impellers and oil contamination of the dry gas seal. In addition, seal clearances across the rotor were measured at 1.2 mm, above the 0.9mm maximum limit, which is indicative of wear, possibly due to rubbing between the rotor and seal. Figure 9 (left) shows the condition of the NDE Stationary seal after one year of operation and corresponding new seal (right); comparing the two conditions it can be seen that the de-swirl chambers in the old seal are fully blocked by the carbon deposit. The de-swirl chambers are used to reduce the circumferential gas velocity at seal entrance which can cause the generation of destabilisation cross coupling forces at the location of the labyrinth seals [1].

Fig. 9. Fouled Stationary NDE Seal (Left), New NDE Seal (Right)

Figure 10 shows both DE and NDE long span Labyrinth seals with carbon deposits evident at both ends especially at the impeller eye. This observation explains the HP compressor performance degradation as well as the high vibration amplitude. The formation of the carbon deposit within the impeller eye would have restricted the inlet flow and this in turn will affect the efficiency of the unit. In addition, the carbon deposit affected the rotor balance state, which resulted in high vibration amplitude. Typically, fouling occurs mainly in the inlet of the 3rd and 4th stage impellers and both are located close to the long span end labyrinth seals.

Using the observations discussed above, the sequence of the compressor degradation and failure becomes clearer. At the earliest stages of the degradation event, the initial high vibration amplitude was caused by unbalance and cross coupling forces generated by the labyrinth seals (on the basis that the seal clearance was optimised at 0.2mm). At this initial stage, the compressor operators had to engage the 'trip multiplier' to pass the critical speed. The trip multiplier essentially allows the operators to increase the vibration trip limit. Furthermore, with time, the degradation of performance continued to such an extent that the compressor went in and out of surge and to overcome this process instability the compressor speed automatically varied from maximum continuous operating speed to 3000 rpm, i.e., traversing the critical speed numerous times (about twice a week) in an attempt to overcome the surge problem. However, by continuously traversing the critical speed the compressor experienced the high vibration levels described and this in turn contributed to the gradual wear of the labyrinth seals along the rotor span. As the seals gradually lost their optimum clearances the cross coupling forces were altered and this in turn reduced the rotor vibration at critical speed, allowing the operators to travers the critical speed without using the 'trip-multiplier'. However, the continuous accumulation of deposits on the impellers caused an increase in the unbalance amount that eventually reached a point where at MCOS the vibration amplitude exceeded the allowable limit.

Fig. 10. Worn DE (left) & NDE (right) Labyrinth Seal and carbon deposit formation on the Impeller Eye

7. Conclusion

A variety of models were used to accurately predict the high-pressure centrifugal compressor synchronous response, where the effect of rotor unbalance due to natural gas fouling and seal clearances were considered. The numerical model developed compared well with the compressor field data for the variety of conditions analysed. The results show that rotor unbalance response behaviour depends on the unbalance amount, unbalance locations, as well as the labyrinth seals clearance. The seal wear rate plays a vital role in increasing or decreasing the vibration amplitude particularly when traversing the compressor critical speed. The 16% efficiency drop was attributed to process fouling at both stages inlet stages as well as to the seal degradation, which was observed during the major overhaul. The high level of vibration amplitude at MCOS was attributed to the unbalance mass due to fouling at the inlet stages. Finally, this study shows that the sensitivity of the location of the inlet stages to rotor vibration, particularly when fouling is evident. Therefore, from a design point of view, locating the inlet stages at node locations will help manage vibration levels on rotors that are susceptible to high amounts of unbalance from fouling. A root cause analysis for a centrifugal compressor was successfully accomplished with a rotor dynamic model.

References

[1]. Ertas, B. H., Delgado, A. and Vannini, G. (2012), "Rotordynamic force coefficients for three types of annular gas seals with inlet preswirl and high differential pressure ratio", Journal of Engineering for Gas Turbines and Power, vol. 134, no. 4.

[2]. API684, 5. E. (2005), Rotordynamic tutorial: lateral critical speeds, unbalance response, stability, train torsionals, and rotor balancing, 5th ed, American Petroleum Institute, USA.

[3]. API 617, 7. E. (2002), Axial and Centrifugal Compressors and Expander-Compressors for Petroleum Chemical and Gas Industry Services, 7th ed, American Petroleum Institute, USA.

[4]. Collins, D., Teixeira, J. and Crudgington, P. (2008), "The degradation of abradable honeycomb labyrinth seal performance due to wear", Sealing Technology, vol. 2008, no. 8, pp. 7-10.

[5]. Lees, A. W. (2006), "Agnes Muszynska, ,Rotordynamics (2005) CRC Press,Boca Raton 0-8247-2399-6 (1075pp., £97).", Journal of Sound and Vibration, vol. 298, no. 1–2, pp. 477.

[6]. Vance, J., Zeidan, F. and Murphy, B. (2010), machinery vibration and rotordynamics, John Wiley & Sons, Inc, Hoboken, New Jersey

[7]. Heinrich, H., (1921), Die Berechnung der Drehschwingungen und ihre Anwendung im Maschinenbau, J, Springer.

[8]. Myklestad, N. (1945), "New method of calculating natural modes of coupled bending-torsion vibration of beams", Trans.ASME, vol. 67, no. 1, pp. 61-67.
[9]. Prohl, M. (1945), "A general method for calculating critical speeds of flexible rotors", Journal of Applied Mechanics, vol. 12, no. 3, pp. 142-148.
[10]. Murphy, B. T. and Vance, J. M. (1982), "An improved method for calculating critical speeds and rotordynamic stability of turbomachinery." .
[11]. Wachel, J. and WW, v. (1981), "Ensuring the Reliability of Offshore Gas Compression Systems", Journal of Petroleum Technology, vol. 33, no. 11, pp. 2252-2260.
[12]. Murphy, B. T. (2013), XLHydroDyn, available at: http://www.xlrotor.com/xlrotor/xlhydrodyn.shtml (accessed 06/13).
[13]. Childs, D. (1993), Turbomachinery Rotordynamics: Phenomena, Modeling and Analysis, John Wiley & Sons, INC, United states of America.

Detection and Diagnosis of Compound Faults in a Reciprocating Compressor based on Motor Current Signatures

Usama Haba[1], Guojin Feng[1], Abdulkarim Shaeboub[1], Xinyu Peng[2], Fengshou Gu[1] and Andrew D. Ball[1]

Centre of Efficiency and Performance Engineering, University of Huddersfield, Queensgate, Huddersfield HD1 3DH, UK
Shanxi key Laboratory of Fully Mechanised Coal Mining Equipment at Taiyuan University of Technology, P.R.China

E-mail: usama.haba@hud.ac.uk

Abstract. Induction motors are the most common driver in the industry and consume tremendous energy every year. Monitoring the status of a motor and its downstream equipment and diagnosing faults in time not only avoids great damage to mechanical systems but also allows the motor to run at optimal efficiency. This paper studies the use of information from motor current signals to detect and diagnose faults of a reciprocating compressor (RC) and its upstream three-phase motor. The motor is applied by the RC with an oscillator torque which induces additional components in measured current signals. Moreover, the current signatures contain changes with the torque profiles due to different types of faults. Based on these analytical studies, experimental studies were carried out for different common RC faults, such as valve leakage, intercooler leakage, stator asymmetries and the compounds of them. The envelope analysis of current signals allows accurate demodulation of the torque profiles and thereby it can be combined with overall current levels for implementing model based detections and diagnosis. The results show these simulated faults can be separated under all operating pressures.

Keywords: Reciprocating compressor, induction motor, stator asymmetries faults, envelope analysis, motor current signatures analysis.

1 Introduction

Reciprocating compressors (RC) and their associated induction motor plays an important part in various industrial systems because of their powerful design, flexibility in use and high thermo-efficiency offering high value of reliability. Due to the RC working condition requirement, some of its components work under harsh conditions such as high pressure and high temperature [1]. Therefore, failures occurring in RC not only degrade performance, consume additional energy, but also possibly even result in system shut-down [2]. According to [3], damages to RC valves can lead to serious machine breakdown, which represents the biggest source of compressor failures and accounts for 31% of machine breakdowns. Additionally, failures in its associated motor such as broken rotor bar or stator asymmetry winding could also affect the compressor performance. Increases in stator resistance of a three phase induction motor can lead to voltage imbalances in the motor causing a reduction of motor efficiency, increase in motor temperature and oscillatory running conditions. Therefore, efficient and effective condition monitoring techniques are actively studied to detect and diagnose RC and its associated motor faults at an early stage in order to prevent any major failures on motors [4, 5].

In practice, faults in machines may develop simultaneously and the combined faults may create more damage or performance reductions to the machine. It has been found that the exposure of the motor to asymmetry supply and combined fault significantly increases the temperature and reduces the efficiency of the machine [6, 7]. However, the combined faults make the fault features more complicated to interpret and the fault detection even more challenging. Recently, research has been carried out in this area. Ballal et al.[8] improved a combined method with neural fuzzy inference system logic to detect stator inter-turn insulation and bearing wear faults in a single phase induction motor. In [9], Garcia-Perez successfully used high-resolution spectral analysis for identifying multiple combined faults in induction motors.

Motor current signature analysis (MCSA) has been proven as an effective and efficient technique for monitoring different motor faults, including air-gap eccentricity, broken rotor bar [10, 11] and turn to turn fault in the stator [12]. In [13], Haynes and Kryter showed that MCSA can also be employed for detecting abnormalities in the downstream machines such as compressors, pumps, rolling mills, air blowers and air conditioning systems. Particularly, Gu et al. [14] has investigated using motor current signals for diagnosing individual faults of RC machines. These all show the great benefit of MCSA in that it does not require any additional systems for measurements [10, 11].

This paper studies further the use of information from motor current signals to detect and diagnose faults of a RC and its associated three-phase motor. Three types of common single faults including intercooler leakage, valve leakage and stator asymmetries and one compound fault are introduced on the RC. The experimental results showed that all the simulated faults can be successfully separated through the analysis of the motor current signal.

2 Reciprocating Compressor Fault Features

2.1 Valve Leakage

Valve leakages are considered as the most common fault in RC and the most frequent root cause for unplanned compressors shutdowns [15]. In multi-stage compressors, especially at high pressure stages, both suction and discharge valves are exposed to a high number of impacts per second and also to high speed flow and high temperature. This harsh working environment often causes non-uniform wear of the sealing surfaces between the valve seat and its plates, which thereafter causes leakage in the valve system. Leakage within the valve allows high temperature gases to be pushed through the valve and accelerates valve system deterioration which eventually affects the compressor efficiency [16].

Under these conditions, when the compressor starts the compression process it requires more power whereas expansion needs less power, hence the motor speed will vary accordingly. When the RC runs under normal conditions, the fluctuation increases as the discharge pressure increases which means the motor current signals change with this fluctuation and the measured current signal will exhibit amplitude modulation. Thus, if the compressor produces a fault such as valve leakage, the load fluctuation will increase and hence cause a high degree of amplitude modulation [17].

In the motor current spectrum, the modulation will exhibit a series of sidebands f_{sb} at the supply frequency, which can be measured as [17]:

$$f_{sb} = f_c \pm mf_x \tag{1}$$

where f_c denotes the frequency of supply, f_x is the working frequency of the compressor and $m = 1,2,3,...$

2.2 Stator Asymmetries Faults

A small amount of resistance imbalance will cause an increase in the stator winding temperature by a large amount. As a general rule the temperature rises by 25% (in °C) for every 3.5% voltage imbalance [18]. For every 1% voltage imbalance, there is a 7% current imbalance expected, which is equivalent to the negative sequence voltage that causes negative braking torque. Rotating flux opposes the main flux asymmetry-voltage operation that will also create a pulsating torque which produces speed pulsation, mechanical vibration and consequently acoustic noise [19]. The voltage imbalance can be defined as[20]:

$$\text{Voltage imbalance} = \left(1 - \frac{3v_{\min}}{\sum \text{V}}\right) \times 100 \tag{2}$$

where, v_{min} is the lowest phase voltage (V), and V is the voltage cross each phase (V). This asymmetric voltage would also lead to changes in motor speed-torque behaviour in the form of current signature changes.

3 Experimental Setup

A two-stage, single action V-shape Broom Wade TS9 reciprocating compressor is employed for testing. It can deliver compressed air up to 0.8MPa (120 psi) to a horizontal air tank. The compressor is derived by a three phase KX-C184, 2.5kw induction motor. In the test, five working conditions were performed one by one: healthy condition (BL), discharge valve leakage on the second stage (DVL), leakage on the intercooler (ICL), asymmetry stator winding (Stator) and a combined fault with both DVL and Stator (DVL and Stator).

From the first and second cylinder pressure comparison in Fig. 1, it can be seen the simulated faults have caused observable changes on the compressor performance.

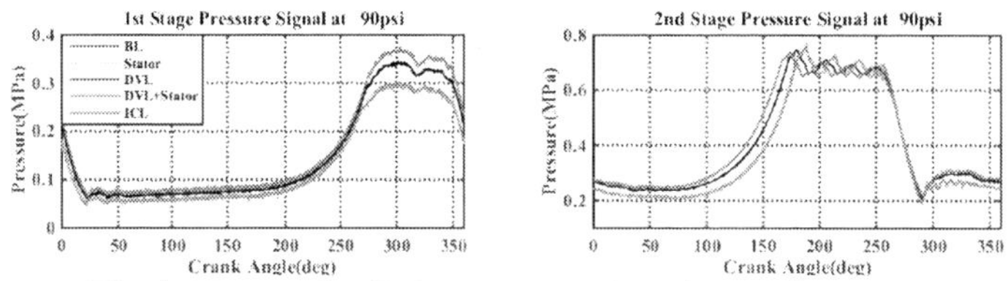

Fig. 1. Measured cylinder pressures for 1st and 2nd stage

4 Results and Discussion

4.1 Motor Current Signal Analysis

The motor current signals for all test cases under pressure of 80psi are presented in Fig. 2 in both time domain and frequency domain. A clear amplitude modulation (AM) effect can be clearly observed from the waveform and the current amplitude under faulty conditions (DVL and DVL+Stator) is slightly higher than that of healthy conditions (BL). From the spectra in Fig. 2 (b), a clear AM effect can also be seen. A series of sidebands appear around the 50 Hz carrier component with a fundamental frequency at 7.5Hz (working frequency of the compressor). These sidebands are caused by an asymmetry between three phases, power supply imbalance and machine operation load fluctuation, with sufficient information for fault diagnosis. Note that the sideband amplitude of the motor current for the leaking valve and combined fault seems slightly higher than that the healthy conditions.

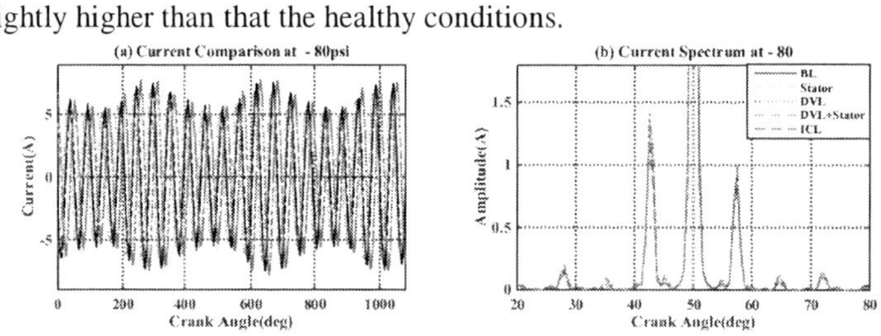

Fig. 2. Current signals for all test cases: (a) time domain and (b) frequency domain

By calculating the average amplitude of the first lower and upper sidebands, i. e. 50±7.5Hz, as symbol sideband amplitude for the corresponding pressure condition, a trend for the symbol sideband changing under pressure is presented in Fig. 3.

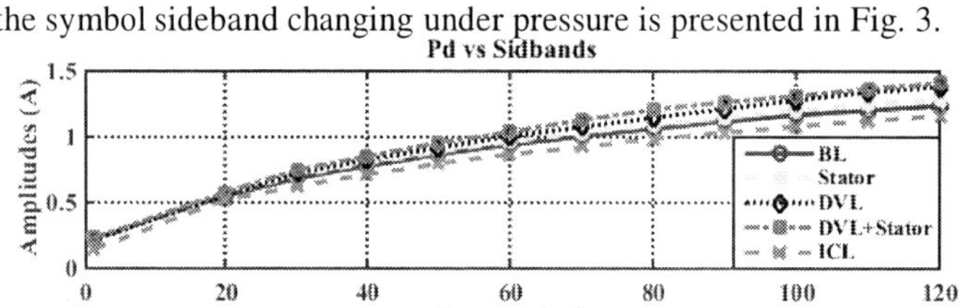

Fig. 3. Amplitudes of sidebands average for all cases.

It can be seen very clearly that the sideband amplitude increases as the pressure increases. Furthermore, the trends for these five test cases can be roughly seperated, with the combined fault being the highest one, followed by valve leakage fault then stator asymmetry fault; the healthy condition and the intercooler leakage being the lowest one.

4.2 Envelope Signal Features

Envelope analysis is suitable for diagnostics of machinery where faults have an amplitude-modulating effect on the characteristic frequencies of the machinery [21]. Therefore, the envelope of the motor current signal is calculated using Hilbert transform [22] and the envelope signal for pressure at 40 psi and 100 psi are given in Fig. 4. It can be observed that the modulation effect becomes more obvious as the pressure increases. It is difficult to assess any change in the pattern, the shape and amplitude between healthy and faulty cases.

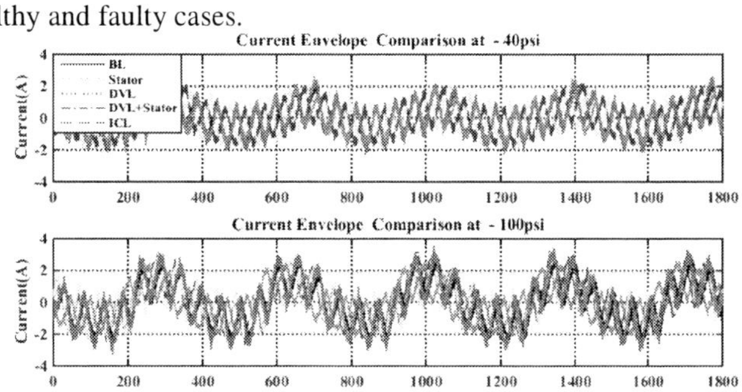

Fig. 4. Envelope signal for normal and faulty operation and envelope values for different discharge pressure

4.3 Diagnosis based on Combined Features

Take the current RMS value as a function of discharge pressure and envelope amplitude as shown in Fig. 5 (a) and (b), it clearly shows that both the healthy and faulty conditions have an increasing trend. The relationship between the motor current RMS values and discharge pressure is shown in Fig. 5 (a) clearly shows a difference between healthy and various fault cases. However it is unable to separate discharge valve leakage, stator asymmetric and combined fault. The envelope analysis in Fig. 5 (b) shows a substantial difference between healthy and faulty cases, especially between discharge valve leakage and combined fault but not for intercooler leakage. Stator asymmetry can be clearly viewed in the RMS values.

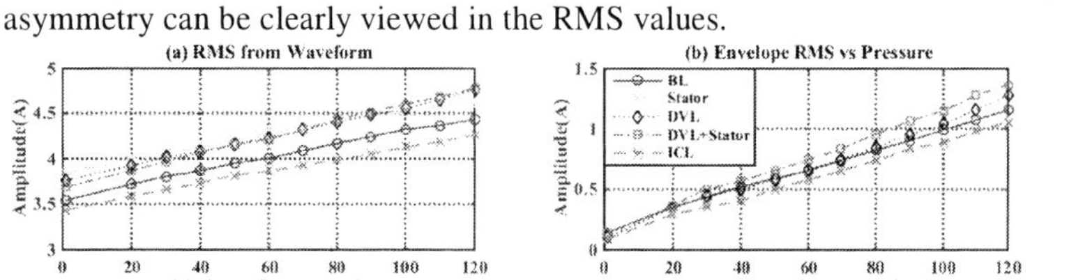

Fig. 5. (a) Current RMS values and (b) envelope amplitude values of different discharge pressure

The envelope analysis combined with overall current levels for implementing a model based approach. Fig. 6 (a) illustrates the combined envelope and current RMS. One can see that simulated faults can be separated under all operating pressures, furthermore, the differences spread over compressor discharge pressure range from 1 psi to 120 psi. It can be concluded that either too low or too high feature values will indicate a faulty or abnormal compressor operation.

On this basis, compressor fault diagnosis can be carried out at its operation pressure range (1psi to 120psi). The upper feature values can be used to differentiate the intercooler leakage from other faulty cases. In addition, all compressor faults can be classified though model based detections, shown as Eq. (3) and (4). The diagnosis results are presented in Fig. 6 (b)

$$\text{Model: } \hat{A}_e = 1.1073A - 3.7806 \tag{3}$$

$$\text{Residual: } R = A_e - \hat{A}_e \tag{4}$$

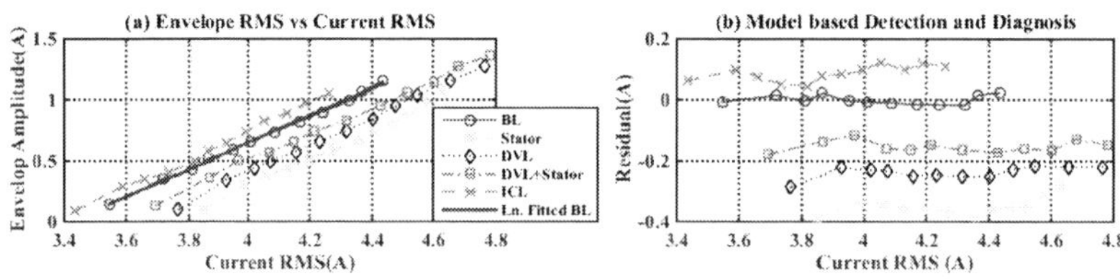

Fig. 6. (a). Envelop RMS vs Current RMS and (b) model based approach

5 Conclusion

This paper studies the usage of motor current signal for the diagnosis of faults on the induction motor, faults on its downstream reciprocating compressor and also the compound faults. The experimental results have shown the motor current signal contains sufficient information (average torque and dynamic torques) for diagnosing the simulated fault cases. A model based fault detection based on the relationship between current RMS and envelope RMS value are proposed. The proposed model is effective for a wide pressure range.

References

1. AlThobiani, F. and A. Ball, An approach to fault diagnosis of reciprocating compressor valves using Teager–Kaiser energy operator and deep belief networks. Expert Systems with Applications, 2014. 41(9): p. 4113-4122.
2. Ahmed, M., et al. Fault diagnosis of reciprocating compressors using revelance vector machines with a genetic algorithm based on vibration data. in Automation and Computing (ICAC), 2014 20th International Conference on. 2014. IEEE.
3. Bardou, O. and M. Sidahmed, Early detection of leakages in the exhaust and discharge systems of reciprocating machines by vibration analysis. Mechanical Systems and Signal Processing, 1994. 8(5): p. 551-570.

4. N. A. Hussein, D.Y.M., E. M. Abdul-Baki, 3-phase Induction Motor Bearing Fault Detection and Isolation using MCSA Technique based on neural ne22twork Algorithm. Journal of Engineering and Development, 2012. 16(3): p. 15.
5. Alwodai, A., F. Gu, and A. Ball. A comparison of different techniques for induction motor rotor fault diagnosis. in Journal of Physics: Conference Series. 2012. IOP Publishing.
6. Sridhar, S. and K.U. Rao. Detection of simultaneous unbalanced under-voltage and broken rotor fault in induction motor. in Condition Assessment Techniques in Electrical Systems (CATCON), 2013 IEEE 1st International Conference on. 2013. IEEE.
7. Constantine, R.A.E., Effects of Mixed Faults on the Stator Current Spectrum of the Induction Machine.
8. Ballal, M.S., et al., Adaptive neural fuzzy inference system for the detection of inter-turn insulation and bearing wear faults in induction motor. Industrial Electronics, IEEE Transactions on, 2007. 54(1): p. 250-258.
9. Garcia-Perez, A., et al., The application of high-resolution spectral analysis for identifying multiple combined faults in induction motors. Industrial Electronics, IEEE Transactions on, 2011. 58(5): p. 2002-2010.
10. El Hachemi Benbouzid, M., A review of induction motors signature analysis as a medium for faults detection. Industrial Electronics, IEEE Transactions on, 2000. 47(5): p. 984-993.
11. Ashari, D., et al., Detection and Diagnosis of Broken Rotor Bar Based on the Analysis of Signals from a Variable Speed Drive. 2014.
12. Sharifi, R. and M. Ebrahimi, Detection of stator winding faults in induction motors using three-phase current monitoring. ISA transactions, 2011. 50(1): p. 14-20.
13. Haynes, H. and R. Kryter, Condition monitoring of machinery using motor current signature analysis, 1989, Oak Ridge National Lab., TN (USA).
14. Gu, F., et al., Electrical motor current signal analysis using a modified bispectrum for fault diagnosis of downstream mechanical equipment. Mechanical Systems and Signal Processing, 2011. 25(1): p. 360-372.
15. Goebel, D. Reciprocating compressor suction and discharge valve monitoring. 2013.
16. Elhaj, M., Condition monitoring of reciprocating compressor valves, 2005, University of Manchester: Manchester.
17. Naid, A., et al. Bispectrum Analysis of Motor Current Signals for Fault Diagnosis of Reciprocating Compressors. in Key Engineering Materials. 2009. Trans Tech Publ.
18. Payne, B.S., Condition Monitoring of Electrical Motors For Improved Asset Management, in School of Engineering2003, University of Manchester: Manchester.
19. Gaeid, K.S., et al., Fault Diagnosis of Induction Motor Using MCSA and FFT. Electrical and Electronic Engineering, 2011. 1(2): p. 85-92.
20. Saleh, A.F., Detection And Diagnosis of Electrical Faults In Induction Motors Using Instantaneous Phase Variation, in School of Engineering2005, University of Manchester: Manchester.
21. Feng, G., et al. The real-time implementation of envelope analysis for bearing fault diagnosis based on wireless sensor network. in Automation and Computing (ICAC), 2013 19th International Conference on. 2013. IEEE.
22. Obaid, R.R., T.G. Habetler, and R.M. Tallam. Detecting load unbalance and shaft misalignment using stator current in inverter-driven induction motors. in Electric Machines and Drives Conference, 2003. IEMDC'03. IEEE International. 2003. IEEE.

AUTHOR INDEX

AUTHOR INDEX